COHERENCE OF LIGHT

Second completely revised edition

COHERENCE OF LIGHT

Second completely revised edition

by

JAN PEŘINA
Palacký University, Olomouc, Czechoslovakia

D. REIDEL PUBLISHING COMPANY
A MEMBER OF THE KLUWER ACADEMIC PUBLISHERS GROUP
DORDRECHT / BOSTON / LANCASTER

Library of Congress Cataloging in Publication Data

CIP

Peřina, Jan, 1936
Coherence of light

Bibliography: p.
Includes index
1. Coherence (Optics) I. Title
QC476.C6P47 1984 535′.2 85-2441
ISBN 90-277-2004-5

Published by D. Reidel Publishing Company,
P.O. Box 17, 3300 AA Dordrecht, Holland
in co-edition with SNTL-Publishers of Technical Literature – Prague, Czechoslovakia.

Distributed in Albania, Bulgaria, Chinese People's Republic, Cuba, Czechoslovakia, German Democratic Republic, Hungary, Korean People's Democratic Republic, Mongolia, Poland, Rumania, the U.S.S.R., Vietnam and Yugoslavia by Artia, Prague

Sold and distributed in the U.S.A. and Canada
by Kluwer Academic Publishers,
190 Old Derby Street, Hingham, MA 02043, U.S.A.

Sold and distributed in all other countries by
Kluwer Academic Publishers Group,
P.O. Box 322, 3300 AH Dordrecht, Holland

Printed in Czechoslovakia by SNTL, Prague

PREFACE TO THE SECOND EDITION

This book deals with the theory of coherence in both classical and quantum terms. Compared to the first edition of this book, it is substantially revised and extended. The classical treatment now includes a general review of research in the radiometry of light of arbitrary states of coherence, the theory developed in the group of Professor E. Wolf, and by other authors. In the quantum part of this book the coherent-state technique is further demonstrated, using the generalized Fokker-Planck equation and the Heisenbreg-Langevin equations, by the quantum statistical properties of light in random and nonlinear media. For this purpose the original treatment is extended to include general formulations of the problem of the interaction of radiation with matter.

While the previous monograph on Quantum Statistics of Linear and Nonlinear Optical Phenomena by this author contains only a basic account of the general methods in order to make it self-contained, the main attention being concentrated on detailed treatments of linear and nonlinear phenomena, this book mainly provides general classical and quantum methods, using stochastic and nonlinear phenomena as brief demonstrations (Chapters 19 – 22). Many new results are presented and most up-to-date references are included.

The author thanks his collaborators Drs. R. Horák, L. Mišta and V. Peřinová for co-operation for many years. Fruitful co-operation with Professors M. C. Teich and P. Diament of Columbia University, Professor G. Lachs of Pennsylvania State University and Professor S. Kielich and Dr. P. Szlachetka of A. Mickiewicz University is acknowledged. The author is obliged to Professor Bedřich Havelka of Palacký University and Professor Emil Wolf of Rochester University for their kind support, and to many scientists for regularly sending him preprints and reprints for many years. He is very indebted to Prof. R. M. Sillitto for his careful reading of the manuscript and for its improvement. The author highly appreciated the opportunity to be the I. I. Rabi Visiting Scientist at Columbia University in 1983. He also thanks very much Mrs. M. Rozsypalová for careful preparation of the figures.

Permission to reproduce figures is acknowledged from: The American Physical Society, Fig. 10.2 by B. L. Morgan and L. Mandel, Fig. 17. 1 by A. W. Smith and J. A. Armstrong, Fig. 17.2 by R. F. Chang, V. Korenman, C. O. Alley and R. W. Detenbeck, Fig. 17.4 by F. T. Arecchi, V. Degiorgio and B. Querzola, Fig. 22.8 by M. S. Zubairy and J. J. Yeh, Fig. 22.9 by M. Dagenais and L. Mandel; North-Holland Publishing Company, Fig. 4.1 by E. Wolf and W. H. Carter,

Fig. 10.1 by F. T. Arecchi, E. Gatti and A. Sona, Fig. 19.5 by R. E. Slusher; Optical Society of America, Fig. 17.7 by M. C. Teich and G. Vannucci;Springer Publishing House, Fig. 22.4 by J. Wagner, P. Kurowski and W. Martienssen; The Institute of Electrical and Electronics Engineers, Fig. 17. 3 by F. T. Arecchi, A. Berné, A. Sona and P. Burlamacchi; J. A. Barth Publishing House, Fig. 22.7 by U. Mohr and H. Paul.

Olomouc, May 1985 J. P.

TABLE OF CONTENTS

INTRODUCTION

Interference and diffraction phenomena of electromagnetic waves are usually described with the aid of *ideally coherent* or *ideally incoherent* light beams. In the first case the amplitudes of beams are superimposed and the superposition of such beams gives an observable interference pattern on the screen. In the second case the intensities are superimposed and the interference pattern is not observable. In fact both these cases are a mathematical idealization since real beams partially influence one another, i.e. they are correlated. Thus the actual case is an intermediate one, involving partially coherent light beams. A superposition of such beams gives an interference pattern whose visibility is less than the visibility of the interference pattern formed by coherent beams. The inadequacy of the description of light by ideally coherent or ideally incoherent beams was first proved by Verdet (1869) who showed that light from two pinholes illuminated by the Sun creates an observable interference pattern on the screen if the separation of the pinholes is less than about 1/20 mm. The Sun must be considered as an incoherent source composed of many elementary radiators (atoms) which practically do not influence one another. Since coherence is the property of coherent radiators (which are mutually synchronized), it is obvious that states intermediate to the states of coherence and incoherence, i.e. states of *partial coherence*, must be considered.

Nevertheless, little attention was devoted to partial coherence for a long time. However, from about 1960 the concept of partial coherence became important in many branches of physics: in the theory of the electromagnetic field in all spectral regions, but especially in optics (image formation, interferometry, spectroscopy), in radioastronomy, in the theory of masers and lasers, in optical communication (see for instance Helstrom (1972, 1976), Sheremetyev (1971), Kuriksha (1973), Gagliardi and Karp (1976)), but also in cross-disciplinary sciences such as biophysics and psychophysics (see Teich and McGill (1976), Teich, Matin and Cantor (1978), Teich, Prucnal, Vannucci, Breton and McGill (1982a, b), Prucnal and Teich (1982)).

Every optical field found in nature has certain statistical features because such an electromagnetic field is generated by many mostly uncorrelated elementary radiators (atoms) and so it represents, in general, a *statistical dynamic system.* Thus the theory of coherence, from a general point of view, is connected to the *statistical description* of the electromagnetic field, just as the *statistical properties* of optical fields manifest themselves as the *coherence properties* of these fields.

The concept of optical coherence was first introduced in connection with the description of interference and diffraction phenomena. However new possibilities of detecting optical fields as well as the preparation of man-made sources such as masers and lasers led to the need for a systematic and complete description and classification of optical coherence phenomena including coherence effects of all orders. This problem can be solved either from a *classical* (wave) or a *quantum* (particle) point of view. In the first case the theory of coherence is based upon the Maxwell wave theory and the theory of stochastic functions, whereas in the second case it is necessary to formulate the theory of coherence in terms of quantum electrodynamics. From this it is obvious that the theory of coherence may be regarded as a component of a more general theory – information theory – which investigates the ability of the field, considered as a statistical dynamic system, to transfer information. Consequently the theory of coherence is closely related to noise theory (in classical terms) and to the theory of quantum fluctuations of the field (in quantum terms).

The earliest investigations of partial coherence and polarization, limited to effects of the second order in the complex amplitudes, were carried out by Verdet (1869), von Laue (1907), Berek (1926a – c), Michelson (1890), Van Cittert (1934, 1939), Zernike (1938), Wiener (1928, 1929, 1930), Hopkins (1951, 1953, 1957a, b), Wolf (1954a, b, 1955, 1956), Blanc-Lapierre and Dumontet (1955) and Pancharatnam (1963). Their results were further continued in papers by Wolf (1959), Mandel (1961a, b, 1962), Mandel and Wolf (1961a), Parrent (1959a, b), Beran and Parrent (1962, 1963), Roman and Wolf (1960), Roman (1961a, b), Parrent and Roman (1960), Barakat (1963), Gamo (1964) and Gabor (1956, 1961). These considerations have been extended to the frequency domain by Mandel and Wolf (1976, 1981), Bastiaans (1977, 1981) and Wolf (1981, 1982). Recently, great effort has been devoted to the radiation from partially coherent sources and to the radiometry of fields of any state of coherence (Marchand and Wolf (1972, 1974a, b), Wolf and Carter (1975, 1976), Carter and Wolf (1975, 1977, 1981a, b), Wolf (1978), Friberg (1978a, b, 1979a, b, 1981a, b), Baltes, Geist and Walther (1978), Collet and Wolf (1979, 1980)).

A new period in the development of the theory of coherence began after the experiments of Hanbury Brown and Twiss (1956a – c, 1957a, b, 1958) in which the fourth-order correlation effects were measured. In principle experiments of this kind make it possible to consider correlation effects of arbitrary order. This is particularly important for the complete statistical description of non-chaotic (e.g. laser) light, chaotic light being completely described by the second-order moment – correlation function. Such a general description of the statistical properties of optical fields, in which coherence effects of all orders are included, was proposed in the classical terms of the theory of stochastic processes by Wolf (1963a).

As the information about the statistics of an optical field is obtained mainly using photodetectors (which are a special type of a more general class of quadratic detectors) it was necessary to find a relation between the statistics of photons in the field and the statistics of photoelectrons emitted by a photodetector exposed in this field. This was done by Mandel (1958, 1959, 1963a), who derived classically the so-called photodetection equation relating the probability distribution of the integrated intensity to the probability distribution of emitted photoelectrons.

In 1963 Glauber (1963a – c, 1964, 1965, 1966a, b, 1967, 1969, 1970, 1972) developed the quantum theory of coherence based on quantum electrodynamics. Quantum correlation functions introduced by him represent expectation values of normally ordered products of field operators and they are closely related to the quantities measured by means of photoelectric detectors. The "diagonal" representation of the density matrix obtained by introducing the so-called coherent states, obtained by Glauber (1963a – c) and Sudarshan (1963a, b), enables us to study the relation between the quantum and classical descriptions of the statistical properties of optical fields. In this representation the classical and quantum correlation functions are formally equivalent if a generalized phase-space distribution is introduced. In the quantum theory of coherence the problem of correspondence between functions of operators (q-numbers) and functions of classical quantities (c-numbers) plays an important role. This correspondence was investigated by Agarwal and Wolf (1968a – c, 1970), Lax (1968b) and Cahill and Glauber (1969). In particular it was shown that antinormally ordered products of field operators are closely related to quantities measured by the so-called quantum counters (Mandel (1966b)), which operate by stimulated emission rather than by absorption.

The mathematical technique used in connection with partial coherence is also useful in the analysis of partial polarization. Both these phenomena may be unified in the correlation theory using the correlation tensors.

An interesting feature of the theory of coherence is that it operates with measurable quantities only. The classical Maxwell theory of the electromagnetic field assumes that the electric and magnetic fields are measurable functions in space and time. In fact electromagnetic vibrations, such as light, are represented by rapidly oscillating quantities and no real detector can follow such rapid changes. Apart from this, the field represents a statistical dynamic system, as has already been mentioned. Therefore it is necessary to introduce an averaging process for physical quantities and only such averaged quantities can be measured. Thus the laws of the electromagnetic field are formulated in the theory of coherence in terms of measurable quantities, including fields in interaction.

The purpose of this book is to investigate the statistical and coherence properties of free electromagnetic fields as well as of fields in interaction with matter and their detection (Scully and Whitney (1972), Agarwal (1973), Haake (1973), Davies (1976)), including the interaction with random and nonlinear media, particularly in the laser (Risken (1968, 1970), Lax (1967a, b, 1968a, b), Haken (1967, 1970a, b, 1972), Lax and Zwanziger (1970, 1973), Louisell (1970, 1973), Sargent and Scully (1972), Sargent, Scully and Lamb (1974)), in nonlinear optical processes (Shen (1967), Loudon (1973), Nussenzveig (1973), Graham (1973, 1974), Schubert and Wilhelmi (1976, 1978, 1980), Peřina (1980, 1984), Peřina, Peřinová, Sibilia and

Bertolotti (1984), Peřina, Peřinová and Koďousek (1984)) and in scattering processes (Crosignani, Di Porto and Bertolotti (1975)). Many experimental results have been reported in the literature (Armstrong and Smith (1967), Pike (1969, 1970), Pike and Jakeman (1974), Arecchi and Degiorgio (1972), Akhmanov, Dyakov and Tchirkin (1981), Demtchuk and Ivanov (1981)). Optical pumping has been discussed for example by Cohen-Tannoudji and Kastler (1966) and by Series (1970). The main subject of our investigations is light but most of our results are valid for the electromagnetic field regardless of the spectral region.

Compared to the first edition of this book, questions of the radiometry of light of any state of coherence and of the quantum statistical properties of radiation in interaction with random and nonlinear media are additionally treated, making use of the Heisenberg-Langevin method and the master equation and the generalized Fokker-Planck equation methods. The most interesting property of the nonlinear optical processes is that, under certain conditions, they provide radiation having no classical analogue exhibiting the so-called *anticorrelation* or *antibunching of photons* whose statistics can be sub-poissonian (Loudon (1980), Schubert and Wilhelmi (1980), Peřina (1980), Paul (1982)), i.e. they have less uncertainty in the number of photons than the field in the coherent state and the variance of their intensity is negative. Recently a great success has been achieved in this field when these quantum aspects of quantum optical systems have been experimentally observed in pioneering experiments by Kimble, Dagenais and Mandel (1977, 1978), Dagenais and Mandel (1978) and by Leuchs, Rateike and Walther (1979) in resonance fluorescence of sodium atoms. A simulation experiment based on second-harmonic generation has been performed by Wagner, Kurowski and Martienssen (1979). Stationary sub-poissonian light has been observed for the first time by Teich and Saleh (1985) using the Franck-Hertz experiment (for the theory, see Teich, Saleh and Peřina (1984)).

There exist several sources on the subject of the theory of coherence. The classical theory has been treated in the following books and reviews: "Principles of Optics" by M. Born and E. Wolf (1965), "Diffraction. Structure des images" by A. Maréchal and M. Francon (1960), "Theory of Partial Coherence" by M. Beran and G. B. Parrent (1964), "Cohérence en optique" by M. Francon and S. Slansky (1965), "Introduction to Statistical Optics" by E. L. O'Neill (1963), "Diffraction. Coherence in Optics" by M. Francon (1966), "Image Formation with Partially Coherent Light" by B. J. Thompson (1969), "Theory of the Electromagnetic Field" by V. A. Potechin and V. N. Tatarinov (1978) and "Elements of Optical Coherence Theory" by A. S. Marathay (1982). A survey of the quantum theory of coherence can be found in lectures by R. J. Glauber (1965, 1969, 1970, 1972) given at Summer Schools. Both the classical and quantum theories of coherence and their interrelations have been treated by L. Mandel and E. Wolf (1965), J. Peřina (1972, 1974, 1975) and by J. R. Klauder and E. C. G. Sudarshan (1968) in their book "Fundamentals of Quantum Optics". Brief reviews of the theory of coherence can be found in the books "Optical Coherence Theory" by

G. J. Troup (1967) and "Optical Coherence" by J. F. Vinson (1971). Particular problems have been also discussed, such as photoelectron statistics by Saleh (1978), coherence and correlation in scattering by Chu (1974) and Crosignani, Di Porto and Bertolotti (1975) and quantum statistics in nonlinear optics by Schubert and Wilhelmi (1980), Peřina (1980, 1984), Kielich (1981) and Klyshko (1980). The generalized coherent states have been systematically discussed by Malkin and Man'ko (1979). Interesting historical remarks can be found in the book "Masers and Lasers" by M. Bertolotti (1983).

DEFINITIONS AND MATHEMATICAL PRELIMINARIES

2.1 Complex value representation of real polychromatic fields

It is well known that the electromagnetic field is a real physical field in contrast to the electron-positron field which is naturally complex. However in classical coherence theory it is advantageous to represent the real electromagnetic field by a complex quantity because of its mathematical simplicity and also because it serves to emphasize that the coherence theory deals with phenomena which are sensitive to the "envelope" or to the "average intensity" of the field. In spite of the fact that such a complex representation is rather artificial in the classical theory it has a deep physical meaning in the quantum theory, providing insight into the detection process. This complex representation provides a bridge between the classical and quantum formulations of coherence phenomena, and the complex representation of the real polychromatic field, introduced in the following, represents a natural generalization of the well-known complex representation of monochromatic fields used in classical optics.

Let us consider the scalar quantity $V^{(r)}(t)$ which may represent for example a Cartesian component of the electric vector $\mathbf{E}$ of the electromagnetic field. If $V^{(r)}(t)$ possesses a Fourier transform, we can write

$$V^{(r)}(t) = \int_{-\infty}^{+\infty} \tilde{V}^{(r)}(\nu) \exp(-\mathrm{i}\, 2\pi\nu t)\, \mathrm{d}\nu =$$

$$= 2\int_{0}^{\infty} |\tilde{V}^{(r)}(\nu)| \cos(2\pi\nu t - \arg \tilde{V}^{(r)}(\nu))\, \mathrm{d}\nu, \tag{2.1}$$

where

$$\tilde{V}^{(r)}(\nu) = \int_{-\infty}^{+\infty} V^{(r)}(t) \exp(\mathrm{i}\, 2\pi\nu t)\, \mathrm{d}t \tag{2.2}$$

and the condition $\tilde{V}^{(r)*}(\nu) = \tilde{V}^{(r)}(-\nu)$ has been used. This condition is a consequence of the reality of $V^{(r)}(t)$ (the asterisk denotes complex conjugation). Hence, the negative frequency components ($\nu < 0$) do not carry any physical information which is not already contained in the positive frequency ones ($\nu > 0$) and consequently they can be omitted. Thus we may employ the complex function

$$V(t) = \int_{0}^{\infty} \tilde{V}^{(r)}(\nu) \exp(-\mathrm{i}\, 2\pi\nu t)\, \mathrm{d}\nu, \tag{2.3}$$

where

$$\tilde{V}^{(r)}(\nu) = \int_{-\infty}^{+\infty} V(t) \exp(\mathrm{i}\, 2\pi\nu t)\, \mathrm{d}t, \qquad \nu \geqq 0,$$
$$= 0, \qquad \nu < 0. \tag{2.4}$$

The function $V(t)$, called the *complex analytic signal*, was introduced by Gabor (1946) (see also Born and Wolf (1965), Sec. 10.2 and Beran and Parrent (1964), Chap. 2). The function $V(t)$ can be written in the form

$$V(t) = \tfrac{1}{2}[V^{(r)}(t) + \mathrm{i}V^{(i)}(t)], \tag{2.5}$$

where

$$V^{(i)}(t) = -2 \int_0^{\infty} |\tilde{V}^{(r)}(\nu)| \sin(2\pi\nu t - \arg \tilde{V}^{(r)}(\nu))\, \mathrm{d}\nu. \tag{2.6}$$

As the total energy $\int_0^{\infty} |\tilde{V}^{(r)}(\nu)|^2 \mathrm{d}\nu$ (which, using Parseval's equality, equals $\int_{-\infty}^{+\infty} |V(t)|^2 \mathrm{d}t$) is assumed to be finite, the function (2.3) is (using the Schwarz inequality) analytic in the lower half of the complex z-plane ($z = t + \mathrm{i}\vartheta$) and $V(t)$ is the boundary value of such an analytic function for $\vartheta \to -0$. Hence, making use of the Cauchy theorem on analytic functions, we obtain the *dispersion relations*, which are also called *Hilbert transforms* (Titchmarsh (1948))

$$V^{(i)}(t) = \frac{1}{\pi} P \int_{-\infty}^{+\infty} \frac{V^{(r)}(t')}{t' - t} \mathrm{d}t', \qquad V^{(r)}(t) = -\frac{1}{\pi} P \int_{-\infty}^{+\infty} \frac{V^{(i)}(t')}{t' - t} \mathrm{d}t', \tag{2.7}$$

where P denotes the Cauchy principal value of the integral at $t = t'$. (A more detailed derivation of dispersion relations is contained in Sec. 4.4.) Equations (2.7) show that the functions $V^{(r)}(t)$ and $V^{(i)}(t)$ are connected to each other and we see again that $V(t)$ given by (2.5) cannot contain more physical information than $V^{(r)}(t)$.

Defining the *generalized functions*

$$\delta^{(\pm)}(t) = \int_0^{\infty} \exp(\pm \mathrm{i}\, 2\pi\nu t)\, \mathrm{d}\nu = \lim_{\varepsilon \to +0} \left[-\frac{1}{2\pi\mathrm{i}(\pm t + \mathrm{i}\varepsilon)} \right] = \frac{1}{2}\left[\delta(t) \pm \frac{\mathrm{i}}{\pi} \frac{P}{t} \right],$$

where $\delta(t)$ is the Dirac function, we obtain by substituting (2.2) into (2.3)

$$V(t) = \int_{-\infty}^{+\infty} V^{(r)}(t')\, \delta^{(-)}(t - t')\, \mathrm{d}t' = \lim_{\varepsilon \to +0} \frac{1}{2\pi\mathrm{i}} \int_{-\infty}^{+\infty} \frac{V^{(r)}(t')}{t - t' - \mathrm{i}\varepsilon} \mathrm{d}t' =$$
$$= \frac{1}{2}\left[V^{(r)}(t) - \frac{\mathrm{i}}{\pi} P \int_{-\infty}^{+\infty} \frac{V^{(r)}(t')}{t - t'} \mathrm{d}t' \right] \tag{2.8a}$$

and

$$V^*(t) = \int_{-\infty}^{+\infty} V^{(r)}(t')\, \delta^{(+)}(t - t')\, \mathrm{d}t' = \lim_{\varepsilon \to +0} \frac{1}{2\pi \mathrm{i}} \int_{-\infty}^{+\infty} \frac{V^{(r)}(t')}{t' - t - \mathrm{i}\varepsilon}\, \mathrm{d}t' =$$
$$= \frac{1}{2}\left[V^{(r)}(t) + \frac{\mathrm{i}}{\pi} P \int_{-\infty}^{+\infty} \frac{V^{(r)}(t')}{t - t'}\, \mathrm{d}t' \right]. \tag{2.8b}$$

Thus if the function $V^{(r)}(t)$ fulfils the wave equation in vacuo,

$$\nabla^2 V^{(r)}(t') - \frac{1}{c^2} \frac{\partial^2 V^{(r)}(t')}{\partial t'^2} = 0, \tag{2.9}$$

where ∇^2 is the Laplace operator, we obtain the same equation for the complex function $V(t)$ by multiplying (2.9) by $\delta^{(-)}(t - t')$ and integrating over t',

$$\nabla^2 V(t) - \frac{1}{c^2} \int_{-\infty}^{+\infty} \frac{\partial^2 V^{(r)}(t')}{\partial t'^2}\, \delta^{(-)}(t - t')\, \mathrm{d}t' =$$
$$= \nabla^2 V(t) - \frac{1}{c^2} \int_{-\infty}^{+\infty} V^{(r)}(t') \frac{\partial^2}{\partial t'^2}\, \delta^{(-)}(t - t')\, \mathrm{d}t' =$$
$$= \nabla^2 V(t) - \frac{1}{c^2} \frac{\partial^2}{\partial t^2} V(t) = 0, \tag{2.10}$$

where we have twice integrated by parts; c in (2.9) and (2.10) denotes the velocity of light *in vacuo*.

2.2 Correlation functions and their properties

Let the electromagnetic field in the complex representation with polarization μ at a point $\mathbf{x}$ at time t be $V_\mu(\mathbf{x}, t)$. We define the $(m + n)$th-order *correlation function* as an ensemble average of the product of these field functions considered at different space-time points and for different polarizations as follows (Wolf (1963a, 1964, 1965, 1966))

$$\Gamma^{(m,n)}_{\mu_1 \ldots \mu_{m+n}}(\mathbf{x}_1, \ldots, \mathbf{x}_{m+n}, t_1, \ldots, t_{m+n}) = \langle \prod_{j=1}^{m} V^*_{\mu_j}(\mathbf{x}_j, t_j) \prod_{k=m+1}^{m+n} V_{\mu_k}(\mathbf{x}_k, t_k) \rangle, \tag{2.11}$$

where the brackets $\langle \ldots \rangle$ denote an ensemble average. Quantum correlation functions corresponding to (2.11) suitable for the description of the coherence properties of fields determined using photodetectors can be written in the form

$$\Gamma^{(m,n)}_{\mathcal{N}, \mu_1 \ldots \mu_{m+n}}(x_1, \ldots, x_{m+n}) = \mathrm{Tr}\, \{\hat{\varrho} \prod_{j=1}^{m} \hat{A}^{(-)}_{\mu_j}(x_j) \prod_{k=m+1}^{m+n} \hat{A}^{(+)}_{\mu_k}(x_k)\}, \tag{2.12}$$

where $x = (\mathbf{x}, t)$, $\hat{\varrho}$ is the density matrix, Tr denotes the trace and $\hat{A}^{(+)}_\mu(x)$ and $\hat{A}^{(-)}_\mu(x)$ are the annihilation and creation operators of a photon of the polarization μ at x

respectively (Glauber (1963a, b, 1964)). The suffix $\mathcal{N}$ indicates that the product of operators $\hat{\mathbf{A}}^{(+)}$ and $\hat{\mathbf{A}}^{(-)}$ is in *normal form*, i.e. all annihilation operators $\hat{\mathbf{A}}^{(+)}$ stand to the right of all creation operators $\hat{\mathbf{A}}^{(-)}$. The correspondence between the correlation functions (2.11) and (2.12) is given by substitutions $\langle\ldots\rangle \rightleftarrows \mathrm{Tr}\,\{\hat{\varrho}\ldots\}$, $\mathbf{V} \rightleftarrows \hat{\mathbf{A}}^{(+)}$ and $\mathbf{V}^* \rightleftarrows \hat{\mathbf{A}}^{(-)}$. Of course, this correspondence is not unique since the operators $\hat{\mathbf{A}}^{(+)}$ and $\hat{\mathbf{A}}^{(-)}$ do not commute. A rigorous definition of the correlation functions (2.11) will be given in Chapter 8. The correlation functions (2.12) and quantum correlation functions involving other orderings of the field operators $\hat{\mathbf{A}}^{(+)}$ and $\hat{\mathbf{A}}^{(-)}$ will be discussed in Chapters 12, 14 and 16, together with their correspondence to the classical correlation functions (2.11) and to various types of physical measurements.

An important class of optical fields appearing in nature is the class of *stationary* and *ergodic* fields. For a stationary field the correlation functions (as well as the probability densities describing the statistics of the field, Sec. 8.1) are independent of translations of the time origin. For an ergodic field the *ensemble* average can be replaced by the *time* average,

$$\langle\ldots\rangle = \lim_{T\to\infty} \frac{1}{2T} \int_{-T}^{+T} \ldots \,\mathrm{d}t. \tag{2.13}$$

For this class of fields the correlation function (2.11) can be rewritten in the form

$$\Gamma^{(m,n)}(\mathbf{x}_1, \ldots, \mathbf{x}_{m+n}, t_1, \ldots, t_{m+n}) \equiv \Gamma^{(m,n)}(\mathbf{x}_1, \ldots, \mathbf{x}_{m+n}, \tau_2, \ldots, \tau_{m+n}) =$$
$$= \lim_{T\to\infty} \frac{1}{2T} \int_{-T}^{+T} \prod_{j=1}^{m} V^*(\mathbf{x}_j, t + \tau_j) \prod_{k=m+1}^{m+n} V(\mathbf{x}_k, t + \tau_k)\,\mathrm{d}t; \tag{2.14}$$

τ_1 can be put to zero and then $\tau_j = t_j - t_1$ for $j = 1, 2, \ldots, m + n$. In (2.14) we have omitted the polarization indices for simplicity; or we can assume that light at a space-time point $x_j \equiv (\mathbf{x}_j, t_j)$ is of polarization μ_j without explicitly writing the polarization indices (i.e. $x_j \equiv (\mathbf{x}_j, t_j, \mu_j)$)†.

The correlation function of the second order ($m = n = 1$) for stationary and ergodic fields is called the *mutual coherence function* and it can be written as

$$\Gamma(\mathbf{x}_1, \mathbf{x}_2, \tau) \equiv \Gamma^{(1,1)}(\mathbf{x}_1, \mathbf{x}_2, \tau) = \lim_{T\to\infty} \frac{1}{2T} \int_{-T}^{+T} V^*(\mathbf{x}_1, t)\, V(\mathbf{x}_2, t + \tau)\,\mathrm{d}t. \tag{2.15}$$

The correlation functions for stationary fields depend on time differences $\tau_j = t_j - t_1$ and the mutual coherence function depends on the time difference $\tau = t_2 - t_1$ only.

The mutual coherence function, which can be concisely denoted as $\Gamma_{12}(\tau)$, will play an important role in the description of the second-order coherence effects

† For another definition of the time average using the so-called truncated functions ($V_T^{(r)}(t) = V^{(r)}(t)$ for $|t| < T$ and $V_T^{(r)}(t) = 0$ for $|t| > T$) we refer the reader to discussions by Born and Wolf (1965) and Beran and Parrent (1964).

connected with classical interference and diffraction phenomena. The higher-order correlation functions will describe higher-order coherence phenomena observed with the help of a number of quadratic detectors (photoelectric detectors) whose photocurrents are correlated, or employing nonlinear media. The earlier investigations of the coherence properties of light used only the function $\Gamma_{12}(\tau)$ (Born and Wolf (1965), Beran and Parrent (1964), Francon and Slansky (1965), Francon (1966), Thompson (1969)).

Let us note here that the difference between the classical and quantum correlation functions (2.11) and (2.12) arises from the averaging, i.e. from the character of the angle brackets. Consequently results concerning the space-time-polarization behaviour of the correlation functions, such as

(i) the spectral decomposition of correlation functions,
(ii) the interference law,
(iii) wave equations,
(iv) the propagation laws of the correlation functions,
(v) the concept of cross-spectral purity of light,
(vi) the conservation laws for correlation functions,
(vii) the matrix formulation of the theory and
(viii) the formulation of the polarization properties of light, etc.,

will be independent of the character of the angle brackets and therefore they are valid for classical as well as quantum correlation functions. This will be demonstrated using the Glauber-Sudarshan diagonal representation of the density matrix which allows the quantum average of operators (q-numbers) to be transformed into a "classical" average of classical fields (c-numbers) in a phase space.

Spectral and analytic properties of the correlation functions

Assuming that the field $V(\mathbf{x}, t)$ possesses a Fourier transform

$$V(\mathbf{x}, t) = \int_0^\infty \tilde{V}(\mathbf{x}, \nu) \exp(-\mathrm{i}\, 2\pi\nu t)\, \mathrm{d}\nu, \tag{2.16}$$

(we write $\tilde{V}$ instead of $\tilde{V}^{(r)}$ for simplicity) we can derive the following spectral decomposition of the correlation function. Substituting (2.16) into (2.11), we have

$$\Gamma^{(m,n)}(\mathbf{x}_1, \ldots, \mathbf{x}_{m+n}, \mathbf{t}) = \int_0^\infty G^{(m,n)}(\mathbf{x}_1, \ldots, \mathbf{x}_{m+n}, \boldsymbol{\nu}) \exp[\mathrm{i}\, 2\pi(\boldsymbol{\nu}, \mathbf{t})]\, \mathrm{d}\boldsymbol{\nu}, \tag{2.17}$$

where

$$G^{(m,n)}(\mathbf{x}_1, \ldots, \mathbf{x}_{m+n}, \boldsymbol{\nu}) = \langle \prod_{j=1}^{m} \tilde{V}^*(\mathbf{x}_j, \nu_j) \prod_{k=m+1}^{m+n} \tilde{V}(\mathbf{x}_k, \nu_k)\rangle =$$

$$= \int_{-\infty}^{+\infty} \Gamma^{(m,n)}(\mathbf{x}_1, \ldots, \mathbf{x}_{m+n}, \mathbf{t}) \exp[-\mathrm{i}\, 2\pi(\boldsymbol{\nu}, \mathbf{t})]\, \mathrm{d}\mathbf{t}, \quad (\nu_j \geqq 0 \text{ for all } j) \tag{2.18}$$

is the *spectral correlation function*. Here $\mathbf{t} \equiv (t_1, \ldots, t_{m+n})$,

$$\boldsymbol{\nu} \equiv (\nu_1, \ldots, \nu_m, -\nu_{m+1}, \ldots, -\nu_{m+n}), \qquad \mathrm{d}\mathbf{t} \equiv \prod_{j=1}^{m+n} \mathrm{d}t_j, \qquad \mathrm{d}\boldsymbol{\nu} \equiv \prod_{j=1}^{m+n} \mathrm{d}\nu_j$$

and

$$(\boldsymbol{\nu}, \mathbf{t}) \equiv \sum_{j=1}^{m+n} \varepsilon_j \nu_j t_j, \qquad \varepsilon_j = \begin{cases} +1, & j = 1, \ldots, m, \\ -1, & j = m+1, \ldots, m+n. \end{cases}$$

For a stationary and ergodic field we obtain the corresponding relations using the definition (2.14) of the correlation function and the spectral decomposition (2.16),

$$\Gamma^{(m,n)}(\mathbf{x}_1, \ldots, \mathbf{x}_{m+n}, \boldsymbol{\tau}) = \int_0^\infty G^{(m,n)}(\mathbf{x}_1, \ldots, \mathbf{x}_{m+n}, \boldsymbol{\nu}) \times$$
$$\times\, \delta\Big(\sum_{j=1}^{m+n} \varepsilon_j \nu_j\Big) \exp\left[\mathrm{i}\, 2\pi(\boldsymbol{\nu}, \boldsymbol{\tau})\right] \mathrm{d}\boldsymbol{\nu}, \tag{2.19}$$

where the spectral correlation function is given by

$$G^{(m,n)}(\mathbf{x}_1, \ldots, \mathbf{x}_{m+n}, \boldsymbol{\nu}) = \lim_{T \to \infty} \frac{1}{2T} \Big\langle \prod_{j=1}^{m} \tilde{V}^*(\mathbf{x}_j, \nu_j) \prod_{k=m+1}^{m+n} \tilde{V}(\mathbf{x}_k, \nu_k) \Big\rangle_e \tag{2.20}$$

and

$$\boldsymbol{\tau} \equiv (\tau_2, \tau_3, \ldots, \tau_{m+n}), \qquad (\boldsymbol{\nu}, \boldsymbol{\tau}) \equiv \sum_{j=2}^{m+n} \varepsilon_j \nu_j \tau_j.$$

The brackets $\langle \ldots \rangle_e$ in (2.20) denote an ensemble average ensuring the existence of the above limit for stationary fields (cf. a discussion by Born and Wolf (1965) and Beran and Parrent (1964); a connection of the theories of coherence in t- and ν-domains has been discussed by Wolf (1981, 1982)). Time-dependent spectra have been investigated by Eberly and Wódkiewicz (1977) and Gase and Schubert (1982). Performing the integration in (2.19) over the variable ν_1 and introducing the quantity

$$G_H^{(m,n)}(\mathbf{x}_1, \ldots, \mathbf{x}_{m+n}, \boldsymbol{\nu}) \equiv$$
$$\equiv G^{(m,n)}\Big(\mathbf{x}_1, \ldots, \mathbf{x}_{m+n}, -\sum_{j=2}^{m+n} \varepsilon_j \nu_j, \nu_2, \ldots, \nu_m, -\nu_{m+1}, \ldots, -\nu_{m+n}\Big),$$

which is the spectral correlation function considered on the hypersurface

$$\sum_{j=1}^{m+n} \varepsilon_j \nu_j = 0 \tag{2.21}$$

of the space $\nu_1 \otimes \nu_2 \otimes \ldots \otimes \nu_{m+n}$, we can invert (2.19) giving

$$G_H^{(m,n)}(\mathbf{x}_1, \ldots, \mathbf{x}_{m+n}, \boldsymbol{\nu}) = \int_{-\infty}^{+\infty} \Gamma^{(m,n)}(\mathbf{x}_1, \ldots, \mathbf{x}_{m+n}, \boldsymbol{\tau}) \exp\left[-\mathrm{i}\, 2\pi(\boldsymbol{\nu}, \boldsymbol{\tau})\right] \mathrm{d}\boldsymbol{\tau},$$
$$(\nu_j \geqq 0 \text{ for all } j). \tag{2.22}$$

Equations (2.17) and (2.19) represent the *generalized Wiener-Khintchine theorem.*

From (2.17) and (2.19) it can be seen that

$$G^{(m,n)}(\mathbf{v}) = G_H^{(m,n)}(\mathbf{v})\,\delta\Big(\sum_{j=1}^{m+n} \varepsilon_j v_j\Big), \tag{2.23}$$

where we have suppressed the space dependence. Substituting this expression into (2.17) we obtain (2.19) with $\tau_j = t_j - t_1$. The δ-function dependence implies that $(m+n)$ frequency components $\tilde{V}(v_1), \ldots, \tilde{V}(v_{m+n})$ of a stationary field may be correlated if and only if the frequencies $v_1, \ldots, v_{m+n}$ are coupled by the relation (2.21).

Of course, the same frequency condition (2.21) is a necessary condition for the non-vanishing of the correlation function of a stationary field. This follows from the definitions (2.11) and (2.12) of the correlation functions using the ensemble and quantum average, respectively, taking into account that such correlation functions are independent of the translation of the time origin, i.e.

$$\Gamma^{(m,n)}(t_1 + \tau, t_2 + \tau, \ldots, t_{m+n} + \tau) = \Gamma^{(m,n)}(t_1, t_2, \ldots, t_{m+n}),$$

which follows using the stationary condition and the spectral decomposition (2.16). In this way (2.18) is appropriate with (2.21) involved. The connection between this definition of the stationary property and the quantum-mechanical definition $[\hat{\varrho}, \hat{H}] = \hat{0}$, where $\hat{\varrho}$ is the density matrix, $\hat{H}$ is the Hamiltonian of the field and $[\hat{\varrho}, \hat{H}] = \hat{\varrho}\hat{H} - \hat{H}\hat{\varrho}$ is the commutator, will be discussed in greater detail in Chapter 15.

We note that (2.22) may be useful for getting information about mode coupling in laser light from the measured correlation functions. Some further details about the spectral decompositions of the correlation functions and their applications to optical imagery, spectroscopy, etc., are available (Mandel (1964d), Wolf (1965), Mehta and Mandel (1967), Peřina (1969)).

For the description of second-order coherence and polarization phenomena the mutual coherence function (2.15) will be appropriate. In this case we obtain from (2.19)

$$\Gamma(\mathbf{x}_1, \mathbf{x}_2, \tau) = \int_0^\infty G(\mathbf{x}_1, \mathbf{x}_2, v) \exp(-\mathrm{i}\,2\pi v\tau)\,\mathrm{d}v, \tag{2.24}$$

where

$$\begin{aligned} G(\mathbf{x}_1, \mathbf{x}_2, v) &\equiv G^{(1,1)}(\mathbf{x}_1, \mathbf{x}_2, v) \equiv G^{(1,1)}(\mathbf{x}_1, \mathbf{x}_2, v, -v) = \\ &= \lim_{T\to\infty} \frac{1}{2T} \langle \tilde{V}^*(\mathbf{x}_1, v)\, \tilde{V}(\mathbf{x}_2, v)\rangle_e = \int_{-\infty}^{+\infty} \Gamma(\mathbf{x}_1, \mathbf{x}_2, \tau) \exp(\mathrm{i}\,2\pi v\tau)\,\mathrm{d}\tau, \quad v \geqq 0, \\ &= 0, \qquad v < 0, \end{aligned} \tag{2.25}$$

which follows from (2.20) and (2.22). The function (2.25) is called the *cross-spectral density* or the *mutual spectral density*. Equation (2.24) is known in the theory of stationary stochastic processes as the *Wiener-Khintchine theorem*.

One can see from (2.24) that the mutual coherence function $\Gamma_{12}(\tau) \equiv \Gamma(\mathbf{x}_1, \mathbf{x}_2, \tau)$ is analytic in the lower half-plane of the complex τ-plane as the analytic signal (2.3) is. Therefore the dispersion relations hold for the mutual coherence function, in analogy to (2.7),

$$\Gamma_{12}^{(i)}(\tau) = \frac{1}{\pi} P \int_{-\infty}^{+\infty} \frac{\Gamma_{12}^{(r)}(\tau')}{\tau' - \tau} \mathrm{d}\tau', \quad \Gamma_{12}^{(r)}(\tau) = -\frac{1}{\pi} P \int_{-\infty}^{+\infty} \frac{\Gamma_{12}^{(i)}(\tau')}{\tau' - \tau} \mathrm{d}\tau'; \tag{2.26}$$

the functions $\Gamma_{12}^{(r)}$ and $\Gamma_{12}^{(i)}$ may be determined in the following way. Using (2.5) we have

$$\Gamma_{12}(\tau) = \langle V_1^*(t)\, V_2(t+\tau)\rangle = \frac{1}{4}\{\langle V_1^{(r)}(t)\, V_2^{(r)}(t+\tau)\rangle + \langle V_1^{(i)}(t)\, V_2^{(i)}(t+\tau)\rangle\}$$
$$- \frac{\mathrm{i}}{4}\{\langle V_1^{(i)}(t)\, V_2^{(r)}(t+\tau)\rangle - \langle V_1^{(r)}(t)\, V_2^{(i)}(t+\tau)\rangle\}. \tag{2.27}$$

However, from the direct and inverted dispersion relations we obatin

$$\langle V_1^{(r)}(t)\, V_2^{(r)}(t+\tau'')\rangle = -\frac{1}{\pi^2} P \int_{-\infty}^{+\infty} \frac{\mathrm{d}\tau'}{\tau' - \tau''} P \int_{-\infty}^{+\infty} \frac{\langle V_1^{(r)}(t)\, V_2^{(r)}(t+\tau)\rangle}{\tau - \tau'} \mathrm{d}\tau =$$
$$= -\frac{1}{\pi} P \int_{-\infty}^{+\infty} \frac{\mathrm{d}\tau'}{\tau' - \tau''} \langle V_1^{(r)}(t)\, V_2^{(i)}(t+\tau')\rangle =$$
$$= -\frac{1}{\pi} P \int_{-\infty}^{+\infty} \frac{\mathrm{d}\tau'}{\tau' - \tau''} \langle V_1^{(r)}(t-\tau')\, V_2^{(i)}(t)\rangle = \langle V_1^{(i)}(t)\, V_2^{(i)}(t+\tau'')\rangle, \tag{2.28a}$$

where we have used the invariance property of the mutual coherence function with respect to translations of the time origin. We can obtain in the same way

$$\langle V_1^{(r)}(t)\, V_2^{(i)}(t+\tau)\rangle = -\langle V_1^{(i)}(t)\, V_2^{(r)}(t+\tau)\rangle, \tag{2.28b}$$

so that (2.27) gives for the mutual coherence function

$$\Gamma_{12}(\tau) = \frac{1}{2}\langle V_1^{(r)}(t)\, V_2^{(r)}(t+\tau)\rangle + \frac{\mathrm{i}}{2}\langle V_1^{(r)}(t)\, V_2^{(i)}(t+\tau)\rangle. \tag{2.29}$$

Hence

$$\Gamma_{12}^{(r)}(\tau) = \tfrac{1}{2}\langle V_1^{(r)}(t)\, V_2^{(r)}(t+\tau)\rangle = \tfrac{1}{2}\langle V_1^{(i)}(t)\, V_2^{(i)}(t+\tau)\rangle,$$
$$\Gamma_{12}^{(i)}(\tau) = \tfrac{1}{2}\langle V_1^{(r)}(t)\, V_2^{(i)}(t+\tau)\rangle = -\tfrac{1}{2}\langle V_1^{(i)}(t)\, V_2^{(r)}(t+\tau)\rangle; \tag{2.30}$$

we note that the same result follows from the condition $\langle V_1(t)\, V_2(t+\tau)\rangle = 0$, which is a consequence of the non-negativeness of frequencies (Mandel (1963a)). In particular for $\mathbf{x}_1 = \mathbf{x}_2$ and $\tau = 0$

$$\langle [V^{(r)}(\mathbf{x}, t)]^2\rangle = \langle [V^{(i)}(\mathbf{x}, t)]^2\rangle = \tfrac{1}{2}\langle |V(\mathbf{x}, t)|^2\rangle, \tag{2.31a}$$

$$\langle V^{(r)}(\mathbf{x}, t)\, V^{(i)}(\mathbf{x}, t)\rangle = 0. \tag{2.31b}$$

From the definitions of $\Gamma_{12}(\tau)$ and $G_{12}(\nu)$ the following identities follow immediately

$$\Gamma_{12}^*(\tau) = \Gamma_{21}(-\tau), \tag{2.32a}$$

$$G_{12}^*(\nu) = G_{21}(\nu). \tag{2.32b}$$

The first condition is called the *cross-symmetry condition* and the second one expresses the hermiticity of G as a matrix.

Analogous results for the third- and fourth-order correlation functions have been discussed by Beran and Corson (1965).

One can see from (2.17) that the correlation function $\Gamma^{(m,n)}(t_1, \ldots, t_{m+n})$ is an analytic function in the upper half of the complex t-plane in variables $t_1, \ldots, t_m$ and in the lower half of the complex t-plane in variables $t_{m+1}, \ldots, t_{m+n}$ and consequently the dispersion relations hold,

$$\operatorname{Im} \Gamma^{(m,n)}(t_1, \ldots, t_{m+n}) = -\varepsilon_k \frac{1}{\pi} P \int_{-\infty}^{+\infty} \frac{\operatorname{Re} \Gamma^{(m,n)}(t_1, \ldots, t'_k, \ldots, t_{m+n})}{t'_k - t_k} \mathrm{d}t'_k,$$

$$\operatorname{Re} \Gamma^{(m,n)}(t_1, \ldots, t_{m+n}) = \varepsilon_k \frac{1}{\pi} P \int_{-\infty}^{+\infty} \frac{\operatorname{Im} \Gamma^{(m,n)}(t_1, \ldots, t'_k, \ldots, t_{m+n})}{t'_k - t_k} \mathrm{d}t'_k, \quad (2.33)$$

where $\varepsilon_k = 1$ for $k = 1, \ldots, m$ and $\varepsilon_k = -1$ for $k = m + 1, \ldots, m + n$. In (2.33) Im and Re denote the imaginary and real parts respectively. Analogous relations are valid for stationary fields in the complex τ-plane (cf. (2.19)).

The wave equations *in vacuo*

As a consequence of the wave equation (2.10) for the complex amplitude of the field, the following system of wave equations holds in vacuo for the correlation functions and the spectral correlation functions

$$\Box_j \Gamma^{(m,n)}(\mathbf{x}_1, \ldots, \mathbf{x}_{m+n}, \mathbf{t}) = 0, \qquad j = 1, \ldots, m + n, \quad (2.34)$$

$$\left[\nabla_j^2 + \left(\frac{2\pi\nu_j}{c}\right)^2\right] G^{(m,n)}(\mathbf{x}_1, \ldots, \mathbf{x}_{m+n}, \boldsymbol{\nu}) = 0, \qquad j = 1, \ldots, m + n, \quad (2.35)$$

where the spectral decompositon (2.17) has been used and $\Box_j$ and ∇_j^2 are D'Alembert and Laplace operators respectively, for the j-th coordinates.

In the case of stationary fields the following system of wave equations holds for the correlation functions

$$\left[\nabla_1^2 - \frac{1}{c^2}\left(\frac{\partial}{\partial\tau_2} + \frac{\partial}{\partial\tau_3} + \ldots + \frac{\partial}{\partial\tau_{m+n}}\right)^2\right] \Gamma^{(m,n)}(\mathbf{x}_1, \ldots, \mathbf{x}_{m+n}, \boldsymbol{\tau}) = 0,$$

$$\left[\nabla_j^2 - \frac{1}{c^2}\frac{\partial^2}{\partial\tau_j^2}\right] \Gamma^{(m,n)}(\mathbf{x}_1, \ldots, \mathbf{x}_{m+n}, \boldsymbol{\tau}) = 0, \qquad j = 2, 3, \ldots, m + n. \quad (2.36)$$

These correspond to the following system of equations for the spectral quantities

$$\left[\nabla_j^2 + \left(\frac{2\pi\nu_j}{c}\right)^2\right] G_H^{(m,n)}(\mathbf{x}_1, \ldots, \mathbf{x}_{m+n}, \boldsymbol{\nu}) = 0, \qquad j = 1, 2, \ldots, m + n. \quad (2.37)$$

Here we have used (2.19) and the conditions (2.21) hold for frequencies.

The wave equations for the mutual coherence function $\Gamma_{12}(\tau)$ were first derived by Wolf (1955) (see also Born and Wolf (1965) and Beran and Parrent (1964)).

Degrees of coherence

Before defining the quantities which can be called the degrees of coherence, we summarize some simple inequalities for the correlation functions.

From the definition (2.11) and also from (2.12) it follows immediately that

$$[\Gamma^{(m,n)}(x_1, \ldots, x_{m+n})]^* = \Gamma^{(m,n)}(x_{m+n}, \ldots, x_1) \tag{2.38}$$

and condition (2.32a) represents a special case of this identity ($m = n = 1$ and $t_2 - t_1 = \tau$ for stationary fields).

A further interesting property is that an exchange of arguments $x_1, \ldots, x_m$ and also of arguments $x_{m+1}, \ldots, x_{m+n}$ does not change the value of the correlation function $\Gamma^{(m,n)}$.

We can easily see that

$$\Gamma^{(n,n)}(x_1, \ldots, x_n, x_n, \ldots, x_1) \geqq 0. \tag{2.39}$$

Using Hölder's inequality we can show for fields having classical analogues

$$|\Gamma^{(n,n)}(x_1, \ldots, x_{2n})|^{2n} \leqq \prod_{j=1}^{2n} \Gamma^{(n,n)}(x_j, \ldots, x_j). \tag{2.40}$$

Further it follows that

$$\Gamma^{(n,n)}(x_1, \ldots, x_n, x_n, \ldots, x_1)\, \Gamma^{(n,n)}(x_{n+1}, \ldots, x_{2n}, x_{2n}, \ldots, x_{n+1}) \geqq$$
$$\geqq |\Gamma^{(n,n)}(x_1, \ldots, x_{2n})|^2. \tag{2.41}$$

These as well as a number of further properties of the correlation functions will be derived in Chapter 12 (Glauber (1963a, 1965)).

Now we can define the following quantities

$$\gamma^{(n,n)}(x_1, \ldots, x_{2n}) = \frac{\Gamma^{(n,n)}(x_1, \ldots, x_{2n})}{\{\prod_{j=1}^{2n} \Gamma^{(n,n)}(x_j, \ldots, x_j)\}^{1/2n}}, \tag{2.42a}$$

$${}^{(G)}\gamma^{(n,n)}(x_1, \ldots, x_{2n}) = \frac{\Gamma^{(n,n)}(x_1, \ldots, x_{2n})}{\{\prod_{j=1}^{2n} \Gamma^{(1,1)}(x_j, x_j)\}^{1/2}}, \tag{2.42b}$$

$${}^{(S)}\gamma^{(n,n)}(x_1, \ldots, x_{2n}) =$$
$$= \frac{\Gamma^{(n,n)}(x_1, \ldots, x_{2n})}{\{\Gamma^{(n,n)}(x_1, \ldots, x_n, x_n, \ldots, x_1)\, \Gamma^{(n,n)}(x_{n+1}, \ldots, x_{2n}, x_{2n}, \ldots, x_{n+1})\}^{1/2}}. \tag{2.42c}$$

Each of these we call the *degree of coherence*. The quantity (2.42a) was introduced by Peřina and Peřinová (1965) and Mehta (1966). The quantity (2.42b) was introduced by Glauber (1963a, 1964, 1965) and the quantity (2.42c) was introduced by Sudarshan (see Klauder and Sudarshan (1968)). One can see from (2.40) and (2.41) that

$$|\gamma^{(n,n)}(x_1, \dots, x_{2n})| \leqq 1 \tag{2.43a}$$

and

$$|{}^{(S)}\gamma^{(n,n)}(x_1, \dots, x_{2n})| \leqq 1 \tag{2.43b}$$

respectively, for fields having classical analogues; but such an inequality does not hold for ${}^{(G)}\gamma^{(n,n)}$. The degree of coherence for stationary fields, which depends on $\tau_j = t_j - t_1, j = 2, 3, \dots, 2n$, can be defined in an identical way.

The correlation functions defined here represent the basic mathematical as well as physical quantities of this book and they will serve as a powerful tool for investigations of the coherence and polarization properties of the electromagnetic field.

Additionally we can define the degree of coherence in the frequency domain (Mandel and Wolf (1976))

$$\gamma(\mathbf{x}_1, \mathbf{x}_2, \nu) = \frac{G(\mathbf{x}_1, \mathbf{x}_2, \nu)}{[G(\mathbf{x}_1, \mathbf{x}_1, \nu)\, G(\mathbf{x}_2, \mathbf{x}_2, \nu)]^{1/2}}, \tag{2.44a}$$

and in the wave-vector and frequency domain we have (Carter and Wolf (1981a))

$$\gamma(\mathbf{k}_1, \mathbf{k}_2, \nu) = \frac{G(\mathbf{k}_1, \mathbf{k}_2, \nu)}{[G(\mathbf{k}_1, \mathbf{k}_1, \nu)\, G(\mathbf{k}_2, \mathbf{k}_2, \nu)]^{1/2}}; \tag{2.44b}$$

both these degrees of coherence satisfy inequalities such as (2.43) (Sec. 4.2) and $G(\mathbf{k}_1, \mathbf{k}_2, \nu)$ is obtained from $G(\mathbf{x}_1, \mathbf{x}_2, \nu)$ by spatial Fourier transformations with respect to $\mathbf{x}_1$ and $\mathbf{x}_2$.

2.3 Elementary ideas of temporal and spatial coherence

Temporal coherence

Suppose that a light beam from a point source σ (Fig. 2.1) is divided into two beams in a Michelson interferometer and that these two beams are unified after a path delay $\Delta s = c\,\Delta t$ is introduced between them (c is velocity of light). If Δs is sufficiently small, interference fringes are formed in the plane $\mathscr{B}$. The appearance of the fringes is a manifestation of *temporal coherence* between the two beams, since the visibility of fringes depends on the time delay Δt introduced between them. In general, interference fringes will be observed if

$$\Delta t\, \Delta \nu \lesssim 1, \tag{2.45a}$$

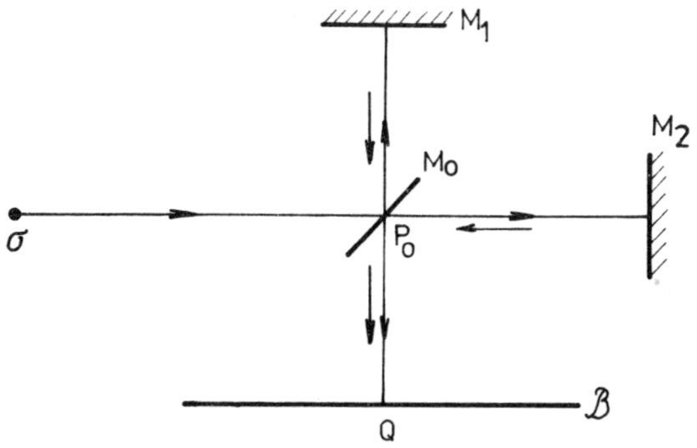

Fig. 2.1 Illustration of temporal coherence by means of the Michelson interferometer; M_1 and M_2 are mirrors, M_0 is a half-silvered mirror, σ is a point source, $\mathscr{B}$ is the observation screen.

where $\Delta\nu$ is the effective bandwidth of the light. The time delay

$$\Delta t \approx \frac{1}{\Delta\nu} \tag{2.45b}$$

is called the *coherence time* of the light and it is denoted as τ_c; the corresponding path $c\tau_c = l_c$ is called the *coherence length.* A more detailed discussion of the terms coherence time and coherence length based on wave-packet considerations was presented by Francon and Slansky (1965).

Using the second-order degree of coherence $\gamma(\mathbf{x}, \mathbf{x}, \tau) \equiv \gamma(\tau)$ of stationary fields we can give a somewhat more precise definition of the coherence time (Wolf (1958), Born and Wolf (1965)).

Considering the simple experiment illustrated in Fig. 2.1 again, where the time delay between two beams is τ, then the visibility of the interference fringes observed in the plane $\mathscr{B}$ is proportional to the modulus of the degree of coherence $\gamma(\mathbf{x}, \mathbf{x}, \tau)$ ($\mathbf{x}$ specifies the position of the point P_0 of the mirror M_0), as we shall discuss in Chapter 3. Therefore one can define the coherence time by means of the moment formula (Wolf (1958))

$$\tau_c^2 = (\Delta t)^2 = N^{-1} \int_{-\infty}^{+\infty} \tau^2 |\gamma(\tau)|^2 \, d\tau = 2N^{-1} \int_0^{\infty} \tau^2 |\gamma(\tau)|^2 \, d\tau, \tag{2.46}$$

where

$$N = \int_{-\infty}^{+\infty} |\gamma(\tau)|^2 \, d\tau = \int_0^{\infty} g^2(\nu) \, d\nu. \tag{2.47}$$

Here $g(\nu)$ denotes the *normalized spectral density* of the light

$$g(\nu) = \frac{G(\nu)}{\int_0^{\infty} G(\nu) \, d\nu}, \tag{2.48}$$

where $G(\nu) \equiv G(\mathbf{x}, \mathbf{x}, \nu)$. Let us note that $\bar{\tau} \equiv \int_{-\infty}^{+\infty} \tau |\gamma(\tau)|^2 \, d\tau = 0$ since $|\gamma(\tau)|$ is an even function. The equality of the two integrals in (2.47) follows from Parseval's theorem applied to the normalized spectral decomposition (2.24),

$$\gamma(\tau) = \int_0^\infty g(\nu) \exp(-\mathrm{i}\, 2\pi\nu\tau)\, \mathrm{d}\nu; \tag{2.49}$$

obviously $\int_0^\infty g(\nu)\, \mathrm{d}\nu = 1$.

If the effective bandwidth of the light is defined by

$$(\Delta\nu)^2 = N^{-1} \int_0^\infty (\nu - \bar{\nu})^2 g^2(\nu)\, \mathrm{d}\nu \tag{2.50}$$

with

$$\bar{\nu} = N^{-1} \int_0^\infty \nu g^2(\nu)\, \mathrm{d}\nu, \tag{2.51}$$

then one may show that (Born and Wolf (1965))

$$\Delta t\, \Delta\nu \geqq \frac{1}{4\pi}. \tag{2.52}$$

For quasi-monochromatic light with a Gaussian spectral profile the equality sign is approximately correct.

Another definition of the coherence time can also be adopted (Mandel (1959, 1963a)),

$$\tau_c = \Delta t = \int_{-\infty}^{+\infty} |\gamma(\tau)|^2\, \mathrm{d}\tau. \tag{2.53}$$

More general definitions of coherence time, appropriate for optical fields of arbitrary statistics, will be given in Sec. 10.5. If the bandwidth is defined as

$$\Delta\nu = \left[\int_0^\infty g^2(\nu)\, \mathrm{d}\nu\right]^{-1}, \tag{2.54}$$

then we have

$$\Delta t\, \Delta\nu = 1, \tag{2.55}$$

applying the Parseval's equality (2.47). This definition is useful in connection with the Hanbury Brown-Twiss experiment measuring the fourth-order correlation function and with the theory of photocount statistics (Mandel (1963a)). For simple types of spectral profiles both these definitions are approximately equivalent (Mandel and Wolf (1962), Mehta (1963)).

For experiments involving the division of light beams at two points (see Young's experiment in Chapter 3) more general definitions can be given (Wolf (1958), Born and Wolf (1965)). Still more general definitions of the coherence time which apply to non-stationary fields have been proposed by Carusotto (1970). The reciprocity relations with partially coherent sources have been discussed by Friberg and Wolf (1983).

Spectra and temporal coherence

Consider now rectangular, Lorentzian and Gaussian forms of spectra and calculate the corresponding degrees of temporal coherence using (2.49).

The normalized rectangular spectrum is defined as

$$g(\nu) = \frac{\pi}{\Gamma}, \qquad |\nu - \nu_0| \leqq \Gamma/2\pi, \tag{2.56a}$$

with the mean frequency ν_0 and the halfwidth Γ. From (2.49)

$$\gamma(\tau) = \exp(-i2\pi\nu_0\tau)\frac{\sin(\Gamma\tau)}{\Gamma\tau}; \tag{2.56b}$$

we see that the degree of coherence (contrast of interference fringes) decreases non-monotonically with increasing τ and Γ.

In the case of the Lorentzian spectrum we have

$$g(\nu) = \frac{1}{2\pi}\frac{2(\Gamma/2\pi)}{(\nu - \nu_0)^2 + (\Gamma/2\pi)^2} \tag{2.57a}$$

and we obtain by means of the residue theorem

$$\gamma(\tau) = \exp(-i2\pi\nu_0\tau - \Gamma|\tau|). \tag{2.57b}$$

For the Gaussian spectrum

$$g(\nu) = \frac{2\pi^{1/2}}{\Gamma}\exp\left[-\frac{(\nu - \nu_0)^2}{(\Gamma/2\pi)^2}\right] \tag{2.58a}$$

we have from (2.49)

$$\gamma(\tau) = \exp\left(-i2\pi\nu_0\tau - \frac{\Gamma^2\tau^2}{4}\right). \tag{2.58b}$$

Since $\Gamma \ll \nu_0$, we have included small non-zero values for $\nu < 0$ in the Lorentzian and Gaussian spectra instead of putting $g(\nu) = 0$ for $\nu < 0$. In the last two cases the contrast of the interference fringes decreases monotonically with increasing τ and Γ.

It is obvious from (2.53) and (2.57b) that $\tau_c = 1/\Gamma$ in agreement with the above considerations. Thus the coherence time increases with decreasing spectral halfwidth.

Spatial coherence

Let us consider another interference experiment of the Young type, illustrated by Fig. 2.2. Assuming quasi-monochromatic light ($\Delta\nu/\bar{\nu} \ll 1$) from an extended chaotic source σ in the form of a square of side Δa, we can observe the

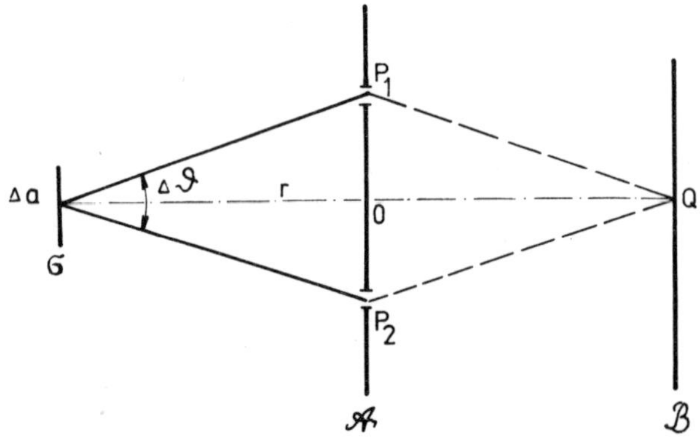

Fig. 2.2 Illustration of spatial coherence by means of the Young's interference experiment; σ is an extended chaotic quasi-monochromatic source, $\mathscr{A}$ is a screen with two pinholes P_1 and P_2 and $\mathscr{B}$ is the observation screen.

interference fringes near the point Q of the screen $\mathscr{B}$ if the pinholes P_1 and P_2 are close enough to each other. The appearance of these fringes is a manifestation of *spatial coherence* between the beams arriving at Q from P_1 and P_2, since the visibility of these fringes depends on the spatial separation of the pinholes. Interference fringes are generally observable if

$$\Delta\vartheta\, \Delta a < \bar{\lambda}, \tag{2.59}$$

where $\bar{\lambda} = c/\bar{\nu}$ is the effective wavelength of the light. In order to observe the fringes near Q the pinholes must be situated within an area around the point 0 of size

$$A_c = \Delta A \approx (r\, \Delta\vartheta)^2 \approx \frac{r^2 \bar{\lambda}^2}{S}, \tag{2.60}$$

where $S = (\Delta a)^2$ is the area of the source. The area A_c is called the coherence area of the light in the plane $\mathscr{A}$ around the point 0. The solid angle $\Delta\Omega$ which the area ΔA subtends at the source is given by

$$\Delta\Omega \approx \frac{\Delta A}{r^2} \approx \frac{\bar{\lambda}^2}{S}. \tag{2.61}$$

Both these kinds of coherence are connected in the degree of coherence $\gamma(\mathbf{x}_1, \mathbf{x}_2, \tau)$; temporal coherence is described by the function $\gamma(\mathbf{x}_1, \mathbf{x}_1, \tau)$ and spatial coherence is described by $\gamma(\mathbf{x}_1, \mathbf{x}_2, 0)$. In general it is not possible to separate temporal and spatial coherence (i.e. to decompose the degree of coherence $\gamma_{12}(\tau)$ into the product of $\gamma_{11}(\tau)$ and $\gamma_{12}(0)$) due to the fact that temporal and spatial coherence are mutually connected, since $\Gamma_{12}(\tau)$ obeys *in vacuo* the wave equations, which relate the space and time variations of $\Gamma_{12}(\tau)$. More precisely, temporal coherence is determined by spatial coherence on the basis of the wave equations (2.36) (Wolf, Devaney and Foley (1981), Wolf and Devaney (1981), see Sec. 4.3).

However, there exists a class of optical fields, for which $\gamma_{12}(\tau)$ may be factorized into the form

$$\gamma_{12}(\tau) = \gamma_{11}(\tau)\gamma_{12}(0), \tag{2.62}$$

i.e. temporal and spatial coherence may be separated in this case. Such fields are said to be *cross-spectrally pure* (Mandel (1961a)). Under certain assumptions it may be shown that temporal coherence is determined by the spectral properties of light while spatial coherence is determined by geometric properties of the source (Secs. 3.3 and 4.3).

Volume of coherence

The *volume of coherence* can be defined as a right-angled cylinder, whose base is the area of coherence and whose height is the coherence length,

$$V_c \equiv \Delta V = c\,\Delta t\,\Delta A \approx \left(\frac{\bar{\lambda}}{|\Delta\lambda|}\right)\left(\frac{r}{\Delta a}\right)^2 \bar{\lambda}^3, \tag{2.63}$$

where $|\Delta\lambda| = |\Delta(c/\bar{\nu})| = (c/\bar{\nu}^2)|\Delta\nu|$. This is just the volume corresponding to one cell of phase space of photons (Kastler (1964)).

A very important parameter is the *degeneracy parameter* δ (Mandel (1961b)), which represents the average number of photons in the same state of polarization in the volume of coherence, i.e. it represents the average number of photons in the same state of polarization which traverse the area of coherence per coherence time. If $\langle n_\nu \rangle$ is the average number of photons emitted per unit area of the source per unit frequency interval, per unit solid angle around the direction normal to the source per unit time, then

$$\delta = \tfrac{1}{2}\langle n_\nu \rangle S\,\Delta\nu\,\Delta\Omega\,\Delta t. \tag{2.64}$$

The factor 1/2 on the right-hand side arises from the fact that the light is assumed to be generated by a chaotic source and so it is unpolarized, i.e. both orthogonal polarizations are present with the same weight. From (2.45) and (2.61) we obtain

$$\delta \approx \frac{c^2}{2\bar{\nu}^2}\langle n_\nu \rangle, \tag{2.65}$$

which is independent of the geometry. For blackbody radiation emerging from an equilibrium enclosure

$$\langle n_\nu \rangle = \frac{2\nu^2}{c^2}\left[\exp\left(\frac{h\nu}{KT}\right) - 1\right]^{-1}, \tag{2.66}$$

where K is the Boltzmann constant, T is the absolute temperature of the radiation and h is the Planck's constant. Equation (2.65) gives

$$\delta \approx \left[\exp\left(\frac{h\nu}{KT}\right) - 1\right]^{-1}, \tag{2.67}$$

which is the expression first derived by Einstein (1912) in connection with his study of radiation in a cavity in thermal equilibrium with the walls of the cavity. This quantity describes the average number of photons in a cell of phase space and in quantum statistics it is called the degeneracy parameter of the radiation.

Of course, the degeneracy parameter can also be defined for non-thermal light. It should be mentioned that for non-thermal (laser) light $\delta \gg 1$ while for thermal light $\delta \ll 1$ (Gabor (1961), Mandel (1961b)).

The concept of the volume of coherence may be shown to correspond to a quantum-mechanically defined cell of phase space, i.e. to $\Delta q_x \Delta q_y \Delta q_z = h^3 (\Delta p_x \Delta p_y \Delta p_z)^{-1} \equiv \Delta V$, where q and p are canonical coordinate and momentum respectively (Mandel and Wolf (1965)).

2.4 Quasi-monochromatic approximation

In cases of practical interest so-called quasi-monochromatic light is frequently used. If the frequency bandwidth of such light is $\Delta\nu$ and the mean frequency is $\bar{\nu}$, then for quasi-monochromatic light the condition

$$\Delta\nu \ll \bar{\nu} \tag{2.68}$$

holds. Writing the correlation function in the form

$$\begin{aligned} \Gamma^{(n,n)}(\mathbf{x}_1, \ldots, \mathbf{x}_{2n}, \boldsymbol{\tau}) &= \left| \Gamma^{(n,n)}(\mathbf{x}_1, \ldots, \mathbf{x}_{2n}, \boldsymbol{\tau}) \right| \times \\ &\times \exp\left[\mathrm{i}\alpha(\mathbf{x}_1, \ldots, \mathbf{x}_{2n}, \boldsymbol{\tau}) + \mathrm{i}\, 2\pi\bar{\nu} \sum_{j=2}^{2n} \varepsilon_j \tau_j \right], \end{aligned} \tag{2.69}$$

where

$$\alpha = \arg \Gamma^{(n,n)} - 2\pi\bar{\nu} \sum_{j=2}^{2n} \varepsilon_j \tau_j$$

and using (2.19), we obtain

$$\begin{aligned} \Gamma^{(n,n)}(\mathbf{x}_1, \ldots, \mathbf{x}_{2n}, \boldsymbol{\tau}) &= \exp\Big(\mathrm{i}\, 2\pi\bar{\nu} \sum_{j=2}^{2n} \varepsilon_j \tau_j\Big) \times \\ &\times \int_0^\infty \ldots \int_0^\infty G_H^{(n,n)}(\mathbf{x}_1, \ldots, \mathbf{x}_{2n}, \boldsymbol{\nu}) \exp\Big[\mathrm{i}\, 2\pi \sum_{j=2}^{2n} \varepsilon_j(\nu_j - \bar{\nu})\, \tau_j\Big] \prod_{j=2}^{2n} \mathrm{d}\nu_j = \\ &= \exp\Big(\mathrm{i}\, 2\pi\bar{\nu} \sum_{j=2}^{2n} \varepsilon_j \tau_j\Big) \int_{-\bar{\nu}}^\infty \ldots \int_{-\bar{\nu}}^\infty G^{(n,n)}\Big(\mathbf{x}_1, \ldots, \mathbf{x}_{2n}, \bar{\nu} - \sum_{j=2}^{2n} \varepsilon_j \nu_j, \bar{\nu} + \\ &+ \nu_2, \ldots, -\bar{\nu} - \nu_{2n}\Big) \exp\Big(\mathrm{i}\, 2\pi \sum_{j=2}^{2n} \varepsilon_j \nu_j \tau_j\Big) \prod_{j=2}^{2n} \mathrm{d}\nu_j, \end{aligned} \tag{2.70}$$

where we have performed the substitutions $\nu_j - \bar{\nu} \to \nu_j (-\sum_{j=2}^{2n} \varepsilon_j \nu_j \to \bar{\nu} - \sum_{j=2}^{2n} \varepsilon_j \nu_j)$. As we consider quasi-monochromatic light, the function $G^{(n,n)}$ in (2.70) will be effectively non-zero only in a region $(\bar{\nu} - \Delta\nu/2, \bar{\nu} + \Delta\nu/2) \otimes \ldots \otimes (\bar{\nu} - \Delta\nu/2, \bar{\nu} +$

$+ \Delta\nu/2$). Thus since $\Delta\nu/\bar{\nu} \ll 1$, the function $|\Gamma^{(n,n)}|\exp(\mathrm{i}\alpha)$ contains only low-frequency components and hence it will vary slowly with τ_j compared with variations arising from the periodic term $\exp(\mathrm{i}\,2\pi\bar{\nu}\sum_{j=2}^{2n}\varepsilon_j\tau_j)$. Considering $\boldsymbol{\tau}$ such that

$$|\tau_j| \ll \frac{1}{\Delta\nu}, \qquad j = 2, 3, \ldots, 2n, \tag{2.71}$$

we see from (2.70) that $|\Gamma^{(n,n)}|\exp(\mathrm{i}\alpha)$ is independent of $\boldsymbol{\tau}$ and so

$$\Gamma^{(n,n)}(\boldsymbol{x}_1, \ldots, \boldsymbol{x}_{2n}, \boldsymbol{\tau}) = \exp\Big(\mathrm{i}\,2\pi\bar{\nu}\sum_{j=2}^{2n}\varepsilon_j\tau_j\Big)\,\Gamma^{(n,n)}(\boldsymbol{x}_1, \ldots, \boldsymbol{x}_{2n}, \boldsymbol{0}). \tag{2.72}$$

Considering two vectors $\boldsymbol{\tau}_1$ and $\boldsymbol{\tau}_2$ for which

$$|\tau_{1j} - \tau_{2j}| \ll \frac{1}{\Delta\nu}, \qquad j = 2, 3, \ldots, 2n, \tag{2.73}$$

we obtain

$$\Gamma^{(n,n)}(\boldsymbol{x}_1, \ldots, \boldsymbol{x}_{2n}, \boldsymbol{\tau}_1) = \exp\Big[\mathrm{i}\,2\pi\bar{\nu}\sum_{j=2}^{2n}\varepsilon_j(\tau_{1j} - \tau_{2j})\Big] \times$$
$$\times\, \Gamma^{(n,n)}(\boldsymbol{x}_1, \ldots, \boldsymbol{x}_{2n}, \boldsymbol{\tau}_2). \tag{2.74}$$

Similar results apply to the quantum correlation function $\Gamma_{\mathcal{N}}^{(n,n)}$ as well as to the quantum correlation functions involving other operator orderings (Chapter 16).

CHAPTER 3

SECOND-ORDER COHERENCE

In this chapter we study the second-order coherence effects. We demonstrate second-order coherence by elementary considerations based on Young's interference experiment and we derive the propagation laws of partial coherence. The description given here is classical. However, we note that the quantum description is formally identical (Sec. 12.2) and thus the results given here also apply to the quantum correlation functions.

3.1 Interference law for two partially coherent beams

In this chapter we assume a stationary and ergodic field. We also assume light to be linearly polarized for simplicity so that it may be represented by a complex scalar function $V(\boldsymbol{x}, t)$. Further we consider quasi-monochromatic light beams.

In the Maxwell theory of the electromagnetic field one assumes the electric field $\boldsymbol{E}$ and magnetic field $\boldsymbol{H}$ to be measurable quantities. However, as has already been mentioned, optical vibrations are so fast that no real detector can follow them, and, moreover, the field represents a statistical dynamic system. Therefore it is necessary to introduce an averaging process for physical quantities. Since we have restricted ourselves to the study of stationary and ergodic fields, the time average can be adopted for the definition of the mutual coherence function. In such a case one realization of the experiment followed in time is sufficient.

Consider the Young's interference experiment with the arrangement illustrated in Fig. 3.1, where σ represents an extended quasi-monochromatic source of

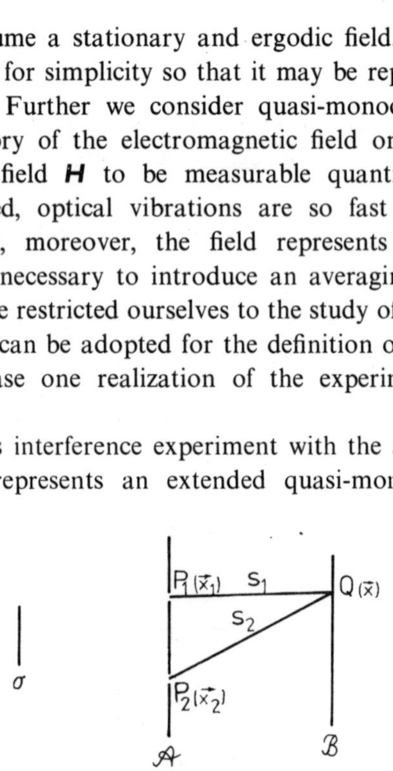

Fig. 3.1 Significance of second-order coherence in the two-beam interference experiment; σ is an extended quasi-monochromatic source, $\mathscr{A}$ is a screen with two pinholes P_1 and P_2 and $\mathscr{B}$ is the observation screen.

light. Light vibrations from the pinholes $P_1(\mathbf{x}_1)$ and $P_2(\mathbf{x}_2)$ of the screen $\mathscr{A}$ interfere at the point $Q(\mathbf{x})$ on the screen $\mathscr{B}$; the point Q is situated at distances s_1 from P_1 and s_2 from P_2 respectively. As we consider linearly polarized light (or we disregard the polarization phenomena in which the vectorial properties of the electromagnetic field manifest themselves), we can describe this phenomenon by the scalar complex function $V(\mathbf{x}, t)$ for which we have at the point $Q(\mathbf{x})$

$$V(\mathbf{x}, t) = a_1 V(\mathbf{x}_1, t - t_1) + a_2 V(\mathbf{x}_2, t - t_2), \tag{3.1}$$

where $t_j = s_j/c$, $j = 1, 2$, c being velocity of light (for simplicity an air medium is assumed). The purely imaginary numbers a_1 and a_2 are transmission factors (propagators) between P_1 and Q and P_2 and Q respectively, defined for the mean frequency; they depend on the size of the pinholes and on the geometry ($a = \bar{\Lambda}/2\pi s = \mathrm{i}\bar{k}\cos(\vartheta)/2\pi s$, $\bar{\Lambda}$ being the inclination factor, ϑ is the inclination angle and $\bar{k}$ is the mean wave number, see (3.36)).

For the mean intensity at the point $Q(\mathbf{x})$ we obtain

$$\begin{aligned} I(\mathbf{x}) &= \langle I(\mathbf{x}, t)\rangle = \langle V^*(\mathbf{x}, t)\, V(\mathbf{x}, t)\rangle = \\ &= |a_1|^2 I(\mathbf{x}_1) + |a_2|^2 I(\mathbf{x}_2) + 2|a_1 a_2| \operatorname{Re} \Gamma(\mathbf{x}_1, \mathbf{x}_2, \tau), \end{aligned} \tag{3.2}$$

where $\tau = t_1 - t_2 = (s_1 - s_2)/c$ and

$$\Gamma_{12}(\tau) \equiv \Gamma(\mathbf{x}_1, \mathbf{x}_2, \tau) = \langle V^*(\mathbf{x}_1, t)\, V(\mathbf{x}_2, t + \tau)\rangle \tag{3.3}$$

is the mutual coherence function. The mean intensities at $\mathbf{x}_1$ and $\mathbf{x}_2$ are related to the mutual coherence function by

$$I(\mathbf{x}_j) = \Gamma(\mathbf{x}_j, \mathbf{x}_j, 0), \qquad j = 1, 2. \tag{3.4}$$

In (3.2) we have also used the stationary condition

$$\langle V^*(\mathbf{x}_1, t - t_1)\, V(\mathbf{x}_2, t - t_2)\rangle = \langle V^*(\mathbf{x}_1, t)\, V(\mathbf{x}_2, t + t_1 - t_2)\rangle, \tag{3.5}$$

which expresses the fact that the correlation function depends on the time difference $t_1 - t_2$.

We introduce the intensities

$$I^{(j)}(\mathbf{x}) = |a_j|^2 I(\mathbf{x}_j), \qquad j = 1, 2 \tag{3.6}$$

at Q arising from P_1 and P_2 alone (if only one pinhole were open) and the degree of coherence according to (2.42a) is for $n = 1$ (all degrees of coherence $\gamma^{(1,1)}$, ${}^{(G)}\gamma^{(1,1)}$ and ${}^{(S)}\gamma^{(1,1)}$ given by (2.42a – c) are identical in this case)

$$\gamma(\mathbf{x}_1, \mathbf{x}_2, \tau) = \frac{\Gamma(\mathbf{x}_1, \mathbf{x}_2, \tau)}{\{\Gamma(\mathbf{x}_1, \mathbf{x}_1, 0)\, \Gamma(\mathbf{x}_2, \mathbf{x}_2, 0)\}^{1/2}}, \tag{3.7}$$

for which $|\gamma(\mathbf{x}_1, \mathbf{x}_2, \tau)| \leqq 1$ as follows from (2.43a) for $n = 1$. We then obtain the final form of the general interference law for stationary quasi-monochromatic optical fields

$$I(\mathbf{x}) = I^{(1)}(\mathbf{x}) + I^{(2)}(\mathbf{x}) + 2[I^{(1)}(\mathbf{x})\, I^{(2)}(\mathbf{x})]^{1/2}\, \mathrm{Re}\, \gamma(\mathbf{x}_1, \mathbf{x}_2, \tau) = \tag{3.8a}$$

$$= I^{(1)}(\mathbf{x}) + I^{(2)}(\mathbf{x}) + 2[I^{(1)}(\mathbf{x})\, I^{(2)}(\mathbf{x})]^{1/2}\, |\gamma(\mathbf{x}_1, \mathbf{x}_2, \tau)| \times$$

$$\times \cos[\alpha(\mathbf{x}_1, \mathbf{x}_2, \tau) - 2\pi\bar{\nu}\tau], \tag{3.8b}$$

where (2.69) for $n = 1$ has also been used. Measuring the total intensity $I(\mathbf{x})$ at Q and the intensities $I^{(1)}(\mathbf{x})$ and $I^{(2)}(\mathbf{x})$ at Q arising from single pinholes, one can determine $\mathrm{Re}\, \gamma_{12}(\tau)$ and hence $\mathrm{Re}\, \Gamma_{12}(\tau)$. Consequently $\gamma_{12}(\tau)$ and $\Gamma_{12}(\tau)$ can be found using the dispersion relations. Hence these are *measurable quantities.* The third term in (3.8) represents an interference term arising if both pinholes are open. Since it is non-zero only if $|\tau| \leqq \tau_c = 1/\Delta\nu$ and from the Heisenberg uncertainty principle it holds that $|\tau| \geqq 1/\Delta\nu$, we see that the interference pattern expresses the impossibility of distinguishing from which beam a given photon comes to the interference pattern.

We can define the usual measure of the sharpness of interference fringes, which can be called the *visibility*, introduced by Michelson†,

$$\mathscr{V}(\mathbf{x}) = \frac{I_{\max} - I_{\min}}{I_{\max} + I_{\min}}, \tag{3.9}$$

where

$$I_{\substack{\max \\ \min}} = I^{(1)}(\mathbf{x}) + I^{(2)}(\mathbf{x}) \pm 2[I^{(1)}(\mathbf{x})\, I^{(2)}(\mathbf{x})]^{1/2}\, |\gamma(\mathbf{x}_1, \mathbf{x}_2, \tau)|, \tag{3.10}$$

which follows from (3.8b), since $\mathrm{Max} \cos x = 1$ (if $\alpha - 2\pi\bar{\nu}\tau = 2\pi m$, $m = 0, \pm 1, \ldots$) and $\mathrm{Min} \cos x = -1$ (if $\alpha - 2\pi\bar{\nu}\tau = (2m + 1)\pi$). Thus for the visibility (3.9) we have

$$\mathscr{V}(\mathbf{x}) = \frac{2[I^{(1)}(\mathbf{x})\, I^{(2)}(\mathbf{x})]^{1/2}}{I^{(1)}(\mathbf{x}) + I^{(2)}(\mathbf{x})}\, |\gamma(\mathbf{x}_1, \mathbf{x}_2, \tau)| =$$

$$= 2\left\{\left[\frac{I^{(1)}(\mathbf{x})}{I^{(2)}(\mathbf{x})}\right]^{1/2} + \left[\frac{I^{(2)}(\mathbf{x})}{I^{(1)}(\mathbf{x})}\right]^{1/2}\right\}^{-1} |\gamma(\mathbf{x}_1 . \mathbf{x}_2, \tau)|. \tag{3.11}$$

In particular, if the intensities of the two beams are equal, then

$$\mathscr{V}(\mathbf{x}) = |\gamma(\mathbf{x}_1, \mathbf{x}_2, \tau)|, \tag{3.12}$$

i.e. the visibility of the fringes is equal to the modulus of the degree of coherence in this case. Hence $|\gamma|$ can be determined from simple measurements.

---

† Another definition of the visibility has been adopted by Francon and Slansky (1965), $\mathscr{V}(\mathbf{x}) = (I_{\max} - I_{\min})/I_{\max}$.

The phase of γ also has a simple operational significance. From (3.8b) the positions of the maxima of the intensity in the fringe pattern are given by

$$\alpha\left(\mathbf{x}_1, \mathbf{x}_2, \frac{s_1 - s_2}{c}\right) - \frac{2\pi\bar{\nu}}{c}(s_1 - s_2) =$$

$$= \alpha\left(\mathbf{x}_1, \mathbf{x}_2, \frac{s_1 - s_2}{c}\right) - \frac{2\pi}{\bar{\lambda}}(s_1 - s_2) = 2m\pi, \qquad m = 0, \pm 1, \ldots. \tag{3.13}$$

The positions of the maxima given by (3.13) coincide with those which would be obtained if the two pinholes were illuminated by monochromatic light of the effective wavelength $\bar{\lambda}$ and the phase of the vibrations at P_1 was retarded with respect to that at P_2 by $\alpha(\mathbf{x}_1, \mathbf{x}_2, (s_1 - s_2)/c)$. Hence $\alpha(\mathbf{x}_1, \mathbf{x}_2, (s_1 - s_2)/c)$ may be regarded as representing the effective retardation of the light at P_1 with respect to the light at P_2. From (3.13) it follows that the phase of γ may be determined in principle from measurements of the positions of the maxima of the fringes.

The determination of γ from measurements enables us to calculate the spectrum of the radiation if the inverse of the Wiener-Khintchine theorem is used. Unfortunately the positional measurements determining the phase of γ are difficult to perform in the optical region. And so we shall use another method of determining the phase of γ based on the analytic properties of γ (Sec. 4.4).

If we restrict ourselves to such time delays τ that (2.71) holds, i.e. $|\tau| \ll (\Delta\nu)^{-1}$ $(\Delta s = |s_1 - s_2| \ll c(\Delta\nu)^{-1})$, it follows from (2.72) for $n = 1$ that $|\Gamma(\mathbf{x}_1, \mathbf{x}_2, \tau)| = |\Gamma(\mathbf{x}_1, \mathbf{x}_2, 0)|$, $\arg \Gamma(\mathbf{x}_1, \mathbf{x}_2, \tau) = \alpha(\mathbf{x}_1, \mathbf{x}_2, 0) - 2\pi\bar{\nu}\tau$ and so the quantities $|\gamma(\mathbf{x}_1, \mathbf{x}_2, \tau)|$ and $\alpha(\mathbf{x}_1, \mathbf{x}_2, \tau)$ appearing in (3.8b) and in the following equations can be replaced by $|\gamma(\mathbf{x}_1, \mathbf{x}_2, 0)|$ and $\alpha(\mathbf{x}_1, \mathbf{x}_2, 0)$. The quantity $\Gamma(\mathbf{x}_1, \mathbf{x}_2, 0) \equiv \Gamma(\mathbf{x}_1, \mathbf{x}_2)$ is called the *mutual intensity*.

Returning to (3.8b) we see that no interference fringes are formed if $\gamma = 0$ $(|\gamma| = 0, \mathscr{V} = 0)$; this is the case when the two beams reaching the point Q from P_1 and P_2 are mutually *incoherent* and their intensities are summed. If $|\gamma| = 1$, the interference fringes have the maximum possible visibility and the beams are completely *coherent*; in this case the complex amplitudes are summed. The intermediate cases with $0 < |\gamma| < 1$ characterize states of *partial coherence*.

Rewriting the interference law in the form (Born and Wolf (1965))

$$I(\mathbf{x}) = |\gamma(\mathbf{x}_1, \mathbf{x}_2, \tau)| \{I^{(1)}(\mathbf{x}) + I^{(2)}(\mathbf{x}) + 2[I^{(1)}(\mathbf{x}) I^{(2)}(\mathbf{x})]^{1/2} \times$$

$$\times \cos[\alpha(\mathbf{x}_1, \mathbf{x}_2, \tau) - 2\pi\bar{\nu}\tau]\} + [1 - |\gamma(\mathbf{x}_1, \mathbf{x}_2, \tau)|] \{I^{(1)}(\mathbf{x}) + I^{(2)}(\mathbf{x})\}, \tag{3.14}$$

we have an interesting result that light which reaches Q from both pinholes P_1 and P_2 may be regarded as a mixture of coherent and incoherent light. The first term in (3.14) may be regarded as arising from the coherent superposition of two beams of intensities $|\gamma(\mathbf{x}_1, \mathbf{x}_2, \tau)|\, I^{(1)}(\mathbf{x})$ and $|\gamma(\mathbf{x}_1, \mathbf{x}_2, \tau)|\, I^{(2)}(\mathbf{x})$ and of relative phase difference $\alpha(\mathbf{x}_1, \mathbf{x}_2, \tau) - 2\pi\bar{\nu}\tau$ while the second term may be regarded as arising from the incoherent superposition of two beams of intensities $(1 - |\gamma(\mathbf{x}_1, \mathbf{x}_2, \tau)|)\, I^{(1)}(\mathbf{x})$ and $(1 - |\gamma(\mathbf{x}_1, \mathbf{x}_2, \tau)|)\, I^{(2)}(\mathbf{x})$.

3.2 Propagation laws of partial coherence

We assume that the mutual coherence function $\Gamma_{12}(\tau) \equiv \Gamma(\mathbf{x}_1, \mathbf{x}_2, \tau)$ is given over a certain area $\mathscr{A}$ of the space. We have to determine this function at pairs of points Q_1, Q_2, which do not lie on $\mathscr{A}$ (Fig. 3.2). We denote $r_1 = \overline{P_1Q_1}$ and $r_2 = \overline{P_2Q_2}$. To solve this problem, we use the Green's function technique for solving the wave equations (2.36) (with $m = n = 1$) for the mutual coherence function (Parrent (1959a), Beran and Parrent (1964)). From (2.36) we have the wave equations for $\Gamma_{12}(\tau)$,

$$\nabla_j^2 \Gamma_{12}(\tau) = \frac{1}{c^2} \frac{\partial^2 \Gamma_{12}(\tau)}{\partial \tau^2}, \qquad j = 1, 2. \tag{3.15}$$

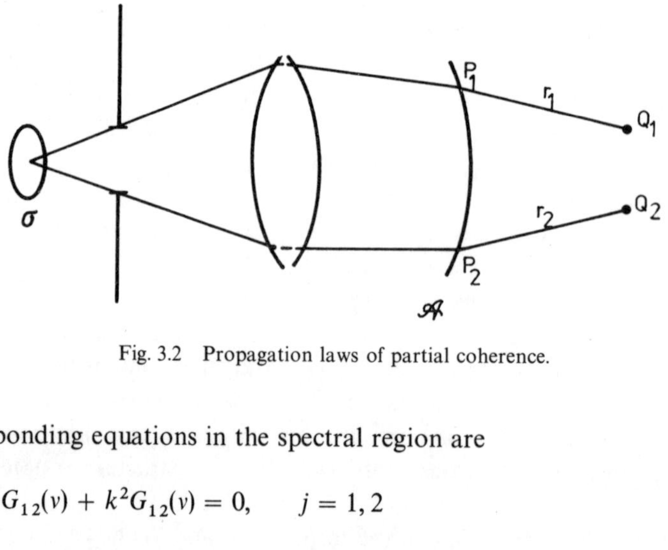

Fig. 3.2 Propagation laws of partial coherence.

The corresponding equations in the spectral region are

$$\nabla_j^2 G_{12}(\nu) + k^2 G_{12}(\nu) = 0, \qquad j = 1, 2 \tag{3.16}$$

for all ν and $k = 2\pi\nu/c$. Let the function $G_{12}(\nu)$ be given on the area $\mathscr{A}$ and let $\mathscr{G}_1$ and $\mathscr{G}_2$ be the Green functions of the equations (3.16), i.e.

$$(\nabla_j^2 + k^2)\, \mathscr{G}_j(Q_j, P_j, \nu) = -\delta(Q_j - P_j), \qquad j = 1, 2, \tag{3.17}$$

where δ is the Dirac function, with the boundary conditions

$$\mathscr{G}_j(Q_j, P'_j, \nu)|_{P'_j \equiv P_j} = 0, \qquad P_j \in \mathscr{A}, \qquad j = 1, 2. \tag{3.18}$$

Using repeatedly the Green formula (Beran and Parrent (1964)), we arrive at the propagation law for the mutual spectral density

$$G(Q_1, Q_2, \nu) = \iint_{\mathscr{A}} \frac{\partial \mathscr{G}_1(Q_1, P_1, \nu)}{\partial n_1} \frac{\partial \mathscr{G}_2(Q_2, P_2, \nu)}{\partial n_2} G(P_1, P_2, \nu) \, \mathrm{d}P_1 \, \mathrm{d}P_2, \tag{3.19}$$

where $\partial/\partial n_j$ denotes the derivative with respect to the normal to the surface $\mathscr{A}$ at the point P_j ($j = 1, 2$). Since $G^*(Q_1, Q_2, \nu) = G(Q_2, Q_1, \nu)$ (see (2.32b)), it follows that

$$\mathscr{G}_1 = \mathscr{G}_2^*. \tag{3.20}$$

Defining $K(Q, P, \nu) \equiv \partial\mathscr{G}(Q, P, \nu)/\partial n$ in (3.19), we obtain

$$G(Q_1, Q_2, \nu) = \iint_{\mathscr{A}} K^*(Q_1, P_1, \nu) \, K(Q_2, P_2, \nu) \, G(P_1, P_2, \nu) \, \mathrm{d}P_1 \, \mathrm{d}P_2, \tag{3.21}$$

where $K(Q, P, \nu)$ represents the *diffraction function* (it is the complex amplitude at Q caused by a unit point source of frequency ν situated at P). For example, in applying (3.21) to all the surfaces of an optical system, such a function $K(Q, P, \nu)$ may again be introduced for the whole system as a consequence of the linearity of (3.21) and this function serves as the kernel of the same integral relation between an object and its image. This result will be used in Chapter 9 for an analysis of optical imaging and that analysis will serve as an example of the application of the propagation laws of partial coherence.

From (3.21) one can determine the *power spectrum* $G(Q, \nu)$ by putting $Q_1 \equiv Q_2 \equiv Q$. This quantity is not determined by the same quantity over the surface $\mathscr{A}$ but by the mutual spectral density $G(P_1, P_2, \nu)$.

The mutual coherence function is obtained by the Fourier transform (2.24),

$$\Gamma(Q_1, Q_2, \tau) = \int_0^\infty \mathrm{d}\nu \exp(-\mathrm{i}\, 2\pi\nu\tau) \iint_{\mathscr{A}} K^*(Q_1, P_1, \nu) \, K(Q_2, P_2, \nu) \times \\ \times G(P_1, P_2, \nu) \, \mathrm{d}P_1 \, \mathrm{d}P_2. \tag{3.22}$$

If $K(Q, P, \nu)$ depends only slightly on the frequency, this function may be replaced by its value $K(Q, P, \bar{\nu})$ at some mean frequency $\bar{\nu}$ and we obtain the following propagation law for the mutual coherence function

$$\Gamma(Q_1, Q_2, \tau) = \iint_{\mathscr{A}} K^*(Q_1, P_1, \bar{\nu}) \, K(Q_2, P_2, \bar{\nu}) \, \Gamma(P_1, P_2, \tau) \, \mathrm{d}P_1 \, \mathrm{d}P_2. \tag{3.23a}$$

From this we obtain for the mutual intensity $\Gamma(Q_1, Q_2)$, by putting $\tau = 0$,

$$\Gamma(Q_1, Q_2) = \iint_{\mathscr{A}} K^*(Q_1, P_1, \bar{\nu}) \, K(Q_2, P_2, \bar{\nu}) \, \Gamma(P_1, P_2) \, \mathrm{d}P_1 \, \mathrm{d}P_2. \tag{3.23b}$$

The intensity $I(Q)$ is given as $I(Q) \equiv \Gamma(Q, Q, 0)$ and the degree of coherence is determined by (3.7) where $\Gamma(Q_j, Q_j, 0) \equiv I(Q_j)$.

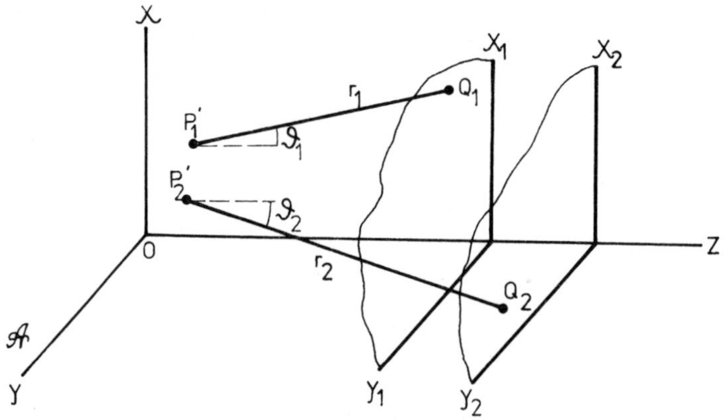

Fig. 3.3 Geometry for describing propagation of partial coherence from a plane surface $\mathscr{A}$.

We now consider propagation from a finite plane surface – a problem having important applications (Parrent (1959a), Beran and Parrent (1964)). The geometry is shown in Fig. 3.3.

The Green functions of this problem satisfying the corresponding Sommerfeld radiation condition at infinity (asymptotic behaviour of the form $\exp(ikr)/r$) and having the property (3.18) are

$$\mathscr{G}_j = \frac{\exp(ikr_j)}{4\pi r_j} - \frac{\exp(ikr_j'')}{4\pi r_j''}, \qquad j = 1, 2. \tag{3.24}$$

Here $r_j = \overline{P_jQ_j}$ and r_j'' is the distance between $P_j' \to P_j$ on $\mathscr{A}$ and the mirror point of Q with respect to the plane (x, y), i.e.

$$\begin{aligned} r_j &= [(x_j - x_j')^2 + (y_j - y_j')^2 + (z_j - z_j')^2]^{1/2}, \\ r_j'' &= [(x_j - x_j')^2 + (y_j - y_j')^2 + (z_j + z_j')^2]^{1/2}. \end{aligned} \tag{3.25}$$

Hence the normal derivatives give us

$$\frac{\partial \mathscr{G}_j}{\partial n_j} = \left(ik - \frac{1}{r_j}\right)\frac{\exp(ikr_j)}{4\pi r_j}\frac{\partial r_j}{\partial n_j} - \left(ik - \frac{1}{r_j''}\right)\frac{\exp(ikr_j'')}{4\pi r_j''}\frac{\partial r_j''}{\partial n_j}. \tag{3.26}$$

However, the following identities hold

$$\begin{aligned} &\frac{\partial r_j}{\partial n_j} = -\frac{\partial r_j''}{\partial n_j} = -\left.\frac{\partial r_j}{\partial z_j'}\right|_{\mathscr{A}} = \left.\frac{\partial r_j''}{\partial z_j'}\right|_{\mathscr{A}} = \left.\frac{z_j}{r_j}\right|_{\mathscr{A}} = \cos(\vartheta_j), \\ &r_j|_{\mathscr{A}} = r_j''|_{\mathscr{A}} \end{aligned} \tag{3.27}$$

and so from (3.26)

$$\frac{\partial \mathscr{G}_j}{\partial n_j} = 2\left(ik - \frac{1}{r_j}\right)\frac{\exp(ikr_j)}{4\pi r_j}\cos(\vartheta_j), \tag{3.28}$$

giving, using (3.19),

$$G(Q_1, Q_2, \nu) = \frac{1}{4\pi^2} \iint_{\mathscr{A}} (1 + ikr_1)(1 - ikr_2) \frac{\exp[-ik(r_1 - r_2)]}{r_1^2 r_2^2} \times$$
$$\times \cos(\vartheta_1)\cos(\vartheta_2)\, G(P_1, P_2, \nu)\, dP_1\, dP_2. \qquad (3.29)$$

The mutual coherence function is obtained (using the Fourier transform) as

$$\Gamma(Q_1, Q_2, \tau) = \frac{1}{4\pi^2} \int_0^\infty \iint_{\mathscr{A}} H(r_1, r_2, \nu, \tau)\, G(P_1, P_2, \nu) \times$$
$$\times \cos(\vartheta_1)\cos(\vartheta_2)\, dP_1\, dP_2\, d\nu, \qquad (3.30)$$

where

$$H(r_1, r_2, \nu, \tau) = \exp\left[-i\,2\pi\nu\left(\tau + \frac{r_1 - r_2}{c}\right)\right]\left\{\frac{1}{r_1^2 r_2^2} + \frac{ik(r_1 - r_2)}{r_1^2 r_2^2} + \frac{k^2}{r_1 r_2}\right\}.$$

This can be written in a concise form

$$\Gamma(Q_1, Q_2, \tau) = \frac{1}{4\pi^2} \iint_{\mathscr{A}} \frac{\cos(\vartheta_1)\cos(\vartheta_2)}{r_1^2 r_2^2}\, \mathscr{D}\Gamma\left(P_1, P_2, \tau + \frac{r_1 - r_2}{c}\right) dP_1\, dP_2, \qquad (3.31)$$

where the operator $\mathscr{D}$ is given by

$$\mathscr{D} \equiv 1 - \frac{r_1 - r_2}{c}\frac{\partial}{\partial\tau} - \frac{r_1 r_2}{c^2}\left(\frac{\partial}{\partial\tau}\right)^2. \qquad (3.32)$$

Equation (3.31) represents the general propagation law for the mutual coherence of a field produced by a plane polychromatic partially coherent primary or secondary source.

In practice it is usual that

$$k = \frac{2\pi}{\lambda} \gg \frac{1}{r} \qquad (3.33)$$

and then we have from (3.31)

$$\Gamma(Q_1, Q_2, \tau) = \frac{1}{4\pi^2} \iint_{\mathscr{A}} \frac{\Gamma\left(P_1, P_2, \tau + \frac{r_1 - r_2}{c}\right)}{r_1 r_2}\, \bar{\Lambda}_1^* \bar{\Lambda}_2\, dP_1\, dP_2, \qquad (3.34)$$

where the Λ_j are inclination factors defined by $\Lambda_j = ik\cos(\vartheta_j)$, describing the change of direction of secondary emitted radiation on $\mathscr{A}$ and $\bar{\Lambda}_j$ is a mean value of Λ_j over the frequency region, or it denotes Λ_j considered at the mean frequency $\bar{\nu}$.

Of course, the same propagation law can be directly obtained by using the Huygens-Fresnel principle for the spectral component $\tilde{V}(\nu)$,

$$\tilde{V}(Q_j, \nu) = \int_{\mathscr{A}} \tilde{V}(P_j, \nu) \frac{\exp(ikr_j)}{2\pi r_j} \Lambda_j\, dP_j, \qquad (3.35)$$

if it is substituted into (2.24) and (2.25).

The intensity at Q can be calculated from (3.34) by putting $Q_1 \equiv Q_2 \equiv Q$ and $\tau = 0$,

$$I(Q) = \iint_{\mathscr{A}} \frac{[I(P_1)\, I(P_2)]^{1/2}}{4\pi^2 r_1 r_2} \gamma\left(P_1, P_2, \frac{r_1 - r_2}{c}\right) \bar{\Lambda}_1^* \bar{\Lambda}_2 \, \mathrm{d}P_1 \, \mathrm{d}P_2, \tag{3.36}$$

where the degree of coherence γ has been introduced from (3.7).

One can see that only measurable quantities appear in the propagation laws, rather than the non-measurable vectors of Maxwell theory, and this is a characteristic feature of this formulation.

Identifying the area $\mathscr{A}$ with the unilluminated side of the screen $\mathscr{A}$ in Fig. 3.1, we again obtain from (3.36) the interference law (3.8).

Identifying the area $\mathscr{A}$ with the plane of an extended primary plane source σ, the elementary radiators of which are mutually incoherent, then only pairs of points $P_1 \equiv P_2 \equiv P$ will contribute to Γ in (3.34) and we may write

$$G(P_1, P_2, \nu) = G(P_1, \nu)\, \delta(P_1 - P_2), \tag{3.37a}$$

$$\Gamma(P_1, P_2, \tau) = \Gamma(P_1, P_1, \tau)\, \delta(P_1 - P_2), \tag{3.37b}$$

where $G(P, \nu) \equiv G(P, P, \nu)$ is the power spectrum. Thus we have from (3.34)

$$\Gamma(Q_1, Q_2, \tau) = \frac{1}{4\pi^2} \int_0^\infty \mathrm{d}\nu \exp(-\mathrm{i}2\pi\nu\tau) \times$$

$$\times \int_\sigma G(P, \nu) \frac{\exp[-\mathrm{i}k(r_1 - r_2)]}{r_1 r_2} \bar{\Lambda}_1^* \bar{\Lambda}_2 \, \mathrm{d}P. \tag{3.38}$$

The degree of coherence $\gamma(Q_1, Q_2, \tau)$ is determined by (3.7), where

$$\Gamma(Q_j, Q_j, 0) \equiv I(Q_j) = \frac{1}{4\pi^2} \int_0^\infty \mathrm{d}\nu \int_\sigma \frac{G(P, \nu)}{r_j^2} |\bar{\Lambda}_j|^2 \, \mathrm{d}P =$$

$$= \frac{1}{4\pi^2} \int_\sigma \frac{I(P)}{r_j^2} |\bar{\Lambda}_j|^2 \, \mathrm{d}P. \tag{3.39}$$

Denoting

$$[G(P, \nu)]^{1/2} \frac{\exp(\mathrm{i}kr_j)}{2\pi r_j} \bar{\Lambda}_j \equiv U(Q_j, P, \nu), \tag{3.40}$$

we may rewrite (3.38) in the form

$$\Gamma(Q_1, Q_2, \tau) = [I(Q_1)\, I(Q_2)]^{1/2} \gamma(Q_1, Q_2, \tau) =$$

$$= \int_0^\infty \mathrm{d}\nu \exp(-\mathrm{i}2\pi\nu\tau) \int_\sigma U^*(Q_1, P, \nu)\, U(Q_2, P, \nu) \, \mathrm{d}P, \tag{3.41}$$

where

$$I(Q_j) = \int_0^\infty \mathrm{d}\nu \int_\sigma |U(Q_j, P, \nu)|^2 \, \mathrm{d}P, \qquad j = 1, 2. \tag{3.42}$$

This formula expresses the mutual coherence function and the degree of coherence in terms of the light distribution arising from an associated fictional source since, according to (3.40), the function $U(Q, P, \nu)$ may be regarded as the complex amplitude at Q due to a monochromatic point source of frequency ν and real amplitude $[G(P, \nu)]^{1/2}$ situated at P.

3.3 Van Cittert-Zernike theorem

In practice quasi-monochromatic light is often used. If moreover the assumption

$$\left| \tau + \frac{r_1 - r_2}{c} \right| \ll \frac{1}{\Delta\nu} \tag{3.43}$$

holds, then (3.38) gives

$$\begin{aligned} \Gamma(Q_1, Q_2, \tau) &= [I(Q_1)\, I(Q_2)]^{1/2}\, \gamma(Q_1, Q_2, \tau) = \\ &= \left(\frac{\bar{k}}{2\pi}\right)^2 \exp(-\mathrm{i}\, 2\pi\bar{\nu}\tau) \int_\sigma I(P) \frac{\exp[-\mathrm{i}\bar{k}(r_1 - r_2)]}{r_1 r_2}\, \mathrm{d}P, \end{aligned} \tag{3.44}$$

where only small angles ϑ_j are considered so that $\cos(\vartheta_j) \approx 1$. The degree of coherence then has the form

$$\gamma(Q_1, Q_2, \tau) \equiv \gamma_{12}(\tau) = \gamma_{12}(0) \exp(-\mathrm{i}\, 2\pi\bar{\nu}\tau), \tag{3.45}$$

with the degree of spatial coherence $\gamma_{12}(0)$ given by

$$\gamma_{12}(0) \equiv \gamma_{12} = \frac{1}{[I(Q_1)\, I(Q_2)]^{1/2}} \int_\sigma I(P) \frac{\exp[-\mathrm{i}\bar{k}(r_1 - r_2)]}{r_1 r_2}\, \mathrm{d}P, \tag{3.46}$$

where

$$I(Q_j) = \int_\sigma \frac{I(P)}{r_j^2}\, \mathrm{d}P, \qquad j = 1, 2. \tag{3.47}$$

This is the Van Cittert-Zernike theorem which enables us to calculate (under the above assumptions) the degree of coherence of an extended quasi-monochromatic incoherent source in terms of the intensity distribution over the source. It says that the degree of spatial coherence of light from this source is equal to the normalized complex amplitude at point Q_2 obtained by the diffraction of a spherical wave centered at point Q_1, at an aperture of the form σ with the real amplitude numerically equal to the source intensity.

It is interesting to note that the incoherent source gives rise in general to a partially coherent field and it means that multiple propagation and diffraction can improve the degree of coherence of the light. This has been demonstrated by Streifer (1966) with the help of a system of lenses and it may play a role in connection with laser light in a cavity, as has been pointed out by Wolf (1963b, 1964) and Wolf and Agarwal (1984).

If (3.45) holds, the visibility of fringes is determined by the geometric properties of the source and it is sufficient to take into account only spatial coherence (described by $\gamma_{12}(0)$), which is independent of the spectral properties of the radiation. However, from (3.38), it is obvious that spatial coherence described by the degree of coherence $\gamma_{12}(0)$ ($\tau = 0$) and temporal coherence described by the degree of coherence $\gamma_{11}(\tau)$ ($P_1 \equiv P_2$) are generally dependent on both the geometrical properties of the source and the spectral properties of the radiation. In general it is not possible to express the degree of coherence $\gamma_{12}(\tau)$ as a product of the degree of spatial coherence $\gamma_{12}(0)$ and the degree of temporal coherence $\gamma_{11}(\tau)$ due to the fact that spatial and temporal coherence are mutually connected. This follows from the fact that $\Gamma_{12}(\tau)$ obeys the wave equations (3.15) *in vacuo* (see Sec. 4.3). In some cases of practical interest however such a decomposition is possible under certain restrictions. For example, assuming that the power spectrum $G(P, \nu)$ is independent of P and that

$$\frac{|r_1 - r_2|}{c} \ll \frac{1}{\Delta\nu}, \tag{3.48}$$

we obtain from (3.38)

$$\begin{aligned} \Gamma(Q_1, Q_2, \tau) &= \frac{1}{4\pi^2}\Gamma(\tau)\int_\sigma \frac{\exp[-i\bar{k}(r_1 - r_2)]}{r_1 r_2}\bar{\Lambda}_1^*\bar{\Lambda}_2\,dP = \\ &= \Gamma(\tau)\,\Gamma(Q_1, Q_2, 0), \end{aligned} \tag{3.49}$$

where $\Gamma(\tau)$ is the Fourier transform of $G(\nu)$ and thus

$$\gamma(Q_1, Q_2, \tau) = \gamma(\tau)\,\gamma(Q_1, Q_2, 0). \tag{3.50}$$

Here $\gamma(\tau) = \Gamma(\tau)/\Gamma(0)$ and $\gamma(Q_1, Q_2, 0) = \Gamma(Q_1, Q_2, 0)/[\Gamma(Q_1, Q_1, 0)\,\Gamma(Q_2, Q_2, 0)]^{1/2}$. In this case temporal coherence is connected with the spectral properties of the radiation and spatial coherence is connected with the geometric properties of the source. This result is a special case of more general considerations about the cross-spectral purity of light (Mandel (1961a)) (Sec. 4.3).

The propagation laws of the mutual coherence function (mutual intensity) for quasi-monochromatic light may be obtained from (3.34) (under the assumption (3.43) and for small angles ϑ_j) in the form

$$\begin{aligned} \Gamma(Q_1, Q_2, \tau) &= \left(\frac{\bar{k}}{2\pi}\right)^2 \exp(-i2\pi\bar{\nu}\tau) \times \\ &\times \iint_\sigma \frac{\exp[-i\bar{k}(r_1 - r_2)]}{r_1 r_2}\Gamma(P_1, P_2, 0)\,dP_1\,dP_2. \end{aligned} \tag{3.51}$$

Putting $\tau = 0$ this becomes

$$\begin{aligned} \Gamma(Q_1, Q_2) &= [I(Q_1)\,I(Q_2)]^{1/2}\,\gamma(Q_1, Q_2) = \\ &= \left(\frac{\bar{k}}{2\pi}\right)^2 \iint_\sigma \frac{\exp[-i\bar{k}(r_1 - r_2)]}{r_1 r_2}\Gamma(P_1, P_2)\,dP_1\,dP_2, \end{aligned} \tag{3.52}$$

which is a special case of (3.23b) when $K(Q, P, \bar{\nu})$ is replaced by $(\bar{k}/2\pi)\exp(i\bar{k}r)/r$.

Using the analogous substitution to (3.40) in (3.44) we obtain the Hopkins formula

$$\Gamma(Q_1, Q_2) = [I(Q_1)\, I(Q_2)]^{1/2}\, \gamma(Q_1, Q_2) = \int_\sigma U^*(Q_1, P)\, U(Q_2, P)\, \mathrm{d}P, \quad (3.53)$$

which is particularly useful in instrumetal optics. This formula gives the degree of spatial coherence without explicit use of any averaging process.

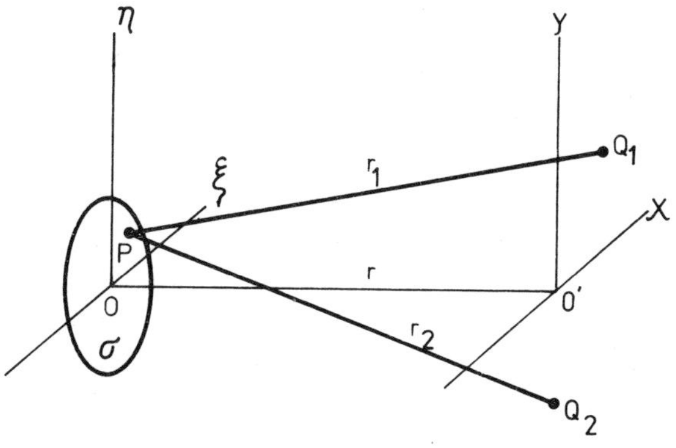

Fig. 3.4 To Van Cittert-Zernike theorem.

Restricting ourselves to small linear dimensions of the source and considering only small angular distances of the points Q_1 and Q_2 (Fig. 3.4), we have

$$r_j = [(x_j - \xi)^2 + (y_j - \eta)^2 + r^2]^{1/2} \approx r + \frac{(x_j - \xi)^2 + (y_j - \eta)^2}{2r}, \quad (3.54)$$

assuming that the linear dimensions of the source are small compared with the distance r and that only small angular distances between OO' and PQ_j are considered; hence

$$r_1 - r_2 = \frac{(x_1^2 + y_1^2) - (x_2^2 + y_2^2)}{2r} - \frac{(x_1 - x_2)\,\xi + (y_1 - y_2)\,\eta}{r}. \quad (3.55)$$

Denoting

$$\psi = -\frac{\bar{k}[(x_1^2 + y_1^2) - (x_2^2 + y_2^2)]}{2r}, \quad p = \frac{x_1 - x_2}{r}, \quad q = \frac{y_1 - y_2}{r}, \quad (3.56)$$

we obtain from (3.46)

$$\gamma(Q_1, Q_2) = \exp(\mathrm{i}\psi) \frac{\int_\sigma I(\xi, \eta) \exp[\mathrm{i}\bar{k}(p\xi + q\eta)]\, \mathrm{d}\xi\, \mathrm{d}\eta}{\int_\sigma I(\xi, \eta)\, \mathrm{d}\xi\, \mathrm{d}\eta}, \quad (3.57)$$

where we have replaced $r_1 r_2$ by r^2. In this case the degree of spatial coherence equals, apart from a phase factor, the *normalized Fourier transform* of the intensity distribution over the source.

For a central, quasi-monochromatic, uniform, spatially incoherent, circular source of radius ϱ we obtain from (3.57)

$$\gamma(Q_1, Q_2) = \frac{2J_1(v)}{v} \exp(\mathrm{i}\psi), \tag{3.58}$$

where $v = \bar{k}\varrho(p^2 + q^2)^{1/2}$ and J_1 is the Bessel function. For a rectangular source with sides $2a$ and $2b$ we obtain

$$\gamma(Q_1, Q_2) = \frac{\sin(\bar{k}pa)}{\bar{k}pa} \frac{\sin(\bar{k}qb)}{\bar{k}qb} \exp(\mathrm{i}\psi). \tag{3.59}$$

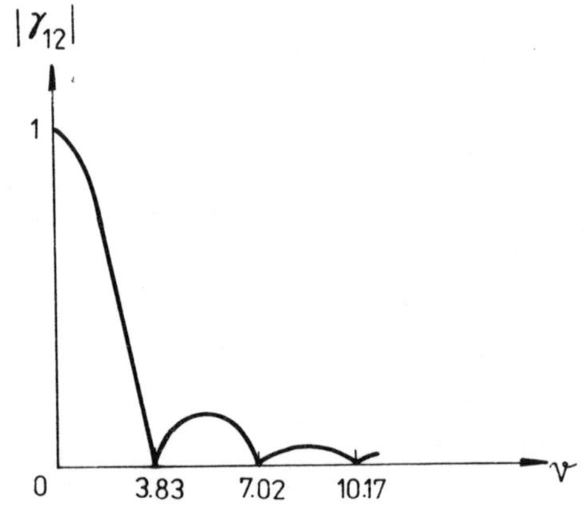

Fig. 3.5 The behaviour of $|\gamma_{12}|$ as a function of $v = (\bar{k}\varrho/r)\,[(x_1 - x_2)^2 + (y_1 - y_2)^2]^{1/2}$ for light from a quasi-monochromatic uniform incoherent circular source.

The typical behaviour of the modulus of the degree of spatial coherence (i.e. of the visibility) is shown in Fig. 3.5, based on (3.58). Using the first zero point of (3.59) to define the coherence area, $\bar{k}pa = \bar{k}qb = \pi$ and we obtain for the coherence area $A_c = (r\bar{\lambda})^2/S$, $S = 4ab$, in agreement with (2.60). Hence the coherence area increases with increasing distance r and with decreasing source area S. More generally, for stationary fields in space the coherence area can be defined as

$$A_c = \iint_{-\infty}^{+\infty} |\gamma(x, y)|^2 \,\mathrm{d}x\,\mathrm{d}y, \tag{3.60}$$

$x = x_1 - x_2$, $y = y_1 - y_2$; the volume of coherence can be defined as

$$V_c = c \iiint_{-\infty}^{+\infty} |\gamma(x, y, \tau)|^2 \,\mathrm{d}x\,\mathrm{d}y\,\mathrm{d}\tau. \tag{3.61}$$

Substituting (3.59) in (3.60) we have $A_c = (r\bar{\lambda})^2/S$ again. More general definitions for fields of arbitrary statistics are given in Sec. 10.5.

Equation (3.58) may be used to determine the angular radius of a star by means of the Michelson stellar interferometer the scheme of which is shown in Fig. 3.6.

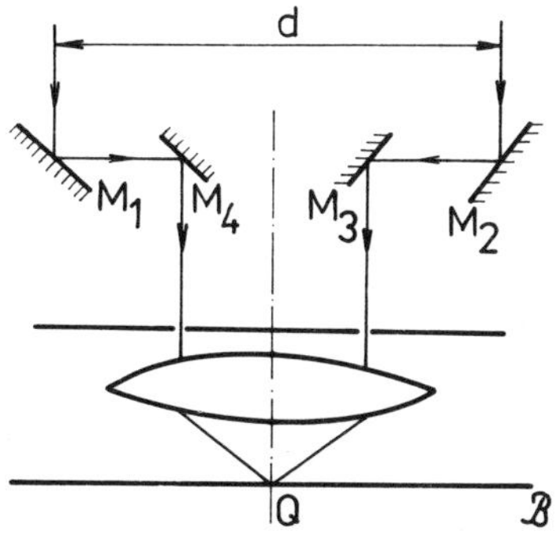

Fig. 3.6 Scheme of the Michelson stellar interferometer; $M_1 - M_4$ are mirrors, the interference is observed on the screen $\mathscr{B}$.

If we change the separation d of the mirrors M_1 and M_2, the visibility of fringes in the neighbourhood of Q will vary according to Fig. 3.5. The first vanishing of the visibility occurs for $v = \bar{k}\varrho d_0/r \doteq 3.83$, i.e. $d_0 = 0.61\bar{\lambda}/\alpha$ ($\psi \approx 0$), $\alpha = \varrho/r$ being the angular radius of the star, and is given by $\alpha = 0.61\bar{\lambda}/d_0$; for the angular diameter we have $2\alpha = 1.22\bar{\lambda}/d_0$.

Numerous methods for measuring the degree of coherence have been developed, mostly based on interferometric measurements with classical types of interferometers, but also using holographic methods or Fourier analysis. The classical interferometric methods are always based on the above description of the formation of interference fringes and they have been reviewed by Francon and Mallick (1967). Further, if $\Gamma(Q_1, Q_2, \tau)$ in (3.34) is a function of $Q_1 - Q_2$, i.e. light is stationary in space, the spatial Fourier transformation of (3.34) provides (Cornacchio and Soni (1965), Cornacchio and Farnham (1966))

$$\gamma(\xi_1 - \xi_2, \eta_1 - \eta_2) = \\ = \frac{\tilde{I}_i(\bar{k}(\xi_1 - \xi_2)/r, \bar{k}(\eta_1 - \eta_2)/r)}{\iint\limits_{-\infty}^{+\infty} I_o^{1/2}(x, y)\, I_o^{1/2}(x - (\xi_1 - \xi_2), y - (\eta_1 - \eta_2))\, \mathrm{d}x\, \mathrm{d}y}, \tag{3.62}$$

where I_o is the object intensity and $\tilde{I}_i$ is the spatial Fourier transform of the image intensity. Interesting methods for measuring the degree of coherence are provided by the *holographic principle* (Soroko (1971), Stasel'ko, Voronin and Smirnov (1973)). Such methods provide the possibility of registering instantaneously an infinite number of degrees of coherence between all points in an area and a reference point. Shearing interferometric measurements of the degree of coherence have been performed by Carter (1977). A review of methods of measuring the degree of coherence has been also given by Gusev and Pojzner (1982).

Note that the second-order coherence concept has been employed also in the x-ray region by Holý (1980, 1982).

CHAPTER 4

GENERAL PROPERTIES OF THE SECOND-ORDER COHERENCE

In this chapter we deal with general properties of the second-order coherence, such as the form of the mutual coherence function for incoherent radiation (Beran and Parrent (1963, 1964)), properties of the coherent radiation as a limiting case of partially coherent radiation (Parrent (1959a, b), Mandel and Wolf (1961a, 1976, 1981), Beran and Parrent (1964), Mehta, Wolf and Balachandran (1966), Barakat (1966), Bastiaans (1977)), the concept of cross-spectral purity of light (Mandel (1961a, 1962)) and the phase problem of interference spectroscopy. A complete solution of the last problem would make it possible to determine the spectrum of radiation from measurements of the visibility of interference fringes (Wolf (1962), Roman and Marathay (1963), Peřina and Tillich (1966), Mehta (1965, 1968), Dialetis and Wolf (1967), Dialetis (1967), Nussenzveig (1967), Burge, Fiddy, Greenaway and Ross (1974, 1976), Barakat (1980)). Finally we treat radiation from partially coherent sources and radiometry with light of any state of coherence, the area systematically developed in recent years (see references in the Introduction).

4.1 Mutual coherence function for incoherent radiation

We return to the propagation law (3.19) and consider the spatially incoherent radiation in the whole space including the surface $\mathscr{A}$, i.e.

$$G(P_1, P_2, \nu) = \begin{cases} G(P, P, \nu) \equiv G(P, \nu), & P_1 \equiv P_2 \equiv P, \\ 0 & , \quad P_1 \not\equiv P_2. \end{cases} \tag{4.1}$$

The integral in (3.19) may be interpreted as an integral in a four-dimensional space while the function under the integral sign is non-vanishing and finite on a two-dimensional hypersurface. Therefore the integral is zero in all cases including $P_1 \equiv P_2$. This is in contradiction to (4.1) and consequently such an incoherent surface does not radiate. As the propagation law (3.19) is a direct consequence of the Helmholtz equations (3.16) we conclude that an incoherent field cannot exist in a free space regardless of the spectral properties of the radiation. However, in agreement with Blanc-Lapierre and Dumontet (1955) we may define a spatially incoherent source by (3.37a, b), i.e.

$$G_{12}(\nu) = G_{11}(\nu)\,\delta(P_1 - P_2) \tag{4.2}$$

for all pairs of points P_1, P_2 of the source. Thus (3.19) gives a non-vanishing result

$$G(Q, \nu) = \int_{\mathscr{A}} \left| \frac{\partial \mathscr{G}(Q, P, \nu)}{\partial n} \right|^2 G(P, \nu)\, \mathrm{d}P, \tag{4.3}$$

where $G(Q, \nu) \equiv G(Q, Q, \nu)$, but the mutual spectral density is non-physical for $P_1 \equiv P_2$ in this case. This difficulty was removed by Beran and Parrent (1963, 1964) who have shown that the δ-function in (4.2) may be replaced by $2J_1(k|P_1 - P_2|)/k|P_1 - P_2|$, where $k = 2\pi/\lambda$, J_1 is the Bessel function and $|P_1 - P_2| = [(x_1 - x_2)^2 + (y_1 - y_2)^2]^{1/2}$ ($P_j \equiv P_j(x_j, y_j)$, $j = 1, 2$), neglecting evanescent waves. Of course, such a function expressing the "smoothed" δ-function is not unique.

4.2 Properties of coherent radiation

We will follow original considerations given by Mehta, Wolf and Balachandran (1966). A fully coherent field (in second-order) may be defined by

$$|\gamma(x_1, x_2)| \equiv 1. \tag{4.4a}$$

If it is stationary, we have

$$|\gamma(\mathbf{x}_1, \mathbf{x}_2, \tau)| \equiv 1. \tag{4.4b}$$

Equations (4.4a, b) must hold for all $\mathbf{x}_1$, $\mathbf{x}_2$, t_1, $t_2(\tau)$. This implies

$$|\Gamma(x_1, x_2)|^2 = \Gamma(x_1, x_1)\, \Gamma(x_2, x_2) \tag{4.5a}$$

and

$$|\Gamma(\mathbf{x}_1, \mathbf{x}_2, \tau)|^2 = \Gamma(\mathbf{x}_1, \mathbf{x}_1, 0)\, \Gamma(\mathbf{x}_2 . \mathbf{x}_2, 0). \tag{4.5b}$$

Therefore the correlation function $\Gamma(x_1, x_2)$ has the form

$$\Gamma(x_1, x_2) = A(x_1)\, A(x_2) \exp[\mathrm{i}\varphi(x_1, x_2)], \tag{4.6}$$

where $A(x_j)$ $(= [\Gamma(x_j, x_j)]^{1/2})$ and $\varphi(x_1, x_2)$ $(= -\varphi(x_2, x_1))$ are real-valued functions (since $\Gamma(x_1, x_2) = \Gamma^*(x_2, x_1)$). Denoting $F = \sum_j^n c_j V(x_j)$, we obtain

$$0 \leqq \langle |F|^2 \rangle = \sum_{j,k}^n c_j^* c_k \langle V^*(x_j)\, V(x_k) \rangle = \sum_{j,k}^n c_j^* c_k \Gamma(x_j, x_k), \tag{4.7a}$$

i.e.

$$\sum_{j,k}^n d_j^* d_k \gamma(x_j, x_k) \geqq 0, \tag{4.7b}$$

which is the *non-negative definiteness condition* for the degree of coherence with arbitrary complex constants $d_k (= c_k [\Gamma(x_k, x_k)]^{1/2})$. This condition gives us successively for $n = 1, 2, 3$

$$\gamma(x, x) = 1 > 0, \tag{4.8a}$$

$$\begin{vmatrix} 1 & \gamma(x_1, x_2) \\ \gamma(x_2, x_1) & 1 \end{vmatrix} = 1 - |\gamma(x_1, x_2)|^2 \geqq 0, \tag{4.8b}$$

$$\begin{vmatrix} 1 & \gamma(x_1, x_2) & \gamma(x_1, x_3) \\ \gamma(x_2, x_1) & 1 & \gamma(x_2, x_3) \\ \gamma(x_3, x_1) & \gamma(x_3, x_2) & 1 \end{vmatrix} = (1 - |\gamma(x_1, x_2)|^2)(1 - |\gamma(x_3, x_1)|^2)$$
$$- |\gamma(x_3, x_2) - \gamma(x_1, x_2)\gamma(x_3, x_1)|^2 \geqq 0. \tag{4.8c}$$

The first condition is evident, the second one gives the normalization inequality for $\gamma(x_1, x_2)$. From the third condition it follows that

$$|\gamma(x_3, x_2) - \gamma(x_1, x_2)\gamma(x_3, x_1)|^2 \leqq$$
$$\leqq (1 - |\gamma(x_1, x_2)|^2)(1 - |\gamma(x_3, x_1)|^2). \tag{4.9}$$

Since (4.4a) holds we have

$$\gamma(x_3, x_2) = \gamma(x_3, x_1)\gamma(x_1, x_2), \tag{4.10a}$$

that is

$$\varphi(x_3, x_2) = \varphi(x_1, x_2) + \varphi(x_3, x_1) = \varphi(x_1, x_2) - \varphi(x_1, x_3) \tag{4.10b}$$

and so putting $x_1 = 0$ we have $\varphi(x_3, x_2) - \varphi(x_3, 0) = \varphi(0, x_2)$ which is independent of x_3. Thus the functional equation (4.10b) defines, up to an additive constant, the function $\alpha(x) = \varphi(0, x)$ and so

$$\varphi(x_1, x_2) = \alpha(x_2) - \alpha(x_1). \tag{4.11}$$

Substituting (4.11) into (4.6) the second-order coherence function $\Gamma(x_1, x_2)$ takes the factorized form

$$\Gamma(x_1, x_2) = V^*(x_1)V(x_2), \tag{4.12}$$

where $V(x) = A(x)\exp[\mathrm{i}\alpha(x)]$. Hence, the conditions of full second-order coherence (4.4a) [(4.5a)] and (4.12) are equivalent because (4.12) also implies (4.4a) [(4.5a)]. Some consequences of this factorization property of the second-order correlation function for the higher-order correlation functions will be dealt with in Chapter 12.

Exactly the same considerations can be used for degrees of coherence (2.44a, b) in the frequency and wave-vector-frequency domains (Mandel and Wolf (1981)). Hence

$$0 \leqq |\gamma(\mathbf{x}_1, \mathbf{x}_2, \nu)| \leqq 1, \tag{4.13a}$$

$$0 \leqq |\gamma(\mathbf{k}_1, \mathbf{k}_2, \nu)| \leqq 1. \tag{4.13b}$$

The limiting values 0 and 1 in these inequalities indicate complete second-order incoherence and coherence, respectively. Moreover, if a stationary field of a frequency ν is spatially coherent, $|\gamma(\mathbf{x}_1, \mathbf{x}_2, \nu)| = 1$ for all $\mathbf{x}_1$, $\mathbf{x}_2$ and by the above arguments the cross-spectral density has the factorized form

$$G(\mathbf{x}_1, \mathbf{x}_2, \nu) = \mathscr{U}^*(\mathbf{x}_1, \nu)\mathscr{U}(\mathbf{x}_2, \nu), \tag{4.13c}$$

where the complex spectral amplitude $\mathscr{U}(\mathbf{x}, \nu)$ satisfies the Helmholtz equation.

A more direct way of obtaining the factorization (4.12) or (4.13c) is to use the cross-symmetry condition $\Gamma(x_1, x_2) = \Gamma^*(x_2, x_1)$, giving, together with (4.5a), $\Gamma(x_1, x_2) = A(x_1)\, B(x_2) = A^*(x_2)\, B^*(x_1)$, so that $A(x_1)/B^*(x_1) = A^*(x_2)/B(x_2) = k$; hence $A(x) = kB^*(x)$ and denoting $V(x) = k^{1/2}B(x)$, we just arrive at (4.12) and in a similar way at (4.13c) $(G(\mathbf{x}_1, \mathbf{x}_2, \nu) = G^*(\mathbf{x}_2, \mathbf{x}_1, \nu))$.

Let us note here that a field coherent to all orders may be defined by means of analogous factorization properties imposed upon the correlation functions $\Gamma^{(m,n)}$. The classical field fulfilling such conditions is obviously a deterministic field without any fluctuations. These questions will be discussed in Sec. 12.2.

Equation (4.12) gives in the case of stationary fields

$$\Gamma(\mathbf{x}_1, \mathbf{x}_2, \tau) = U^*(\mathbf{x}_1)\, U(\mathbf{x}_2) \exp(-i 2\pi \nu_0 \tau), \tag{4.14}$$

where $\tau = t_2 - t_1$, ν_0 is a positive constant and $U(\mathbf{x}_j)$ is a function depending on $\mathbf{x}_j$ only. The factorization property (4.14) of the mutual coherence function was derived for quasi-monochromatic light using other methods by Parrent (1959a, b), Beran and Parrent (1964) and Mandel and Wolf (1961a).

Thus the condition (4.4b) for the second-order fully coherent field leads to the result that the coherent field in this sense must be monochromatic with frequency ν_0. On the other hand, substituting the monochromatic field

$$V(\mathbf{x}_j, t) = U(\mathbf{x}_j) \exp(-i 2\pi \nu_0 t), \qquad j = 1, 2 \tag{4.15}$$

into the definition of $\Gamma(\mathbf{x}_1, \mathbf{x}_2, \tau)$, we arrive at (4.14) again. Hence, the stationary field is completely coherent (in second order) in the sense of (4.4) if and only if it is monochromatic. The result follows from restricting ourselves to the class of stationary fields; indeed in this case $\Gamma(t_1, t_2) = \Gamma(t_1 - t_2) = V^*(t_1)\, V(t_2)$, and this is a functional equation which may be fulfilled only by an exponential function, i.e. $V(t) \sim \exp(-i 2\pi \nu_0 t)$. However, considering a class of general non-stationary fields the factorization condition (4.12) and analogous conditions for $\Gamma^{(n,n)}$ (expressing complete coherence) do restrict the statistical properties of the field rather than its spectral properties; such a field $V(t)$ may be of arbitrary spectral composition.

The conditions (4.4) are too strong to be satisfied by real physical fields. Therefore Mandel and Wolf (1961a) considered a weaker, but more realistic, definition of the second-order fully coherent field as fulfilling

$$\operatorname*{Max}_{\tau} |\gamma(\mathbf{x}_1, \mathbf{x}_2, \tau)| \equiv 1 \tag{4.16a}$$

for all $\mathbf{x}_1$, $\mathbf{x}_2$. They have shown that the mutual coherence function is then factorized in the form

$$\Gamma(\mathbf{x}_1, \mathbf{x}_2, \tau) = U^*(\mathbf{x}_1)\, U(\mathbf{x}_2)\, \gamma(\tau), \tag{4.16b}$$

where $\gamma(\tau)$ is a unimodular function. Thus a field coherent in this sense need not be monochromatic and so the class of coherent fields in the sense (4.16) is broader than the class of monochromatic fields. Equation (4.16b) reduces to (4.14) if

$$|\tau| \ll \frac{1}{\Delta\nu}. \tag{4.17}$$

Also other generalizations of the definition of the coherent field have been introduced (Peřina (1965a, b)), with respect to which the monochromatic field need not be even coherent. Such definitions are based on the average using functional integrals in a functional space. The fact that coherence is a manifestation of the statistical properties of the field and that it does not restrict generally the spectrum will be more evident in the quantum description, where the formalism of the so-called coherent states will be used.

It should be mentioned here that (4.14) expresses that for the stationary coherent field the degree of coherence $\gamma(\mathbf{x}_1, \mathbf{x}_2, \tau)$ has no zeros in the complex τ-plane. This is an interesting result in connection with the phase problem of the coherence theory (Sec. 4.4): finding conditions under which the knowledge of the modulus of $\gamma(\mathbf{x}_1, \mathbf{x}_2, \tau)$ – the visibility – for all τ allows a unique determination of its phase and thus of the complete function γ from which the spectrum g (its Fourier transform) can be found. It was shown by Van Kampen (1953) and Edwards and Parrent (1959) that the most general unimodular function, analytic in the lower half of the complex τ-plane, has the form

$$\exp[\mathrm{i}\varphi_{12}(\tau)] = \exp[\mathrm{i}(\beta_{12} - 2\pi\nu_0\tau)] \prod_{j=1}^{\infty} \frac{\tau_j^*}{\tau_j} \frac{\tau_j - \tau}{\tau_j^* - \tau}, \tag{4.18}$$

where β_{12} and $\nu_0 > 0$ are real constants and the τ_j are complex numbers ($\mathrm{Im}\,\tau_j < 0$) determining the positions of the zeros of the function $\exp[\mathrm{i}\varphi_{12}(\tau)]$ in the lower half of the complex τ-plane. For stationary fields (4.6) reduces to

$$\Gamma(\mathbf{x}_1, \mathbf{x}_2, \tau) = A(\mathbf{x}_1)\, A(\mathbf{x}_2) \exp[\mathrm{i}\varphi(\mathbf{x}_1, \mathbf{x}_2, \tau)]. \tag{4.19}$$

By using (4.14) we obtain $\varphi(\mathbf{x}_1, \mathbf{x}_2, \tau) = \beta(\mathbf{x}_1, \mathbf{x}_2) - 2\pi\nu_0\tau = \alpha(\mathbf{x}_2) - \alpha(\mathbf{x}_1) - 2\pi\nu_0\tau$, where $\alpha(\mathbf{x})$ is a function of $\mathbf{x}$ only and $U(\mathbf{x}) = A(\mathbf{x}) \exp[\mathrm{i}\alpha(\mathbf{x})]$. The so-called *Blaschke factors* $(\tau - \tau_j)/(\tau - \tau_j^*)$ in (4.18), in which zeros are included, are not present in (4.14) and consequently $\gamma(\mathbf{x}_1, \mathbf{x}_2, \tau)$ for coherent fields does not contain any zeros. Their absence for coherent fields follows using the fact that $\Gamma(\mathbf{x}_1, \mathbf{x}_2, \tau)$ must satisfy the wave equations (3.15) (Peřina (1966)). The function (4.19), with (4.18) substituted, is of the factorized form $\Gamma(\mathbf{x}_1, \mathbf{x}_2, \tau) = U^*(\mathbf{x}_1)\, U(\mathbf{x}_2)\, f(\tau)$, with $f(\tau) = \exp(-\mathrm{i}2\pi\nu_0\tau) \prod_j \tau_j^*(\tau_j - \tau)/\tau_j(\tau_j^* - \tau)$ and U independent of τ. It follows that $\Gamma(\mathbf{x}_1, \mathbf{x}_2, \tau)$ cannot be a solution of the wave equations unless the zeros are absent and $f(\tau)$ equals $\exp(-\mathrm{i}2\pi\nu_0\tau)$.

Finally let us note that the property of full coherence remains unchanged during propagation, i.e. if a field is coherent on a surface $\mathscr{A}$, then it is coherent in the whole space. This follows by substituting

$$G(\mathbf{x}_1, \mathbf{x}_2, \nu) = U^*(\mathbf{x}_1)\, U(\mathbf{x}_2)\, \delta(\nu - \nu_0) \tag{4.20a}$$

(which is a Fourier transform of (4.14)) into (3.21). We obtain

$$G(Q_1, Q_2, \nu) = \iint_{\mathscr{A}} K^*(Q_1, P_1, \nu)\, K(Q_2, P_2, \nu)\, U^*(P_1)\, U(P_2)\, \delta(\nu - \nu_0)\, \mathrm{d}P_1\, \mathrm{d}P_2$$

and consequently

$$\Gamma(Q_1, Q_2, \tau) = U^*(Q_1)\, U(Q_2) \exp(-\mathrm{i}\, 2\pi \nu_0 \tau), \tag{4.20b}$$

where

$$U(Q) = \int_{\mathscr{A}} K(Q, P, \nu_0)\, U(P)\, \mathrm{d}P, \tag{4.20c}$$

which is again of the form (4.14). By using (4.16b) and (3.34) one can prove (Mandel and Wolf (1961a)), under certain conditions, that the function $\Gamma(Q_1, Q_2, \tau)$ is also of the form (4.16b). Therefore the relation (4.20c), expressing a more general form of the Huygens-Fresnel principle, is valid for a broader class of fields than the class of monochromatic fields, i.e. it is valid for the class of coherent fields (in the sense of (4.16)). All quantities appearing in (4.20c) are measurable.

4.3 Concept of cross-spectral purity of light

As was already mentioned, spatial and temporal coherence are involved in the degree of coherence $\gamma_{12}(\tau)$ and they cannot be separated in general. Wolf, Devaney and Foley (1981) and Wolf and Devaney (1981) have shown that it follows from the wave equations for $\Gamma(\mathbf{x}_1, \mathbf{x}_2, \tau)$ that

$$\Gamma(\mathbf{x}_1, \mathbf{x}_2, \tau) = \int \Gamma(\mathbf{x}_1, \mathbf{x}'_2)\, K(\mathbf{x}_2 - \mathbf{x}'_2, \tau)\, \mathrm{d}^3 x'_2, \tag{4.21}$$

i.e. the mutual coherence function is fully determined by the mutual intensity, describing spatial coherence. In particular for $\mathbf{x}_1 = \mathbf{x}_2$ temporal coherence is fully specified by spatial coherence in the field. Here ($r = |\mathbf{x}|$)

$$K(\mathbf{x}, t) = -\frac{1}{2\pi r} \frac{\partial}{\partial r} [\delta^{(+)}(r - ct) + \delta^{(-)}(r + ct)]. \tag{4.22}$$

From (4.21) the cross-spectral density of the field can be expressed in terms of its mutual intensity

$$\begin{aligned} G(\mathbf{x}_1, \mathbf{x}_2, \nu) &= \frac{k^2}{2\pi^2 c} \int \Gamma(\mathbf{x}_1, \mathbf{x}'_2) \frac{\sin(k|\mathbf{x}_2 - \mathbf{x}'_2|)}{k|\mathbf{x}_2 - \mathbf{x}'_2|} \mathrm{d}^3 x'_2 = \\ &= \frac{k^2}{2\pi^2 c} \int \Gamma(\mathbf{x}'_1, \mathbf{x}_2) \frac{\sin(k|\mathbf{x}_1 - \mathbf{x}'_1|)}{k|\mathbf{x}_1 - \mathbf{x}'_1|} \mathrm{d}^3 x'_1, \end{aligned} \tag{4.23}$$

applying (2.32b).

Nevertheless, it may be shown (Mandel (1961a), Mandel and Wolf (1976)) that in cases of practical interest $\gamma_{12}(\tau)$ may be expressed, at least to a good approximation, as the product of the degrees of spatial and temporal coherence. This property is closely related to the spectral properties of light. The notion of the cross-spectral purity of light, introduced by means of this factorization property, leads to an important classification of optical fields. As an example we can cite (3.50), obtained from the Van Cittert-Zernike theorem under the assumption (3.48). Here we will follow a treatment given by Mandel (1961a).

We restrict ourselves to linearly polarized light and consider two points $P_1(\mathbf{x}_1)$ and $P_2(\mathbf{x}_2)$ at which the complex field amplitudes are $V(\mathbf{x}_1, t)$ and $V(\mathbf{x}_2, t)$ respectively. Let the normalized mutual spectral density corresponding to $\gamma(\mathbf{x}_1, \mathbf{x}_2, \tau)$ be $g(\mathbf{x}_1, \mathbf{x}_2, \nu) \equiv g_{12}(\nu)$ and let $g_{11}(\nu) \equiv g_{22}(\nu)$. Further let the mean frequency be $\bar{\nu}$ and the width of the spectrum be $\Delta\nu$. Suppose that the light from P_1 and P_2 is superimposed at a point $P_3(\mathbf{x}_3)$ with the corresponding normalized spectral density $g_{33}(\nu)$. In general $g_{33}(\nu)$ will not be simply related to $g_{11}(\nu)$ and $g_{22}(\nu)$, even if $g_{11}(\nu)$ and $g_{22}(\nu)$ are identical. If $g_{11}(\nu) \equiv g_{22}(\nu)$ and a region of neighbouring points P_3 exists such that $g_{33}(\nu) \equiv g_{11}(\nu)$, then $V_1(t)$ and $V_2(t)$ are said to be cross-spectrally pure.

Performing the superposition of $V_1(t)$ and $V_2(t)$ at P_3 with path differences $c\eta_1$ and $c\eta_2$ and the corresponding propagators a_1 and a_2 between P_1, P_3 and P_2, P_3 respectively, we arrive at

$$V(\mathbf{x}_3, t) = a_1 V(\mathbf{x}_1, t - \eta_1) + a_2 V(\mathbf{x}_2, t - \eta_2) \tag{4.24}$$

and we obtain for the autocorrelation function $\Gamma_{33}(\tau)$

$$\begin{aligned}\Gamma_{33}(\tau) = |a_1|^2 \Gamma_{11}(\tau) + |a_2|^2 \Gamma_{22}(\tau) + \\ + a_1^* a_2 \Gamma_{12}(\tau + \eta_1 - \eta_2) + a_1 a_2^* \Gamma_{21}(\tau + \eta_2 - \eta_1).\end{aligned} \tag{4.25}$$

The normalization gives us

$$\gamma_{33}(\tau) = \frac{\gamma_{11}(\tau) + \frac{1}{2}K_{12}\gamma_{12}(\tau + \eta_1 - \eta_2) + \frac{1}{2}K_{12}^*\gamma_{21}(\tau + \eta_2 - \eta_1)}{1 + \mathrm{Re}\,\{K_{12}\gamma_{12}(\eta_1 - \eta_2)\}}, \tag{4.26}$$

since $g_{11}(\nu) \equiv g_{22}(\nu)$ and so $\gamma_{11}(\tau) \equiv \gamma_{22}(\tau)$ and

$$K_{12} = \frac{2(I_1 I_2)^{1/2} a_1^* a_2}{|a_1|^2 I_1 + |a_2|^2 I_2}; \tag{4.27}$$

here $I_j \equiv \Gamma_{jj}(0)$ and obviously $|K_{12}| \leqq 1$. Performing the Fourier transformation, we obtain

$$g_{33}(\nu) = \frac{g_{11}(\nu) + \mathrm{Re}\,\{K_{12} g_{12}(\nu) \exp[-\mathrm{i}2\pi\nu(\eta_1 - \eta_2)]\}}{1 + \mathrm{Re}\,\{K_{12}\gamma_{12}(\eta_1 - \eta_2)\}} \tag{4.28}$$

and so

$$\begin{aligned}g_{33}(\nu) - g_{11}(\nu) = \\ = \frac{\mathrm{Re}\,\{K_{12}[g_{12}(\nu)\exp[-\mathrm{i}2\pi\nu(\eta_1 - \eta_2)] - g_{11}(\nu)\gamma_{12}(\eta_1 - \eta_2)]\}}{1 + \mathrm{Re}\,\{K_{12}\gamma_{12}(\eta_1 - \eta_2)\}}.\end{aligned} \tag{4.29}$$

We can now try to find conditions under which the light is cross-spectrally pure, i.e. $g_{11}(\nu) \equiv g_{22}(\nu) \equiv g_{33}(\nu)$. For this the right-hand side of (4.29) must vanish, but this cannot generally be so (except for incoherent beams for which $\gamma_{12}(\tau) \equiv g_{12}(\nu) \equiv 0$). This is readily seen for sufficiently large $|\eta_1 - \eta_2|$ when the function $\gamma_{12}(\eta_1 - \eta_2)$ in (4.29) vanishes and the spectrum is modulated by a cosine function through the first term on the right-hand side. However, if the path difference $c\,|\eta_1 - \eta_2|$ is sufficiently small in the sense that $|\eta_1 - \eta_2| = \eta_0 + \Delta\eta$, where $|\Delta\eta| \ll 1/\Delta\nu$ and η_0 is fixed, we obtain

$$\gamma_{12}(\eta_1 - \eta_2) = \gamma_{12}(\eta_0 + \Delta\eta) = \gamma_{12}(\eta_0)\exp(-\mathrm{i}\,2\pi\bar{\nu}\,\Delta\eta). \tag{4.30}$$

Substituting this into (4.29), we obtain

$$g_{12}(\nu)\exp(-\mathrm{i}\,2\pi\nu\eta_0) = g_{11}(\nu)\,\gamma_{12}(\eta_0) \tag{4.31}$$

provided that $g_{33}(\nu) \equiv g_{11}(\nu)$ for all K_{12} and $|\Delta\eta| \ll 1/\Delta\nu$. From this we have, making a Fourier transformation, the required factorization property

$$\gamma_{12}(\tau + \eta_0) = \gamma_{11}(\tau)\,\gamma_{12}(\eta_0) \tag{4.32a}$$

or

$$\gamma_{12}(\tau) = \gamma_{11}(\tau - \eta_0)\,\gamma_{12}(\eta_0). \tag{4.32b}$$

As $|\gamma_{11}(\tau)| \leqq |\gamma_{11}(0)| = 1$, it follows that $|\gamma_{12}(\tau + \eta_0)| \leqq |\gamma_{12}(\eta_0)|$, i.e. $c\eta_0$ is the path difference which must be introduced between the light beams from P_1 and P_2 to reach the maximum of $|\gamma_{12}|$. In practice η_0 equals $\eta_1 - \eta_2 = (\overline{P_3P_1} - \overline{P_3P_2})/c$. The first factor on the right-hand side of (4.32) represents temporal coherence, the second one represents spatial coherence. Summarizing we have shown (under the above assumptions) that the degree of coherence between points P_1 and P_2 in an optical stationary field is reducible if and only if the corresponding light beams are cross-spectrally pure.

It has been further shown (Mandel (1961a)) with the aid of the propagation law for the mutual coherence function that the reduction formula (4.32), i.e. the cross-spectral purity of light, is conserved during propagation provided that the path differences involved are sufficiently small and that the normalized mutual spectral density is constant on the surface from which it propagates.

From (4.2) it can be seen that a spatially incoherent source, for which

$$\gamma(\mathbf{x}_1, \mathbf{x}_2, \tau) = \gamma(\mathbf{x}_1, \mathbf{x}_1, \tau)\,\delta(\mathbf{x}_1 - \mathbf{x}_2) \tag{4.33}$$

holds (cf. (3.37)), produces a field which is always cross-spectrally pure provided the normalized mutual spectral density is the same for all points of the source (under the restrictions on path differences involved).

The behaviour of cross-spectrally pure beams has been further investigated by Mandel (1962) and Wolf (1983).

4.4 Determination of the spectrum from the visibility of interference fringes – phase problem of coherence theory

We now consider the problem in interference spectroscopy of determining the spectrum from measurements of the visibility of interference fringes. This was first considered by Michelson, who showed that one may obtain information about the energy spectrum of the light from the measured visibility $\mathscr{V} = |\gamma(\mathbf{x}, \mathbf{x}, \tau)| \equiv |\gamma(\tau)|$ as a function of τ.

This can be seen very simply for quasi-monochromatic light obtained from a spectral line of mean frequency $\bar{\nu}$, bandwidth $\Delta\nu\,(\ll \bar{\nu})$ and with symmetric profile, so that the normalized spectral density g satisfies $g(\bar{\nu} + \nu) = g(\bar{\nu} - \nu)$. As (2.49) holds,

$$\gamma(\tau) = \int_0^\infty g(\nu) \exp(-\mathrm{i}\,2\pi\nu\tau)\,\mathrm{d}\nu, \tag{4.34}$$

and the spectrum $g(\nu)$ may be determined by an inverse Fourier transformation,

$$\begin{aligned} g(\nu) &= \int_{-\infty}^{+\infty} \gamma(\tau) \exp(\mathrm{i}\,2\pi\nu\tau)\,\mathrm{d}\tau, \quad \nu \geqq 0, \\ &= 0 \qquad\qquad\qquad\qquad\qquad\quad, \quad \nu < 0. \end{aligned} \tag{4.35}$$

In order to determine $g(\nu)$ we must know not only the modulus of $\gamma(\tau)$ but also the phase $\varphi(\tau)$ of $\gamma(\tau)$. However, in the case of a symmetric spectral profile we obtain from (4.34), by substituting $\nu \to \bar{\nu} + \nu$,

$$\gamma(\tau) = \exp(-\mathrm{i}\,2\pi\bar{\nu}\tau) \int_{-\Delta\nu/2}^{+\Delta\nu/2} g(\bar{\nu} + \nu) \exp(-\mathrm{i}\,2\pi\nu\tau)\,\mathrm{d}\nu. \tag{4.36}$$

But this integral is real since $g(\bar{\nu} + \nu)$ is symmetric with respect to $\bar{\nu}$. Taking the absolute value of (4.36), we see that the integral equals $|\gamma(\tau)|$ and we finally have

$$\gamma(\tau) = |\gamma(\tau)| \exp(-\mathrm{i}\,2\pi\bar{\nu}\tau), \tag{4.37}$$

the phase $\varphi(\tau)$ of $\gamma(\tau)$ being equal to $(-2\pi\bar{\nu}\tau)$ in this case. The spectrum is determined by (4.35), i.e.

$$\begin{aligned} g(\nu) &= \int_{-\infty}^{+\infty} \mathscr{V}(\tau) \exp[\mathrm{i}\,2\pi\tau(\nu - \bar{\nu})]\,\mathrm{d}\tau = \\ &= 2\int_0^\infty \mathscr{V}(\tau) \cos[2\pi\tau(\nu - \bar{\nu})]\,\mathrm{d}\tau, \end{aligned} \tag{4.38}$$

since $\mathscr{V}(\tau) \equiv |\gamma(\tau)|$ is an even function of τ.

If the spectrum is non-symmetric the phase of $\gamma(\tau)$ does not equal $(-2\pi\bar{\nu}\tau)$, but we have mentioned in Sec. 3.1 that this phase can in principle be determined from shifts of interference fringes in the Young experiment. Unfortunately, such measurements are difficult to perform in the optical region. However, it has been pointed out by Wolf (1962) that information about the phase of γ may be deduced from the analytic properties of $\gamma(\tau)$ if the positions of the zeros of γ in the region of analyticity are known or if the zeros are absent.

As the function $\gamma(\tau)$ given by (4.34) is analytic and regular in the lower half, $\Pi^{(-)}$, of the complex τ-plane, the function

$$\ln \gamma(\tau) = \ln |\gamma(\tau)| + i\varphi(\tau) \tag{4.39}$$

will also be analytic in $\Pi^{(-)}$, but it will have logarithmic branch points at zeros of $\gamma(\tau)$. First we assume that there are no zeros of $\gamma(\tau)$ in $\Pi^{(-)}$ and, applying the dispersion relation of the form (2.26), we can obtain a relation between the modulus $|\gamma(\tau)|$ and the phase $\varphi(\tau)$ (Toll (1956), Wolf (1962), Roman and Marathay (1963)). In deriving the dispersion relations for the function $\gamma(\tau)$ we require (Bogolyubov, Medvedev and Polivanov (1958)) that $|\gamma(\tau)| \to 0$ as $|\tau| \to \infty$ at least as $|\tau|^{-1}$ and that $|\gamma(\tau)|$ is square integrable (for this case $\int_{-\infty}^{+\infty} |\gamma(\tau)|^2 \,d\tau = \int_0^{\infty} g^2(\nu)\,d\nu < \infty$). One can then write for an analytic and regular function $\gamma(\tau)$ the Cauchy integral with the contour of integration composed of the real axis and the semi-circle of infinite radius lying in $\Pi^{(-)}$. The integral vanishes over the semi-circle as a consequence of the above assumptions so that

$$\gamma(\tau) = -\frac{1}{2\pi i}\int_{-\infty}^{+\infty} \frac{\gamma(\tau')}{\tau' - \tau}\,d\tau', \qquad \operatorname{Im}\tau < 0. \tag{4.40}$$

Putting $\operatorname{Im}\tau \to 0$ and using the identity

$$\lim_{\substack{\varepsilon\to 0\\ \varepsilon \mp 0}} \frac{1}{\tau > i\varepsilon} = P\frac{1}{\tau} \pm \pi i\delta(\tau), \tag{4.41}$$

where P denotes the principal value in the Cauchy sense, and separating the real and the imaginary parts, we arrive at the dispersion relations of the type (2.26) as follows

$$\operatorname{Im}\gamma(\tau) = \frac{1}{\pi} P \int_{-\infty}^{+\infty} \frac{\operatorname{Re}\gamma(\tau')}{\tau' - \tau}\,d\tau', \tag{4.42a}$$

$$\operatorname{Re}\gamma(\tau) = -\frac{1}{\pi} P \int_{-\infty}^{+\infty} \frac{\operatorname{Im}\gamma(\tau')}{\tau' - \tau}\,d\tau'. \tag{4.42b}$$

The situation is different if the function $\ln\gamma(\tau)$ is considered. Since $|\gamma(\tau)| \approx |\tau|^{-1}$ for $|\tau| \to \infty$, $\ln|\gamma(\tau)| \approx -\ln|\tau|$ and the assumption necessary for the vanishing of the Cauchy integral over the semi-circle is not satisfied. In this case one must consider the function $\ln\gamma(\tau)/\tau$ for which $|\ln\gamma(\tau)/\tau| = |\gamma'(\tau)/\gamma(\tau)| \approx |\tau|^{-1}$ for $|\tau| \to \infty$. Here we adopt another method of solving the phase problem based on the solution of a singular integral equation of the Cauchy type using the method of Muskhelishvili (1953); this method has to be useful also in other branches of physics. We wish also to point out the elegance of this method. We assume the validity of the Lipschitz condition for all the functions used ($|\gamma(\tau_1) - \gamma(\tau_2)| \leqq \alpha |\tau_1 - \tau_2|^\beta$, where α and β are positive constants). The modulus and the phase of $\gamma(\tau)$ may be expressed as

$$|\gamma(\tau)| = [(\mathrm{Re}\,\gamma(\tau))^2 + (\mathrm{Im}\,\gamma(\tau))^2]^{1/2}, \tag{4.43a}$$

$$\tan\varphi(\tau) = \frac{\mathrm{Im}\,\gamma(\tau)}{\mathrm{Re}\,\gamma(\tau)}. \tag{4.43b}$$

First we assume that $\tan\varphi(\tau) \neq 0$ and $\mathrm{Re}\,\gamma(\tau) \neq 0$ for all real τ. This is equivalent to the assumption that $\gamma(\tau)$ has no zeros in $\Pi^{(-)}$, since then $\mathrm{Im}\,\gamma(\tau) \neq 0$ for all τ and the function $\gamma(\tau)$ has indeed no zeros for real τ. Applying a theorem given by Landau and Lifshitz (1964) (§125) the function $\gamma(\tau)$ indeed has no zeros in $\Pi^{(-)}$. Denoting $\mathrm{Im}\,\gamma(\tau) = f(\tau)$ and substituting (4.43b) into (4.42b), we obtain a singular integral equation of the type

$$\frac{f(\tau)}{\tan\varphi(\tau)} = -\frac{1}{\pi} P \int_{-\infty}^{+\infty} \frac{f(\tau')}{\tau' - \tau}\,\mathrm{d}\tau'. \tag{4.44}$$

To solve this equation we use the Muskhelishvili method.

We first derive the formulae of Sokhotski and Plemelj. Let $\psi(\tau)$ be defined as the Cauchy integral

$$\psi(\tau) = \frac{1}{2\pi\mathrm{i}} \int_{-\infty}^{+\infty} \frac{f(\tau')}{\tau' - \tau}\,\mathrm{d}\tau', \qquad \mathrm{Im}\,\tau \neq 0. \tag{4.45}$$

Using (4.41), we arrive at the formulae of Sokhotski and Plemelj,

$$\psi^+(\tau) - \psi^-(\tau) = f(\tau), \tag{4.46a}$$

$$\psi^+(\tau) + \psi^-(\tau) = \frac{1}{\pi\mathrm{i}} P \int_{-\infty}^{+\infty} \frac{f(\tau')}{\tau' - \tau}\,\mathrm{d}\tau', \tag{4.46b}$$

where $\psi^+(\tau)$ and $\psi^-(\tau)$ are boundary values of $\psi(\tau)$ on the real τ-axis from the upper and lower half planes respectively.

Denoting $A(\tau) = \mathrm{i}/\tan\varphi(\tau)$, we obtain from (4.44)

$$A(\tau) f(\tau) - \frac{1}{\pi\mathrm{i}} P \int_{-\infty}^{+\infty} \frac{f(\tau')}{\tau' - \tau}\,\mathrm{d}\tau' = 0, \tag{4.47}$$

and using (4.46a, b) we have

$$A(\tau)\{\psi^+(\tau) - \psi^-(\tau)\} - \{\psi^+(\tau) + \psi^-(\tau)\} = 0. \tag{4.48}$$

Since $\gamma(\tau)$ has no zeros in $\Pi^{(-)}$ (that is $\ln\gamma(\tau)/2\pi\mathrm{i}\,|_C = \varphi(\tau)/2\pi\,|_C = 0$ for any closed contour C in $\Pi^{(-)}$), the relation

$$[\ln\psi(\tau)]^+ - [\ln\psi(\tau)]^- = \ln\frac{A(\tau)+1}{A(\tau)-1} = -\mathrm{i}\,2\varphi(\tau), \tag{4.49}$$

following from (4.48), determines the unique function $\ln \psi(\tau)$ for which, according to (4.46a) and (4.45),

$$\ln \psi(\tau) = \frac{1}{2\pi i} \int_{-\infty}^{+\infty} \frac{-i\,2\varphi(\tau')}{\tau' - \tau} d\tau' = -\frac{1}{\pi} \int_{-\infty}^{+\infty} \frac{\varphi(\tau')}{\tau' - \tau} d\tau'. \tag{4.50}$$

That is

$$\psi(\tau) = \exp\left[-\frac{1}{\pi} \int_{-\infty}^{+\infty} \frac{\varphi(\tau')}{\tau' - \tau} d\tau' \right]. \tag{4.51}$$

Using (4.46a) again we have

$$\begin{aligned} f(\tau) &= \operatorname{Im} \gamma(\tau) = \psi^+(\tau) - \psi^-(\tau) = \\ &= 2i \sin \varphi(\tau) \exp\left[-\frac{1}{\pi} P \int_{-\infty}^{+\infty} \frac{\varphi(\tau')}{\tau' - \tau} d\tau' \right]. \end{aligned} \tag{4.52}$$

Making use of the fact that the solution of a homogeneous equation is determined up to a multiplicative constant and omitting the constant 2i in (4.52) (that is $f(\tau) \equiv 1$ for $\varphi = \pi/2$), we obtain from (4.43b)

$$\operatorname{Re} \gamma(\tau) = \cos \varphi(\tau) \exp\left[-\frac{1}{\pi} P \int_{-\infty}^{+\infty} \frac{\varphi(\tau')}{\tau' - \tau} d\tau' \right] \tag{4.53}$$

and from (4.43a)

$$|\gamma(\tau)| = \exp\left[-\frac{1}{\pi} P \int_{-\infty}^{+\infty} \frac{\varphi(\tau')}{\tau' - \tau} d\tau' \right]. \tag{4.54}$$

From this relation the modulus of $\gamma(\tau)$ may be calculated if the phase $\varphi(\tau)$ is known. However, by using the Sokhotski-Plemelj formulae, one can prove that if two functions G and H are related by the Cauchy integral

$$G(\tau) = \frac{1}{\pi i} P \int_{-\infty}^{+\infty} \frac{H(\tau')}{\tau' - \tau} d\tau', \tag{4.55a}$$

then

$$H(\tau) = \frac{1}{\pi i} P \int_{-\infty}^{+\infty} \frac{G(\tau')}{\tau' - \tau} d\tau'. \tag{4.55b}$$

Hence, from (4.54)

$$i \ln |\gamma(\tau)| = \frac{1}{\pi i} P \int_{-\infty}^{+\infty} \frac{\varphi(\tau')}{\tau' - \tau} d\tau', \tag{4.56}$$

and finally we obtain

$$\varphi(\tau) = \frac{1}{\pi} P \int_{-\infty}^{+\infty} \frac{\ln |\gamma(\tau')|}{\tau' - \tau} d\tau', \tag{4.57}$$

which is the required expression relating the phase $\varphi(\tau)$ to $|\gamma(\tau)|$.

We can now consider the general case when $\gamma(\tau)$ has zeros in $\Pi^{(-)}$ at points τ_j ($j = 1, 2, \ldots$). Making use of the most general form of an analytic unimodular function (4.18), we can write

$$\gamma(\tau) = \gamma_0(\tau) \exp(-\mathrm{i}\, 2\pi\nu_0\tau) \prod_j \frac{\tau - \tau_j}{\tau - \tau_j^*}, \tag{4.58}$$

where $\nu_0 > 0$ and $\operatorname{Im} \tau_j < 0$. The positions of zeros of $\gamma(\tau)$ are described by the Blaschke factors $(\tau - \tau_j)/(\tau - \tau_j^*)$ and $|\gamma(\tau)| \equiv |\gamma_0(\tau)|$ on the real axis; the function $\gamma_0(\tau)$ does not contain any zeros in $\Pi^{(-)}$. Therefore the complete phase $\varphi(\tau)$ of $\gamma(\tau)$ is obtained from (4.58) (Toll (1956))

$$\varphi(\tau) = \frac{1}{\pi} P \int_{-\infty}^{+\infty} \frac{\ln|\gamma(\tau')|}{\tau' - \tau} \mathrm{d}\tau' + \sum_j \arg \frac{\tau - \tau_j}{\tau - \tau_j^*} - 2\pi\nu_0\tau. \tag{4.59}$$

The physical significance of the last term is clear – it represents a term causing the shift of the spectrum to ν_0 and so it can be omitted in the following. Further it can be shown that the second term in (4.59) – expressing the so-called *Blaschke phase* – is non-negative, so that the first term given by means of the dispersion relation may be called the *minimal phase.*

Thus, for the unique determination of the phase $\varphi(\tau)$ of $\gamma(\tau)$ from its modulus and hence for the unique determination of the spectrum from the visibility one must know the positions of the zeros of $\gamma(\tau)$ in $\Pi^{(-)}$; in general the knowledge of only the visibility $\mathscr{V}(\tau) \equiv |\gamma(\tau)|$ is not sufficient to determine $g(\nu)$ uniquely. By using the residue theorem, we can calculate $g(\nu)$ from (4.58) with the explicit contribution of the Blaschke factors to the spectrum $g(\nu)$ as follows (Peřina and Tillich (1966))

$$g(\nu) = g_0(\nu) + 2\pi\mathrm{i} \sum_j (\tau_j^* - \tau_j) \exp(\mathrm{i}\, 2\pi\nu\tau_j^*) \times \\ \times \prod_{k \neq j} \frac{\tau_j^* - \tau_k}{\tau_j^* - \tau_k^*} \int_0^{\nu} g_0(\mu) \exp(-\mathrm{i}\, 2\pi\mu\tau_j^*)\, \mathrm{d}\mu, \tag{4.60}$$

where the minimum spectrum $g_0(\nu)$ is the Fourier transform of $\gamma_0(\tau)$.

Unfortunately very little is known about the physical significance of the zeros and their location, but some restrictions follow from the cross-symmetry condition $\gamma(\tau) = \gamma^*(-\tau)$ expressing the reality of the spectrum $g(\nu)$, $g^*(\nu) = g(\nu)$, and from the non-negativeness of $g(\nu)$. It can easily be shown (Roman and Marathay (1963), Peřina and Tillich (1966)) that the cross-symmetry condition for $\gamma(\tau)$, which has the form (4.58), leads to a symmetrical distribution of zeros with respect to the imaginary τ-axis, i.e. with every factor $(\tau - \tau_j)/(\tau - \tau_j^*)$ the factor $(\tau + \tau_k^*)/(\tau + \tau_k)$ occurs (for every j a k exists such that $\tau_j = -\tau_k^*$). Then (4.59) may be written in the form (omitting the trivial term $(-2\pi\nu_0\tau)$)

$$\varphi(\tau) = \frac{2\tau}{\pi} P \int_0^{\infty} \frac{\ln|\gamma(\tau')|}{\tau'^2 - \tau^2} \mathrm{d}\tau' + \sum_j \left[\arg \frac{\tau - \tau_j}{\tau - \tau_j^*} + \arg \frac{\tau + \tau_j^*}{\tau + \tau_j} \right], \tag{4.61}$$

since $|\gamma(\tau)|$ is an even function ($\gamma^*(\tau) = \gamma(-\tau)$ implies that $|\gamma(-\tau)| = |\gamma(+\tau)|$ and $\varphi(-\tau) = -\varphi(+\tau)$). The condition $g(\nu) \geqq 0$ leads to the elimination of the zeros from the imaginary τ-axis, since if $\tau = ia$ ($a < 0$ and real), then $\gamma(ia) = \int_0^\infty g(\nu) \times \times \exp(2\pi\nu a)\,d\nu > 0$, i.e. $\gamma(ia) \neq 0$.

Some physical conditions may be found under which $\gamma(\tau)$ cannot have zeros in $\Pi^{(-)}$ at all. Then $\varphi(\tau)$ and $g(\nu)$ may be uniquely determined from the dispersion relations. It was shown that (Peřina and Tillich (1966))

$$\frac{1}{\pi}\int_0^\infty \frac{\ln|\gamma(\tau)|}{\tau^{2n}}\,d\tau = (-1)^n \frac{\mu_{2n-1}}{2(2n-1)!} - \frac{1}{2n-1}\sum_j \frac{\operatorname{Im}\tau_j^{2n-1}}{|\tau_j|^{2(2n-1)}}, \tag{4.62a}$$

$$\sum_j \frac{\operatorname{Re}\tau_j^{2n-1}}{|\tau_j|^{2(2n-1)}} = 0, \tag{4.62b}$$

where $n \geqq 1$ and the conditions $|\ln\gamma(\tau)| \leqq A|\tau|^{l_1}$, $l_1 < 2n-1$ for $|\tau| \to \infty$ and $|\ln\gamma(\tau)| \leqq B|\tau|^{l_2}$, $l_2 > 2n-1$ for $|\tau| \to 0$ hold (A and B are real positive constants). As the zeros are distributed symmetrically with respect to the imaginary axis, the condition (4.62b) is an identity. The condition (4.62a) relates the behaviour of $|\gamma(\tau)|$ in the neighbourhood of zero and infinity to the moments μ_{2n-1} of the spectrum of the function $\ln\gamma(\tau)$ and to the distribution of zeros in $\Pi^{(-)}$. For $n = 1$ we obtain (Khalfin (1960))

$$\frac{1}{\pi}\int_0^\infty \frac{\ln|\gamma(\tau)|}{\tau^2}\,d\tau = -\frac{\mu_1}{2} - \sum_j \frac{\operatorname{Im}\tau_j}{|\tau_j|^2}, \tag{4.63}$$

where μ_1 is the first moment of the spectrum of $\gamma(\tau)$, i.e. of $g(\nu)$. As $0 \leqq |\gamma(\tau)| \leqq 1$, $\gamma(0) = 1$ and $\operatorname{Im}\tau_j < 0$, the minimum value of μ_1 is

$$\mu_{1\min} = -\frac{2}{\pi}\int_0^\infty \frac{\ln|\gamma(\tau)|}{\tau^2}\,d\tau > 0 \tag{4.64}$$

and

$$\mu_1 = -\frac{2}{\pi}\int_0^\infty \frac{\ln|\gamma(\tau)|}{\tau^2}\,d\tau - 2\sum_j \frac{\operatorname{Im}\tau_j}{|\tau_j|^2} \geqq \mu_{1\min}. \tag{4.65}$$

Hence the reconstruction of the phase under the additional condition of the minimum of the moment of the spectrum is unique. Another kind of elimination of zeros for a fully coherent stationary field has been discussed in Sec. 4.2. As was shown by Kano and Wolf (1962) $\gamma(\tau)$ has no zeros in $\Pi^{(-)}$ for blackbody radiation so that its phase as well as the spectrum may be reconstructed uniquely from measurements of the visibility of the interference fringes in spite of the fact that the spectrum of this radiation is not symmetric. Some interesting theorems for quasi-monochromatic spectra, particularly of Gaussian and Lorentzian forms, concerning the position of zeros were derived by Nussenzveig (1967). In particular it was shown that the contribution of zeros to the phase $\varphi(\tau)$ is significant so that the minimal phase cannot be considered as a good approximation to the real phase.

Also some alternative experimental approaches to the recovery of the phase $\varphi(\tau)$ from $|\gamma(\tau)|$ have been suggested. Gamo (1963a) (see also Mandel (1963a)) suggested that the phase can be measured by means of triple intensity correlation measurements. Another method was proposed by Mehta (1965) on the basis of the Cauchy-Riemann conditions

$$\frac{\partial |\gamma(\tau_r, \tau_i)|}{\partial \tau_r} = -|\gamma(\tau_r, \tau_i)| \frac{\partial \varphi(\tau_r, \tau_i)}{\partial \tau_i}, \tag{4.66a}$$

$$\frac{\partial |\gamma(\tau_r, \tau_i)|}{\partial \tau_i} = |\gamma(\tau_r, \tau_i)| \frac{\partial \varphi(\tau_r, \tau_i)}{\partial \tau_r}, \tag{4.66b}$$

which must hold in $\Pi^{(-)}$; here $\tau = \tau_r + i\tau_i$. Integrating (4.66b) we may determine the phase φ. To determine $\partial\varphi(\tau_r, \tau_i)/\partial\tau_r|_{\tau_i=0}$ we must also know $|\gamma(\tau)|$ in $\Pi^{(-)}$ for $\tau_i < 0$, but the physical significance of $\gamma(\tau_r, \tau_i)$ is clear as can be seen from (4.34),

$$\gamma(\tau_r, \tau_i) = \int_0^\infty g(\nu) \exp(2\pi\nu\tau_i) \exp(-i\, 2\pi\nu\tau_r)\, d\nu, \tag{4.67}$$

i.e. $\gamma(\tau_r, \tau_i)$ for $\tau_i < 0$ may be obtained from the spectrum $g(\nu)$ by using the exponential filter $\exp[-2\pi\nu(-\tau_i)]$. Yet another method proposed by Mehta (1968) is based on using a reference beam with the known degree of coherence $\gamma_r(\tau)$ and spectrum $g_r(\nu)$. If we superimpose the light from the reference source on light with an unknown spectrum $g(\nu)$, we obtain light with the spectrum

$$\bar{g}(\nu) = g(\nu) + g_r(\nu), \tag{4.68}$$

since the sources are statistically independent. Thus we have

$$\bar{\gamma}(\tau) = \gamma(\tau) + \gamma_r(\tau) \tag{4.69}$$

for the corresponding degree of coherence. From two separate experiments with light of the unknown spectrum and of the superimposed spectrum we obtain $|\gamma(\tau)|$ and $|\bar{\gamma}(\tau)|$ respectively. Taking the squared modulus of (4.69), we arrive at

$$|\bar{\gamma}(\tau)|^2 = |\gamma(\tau)|^2 + |\gamma_r(\tau)|^2 + 2|\gamma(\tau)|\,|\gamma_r(\tau)| \cos[\varphi(\tau) - \varphi_r(\tau)]. \tag{4.70}$$

Here all the quantities are known except $\varphi(\tau)$ which can be determined.

The problem of finding zeros of $\gamma(\tau)$ was transferred by Roman and Marathay (1963) to a nonlinear eigenvalue problem and by Dialetis and Wolf (1967) and Dialetis (1967) to a certain inhomogeneous eigenvalue problem of the Sturm-Liouville type. It was shown that this eigenvalue problem is equivalent to a certain stability problem in mechanics.

The use of the Cauchy-Riemann conditions (4.66) for the experimental reconstruction of the phase has been demonstrated by Kohler and Mandel (1973) (see also Ablekov, Zubkov and Frolov (1976)). A holographic approach to the phase problem of optical coherence has been developed by Kohler and Mandel (1970). The use of exponential filters has been discussed by Nakajima and Asakura (1982) with the help of computer simulation.

Much effort has been devoted to solving the phase problem in electron microscopy (Misell (1973), Misell, Burge and Greenaway (1974), Misell and Greenaway (1974), Spence (1974), Ferwerda and Hoenders (1975), Ferwerda (1978)), particularly in bright-field microscopy, where a complex function $\gamma(\tau) + C \equiv \gamma'(\tau)$ ($C > 0$ being a constant determined by the bright background), having no zeros, can be considered. Burge, Fiddy, Greenaway and Ross (1974) proved, using the above singular equation method, that the phase φ of γ is determined in this case by

$$\tan \varphi(\tau) = \frac{|\gamma'(\tau)| \sin \alpha(\tau)}{|\gamma'(\tau)| \cos \alpha(\tau) - C}, \tag{4.71}$$

where $\alpha(\tau)$ is the phase of $\gamma'(\tau)$, which is equal to the minimal phase determined from the dispersion relation (4.57), since $\gamma'(\tau)$ has no zeros in $\Pi^{(-)}$. The most complex study of the phase problem has been provided by Burge, Fiddy, Greenaway and Ross (1976), who also employed the dispersion relations with subtractions (Bogolyubov, Medvedev and Polivanov (1958), Nussenzveig (1972)) and some algorithms have been given to determine the positions of the zeros. Burge et al. also used the theory of entire functions, which has been further applied in a number of papers (e.g. Ross, Fiddy, Nieto-Vesperinas and Wheeler (1977)). As usually the first few zeros of the degree of coherence are sufficient to determine the spectrum with good accuracy, Barakat (1980) suggested the moment estimator approach to the phase retrieval problem in coherence theory. The two-dimensional phase problem has been considered by Nieto-Vesperinas (1980, 1982) and Kiedron (1981). Nieto-Vesperinas (1980) in particular proved that in general the two-dimensional phase problem has no solution unless it cannot be reduced to the one-dimensional case. Some hybrid phase measurements were discussed by Přikryl and Vest (1983).

A number of further interesting papers have been devoted to the phase problem (Thomas (1971/72), Saxton (1974), Greenaway (1977), Fiddy and Greenaway (1979), Kiedroń (1980), Ablekov, Avdyrevskii, Babaev, Koljadin, Frolov and Fulov (1980), Ablekov, Babaev, Koljadin and Frolov (1980), Ablekov, Babaev, Koljadin, Syrich and Frolov (1981), Bakut, Sviridov and Ustinov (1981), Lawton (1981), Walker (1981)).

Hence in the optical region, where the function $\operatorname{Re} \gamma(\tau)$ is rapidly oscillating quantity and so cannot be measured directly, there are some unsolved problems as to the determination of the spectrum $g(\nu)$ from the visibility $|\gamma(\tau)|$. Such measurements of $\operatorname{Re} \gamma(\tau)$ are possible in the far infra-red region (Strong and Vanasse (1959), Jacquinot (1960), Vanasse and Sakai (1967)). As $\operatorname{Im} \gamma(\tau)$ is determined uniquely from $\operatorname{Re} \gamma(\tau)$ by means of the dispersion relations, the spectrum $g(\nu)$ can in principle be determined uniquely in this case.

Finally let us note that the phase problem arises in other branches of physics too; we can mention the quantum theory of decay (Khalfin (1960)), scattering theory (Goldberger, Lewis and Watson (1963)) and the diffraction theory of image formation (O'Neil and Walther (1963), Walther (1963), Peřina (1963b), Nieto-Vesperinas and Hignette (1979)).

4.5 Radiation from partially coherent planar sources and radiometry with light of any state of coherence

In recent years much effort has been devoted to including the traditional radiometry of incoherent sources as a special case of a more general radiometry with light from partially coherent sources, i.e. with light of arbitrary states of coherence. Walther (1968, 1973) suggested to define the generalized radiance

$$B_\omega(\mathbf{r}, \mathbf{s}) = \left(\frac{k}{2\pi}\right)^2 \cos(\Theta) \int_{(z'=0)} G\left(\mathbf{r} + \frac{\mathbf{r}'}{2}, \mathbf{r} - \frac{\mathbf{r}'}{2}, \omega\right) \exp(-ik\mathbf{s}\,.\,\mathbf{r}')\, \mathrm{d}^2 r', \tag{4.72}$$

where $\omega = 2\pi\nu$, Θ is the angle between the normal to the plane source and the position vector $\mathbf{R} = R\mathbf{s}$ of a typical point. The integration is taken over the plane $z' = 0$ of the source. The generalized radiance can also be written in terms of the four-dimensional spatial Fourier transform,

$$\tilde{G}(\mathbf{f}_1, \mathbf{f}_2, \omega) = \frac{1}{(2\pi)^4} \iint_{(z=0)} G(\mathbf{r}_1, \mathbf{r}_2, \omega) \exp[-\mathrm{i}(\mathbf{f}_1 \cdot \mathbf{r}_1 + \mathbf{f}_2 \cdot \mathbf{r}_2)]\, \mathrm{d}^2 r_1\, \mathrm{d}^2 r_2, \tag{4.73}$$

in the form (Marchand and Wolf (1974a))

$$B_\omega(\mathbf{r}, \mathbf{s}) = k^2 \cos(\Theta) \int_{(\mathbf{f}\text{-plane})} \tilde{G}\left(k\mathbf{s} + \frac{\mathbf{f}}{2}, -k\mathbf{s} + \frac{\mathbf{f}}{2}, \omega\right) \exp(\mathrm{i}\mathbf{f}\,.\,\mathbf{r})\, \mathrm{d}^2 f. \tag{4.74}$$

The generalized radiance $B_\omega(\mathbf{r}, \mathbf{s})$ is always real: taking into account the identity (2.32b) and changing $\mathbf{r}' \to -\mathbf{r}'$ in (4.72) we see at once that $B^*_\omega(\mathbf{r}, \mathbf{s}) = B_\omega(\mathbf{r}, \mathbf{s})$. However, the generalized radiance $B_\omega(\mathbf{r}, \mathbf{s})$ may take on negative values in some cases, being the function of the Fourier conjugated variables $\mathbf{r}$ and $\mathbf{s}$. Hence it cannot be interpreted in general as a true flux density. This property is analogous to the property of the Wigner function in quantum mechanics (Sec. 13.2). However, the generalized radiance is a non-negative valued function for a spatially incoherent source. These questions are discussed in great detail by Friberg (1978a, b, 1979a, 1981a).

For any state of coherence of the source from (4.72) it holds that

$$\int_{(z=0)} B_\omega(\mathbf{r}, \mathbf{s})\, \mathrm{d}^2 r = (2\pi k)^2 \cos(\Theta)\, \tilde{G}(k\mathbf{s}, -k\mathbf{s}, \omega). \tag{4.75}$$

Further we may define the generalized radiant emittance

$$E_\omega(\mathbf{r}) = \int_{(2\pi)} B_\omega(\mathbf{r}, \mathbf{s}) \cos(\Theta)\, \mathrm{d}\Omega \tag{4.76}$$

($\mathrm{d}\Omega$ being the solid angle differential around $\mathbf{s}$), giving (Marchand and Wolf (1974a))

$$E_\omega(\mathbf{r}) = \int_{(z'=0)} G\left(\mathbf{r} + \frac{\mathbf{r}'}{2}, \mathbf{r} - \frac{\mathbf{r}'}{2}, \omega\right) K_\omega(\mathbf{r}')\, \mathrm{d}^2 r', \tag{4.77}$$

where

$$K_\omega(\mathbf{r}') = \frac{k^2}{2(2\pi)^{1/2}} \frac{J_{3/2}(kr')}{(kr')^{3/2}} \tag{4.78}$$

is expressed in terms of the Bessel function $J_{3/2}$.

From (4.75) we have for the generalized radiant intensity

$$J_\omega(\mathbf{s}) = \cos(\Theta) \int_{(z=0)} B_\omega(\mathbf{r}, \mathbf{s})\, d^2r = (2\pi k)^2 \cos^2(\Theta)\, \tilde{G}(k\mathbf{s}, -k\mathbf{s}, \omega) =$$
$$= k^2 \cos^2(\Theta)\, \tilde{C}(k\mathbf{s}, \omega), \tag{4.79}$$

where

$$C(\mathbf{r}, \omega) = \int_{-\infty}^{+\infty} G\left(\mathbf{r}' + \frac{\mathbf{r}}{2}, \mathbf{r}' - \frac{\mathbf{r}}{2}, \omega\right) d^2r' \tag{4.80}$$

and the integration extends over the source plane $z' = 0$. The quantity $C(\mathbf{r}, \omega)$ is called the source-averaged cross-spectral density function of light in the source plane. In view of the physical significance of $G(\mathbf{r}_1, \mathbf{r}_2, \omega)$, the function $C(\mathbf{r}, \omega)$ is clearly proportional to the average value of the correlations of light fluctuations at frequency ω for all pairs of points $\mathbf{r}_1$ and $\mathbf{r}_2$ in the source plane whose relative separation is $\mathbf{r} = \mathbf{r}_1 - \mathbf{r}_2$ and the average is taken over the whole source. Further one can define the coefficient of directionality (Collet and Wolf (1979))

$$\eta(\mathbf{r}, \omega) = \frac{C(\mathbf{r}, \omega)}{C(\mathbf{0}, \omega)}, \tag{4.81}$$

satisfying $\eta(\mathbf{0}, \omega) = 1$ and $|\eta(\mathbf{r}, \omega)| \leqq 1$. For $kR \gg 1$

$$J_\omega(\mathbf{s}) \simeq R^2 G(R\mathbf{s}, R\mathbf{s}, \omega). \tag{4.82}$$

Since $G(\mathbf{R}, \mathbf{R}, \omega) \geqq 0$,

$$J_\omega(\mathbf{s}) \geqq 0 \tag{4.83}$$

irrespective of the state of coherence of the source.

Spatially incoherent sources

For spatially incoherent sources (3.37a) is appropriate with the power spectrum $G(\mathbf{r}_1, \omega) \geqq 0$ ($G(\mathbf{r}_1, \omega) = 0$ if $\mathbf{r}_1$ lies outside the source area) and from (4.72)

$$B_\omega(\mathbf{r}, \mathbf{s}) = \left(\frac{k}{2\pi}\right)^2 \cos(\Theta)\, G(\mathbf{r}, \omega), \tag{4.84}$$

which is non-negative ($-\pi/2 \leqq \Theta \leqq \pi/2$). From (4.79) and (4.84)

$$J_\omega(\mathbf{s}) = \left(\frac{k}{2\pi}\right)^2 \cos^2(\Theta) \int_{(z=0)} G(\mathbf{r}, \omega)\, d^2r = J_\omega(0) \cos^2(\Theta), \tag{4.85}$$

where $J_\omega(0)$ is the radiant intensity at $\Theta = 0$ ($s_x = s_y = 0$), i.e. at the direction perpendicular to the plane of the source. This implies that a spatially incoherent plane source of finite extent cannot radiate isotropically into the half-space $z > 0$. Such isotropic radiation with $\cos(\Theta)$-dependence is produced by blackbody sources, exhibiting correlations over distances of the order of the mean wavelength of radiation.

Coherent sources

For radiation from coherent sources, the factorization (4.13c) for the cross-spectral density holds and consequently

$$B_\omega(\mathbf{r}, \mathbf{s}) = \left(\frac{k}{2\pi}\right)^2 \cos(\Theta) \int_{(z'=0)} \mathscr{U}^*\left(\mathbf{r} + \frac{\mathbf{r}'}{2}\right) \mathscr{U}\left(\mathbf{r} - \frac{\mathbf{r}'}{2}\right) \exp(-\mathrm{i}k\mathbf{s}\,.\,\mathbf{r}')\,\mathrm{d}^2 r'. \tag{4.86}$$

From (4.79)

$$J_\omega(\mathbf{s}) = (2\pi k)^2 \cos^2(\Theta)\,|\,\tilde{\mathscr{U}}(k\mathbf{s})\,|^2, \tag{4.87}$$

where $\tilde{\mathscr{U}}(\mathbf{f})$ is the two-fold Fourier transform to $\mathscr{U}(\mathbf{r})$ and thus $\tilde{G}(\mathbf{f}_1, \mathbf{f}_2, \omega) = \tilde{\mathscr{U}}^*(-\mathbf{f}_1)\,\tilde{\mathscr{U}}(\mathbf{f}_2)$,

$$\mathscr{U}(\mathbf{f}) = \frac{1}{(2\pi)^2} \int_{(z=0)} \mathscr{U}(\mathbf{r}) \exp(-\mathrm{i}\mathbf{f}\,.\,\mathbf{r})\,\mathrm{d}^2 r. \tag{4.88}$$

As the generalized radiance $B_\omega(\mathbf{r}, \mathbf{s})$ as well as the generalized radiant emittance $E_\omega(\mathbf{r})$ may take on negative values, they are not measurable quantities. However, the generalized radiant intensity $J_\omega(\mathbf{s})$ defined by (4.79) is always non-negative and it represents the basic measurable quantity of the radiometry of fields produced by sources of any state of coherence.

Quasi-homogeneous sources

Quasi-homogeneous planar sources are characterized by

$$G(\mathbf{r}_1, \mathbf{r}_2, \omega) = I(\tfrac{1}{2}(\mathbf{r}_1 + \mathbf{r}_2), \omega)\,\gamma(\mathbf{r}_1 - \mathbf{r}_2, \omega), \tag{4.89}$$

where the spectral intensity distribution $I(\mathbf{r}, \omega)$ and the degree of spatial coherence $\gamma(\mathbf{r}, \omega)$ are considered in the plane of the source. Further it is assumed that $I(\mathbf{r}, \omega)$ varies with $\mathbf{r}$ much more slowly than $\gamma(\mathbf{r}, \omega)$, that the linear dimensions of the source are large compared with the wavelength of the light, and that $|\gamma(\mathbf{r}, \omega)|$ is substantially non-zero only within a domain small compared to the size of the source. These assumptions are usually valid in practice.

In this case we obtain (Carter and Wolf (1977))

$$B_\omega(\mathbf{r}, \mathbf{s}) = \left(\frac{k}{2\pi}\right)^2 \cos(\Theta)\, I(\mathbf{r}, \omega) \int_{-\infty}^{+\infty} \gamma(\mathbf{r}', \omega) \exp(-\mathrm{i}k\mathbf{s}\,.\,\mathbf{r}')\,\mathrm{d}^2 r', \tag{4.90a}$$

$$E_\omega(\boldsymbol{r}) = I(\boldsymbol{r}, \omega) \int_{-\infty}^{+\infty} \gamma(\boldsymbol{r}', \omega)\, K_\omega(\boldsymbol{r}')\, \mathrm{d}^2 r', \tag{4.90b}$$

$$J_\omega(\boldsymbol{s}) = k^2 \cos^2(\Theta)\, \tilde{I}(\boldsymbol{0}, \omega) \int_{-\infty}^{+\infty} \gamma(\boldsymbol{r}', \omega) \exp(-ik\boldsymbol{s}\,.\,\boldsymbol{r}')\, \mathrm{d}^2 r', \tag{4.90c}$$

where $\tilde{I}(\boldsymbol{0}, \omega)$ is the two-dimensional spatial Fourier transform to $I(\boldsymbol{r}, \omega)$ at $\boldsymbol{0}$ and $K_\omega(\boldsymbol{r}')$ is given by (4.78).

From (4.90c) we see (Carter and Wolf (1977)) that the angular distribution of the radiant intensity $J_\omega(\boldsymbol{s})$ is proportional to the two-dimensional spatial Fourier transform of the degree of spatial coherence of light across the source, and to the square of the cosine of the angle that the s-direction makes with the normal to the source plane. Hence the coherence properties of a quasi-homogeneous source completely determine the angular distribution of the radiant intensity generated by the source. For these sources Carter and Wolf (1977) have also derived a generalization of the Van Cittert-Zernike theorem, giving the degree of spatial coherence in the far zone apart from a simple geometrical factor, as the normalized spatial Fourier transform of the optical intensity across the source.

For a Gaussian correlated source

$$\gamma(\boldsymbol{r}', \omega) = \exp\left(-\frac{\boldsymbol{r}'^2}{2\sigma^2}\right) \tag{4.91}$$

and from (4.90c)

$$J_\omega(\boldsymbol{s}) = J_\omega(0) \cos^2(\Theta) \exp\left[-\tfrac{1}{2}(k\sigma)^2 \sin^2(\Theta)\right], \tag{4.92}$$

where

$$J_\omega(0) = 2\pi(k\sigma)^2\, \tilde{I}(\boldsymbol{0}, \omega). \tag{4.93}$$

This is illustrated in Fig. 4.1. It is seen from this figure that there is a strong modification of the directionality of the radiant intensity, when the correlation

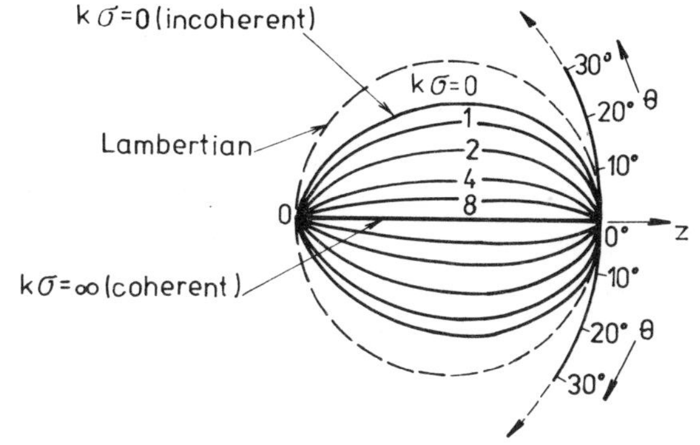

Fig. 4.1 Angular distribution of the normalized radiant intensity $J_\omega(\Theta)/J_\omega(0)$, from a Gaussian correlated statistically homogeneous source, computed from equation (4.92) (after E. Wolf and W. H. Carter, 1975, *Opt. Comm.* **13**, 205).

distance σ increases from zero (incoherent source) to one wavelength ($k\sigma = 8$ corresponds to $\sigma \approx 1.28\lambda$); $k\sigma = \infty$ corresponds to the spatially coherent source (Wolf and Carter (1975)). For comparison the radiant intensity from a Lambertian source is shown (for instance for a blackbody source $\gamma(\mathbf{r}', \omega) = \sin(kr')/kr'$, $J_\omega(\mathbf{s}) \sim \cos(\Theta)$).

Other model sources

In the literature other model sources have been investigated, such as Schell model sources specified by

$$G(\mathbf{r}_1, \mathbf{r}_2, \omega) = [I(\mathbf{r}_1, \omega)\, I(\mathbf{r}_2, \omega)]^{1/2}\, \gamma(\mathbf{r}_1 - \mathbf{r}_2, \omega). \tag{4.94}$$

These sources reduce, in fact, to quasi-homogeneous sources if the optical intensity distribution across the source varies sufficiently slowly.

For Bessel-correlated sources it holds that (Baltes, Steinle and Antes (1976))

$$\gamma(\mathbf{r}, \omega) = \frac{\Gamma\left(\dfrac{n}{2} + 1\right)}{(kr/2)^{n/2}}\, J_{n/2}(kr), \tag{4.95}$$

with

$$J_\omega(\mathbf{s}) = J_\omega(0)\cos^n(\Theta), \tag{4.96}$$

giving the Lambertian source for $n = 1$ ($J_{n/2}(kr)$ in (4.95) is the Bessel function and Γ is the gamma function).

Concluding, we can point out that sources with entirely different coherence properties may produce identical distributions of the radiant intensity; and a class of partially coherent sources may give rise to radiation fields with the same directionality as a fully coherent laser beam (Wolf and Collet (1978)). Recent experimental work (De Santis, Gori, Quattari and Palma (1979), Farina, Narducci and Collet (1980), and see also discussions by Carter and Bertolotti (1978)) demonstrated that practical realization of partially coherent sources is possible with controllable coherence and intensity properties. It was also demonstrated that complete spatial coherence is not a necessary requirement for the generation of highly directional light beams. A great advantage of such beams is that the speckle phenomenon, characteristic of coherent laser beams, is missing.

The inverse problems, i.e. determining the distributions of the optical intensity and the degree of spatial coherence for the source from the measured angular distributions of the radiant intensity, have been reviewed by Baltes, Geist and Walther (1978) who also outlined the fourth-order radiometric theory. The radiance transfer function for space-invariant linear systems has been introduced and investigated by Jannson (1980).

Further interesting discussions of this subject have been provided by Baltes, Steinle and Antes (1976), Baltes (1977), Steinle and Baltes (1977) and Carter and Wolf (1981a, b). A generalization to real optical systems has been suggested by Martínez-Herrero and Mejías (1982b).

MATRIX DESCRIPTION OF PARTIAL COHERENCE

5.1 Sampling theorem

The theory of coherence we have developed in the previous chapters may be regarded as a functional theory. Now we follow another approach to the formulation of the coherence properties of the electromagnetic field using the „quantization" of field functions (e.g. of object and image functions) (Gabor (1956)) based on the so-called *sampling theorem.*

If a square integrable function $f(x)$ has a spectrum non-vanishing only in the finite interval $(-W, +W)$, then this function is fully determined by its values on a countably infinite set of points and it holds that

$$f(x) = \sum_{n=-\infty}^{+\infty} f\left(\frac{n}{2W}\right) u_n(2\pi Wx), \tag{5.1}$$

where

$$u_n(2\pi Wx) = \frac{\sin\left[\pi(2Wx - n)\right]}{\pi(2Wx - n)}. \tag{5.2}$$

It is not difficult to prove this theorem. Such a function may be written as

$$f(x) = \int_{-W}^{+W} g(v) \exp(-\mathrm{i}\, 2\pi v x)\, \mathrm{d}v, \tag{5.3}$$

where $g(v)$ is the spectrum of $f(x)$. Further

$$g(v) = \sigma(v, W) \sum_{n=-\infty}^{+\infty} a_n \exp\left(\mathrm{i}\pi n \frac{v}{W}\right), \tag{5.4}$$

where

$$a_n = \frac{1}{2W} \int_{-W}^{+W} g(v) \exp\left(-\mathrm{i}\pi n \frac{v}{W}\right) \mathrm{d}v \tag{5.5}$$

and

$$\sigma(v, W) = \frac{\operatorname{sgn}(W - v) + \operatorname{sgn}(W + v)}{2}; \tag{5.6}$$

here sgn is the sign function ($\operatorname{sgn} x = 1$ for $x > 0$ and $\operatorname{sgn} x = -1$ for $x < 0$). Comparing (5.3) and (5.5) we see that

$$a_n = \frac{1}{2W} f\left(\frac{n}{2W}\right) \tag{5.7}$$

and substituting (5.7) into (5.4) and performing the Fourier transformation, we finally obtain

$$f(x) = \sum_{n=-\infty}^{+\infty} f\left(\frac{n}{2W}\right) \frac{1}{2W} \int_{-W}^{+W} \exp\left(i\pi n \frac{\nu}{W} - i\,2\pi\nu x\right) d\nu =$$

$$= \sum_{n=-\infty}^{+\infty} f\left(\frac{n}{2W}\right) \frac{\sin(\pi n - 2\pi x W)}{\pi n - 2\pi x W}, \tag{5.8}$$

which is just (5.1). The optical information theory worked out in papers by Shannon (1949), Gabor (1956, 1961), Linfoot (1956) and Toraldo di Francia (1955) (see also Khurgyn and Yakovlev (1971)) among others is based upon this theorem. A characteristic feature of this theorem is the "quantization" of the function $f(x)$ and we now use it to obtain a *matrix formulation* of partial coherence, the relation of which to the usual functional theory is analogous to the relation between the Schrödinger wave and Heisenberg-Dirac matrix formulations of quantum mechanics. Of course, these two formulations are equivalent as a consequence of the isomorphism of the spaces L_2 (the space of functions which are square integrable in the Lebesgue sense) and l_2 (the space of generalized Fourier coefficients of functions from L_2).

We restrict ourselves to quasi-monochromatic light since this will be sufficient to demonstrate this matrix method. For a more complete treatment of the matrix formulation including light of arbitrary spectral composition we refer the reader to the review by Gamo (1964), to the paper by Gabor (1956) and also to recent papers by Wolf (1981, 1982).

First let us summarize some properties of the functions u_n given by (5.2). Obviously

$$u_n(m\pi) = \delta_{mn} = \begin{cases} 1, & m = n, \\ 0, & m \neq n. \end{cases} \tag{5.9}$$

Further we obtain

$$\int_{-\infty}^{+\infty} u_n(2\pi Wx)\, u_m(2\pi Wx)\, dx = \frac{1}{2W} \delta_{nm}, \tag{5.10}$$

which is the orthogonality condition for the complete set of functions u_n.

Inverting (5.1) making use of this condition, we obtain

$$f\left(\frac{n}{2W}\right) = 2W \int_{-\infty}^{+\infty} f(x)\, u_n(2\pi Wx)\, dx \tag{5.11}$$

and further we readily see that

$$\int_{-\infty}^{+\infty} |f(x)|^2\, dx = \frac{1}{2W} \sum_{n=-\infty}^{+\infty} \left| f\left(\frac{n}{2W}\right)\right|^2. \tag{5.12}$$

From (5.1) we obtain, putting $f(x) \equiv 1$ $(z = 2\pi Wx)$,

$$\sum_{n=-\infty}^{+\infty} u_n(z) = \sum_{n=-\infty}^{+\infty} \frac{\sin(z - \pi n)}{z - \pi n} = 1 \tag{5.13}$$

and putting $f(x) = \sin(z - z_2)/(z - z_2)$

$$\sum_{n=-\infty}^{+\infty} \frac{\sin(z_1 - \pi n)}{z_1 - \pi n} \frac{\sin(z_2 - \pi n)}{z_2 - \pi n} = \frac{\sin(z_1 - z_2)}{z_1 - z_2}. \tag{5.14}$$

If $z_1 = z_2 = z$, then

$$\sum_{n=-\infty}^{+\infty} u_n^2(z) = 1. \tag{5.15}$$

5.2 Interference law in matrix form

As we have restricted ourselves to quasi-monochromatic light we may consider the mutual intensity $\Gamma(x_1, x_2) \equiv J_{12}$ in describing the coherence properties of light. Denoting $\Gamma_{11}(0) \equiv J_{11}$ and $\Gamma_{22}(0) \equiv J_{22}$ we can write the interference law in the form

$$I = |a_1|^2 J_{11} + |a_2|^2 J_{22} + a_1^* a_2 J_{12} + a_1 a_2^* J_{21}, \tag{5.16}$$

where the a_j are the propagators introduced in Sec. 3.1. This is a positive semi-definite quadratic form, since $I \geqq 0$. The elements of the matrix $\hat{J} \equiv (J_{ij})$ can be expressed in terms of the eigenvalues λ_1 and λ_2 of the matrix $\hat{J}$, defined by

$$\begin{vmatrix} J_{11} - \lambda & J_{12} \\ J_{21} & J_{22} - \lambda \end{vmatrix} = \lambda^2 - (J_{11} + J_{22})\lambda + J_{11}J_{22} - |J_{12}|^2 = 0, \tag{5.17}$$

and eigenstates

$$\boldsymbol{\varphi}^{(1)} = \begin{pmatrix} U_{11} \\ U_{21} \end{pmatrix}, \qquad \boldsymbol{\varphi}^{(2)} = \begin{pmatrix} U_{12} \\ U_{22} \end{pmatrix} \tag{5.18}$$

in the form

$$J_{ij} = \lambda_1 U_{i1} U_{j1}^* + \lambda_2 U_{i2} U_{j2}^*, \qquad i, j = 1, 2. \tag{5.19}$$

The elements of the matrix $\hat{U} \equiv (\boldsymbol{\varphi}^{(1)}, \boldsymbol{\varphi}^{(2)})$ satisfy the homogeneous equations

$$\begin{pmatrix} J_{11} - \lambda_j & J_{12} \\ J_{21} & J_{22} - \lambda_j \end{pmatrix} \begin{pmatrix} U_{1j} \\ U_{2j} \end{pmatrix} = \hat{0}. \tag{5.20}$$

As $\hat{U}$ is a unitary matrix ($\hat{U}^+ = \hat{U}^{-1}$, $\hat{U}^+$ is the Hermitian conjugate to $\hat{U}$) we have

$$U_{1i}^* U_{1j} + U_{2i}^* U_{2j} = U_{i1} U_{j1}^* + U_{i2} U_{j2}^* = \delta_{ij}, \qquad i, j = 1, 2. \tag{5.21}$$

Since $|\gamma_{12}| \leqq 1$, $|J_{12}|^2 \leqq J_{11}J_{22}$ and the equality occurs for coherent light; in this case $\mathrm{Det}\,\hat{J} = J_{11}J_{22} - |J_{12}|^2 = 0$ (Det denotes the determinant) and so, from (5.17), $\lambda_2 = 0$. Then (5.19) gives $J_{ij} = \lambda_1 U_{i1} U_{j1}^*$, i.e. the elements of the matrix $\hat{J}$ are factorized in agreement with the result of Sec. 4.2.

5.3 Intensity matrix and its properties

We now start a more general treatment of the matrix formulation of partial coherence. First we introduce the intensity matrix $\hat{A}$ corresponding to the mutual intensity. Consider an object plane which is imaged by an optical system with the numerical apertures $\alpha = n_1 \sin(\vartheta_1)$ and $\beta = n_2 \sin(\vartheta_2)$ (n_1, ϑ_1 and n_2, ϑ_2 are the refractive indices and aperture angles in the object and image spaces respectively – see Fig. 5.1). We can show by elementary diffraction considerations that the diffraction function $K(x, \xi)$ (we consider a one-dimensional case for simplicity) is a spatial frequency limited function in the object and image spaces within the domains $(-W, +W) = (-k\alpha/2\pi, +k\alpha/2\pi)$ and $(-k\beta/2\pi, +k\beta/2\pi)$ respectively (Born and Wolf (1965), p. 480), where k is the mean wave number.

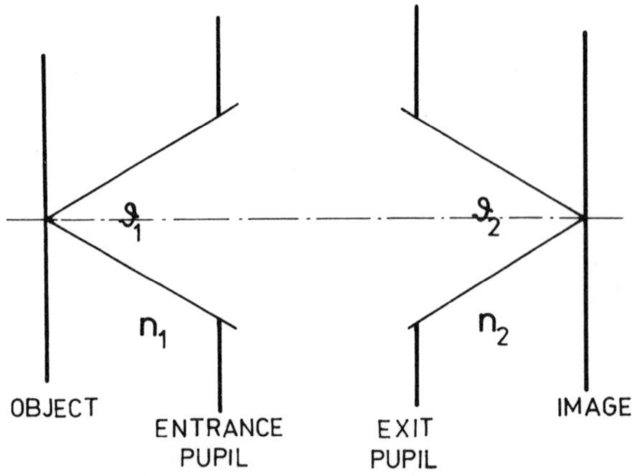

Fig. 5.1 Numerical apertures for entrance and exit pupils.

The mutual intensity $J(x_1, x_2)$ may be decomposed into the complete set of functions u_n in the form

$$J(x_1, x_2) = \sum_{n=-\infty}^{+\infty} \sum_{m=-\infty}^{+\infty} A_{nm} u_n(k\alpha x_1)\, u_m(k\alpha x_2), \tag{5.22}$$

where, using the orthogonality condition (5.10),

$$A_{nm} = \left(\frac{k\alpha}{\pi}\right)^2 \iint_{-\infty}^{+\infty} J(x_1, x_2)\, u_n(k\alpha x_1)\, u_m(k\alpha x_2)\, \mathrm{d}x_1\, \mathrm{d}x_2. \tag{5.23}$$

As $J^*(x_1, x_2) = J(x_2, x_1)$ the matrix $\hat{A} \equiv (A_{nm})$ is Hermitian, i.e. $A_{nm}^* = A_{mn}^+ = A_{mn}$.

The condition of positive semi-definiteness for $J(x_1, x_2)$,

$$\iint_{-\infty}^{+\infty} J(x_1, x_2) f^*(x_1) f(x_2)\, dx_1\, dx_2 \geqq 0, \tag{5.24a}$$

can be transformed to the matrix form

$$\sum_{n=-\infty}^{+\infty} \sum_{m=-\infty}^{+\infty} A_{nm} z_n^* z_m \geqq 0, \tag{5.24b}$$

where $z_n = \int_{-\infty}^{+\infty} f(x)\, u_n(k\alpha x)\, dx$ and $f(x)$ is an arbitrary function. Writing the mutual intensity with the help of the Hopkins formula (3.53)†,

$$J(x_1, x_2) = \int_{\sigma} U^*(x_1, \xi)\, U(x_2, \xi)\, d\xi, \tag{5.25}$$

where the integration is taken over the surface of the source of light, and using the sampling theorem (5.1), we obtain another form of A_{nm}

$$A_{nm} = \int_{\sigma} U^*\left(\frac{n\pi}{k\alpha}, \xi\right) U\left(\frac{m\pi}{k\alpha}, \xi\right) d\xi. \tag{5.26}$$

From this we see that $A_{nm} = J(n\pi/k\alpha, m\pi/k\alpha)$ is the mutual intensity between the wave amplitudes at the nth and mth sampling points; $A_{nn} = J(n\pi/k\alpha, n\pi/k\alpha)$ is the intensity at the nth sampling point. Note that both these quantities are measurable. From $|J_{12}|^2 \leqq J_{11} J_{22}$ it follows that

$$|A_{nm}|^2 \leqq A_{nn} A_{mm}. \tag{5.27}$$

From (5.27), summing over n and m,

$$\sum_{n=-\infty}^{+\infty} \sum_{m=-\infty}^{+\infty} A_{nm} A_{mn} \leqq \Big(\sum_{n=-\infty}^{+\infty} A_{nn}\Big)^2, \tag{5.28a}$$

where we have used the Hermiticity of $\hat{A}$, $A_{nm}^* = A_{mn}$, and so

$$\operatorname{Tr} \hat{A}^2 \leqq (\operatorname{Tr} \hat{A})^2, \tag{5.28b}$$

where Tr denotes the trace of the matrix. The equality occurs for the coherent field.

The integrated intensity may be obtained from (5.22) by putting $x_1 \equiv x_2 \equiv x$, which gives the intensity $J(x, x)$ as a Hermitian, real and non-negative form††, and integrating over x

$$I_{\text{integrated}} = \int_{-\infty}^{+\infty} J(x, x)\, dx = \frac{\pi}{k\alpha} \operatorname{Tr} \hat{A}, \tag{5.29}$$

† Note that this form of the mutual intensity closely corresponds to the form of the density matrix $\varrho(x_1, x_2)$ in quantum mechanics defined by means of the wave function $\psi(x, q)$ as $\varrho(x_1, x_2) = \int \psi^*(x_1, q)\, \psi(x_2, q)\, dq$.

†† Note that the intensity matrix $\hat{A}$ contains the phase information while the intensity $J(x, x)$ does not. From this point of view the intensity matrix $\hat{A} \equiv (A_{nm})$ is a more general physical quantity than the intensity $J(x, x)$.

where the orthogonality condition (5.10) has also been used. We can also show that

$$\iint\limits_{-\infty}^{+\infty} |J(x_1, x_2)|^2 \,dx_1\, dx_2 = \left(\frac{\pi}{k\alpha}\right)^2 \operatorname{Tr} \hat{A}^2. \tag{5.30}$$

By putting $f(x) \equiv 1$ in (5.24a), with (5.22) substituted, we have

$$\iint\limits_{-\infty}^{+\infty} J(x_1, x_2) \,dx_1\, dx_2 = \left(\frac{\pi}{k\alpha}\right)^2 \sum_{n=-\infty}^{+\infty} \sum_{m=-\infty}^{+\infty} A_{nm} \geqq 0, \tag{5.31}$$

where the well-known integral $\int_{-\infty}^{+\infty} \sin(ax)/x \,dx = \pi \operatorname{sgn}(a)$ has been used.

Equation (5.22) may be written in the matrix form as a scalar product

$$J(x_1, x_2) = (\boldsymbol{\Phi}(x_1), \hat{A}\boldsymbol{\Phi}(x_2)), \tag{5.32}$$

where $\boldsymbol{\Phi}$ is a column vector in the Hilbert space created from u_n. Using the Schwarz inequality, we obtain

$$\begin{aligned} |(\boldsymbol{\Phi}(x_1), \hat{A}\boldsymbol{\Phi}(x_2))|^2 &\leqq \|\boldsymbol{\Phi}(x_1)\|^2 \,\|\hat{A}\boldsymbol{\Phi}(x_2)\|^2 = \\ &= \|\hat{A}\boldsymbol{\Phi}(x_2)\|^2 \leqq \sum_{n=-\infty}^{+\infty} \sum_{m=-\infty}^{+\infty} |A_{nm}|^2 = \operatorname{Tr} \hat{A}^2 = \\ &= \left(\frac{k\alpha}{\pi}\right)^2 \iint\limits_{-\infty}^{+\infty} |J(x_1, x_2)|^2 \,dx_1\, dx_2, \end{aligned} \tag{5.33}$$

where the norms $\|\boldsymbol{\Phi}\|$ and $\|\hat{A}\boldsymbol{\Phi}\|$ of $\boldsymbol{\Phi}$ and $\hat{A}\boldsymbol{\Phi}$ fulfil the following equations

$$\|\boldsymbol{\Phi}\|^2 = (\boldsymbol{\Phi}, \boldsymbol{\Phi}) = \sum_{n=-\infty}^{+\infty} u_n^2 = 1 \tag{5.34}$$

according to (5.15) and

$$\|\hat{A}\boldsymbol{\Phi}\|^2 \leqq \|A\|^2 \,\|\boldsymbol{\Phi}\|^2 = \sum_{n=-\infty}^{+\infty} \sum_{m=-\infty}^{+\infty} |A_{nm}|^2 = \operatorname{Tr} \hat{A}^2. \tag{5.35}$$

In (5.33) equation (5.30) has also been used. Equation (5.33) shows that the form (5.22) is bounded and uniformly convergent if

$$\iint\limits_{-\infty}^{+\infty} |J(x_1, x_2)|^2 \,dx_1\, dx_2 < \infty. \tag{5.36}$$

The bilinear form (5.22) may be transformed to the normal (diagonal) form using the standard methods. Let λ_n be the eigenvalues of the matrix $\hat{A}$ determined from the condition

$$\operatorname{Det}(\hat{A} - \lambda\hat{1}) = 0, \tag{5.37}$$

where $\hat{1}$ is the unit matrix (the λ_n are real and non-negative since $\hat{A}$ is Hermitian and non-negative), and let $\boldsymbol{\varphi}^{(n)}$ be the eigenvectors of the matrix $\hat{A}$ obeying the homogeneous equations

$$\hat{A}\boldsymbol{\varphi}^{(n)} = \lambda_n \boldsymbol{\varphi}^{(n)}. \tag{5.38}$$

Define a matrix $\hat{U}$ such that U_{mn} is the mth component of the vector $\boldsymbol{\varphi}^{(n)}$, i.e.

$$\hat{U} \equiv (\boldsymbol{\varphi}^{(1)}, \boldsymbol{\varphi}^{(2)}, \ldots).$$

The matrix $\hat{U}$ is unitary since

$$\hat{U}^+ \hat{U} = \hat{U}\hat{U}^+ = \hat{1}, \tag{5.39}$$

that is

$$\sum_{j=-\infty}^{+\infty} U^*_{jm} U_{jn} = \sum_{j=-\infty}^{+\infty} U_{mj} U^*_{nj} = \delta_{mn}, \tag{5.40}$$

which is a more general form of (5.21). Introducing a new vector

$$\boldsymbol{\Psi} = \hat{U}^+ \boldsymbol{\Phi}, \qquad \boldsymbol{\Phi} = \hat{U}\boldsymbol{\Psi} \tag{5.41}$$

in (5.32), we arrive at the diagonal form of $J(x_1, x_2)$ in terms of $\boldsymbol{\Psi}$

$$J(x_1, x_2) = (\hat{U}\boldsymbol{\Psi}(x_1), \hat{A}\hat{U}\boldsymbol{\Psi}(x_2)) = (\boldsymbol{\Psi}(x_1), \hat{U}^+\hat{A}\hat{U}\boldsymbol{\Psi}(x_2)). \tag{5.42}$$

As $\hat{U}^+\hat{A}\hat{U} = \hat{\Lambda}$, where $\hat{\Lambda}$ is the diagonal matrix with elements λ_n, we obtain from (5.42)

$$J(x_1, x_2) = \sum_{n=-\infty}^{+\infty} \lambda_n \psi^*_n(x_1)\, \psi_n(x_2), \tag{5.43a}$$

which is the required diagonal form of the mutual intensity $J(x_1, x_2)$. For polychromatic light it analogously holds for the cross-spectral density (Gamo (1964), Wolf (1981, 1982))†

$$G(x_1, x_2, \nu) = \sum_{n=-\infty}^{+\infty} \lambda_n(\nu)\, \psi^*_n(x_1, \nu)\, \psi_n(x_2, \nu). \tag{5.43b}$$

Further we obtain

$$\hat{A} = \hat{U}\hat{\Lambda}\hat{U}^+, \tag{5.44a}$$

that is

$$A_{jl} = \sum_{n=-\infty}^{+\infty} \lambda_n U_{jn} U^*_{ln}, \tag{5.44b}$$

---

† It is $G(x_1, x_2, \nu) = \langle \tilde{V}^*(x_1, \nu)\, \tilde{V}(x_2, \nu)\rangle$, if $\tilde{V}(x, \nu) = \sum_n a_n(\nu)\, \psi_n(x, \nu)$ and $\langle a^*_m(\nu)\, a_n(\nu)\rangle = \lambda_n(\nu)\, \delta_{mn}$, λ_n and ψ_n being eigenvalues and eigenfunctions of the homogeneous Fredholm integral equation with the kernel $G(x_1, x_2, \nu)$ (see (5.52)). This provides a new definition of the cross-spectral density $G(x_1, x_2, \nu)$ compared to (2.25) (Wolf (1981, 1982)). In this new theory of coherence due to Wolf, using the eigenvalues and eigenfunctions of a Fredholm integral equation instead of the Fourier integral, the Fourier transformations between $V(x, t)$ and $\tilde{V}(x, \nu)$ are forbidden in ordinary functions and they are admissible only between $\Gamma(x_1, x_2, \tau)$ and $G(x_1, x_2, \nu)$. A clear definition of the cross-spectral density $G(x_1, x_2, \nu)$ as $\langle \tilde{V}^*(x_1, \nu)\, \tilde{V}(x_2, \nu)\rangle$, involving the Fourier transformation, and of its generalization to an arbitrary order, is given in (2.18) involving the stationary frequency condition (2.21), which follows from the time invariance of the correlation functions with respect to translations of the time origin. In this case the angle brackets mean the ensemble or the quantum average.

and from (5.41)

$$\psi_n(x) = \sum_{j=-\infty}^{+\infty} U_{jn}^* u_j(k\alpha x), \qquad u_n(k\alpha x) = \sum_{j=-\infty}^{+\infty} U_{nj}\psi_j(x). \tag{5.45}$$

It is obvious that

$$\| \boldsymbol{\Psi} \|^2 = (\boldsymbol{\Psi}, \boldsymbol{\Psi}) = \sum_{n=-\infty}^{+\infty} \sum_{j=-\infty}^{+\infty} \sum_{l=-\infty}^{+\infty} U_{jn} U_{ln}^* u_j(k\alpha x)\, u_l\,(k\alpha x) =$$

$$= \sum_{j=-\infty}^{+\infty} u_j^2(k\alpha x) = \| \boldsymbol{\Phi} \|^2 = 1, \tag{5.46}$$

where we have used (5.40) and (5.34).

Equation (5.44b) may be written in the form

$$\hat{A} = \sum_{n=-\infty}^{+\infty} \lambda_n \hat{P}^{(n)} \tag{5.47}$$

where

$$(\hat{P}^{(n)})_{jl} = U_{jn} U_{ln}^* \tag{5.48}$$

is the matrix element of the projection operator $\hat{P}^{(n)}$ for which

$$(\hat{P}^{(n)}\hat{P}^{(m)})_{jl} = \sum_r (\hat{P}^{(n)})_{jr}\,(\hat{P}^{(m)})_{rl} = \sum_r U_{jn} U_{rn}^* U_{rm} U_{lm}^* =$$

$$= U_{jn} U_{lm}^* \delta_{nm} = (\hat{P}^{(n)})_{jl}\delta_{nm}, \tag{5.49a}$$

that is

$$\hat{P}^{(n)}\hat{P}^{(m)} = \hat{P}^{(n)}\delta_{nm}\,. \tag{5.49b}$$

Further

$$\mathrm{Tr}\,\hat{P}^{(n)} = 1 \tag{5.50}$$

using (5.40).

Some further properties of the mutual intensity may be obtained if it is considered as a kernel of a homogeneous integral equation instead of considering it as a bilinear form in the Hilbert space. Writing $\varphi_n(x) = u_n(k\alpha x)$, we obtain, if (5.10), (5.41) [(5.45)] and (5.39) [(5.40)] are used,

$$(\varphi_n(x), \varphi_m(x)) = \int_{-\infty}^{+\infty} \varphi_n(x)\,\varphi_m(x)\,\mathrm{d}x =$$

$$= (\psi_n(x), \psi_m(x)) = \int_{-\infty}^{+\infty} \psi_n^*(x)\,\psi_m(x)\,\mathrm{d}x = \frac{\pi}{k\alpha}\,\delta_{nm}\,. \tag{5.51}$$

Multiplying (5.43a) by $\psi_n(x_1)$ and integrating over x_1, we have

$$\int_{-\infty}^{+\infty} J(x_1, x_2)\,\psi_n(x_1)\,\mathrm{d}x_1 = \frac{\pi}{k\alpha}\,\lambda_n\psi_n(x_2). \tag{5.52}$$

Therefore $\psi_n(x)$ are eigenfunctions of the kernel $J(x_1, x_2)$ corresponding to the eigenvalues $\pi\lambda_n/k\alpha$. Equation (5.43a) represents the decomposition of the mutual intensity – the kernel of the integral equation (5.52) – in terms of the eigenfunctions $\psi_n(x)$ (the Mercer theorem).

If $\lambda_j = 0$ for all $j \neq n$ and $\lambda_n > 0$ (all λ_j must be real, since $J^*(x_1, x_2) = J(x_2, x_1)$, and non-negative, since (5.24a) holds), then we obtain from (5.43a)

$$J(x_1, x_2) = \lambda_n \psi_n^*(x_1)\, \psi_n(x_2). \tag{5.53}$$

The integrated intensity (5.29) equals $\pi\lambda_n/k\alpha$ in this case. Thus the mutual intensity is factorized and the field is coherent. A partially coherent field may be regarded, according to (5.43a, b), as the superposition of elementary spatially coherent fields emitted by discrete statistically independent elementary sources. Equation (5.43a) can be written in terms of the projection operator

$$P^{(n)}(x_1, x_2) = \psi_n^*(x_1)\, \psi_n(x_2) \tag{5.54}$$

as

$$J(x_1, x_2) = \sum_{n=-\infty}^{+\infty} \lambda_n P^{(n)}(x_1, x_2), \tag{5.55}$$

for which, in analogy to (5.49b) and (5.50),

$$\frac{k\alpha}{\pi} \int_{-\infty}^{+\infty} P^{(n)}(x_1, x_2)\, P^{(m)}(x_2, x_3)\, \mathrm{d}x_2 = P^{(n)}(x_1, x_3)\, \delta_{nm} \tag{5.56a}$$

and

$$\frac{k\alpha}{\pi} \int_{-\infty}^{+\infty} P^{(n)}(x, x)\, \mathrm{d}x = 1. \tag{5.56b}$$

5.4 Propagation law of partial coherence in matrix form

It is interesting to formulate the propagation law (3.23b) in the matrix form. We can write

$$J'(x_1, x_2) = \iint_{\mathscr{A}_1} K^*(x_1, \xi_1)\, K(x_2, \xi_2)\, J(\xi_1, \xi_2)\, \mathrm{d}\xi_1\, \mathrm{d}\xi_2, \tag{5.57}$$

where $\Gamma(Q_1, Q_2)$ and $\Gamma(P_1, P_2)$ are denoted as $J'(x_1, x_2)$ and $J(\xi_1, \xi_2)$ respectively and $K(x, \xi) \equiv K(Q, P, \bar{\nu})$; $\mathscr{A}_1$ is an object region in which $J(\xi_1, \xi_2)$ is defined while the image region in which the mutual intensity is $J'(x_1, x_2)$ is denoted as $\mathscr{A}_2$. The numerical aperture in the image space is β so that the interval of spatial frequencies present in the image space will be $(-k\beta/2\pi, +k\beta/2\pi)$ as we have mentioned. The transmission matrix $(K_{nm}) \equiv \hat{K}$ may be defined in an identical way to the intensity matrix. We can write

$$K(x, \xi) = \sum_{n=-\infty}^{+\infty} \sum_{m=-\infty}^{+\infty} K_{nm} u_m(k\beta x)\, u_n(k\alpha\xi), \tag{5.58}$$

where

$$K_{nm} = \frac{k^2\alpha\beta}{\pi^2} \int_{\mathscr{A}_2} \int_{\mathscr{A}_1} K(x, \xi)\, u_m(k\beta x)\, u_n(k\alpha\xi)\, \mathrm{d}x\, \mathrm{d}\xi. \tag{5.59}$$

The physical significance of the elements K_{nm} is obtained, similarly to the significance of A_{nm}, by applying (5.11): $K_{nm} = K(m\pi/k\beta, n\pi/k\alpha)$ represents the wave amplitude at the mth sampling point in the image plane caused by the unit amplitude at the nth sampling point of the object plane.

Substituting (5.58) into (5.57), multiplying (5.57) by $u_m(k\beta x_1)\, u_j(k\beta x_2)$, integrating over x_1 and x_2 under the assumptions that $J \equiv 0$ outside $\mathscr{A}_1$ and $J' \equiv 0$ outside $\mathscr{A}_2$† and introducing the elements of the intensity matrix A from (5.23), we arrive at

$$A'_{mj} = \left(\frac{\pi}{k\alpha}\right)^2 \sum_{n=-\infty}^{+\infty} \sum_{l=-\infty}^{+\infty} K^*_{nm} A_{nl} K_{lj}, \tag{5.60a}$$

that is

$$\hat{A}' = \left(\frac{\pi}{k\alpha}\right)^2 \hat{K}^+ \hat{A} \hat{K}. \tag{5.60b}$$

The solution of the problem of the reconstruction of an object from its image by solving the Fredholm integral equation (5.57) corresponds to the solution of the matrix equation (5.60).

5.5 The entropy of light beams

Interpreting (5.43) as the incoherent superposition of elementary coherent beams, we may introduce the probability p_n that a photon belongs to the nth beam

$$p_n = \frac{\lambda_n}{\sum_n \lambda_n}, \qquad \sum_n p_n = 1, \qquad \lambda_n \geqq 0. \tag{5.61}$$

Further we may introduce the entropy as the mean value of the quantity $\ln(1/p_n) = -\ln p_n$ (which varies in the interval $(0, \infty)$ while p_n varies in the interval $(0,1)$) using the relation (Landau and Lifshitz (1964))††

$$H = -\sum_n p_n \ln p_n. \tag{5.62}$$

This quantity is a measure of the disorder in the field and it also reflects the correlation properties of the field. One can easily show that the minimum of the entropy occurs if all $\lambda_j = 0$ except say $\lambda_n > 0$; then $p_j = 0$ for $j \neq n$ and $p_n = 1$ so that $H = H_{\min} = 0$. This is the case of the coherent field. It can

† This assumption is an idealization which can never be fulfilled exactly (see Sec. 9.2).

†† This expression corresponds to the expression for the entropy defined by means of the density matrix $\hat{\varrho}$, $H = -\mathrm{Tr}(\hat{\varrho} \ln \hat{\varrho})$.

also be shown by the method of Lagrange multipliers that the maximum of the entropy occurs if $\lambda_j = 1/N$ ($j = 1, 2, \ldots, N$) when $H_{\max} = \ln N$; this corresponds to the incoherent field. For a partially coherent field the entropy lies between these two values.

Some applications of thermodynamic considerations and of the concept of the entropy to problems of interference were given by Laue (1907) and Gamo (1964).

5.6 Generally covariant formulation of partial coherence

A generally covariant formulation of partial coherence (Peřina (1963a)) makes use of the "quantization" of an object and its image (Gabor (1956)) to obtain an interpretation of the degree of coherence as the metric tensor in a non-Euclidian space, called the optical space. This fact allows a study of partial coherence as a property of the optical space. We consider here real functions and quasi-monochromatic light with time delays $|\tau| \ll 1/\Delta\nu$.

Introducing the degree of coherence $\gamma(\xi_1, \xi_2)/[I(\xi_1)\, I(\xi_2)]^{1/2}$ into (5.57), denoting the real amplitude $I^{1/2}(\xi)$ as $u(\xi)$, considering $K \equiv 1$ for simplicity and performing the "quantization" of the object, we obtain the intensity in the quadratic form

$$I = \sum_{i,j}^{n} \gamma_{ij} u^i u^j, \tag{5.63}$$

where $u^i \equiv u(\xi_i)$, ξ_i are points of "quantization" and $\gamma_{ij} \equiv \gamma(u^i, u^j)$. We may define a non-Euclidian space with the metric tensor $\gamma_{ij} = \gamma_{ji}$ in the usual way (Petrov (1961)) with the use of the quadratic form for the interval $(\mathrm{d}s)^2$,

$$(\mathrm{d}s)^2 = \sum_{i,j}^{n} \gamma_{ij}\, \mathrm{d}u^i\, \mathrm{d}u^j. \tag{5.64}$$

This space is the optical space. As $I \geqq 0$, this quadratic form is positive semidefinite. Thus $(\mathrm{d}s)^2$ may be ragarded as the interval in a Riemann space of n dimensions with the metric tensor γ_{ij}.

One can see that for incoherent light, since $\gamma_{ij} = \delta_{ij}$, the optical space is Euclidian and

$$I = \sum_{i=1}^{n} (u^i)^2. \tag{5.65}$$

For coherent light $\gamma_{ij} \equiv 1$ for all i, j and we obtain

$$I = \Big(\sum_{i=1}^{n} u^i\Big)^2. \tag{5.66}$$

Therefore for coherent and partially coherent light the optical space is non-Euclidian.

This interpretation of the degree of coherence as the metric tensor in the non-Euclidian (Riemann) space makes it possible to apply the formalism of Riemann spaces and to study the coherence properties of optical fields as metric properties of these spaces.

The optical space of maximal homogeneity is invariant with respect to an $n(n+1)/2$ parametric group of movements (n translations and $\binom{n}{2}$ rotations; this is a group of maximal movability). The optical space of maximal homogeneity is the space of ideally incoherent light. The optical space of partially coherent as well as coherent light is invariant with respect to groups with a smaller number of parameters than $n(n+1)/2$.

Some invariants may be constructed in the optical space from the degree of coherence and its derivatives with the help of the Noether theorem.

CHAPTER 6

POLARIZATION PROPERTIES OF LIGHT

So far we have considered light beams as scalar quantities making either a scalar approximation to the vector electromagnetic field or taking one component of the vector field. In the following we shall take into account the vectorial properties of the field. In this chapter we consider the correlation of light beams at one space point with particular attention to the correlation between polarization components of the light beams. A more complete theory including correlations between beams at different space-time points and between different polarization components of the field will be dealt with in the next chapter.

Wiener (1928, 1929, 1930) was the first to introduce the *coherence matrix*† describing the polarization properties of the electromagnetic field. He showed its connection with the density matrix of quantum mechanics and he also showed that the coherence matrix can be decomposed in terms of matrices called now the Pauli matrices with the so-called Stokes parameters as the expansion coefficients. Further he proposed an operational procedure for the experimental determination of the Stokes parameters. These questions were later studied by Fano (1954), Wolf (1954a, 1959) (see also Born and Wolf (1965)), Parrent and Roman (1960) (see also Beran and Parrent (1964)), Barakat (1963, 1977, 1981) and Marathay (1966) (see also O'Neil (1963)). While the papers by Wolf, Parrent and Roman and Marathay treat quasi-monochromatic fields only, results obtained in the papers by Wiener and Barakat are correct for an arbitrary spectral bandwidth of the fields. An analysis by Pancharatnam (1963) has also been made for polychromatic light.

6.1 Definitions of coherence matrix and Stokes parameters

First let us note that the completely deterministic monochromatic wave is also *completely polarized*, i.e. the end point of the electric or the magnetic vector moves periodically with time, in general, over an ellipse. If the end point moves completely chaotically, one speaks of *unpolarized light*. However, light is generally in intermediate states of polarization and then one speaks of *partially polarized light*.

In all considerations here we use the electric vector $\mathbf{E}$ only, but there is no principal difficulty in taking into account the magnetic vector $\mathbf{H}$, too. We assume

† A more appropriate term should be the polarization matrix.

stationary plane waves propagating in the direction of the positive z-axis. The complex analytic signals of a typical wave will be $E_x(\mathbf{x}, t)$ and $E_y(\mathbf{x}, t)$. There are the components of $\mathbf{E}$ in two mutually orthogonal directions perpendicular to the z-direction, with (x, y, z) forming a right-handed triad; $E_z = 0$ far from the source because of the transversality.

We introduce the *coherence matrix* (Wolf (1954a, 1959))†

$$\mathscr{J}(\tau) = \langle \mathscr{E}^+(t) \otimes \mathscr{E}(t + \tau) \rangle = \left\langle \begin{pmatrix} E_x^*(t) \\ E_y^*(t) \end{pmatrix} \otimes (E_x(t + \tau), E_y(t + \tau)) \right\rangle =$$

$$= \begin{pmatrix} \langle E_x^*(t)\, E_x(t + \tau) \rangle & \langle E_x^*(t)\, E_y(t + \tau) \rangle \\ \langle E_y^*(t)\, E_x(t + \tau) \rangle & \langle E_y^*(t)\, E_y(t + \tau) \rangle \end{pmatrix} = \begin{pmatrix} \mathscr{J}_{xx}(\tau) & \mathscr{J}_{xy}(\tau) \\ \mathscr{J}_{yx}(\tau) & \mathscr{J}_{yy}(\tau) \end{pmatrix}, \tag{6.1}$$

where $\otimes$ is the direct (Kronecker) product and

$$\mathscr{J}_{ij} = \langle E_i^*(t)\, E_j(t + \tau) \rangle, \qquad i, j = x, y \tag{6.2}$$

are elements of the coherence matrix and $\mathscr{E} \equiv (E_x, E_y)$ is a row matrix. The explicit dependence on $\mathbf{x}$ is suppressed here. Further we introduce the *spectral coherence matrix* (Wiener (1928, 1929, 1930))

$$\mathscr{R}(\nu) = \begin{pmatrix} \mathscr{R}_{xx}(\nu) & \mathscr{R}_{xy}(\nu) \\ \mathscr{R}_{yx}(\nu) & \mathscr{R}_{yy}(\nu) \end{pmatrix} = \lim_{T \to \infty} \frac{1}{2T} \left\langle \begin{pmatrix} \tilde{E}_x^*(\nu) \\ \tilde{E}_y^*(\nu) \end{pmatrix} \otimes (\tilde{E}_x(\nu), \tilde{E}_y(\nu)) \right\rangle_e , \tag{6.3}$$

where

$$\mathscr{R}_{ij}(\nu) = \int_{-\infty}^{+\infty} \mathscr{J}_{ij}(\tau) \exp(\mathrm{i}\, 2\pi\nu\tau)\, \mathrm{d}\tau = \lim_{T \to \infty} \frac{1}{2T} \langle \tilde{E}_i^*(\nu)\, \tilde{E}_j(\nu) \rangle_e \tag{6.4}$$

and (2.25) has been used.

From these definitions of $\mathscr{J}(\tau)$ and $\mathscr{R}(\nu)$ it is seen that

$$\mathscr{J}^+(\tau) = \mathscr{J}(-\tau) \tag{6.5a}$$

and

$$\mathscr{R}^+(\nu) = \mathscr{R}(\nu), \tag{6.5b}$$

i.e. the spectral coherence matrix $\mathscr{R}(\nu)$ is Hermitian while the coherence matrix $\mathscr{J}(\tau)$ is Hermitian for $\tau = 0$ only; the use of $\mathscr{J}$ is therefore appropriate for quasi-monochromatic light if the time delays τ involved are restricted so that $|\tau| \ll 1/\Delta\nu$.

Writing the matrix $\mathscr{R}(\nu)$ in the form

$$\mathscr{R}(\nu) = \tfrac{1}{2} \begin{pmatrix} s_0 + s_1 & s_2 + \mathrm{i}s_3 \\ s_2 - \mathrm{i}s_3 & s_0 - s_1 \end{pmatrix}, \tag{6.6}$$

† We note that the Wolf's coherence matrix differs slightly from $\mathscr{J}(\tau)$ given by (6.1); his coherence matrix equals $\mathscr{J}^*(0)$.

we see that

$$\begin{aligned} s_0 &= \mathscr{R}_{xx} + \mathscr{R}_{yy}, \\ s_1 &= \mathscr{R}_{xx} - \mathscr{R}_{yy}, \\ s_2 &= \mathscr{R}_{xy} + \mathscr{R}_{yx}, \\ s_3 &= \mathrm{i}(\mathscr{R}_{yx} - \mathscr{R}_{xy}). \end{aligned} \tag{6.7}$$

These quantities are called the *spectral Stokes parameters* (Pancharatnam (1963)).

Further we obtain

$$4 \operatorname{Det} \mathscr{R} = 4(\mathscr{R}_{xx}\mathscr{R}_{yy} - |\mathscr{R}_{xy}|^2) = s_0^2 - s_1^2 - s_2^2 - s_3^2 \geqq 0, \tag{6.8a}$$

and

$$\operatorname{Tr} \mathscr{R} = \mathscr{R}_{xx} + \mathscr{R}_{yy} = s_0 \geqq 0. \tag{6.8b}$$

Therefore the trace of $\mathscr{R}$ equals the spectral intensity $\tilde{I}(\nu)$. The property (6.8a) of the Stokes parameters suggested the use of the Lorentz group of transformations to study the polarization properties of light (Barakat (1963)).

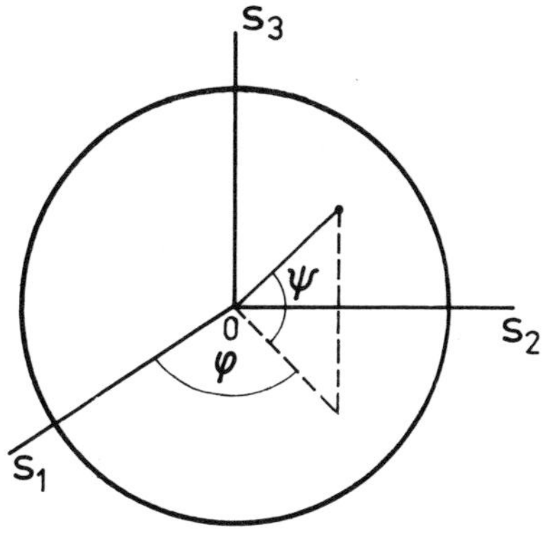

Fig. 6.1 Poincaré sphere.

The relations (6.8a, b) may serve to introduce the Poincaré sphere as follows (Fig. 6.1)

$$\begin{aligned} s_1 &\leqq s_0 \cos(\psi) \cos(\varphi), \\ s_2 &\leqq s_0 \cos(\psi) \sin(\varphi), \\ s_3 &\leqq s_0 \sin(\psi); \end{aligned} \tag{6.8c}$$

the signs of equality in (6.8a, c) are appropriate for the sphere.

Introducing the Pauli matrices

$$\boldsymbol{\sigma}_0 = \begin{pmatrix} 1 & 0 \\ 0 & 1 \end{pmatrix}, \quad \boldsymbol{\sigma}_1 = \begin{pmatrix} 1 & 0 \\ 0 & -1 \end{pmatrix}, \quad \boldsymbol{\sigma}_2 = \begin{pmatrix} 0 & 1 \\ 1 & 0 \end{pmatrix}, \quad \boldsymbol{\sigma}_3 = \begin{pmatrix} 0 & \mathrm{i} \\ -\mathrm{i} & 0 \end{pmatrix}, \tag{6.9}$$

which form an orthogonal system for decompositions of (2×2) Hermitian matrices, and for which

$$\boldsymbol{\sigma}_i\boldsymbol{\sigma}_j = -\boldsymbol{\sigma}_j\boldsymbol{\sigma}_i = -\mathrm{i}\boldsymbol{\sigma}_k, \qquad (i, j, k) = (1, 2, 3) \text{ and cycl,} \tag{6.10a}$$

$$(\boldsymbol{\sigma}_j)^2 = \boldsymbol{\sigma}_0, \qquad j = 0, 1, 2, 3, \tag{6.10b}$$

$$\boldsymbol{\sigma}_j\boldsymbol{\sigma}_0 = \boldsymbol{\sigma}_0\boldsymbol{\sigma}_j = \boldsymbol{\sigma}_j, \qquad j = 0, 1, 2, 3, \tag{6.10c}$$

$$\mathrm{Tr}\,(\boldsymbol{\sigma}_i\boldsymbol{\sigma}_j) = 2\delta_{ij}, \qquad i, j = 0, 1, 2, 3, \tag{6.10d}$$

we can write the spectral coherence matrix in the form

$$\mathscr{R} = \tfrac{1}{2}\sum_{i=0}^{3} s_i\boldsymbol{\sigma}_i. \tag{6.11}$$

Multiplying this relation by $\boldsymbol{\sigma}_j$, taking the trace and using (6.10d), we arrive at

$$s_j = \mathrm{Tr}\,(\boldsymbol{\sigma}_j\mathscr{R}), \tag{6.12}$$

which is a compact form of (6.7).

An identical decomposition to (6.11) holds for the coherence matrix $\mathscr{J}(0)$ (when assuming quasi-monochromatic light and $|\tau| \ll 1/\Delta\nu$) in terms of the Stokes parameters $S_j(0) \equiv S_j$ defined by

$$S_j(\tau) = \int_0^\infty s_j(\nu)\exp(-\mathrm{i}\,2\pi\nu\tau)\,\mathrm{d}\nu. \tag{6.13}$$

The corresponding Poincaré sphere can be introduced in terms of S_j.

6.2 Matrix description of non-image-forming optical elements

Some optical devices not forming an image, such as a compensator or a polarizer, may be described by a matrix $\mathscr{L}(\nu)$ depending on the properties of the device only; this matrix relates the emerging spectral field $\mathscr{F}'(\nu)$ to the incident spectral field $\mathscr{F}(\nu)$,

$$\mathscr{F}' = \mathscr{F}\mathscr{L}. \tag{6.14}$$

The spectral coherence matrix behind the device is given, using (6.3), by

$$\mathscr{R}'(\nu) = \lim_{T\to\infty}\frac{1}{2T}\langle\mathscr{F}'^{+}(\nu)\otimes\mathscr{F}'(\nu)\rangle_e =$$
$$= \lim_{T\to\infty}\frac{1}{2T}\langle\mathscr{L}^{+}\mathscr{F}^{+}\otimes\mathscr{F}\mathscr{L}\rangle_e = \mathscr{L}^{+}(\nu)\,\mathscr{R}(\nu)\,\mathscr{L}(\nu). \tag{6.15}$$

The spectral intensity $\tilde{I}'(\nu)$ behind the device is

$$\tilde{I}'(\nu) = \mathrm{Tr}\,\mathscr{R}'(\nu) = \mathrm{Tr}\,\{\mathscr{L}^{+}(\nu)\,\mathscr{R}(\nu)\,\mathscr{L}(\nu)\} = \mathrm{Tr}\,(\mathscr{R}\mathscr{L}\mathscr{L}^{+}), \tag{6.16}$$

since a cyclic permutation does not change the trace.

It is not difficult to specialize these general results to quasi-monochromatic light under the assumption that $|\tau| \ll 1/\Delta\nu$. In this case the coherence matrix $\mathscr{J}(0) \equiv \mathscr{J}$ must be used instead of the spectral coherence matrix $\mathscr{R}(\nu)$ (Wolf (1959), Parrent and Roman (1960), Born and Wolf (1965), Mandel and Wolf (1965)). It is seen from (6.5a) that the coherence matrix $\mathscr{J}$ is Hermitian in this case and consequently equations (6.6), (6.7), (6.8), (6.11), (6.12), (6.14), (6.15) and (6.16) hold if the spectral quantities are replaced by their corresponding Fourier transform quantities at $\tau = 0$,

$$\begin{aligned} &\mathscr{R}(\nu) \to \mathscr{J}(0), \qquad (\tilde{I}(\nu) \to I = \mathrm{Tr}\,\mathscr{J}), \\ &s_j(\nu) \to S_j(0), \\ &\mathscr{F}(\nu) \to \mathscr{E}(t). \end{aligned} \tag{6.17}$$

The device matrix $\mathscr{L}(\nu)$ must be replaced by $\mathscr{L}(\bar{\nu})$, i.e. this matrix must be taken at the mean frequency. This case was treated for example by Beran and Parrent (1964).

Let us now consider some special kinds of optical devices.

Compensator

When the radiation passes through the compensator, the phases of the $\tilde{E}_x$ – and $\tilde{E}_y$ – components of the field are changed to $\varepsilon_x(\nu)$ and $\varepsilon_y(\nu)$ respectively. Introducing the difference $\delta(\nu) = \varepsilon_x(\nu) - \varepsilon_y(\nu)$, we may describe this device by the matrix

$$\mathscr{L}_c = \begin{pmatrix} \exp(\mathrm{i}\delta/2) & 0 \\ 0 & \exp(-\mathrm{i}\delta/2) \end{pmatrix}. \tag{6.18}$$

This matrix is unitary so that from (6.15)

$$\mathscr{R}' = \mathscr{L}_c^{-1}\mathscr{R}\mathscr{L}_c \tag{6.19a}$$

and

$$\tilde{I}'_c = \tilde{I}_c, \tag{6.19b}$$

since a unitary transform does not change the trace (there is no absorption and no reflection).

Absorber

This device is described by the matrix

$$\mathscr{L}_a = \begin{pmatrix} \exp(-\eta_x/2) & 0 \\ 0 & \exp(-\eta_y/2) \end{pmatrix}, \tag{6.20}$$

where η_x and η_y are real non-negative constants. The spectral coherence matrix $\mathscr{R}'$ is given by (6.15) and the spectral intensity equals

$$\tilde{I}'_a = \mathrm{Tr}\,\{\mathscr{R}\mathscr{L}_a^2\}. \tag{6.21}$$

Rotator

A device which rotates the electric vector through an angle α is described by the real unimodular and unitary matrix

$$\mathscr{L}_r = \begin{pmatrix} \cos\alpha & \sin\alpha \\ -\sin\alpha & \cos\alpha \end{pmatrix} \tag{6.22}$$

and we obtain

$$\mathscr{R}' = \mathscr{L}_r^{-1}\mathscr{R}\mathscr{L}_r = \mathscr{L}_r(-\alpha)\,\mathscr{R}\mathscr{L}_r(\alpha) \tag{6.23a}$$

and

$$\tilde{I}'_r = \tilde{I}_r. \tag{6.23b}$$

Polarizer

A polarizer is a device projecting the electric field onto a direction ϑ and it is characterized by a projection operator $\mathscr{L}_p^{(+)}(\vartheta)$ fulfilling the idempotency condition

$$\mathscr{L}_p^{(+)}(\vartheta)\,\mathscr{L}_p^{(+)}(\vartheta) = \mathscr{L}_p^{(+)}(\vartheta). \tag{6.24a}$$

The conjugate operator $\mathscr{L}_p^{(-)}(\vartheta)$ projecting the field onto the direction perpendicular to ϑ satisfies a condition of the same form and both these operators fulfil the following conditions

$$\mathscr{L}_p^{(+)}\mathscr{L}_p^{(-)} = \mathscr{L}_p^{(-)}\mathscr{L}_p^{(+)} = \hat{0}, \tag{6.24b}$$

$$\mathscr{L}_p^{(+)} + \mathscr{L}_p^{(-)} = \hat{1}. \tag{6.24c}$$

The projecting operators can be represented by the Hermitian matrices

$$\mathscr{L}_p^{(+)}(\vartheta) = \begin{pmatrix} \cos^2\vartheta & \cos(\vartheta)\sin(\vartheta) \\ \sin(\vartheta)\cos(\vartheta) & \sin^2\vartheta \end{pmatrix}, \tag{6.25a}$$

$$\mathscr{L}_p^{(-)}(\vartheta) = \begin{pmatrix} \sin^2\vartheta & -\sin(\vartheta)\cos(\vartheta) \\ -\cos(\vartheta)\sin(\vartheta) & \cos^2\vartheta \end{pmatrix}. \tag{6.25b}$$

It is obvious that $\mathscr{L}_p^{(-)}(\vartheta) = \mathscr{L}_p^{(+)}(\vartheta + \pi/2)$. The spectral coherence matrix transforms according to

$$\mathscr{R}' = \mathscr{L}_p^{(+)}\mathscr{R}\mathscr{L}_p^{(+)} \tag{6.26a}$$

since $\mathscr{L}_p^{(+)+} = \mathscr{L}_p^{(+)}$ and for the spectral intensity we obtain

$$\tilde{I}'_p = \mathrm{Tr}\,\{\mathscr{R}\mathscr{L}_p^{(+)}\}, \tag{6.26b}$$

if (6.24a) is taken into account.

6.3 Interference law for partially polarized beams

Multiplying such matrices successively we may describe cascade systems. As an important example, enabling us to give an operational procedure for defining the elements of the coherence matrix and the Stokes parameters, we consider a combination of a compensator followed by a polarizer. We obtain using (6.18) and (6.25a)

$$\mathscr{L} = \mathscr{L}_c \mathscr{L}_p^{(+)} = \begin{pmatrix} \exp(\mathrm{i}\delta/2) & 0 \\ 0 & \exp(-\mathrm{i}\delta/2) \end{pmatrix} \begin{pmatrix} \cos^2 \vartheta & \cos(\vartheta)\sin(\vartheta) \\ \sin(\vartheta)\cos(\vartheta) & \sin^2 \vartheta \end{pmatrix} =$$

$$= \begin{pmatrix} \exp(\mathrm{i}\delta/2)\cos^2 \vartheta & \exp(\mathrm{i}\delta/2)\cos(\vartheta)\sin(\vartheta) \\ \exp(-\mathrm{i}\delta/2)\sin(\vartheta)\cos(\vartheta) & \exp(-\mathrm{i}\delta/2)\sin^2 \vartheta \end{pmatrix} \tag{6.27}$$

and so

$$\tilde{I}(\nu) = \mathrm{Tr}\,(\mathscr{L}_p^{(+)+} \mathscr{L}_c^{+} \mathscr{R} \mathscr{L}_c \mathscr{L}_p^{(+)}) = \mathrm{Tr}\,(\mathscr{R} \mathscr{L}_c \mathscr{L}_p^{(+)} \mathscr{L}_c^{+}) =$$

$$= \mathrm{Tr}\left\{ \begin{pmatrix} \mathscr{R}_{xx} & \mathscr{R}_{xy} \\ \mathscr{R}_{yx} & \mathscr{R}_{yy} \end{pmatrix} \begin{pmatrix} \exp(\mathrm{i}\delta/2)\cos^2 \vartheta & \exp(\mathrm{i}\delta/2)\cos\vartheta \sin\vartheta \\ \exp(-\mathrm{i}\delta/2)\sin\vartheta\cos\vartheta & \exp(-\mathrm{i}\delta/2)\sin^2 \vartheta \end{pmatrix} \times\right.$$

$$\left. \times \begin{pmatrix} \exp(-\mathrm{i}\delta/2) & 0 \\ 0 & \exp(\mathrm{i}\delta/2) \end{pmatrix} \right\} = \mathscr{R}_{xx}\cos^2 \vartheta + \mathscr{R}_{yy}\sin^2 \vartheta +$$

$$+ \mathscr{R}_{xy}\exp(-\mathrm{i}\delta)\sin(\vartheta)\cos(\vartheta) + \mathscr{R}_{yx}\exp(\mathrm{i}\delta)\sin(\vartheta)\cos(\vartheta) =$$

$$= \mathscr{R}_{xx}\cos^2 \vartheta + \mathscr{R}_{yy}\sin^2 \vartheta + 2\sin(\vartheta)\cos(\vartheta)\,\mathrm{Re}\,\{\mathscr{R}_{xy}\exp(-\mathrm{i}\delta)\}. \tag{6.28}$$

Denoting

$$\tilde{I}^{(1)}(\nu) = \mathscr{R}_{xx}(\nu)\cos^2 \vartheta, \qquad \tilde{I}^{(2)}(\nu) = \mathscr{R}_{yy}(\nu)\sin^2 \vartheta \tag{6.29a}$$

and

$$\mu_{xy}(\nu) = |\mu_{xy}(\nu)|\exp[\mathrm{i}\beta_{xy}(\nu)] = \frac{\mathscr{R}_{xy}(\nu)}{[\mathscr{R}_{xx}(\nu)\mathscr{R}_{yy}(\nu)]^{1/2}}, \tag{6.29b}$$

we can rewrite (6.28) in the form (Pancharatnam (1963))

$$\tilde{I}(\nu) = \tilde{I}^{(1)}(\nu) + \tilde{I}^{(2)}(\nu) + 2[\tilde{I}^{(1)}(\nu)\tilde{I}^{(2)}(\nu)]^{1/2}|\mu_{xy}(\nu)|\cos[\beta_{xy}(\nu) - \delta], \tag{6.30}$$

which is the spectral interference law for partially polarized polychromatic beams; its form is very similar to the form of the interference law (3.8b) for partial coherence.

Considering quasi-monochromatic light and time delays restricted by $|\tau| \ll 1/\Delta\nu$ and also taking all device quantities in (6.30) or (6.28) at the mean frequency we obtain, by means of the Fourier transformation,

$$I(\vartheta, \delta) = \mathscr{I}_{xx}\cos^2 \vartheta + \mathscr{I}_{yy}\sin^2 \vartheta + 2\sin(\vartheta)\cos(\vartheta)\,\mathrm{Re}\,\{\mathscr{I}_{xy}\exp(-\mathrm{i}\delta)\} \tag{6.31a}$$

$$= I^{(1)} + I^{(2)} + 2[I^{(1)}I^{(2)}]^{1/2}|\gamma_{xy}|\cos(\beta_{xy} - \delta), \tag{6.31b}$$

where the correlation between the x- and y-components of the electric field is expressed by

$$\gamma_{xy} = |\gamma_{xy}| \exp(\mathrm{i}\beta_{xy}) = \frac{\mathscr{J}_{xy}}{[\mathscr{J}_{xx}\mathscr{J}_{yy}]^{1/2}}. \tag{6.32}$$

The equations (6.31a, b) represent Wolf's interference law for partially polarized quasi-monochromatic beams (Wolf (1959), Born and Wolf (1965), Mandel and Wolf (1965)), which is analogous to the interference law (3.8b) for quasi-monochromatic partially coherent light beams. The correlation coefficient γ_{xy} plays the role of the second-order degree of coherence $\gamma(\mathbf{x}_1, \mathbf{x}_2, 0)$ of the scalar theory and it expresses the correlation between the components of the electric field at a particular space point. From (6.8a) and the corresponding relation between the elements of $\mathscr{J}$ and S_j we obtain (in analogy to (4.13a) and (2.43a) for $n = 1$)

$$0 \leqq |\mu_{xy}(\nu)| \leqq 1, \tag{6.33a}$$

$$0 \leqq |\gamma_{xy}| \leqq 1. \tag{6.33b}$$

It can be shown from (6.31a) that

$$I_{\substack{\max \\ \min}(\vartheta,\delta)} = \tfrac{1}{2}\operatorname{Tr}(\mathscr{J})\left[1 \pm \left(1 - \frac{4 \operatorname{Det} \mathscr{J}}{(\operatorname{Tr} \mathscr{J})^2}\right)^{1/2}\right] \tag{6.34}$$

and the fringe visibility equals

$$\frac{I_{\max} - I_{\min}}{I_{\max} + I_{\min}} = \left(1 - \frac{4 \operatorname{Det} \mathscr{J}}{(\operatorname{Tr} \mathscr{J})^2}\right)^{1/2}. \tag{6.35}$$

This quantity depends on the rotational invariants $\operatorname{Tr} \mathscr{J}$ and $\operatorname{Det} \mathscr{J}$ only and so it is invariant during a rotation of the (x, y)-system of coordinates.

6.4 Unpolarized and polarized light

For unpolarized (natural) light $I(\vartheta, \delta)$ is a constant and consequently

$$|\gamma_{xy}| = 0, \qquad \mathscr{J}_{xx} = \mathscr{J}_{yy}, \tag{6.36}$$

that is E_x and E_y are mutually incoherent. Therefore $\mathscr{J}_{xy} = \mathscr{J}_{yx} = 0$ and the coherence matrix has the form

$$\mathscr{J} = \tfrac{1}{2}I \begin{pmatrix} 1 & 0 \\ 0 & 1 \end{pmatrix}. \tag{6.37}$$

Consider next coherent light; in this case we have

$$\mathscr{J} = \begin{pmatrix} |E_x|^2 & |E_x E_y| \exp(\mathrm{i}\delta) \\ |E_x E_y| \exp(-\mathrm{i}\delta) & |E_y|^2 \end{pmatrix}, \tag{6.38}$$

where $\delta = \arg E_y - \arg E_x$. Therefore

$$\operatorname{Det} \mathscr{J} = 0 \tag{6.39}$$

and

$$\gamma_{xy} = \exp(i\delta), \tag{6.40a}$$

that is

$$|\gamma_{xy}| = 1 \tag{6.40b}$$

and so the light is also completely polarized. The matrix (6.38) can further be specialized for linearly and circularly polarized light (Born and Wolf (1965)).

Let us now show that every coherence matrix $\mathscr{J}$ may be decomposed into a sum of the coherence matrices for completely polarized and unpolarized light $\mathscr{J}_p$ and $\mathscr{J}_u$ respectively (Born and Wolf (1965)),

$$\mathscr{J} = \mathscr{J}_p + \mathscr{J}_u, \tag{6.41}$$

where

$$\mathscr{J}_u = \begin{pmatrix} A & 0 \\ 0 & A \end{pmatrix}, \qquad \mathscr{J}_p = \begin{pmatrix} B & D \\ D^* & C \end{pmatrix} \tag{6.42}$$

with A, B, C real and non-negative and

$$BC - |D|^2 = 0. \tag{6.43}$$

It follows that

$$A + B = \mathscr{J}_{xx}, \qquad D = \mathscr{J}_{xy},$$
$$D^* = \mathscr{J}_{yx}, \quad A + C = \mathscr{J}_{yy} \tag{6.44}$$

and from (6.43) we obtain

$$(\mathscr{J}_{xx} - A)(\mathscr{J}_{yy} - A) - \mathscr{J}_{xy}\mathscr{J}_{yx} = \begin{vmatrix} \mathscr{J}_{xx} - A & \mathscr{J}_{xy} \\ \mathscr{J}_{yx} & \mathscr{J}_{yy} - A \end{vmatrix} = 0, \tag{6.45}$$

that is the eigenvalues of the coherence matrix are

$$A_{1,2} = \tfrac{1}{2}\,\mathrm{Tr}\,(\mathscr{J})\left[1 \pm \left(1 - \frac{4\,\mathrm{Det}\,\mathscr{J}}{(\mathrm{Tr}\,\mathscr{J})^2}\right)^{1/2}\right], \tag{6.46}$$

which equals (6.34). As the following inequalities hold

$$\mathrm{Det}\,\mathscr{J} \leqq \mathscr{J}_{xx}\mathscr{J}_{yy} \leqq \tfrac{1}{4}(\mathrm{Tr}\,\mathscr{J})^2, \tag{6.47}$$

both these roots are real and non-negative. The quantities B, C, D are determined from (6.44). Taking the negative sign in (6.46) (the positive sign leads to non-physical negative values of B and C), we obtain

$$A = \tfrac{1}{2}(\mathscr{J}_{xx} + \mathscr{J}_{yy}) - \tfrac{1}{2}[(\mathrm{Tr}\,\mathscr{J})^2 - 4\,\mathrm{Det}\,\mathscr{J}]^{1/2},$$
$$B = \tfrac{1}{2}(\mathscr{J}_{xx} - \mathscr{J}_{yy}) + \tfrac{1}{2}[(\mathrm{Tr}\,\mathscr{J})^2 - 4\,\mathrm{Det}\,\mathscr{J}]^{1/2},$$
$$C = \tfrac{1}{2}(\mathscr{J}_{yy} - \mathscr{J}_{xx}) + \tfrac{1}{2}[(\mathrm{Tr}\,\mathscr{J})^2 - 4\,\mathrm{Det}\,\mathscr{J}]^{1/2},$$
$$D = \mathscr{J}_{xy}, \qquad D^* = \mathscr{J}_{yx}; \tag{6.48}$$

$B, C \geqq 0$ since

$$[(\mathrm{Tr}\,\mathscr{I})^2 - 4\,\mathrm{Det}\,\mathscr{I}]^{1/2} = [(\mathscr{I}_{xx} - \mathscr{I}_{yy})^2 + 4\,|\mathscr{I}_{xy}|^2]^{1/2} \geqq |\mathscr{I}_{xx} - \mathscr{I}_{yy}|. \tag{6.49}$$

In this sense the decomposition (6.41) is unique.

6.5 Degree of polarization

The degree of polarization is defined as

$$P = \frac{\mathrm{Tr}\,\mathscr{I}_p}{\mathrm{Tr}\,\mathscr{I}_p + \mathrm{Tr}\,\mathscr{I}_u} = \frac{B + C}{B + C + 2A} = \left(1 - \frac{4\,\mathrm{Det}\,\mathscr{I}}{(\mathrm{Tr}\,\mathscr{I})^2}\right)^{1/2} =$$
$$= \left|\frac{A_1 - A_2}{A_1 + A_2}\right| = \frac{(S_1^2 + S_2^2 + S_3^2)^{1/2}}{S_0}. \tag{6.50a}$$

Introducing the correlation matrix $(\sigma_{ij}) = (\langle \Delta I_i \Delta I_j \rangle)$, ΔI_i being intensity fluctuations $I_i - \langle I_i \rangle$, we can easily prove for chaotic radiation (Carter and Wolf (1973), cf. Chapter 10) that also

$$P = \left(1 - \frac{4\,\mathrm{Det}\,\boldsymbol{\sigma}}{(\mathrm{Tr}\,\boldsymbol{\sigma})^2}\right)^{1/2}. \tag{6.50b}$$

So the degree of polarization equals the visibility (6.35). In (6.50a) we have used (6.48), (6.46) and the corresponding relations to (6.8a, b) for the Stokes parameters S_j. We see that the degree of polarization is rotationally invariant and from (6.47) it holds that

$$0 \leqq P \leqq 1. \tag{6.51}$$

For $P = 1$, $\mathrm{Det}\,\mathscr{I} = 0$, $|\gamma_{xy}| = 1$ and $S_0^2 = S_1^2 + S_2^2 + S_3^2$; in this case the light is completely polarized. For $P = 0$, $(\mathscr{I}_{xx} - \mathscr{I}_{yy})^2 + 4\,|\mathscr{I}_{xy}|^2 = 0$ (i.e. $\mathscr{I}_{xx} = \mathscr{I}_{yy}$ and $\mathscr{I}_{xy} = \mathscr{I}_{yx} = 0$, $S_1 = S_2 = S_3 = 0$) and the light is completely unpolarized ($\gamma_{xy} = 0$). In intermediate cases $0 < P < 1$ one speaks of partially polarized light. Consequently the Poincaré sphere represents completely polarized states, the centre of the Poincaré sphere represents the completely unpolarized state and inner points of the Poincaré ball (excepting the centre) represent states of partial polarization.

It may be shown (Born and Wolf (1965), Parrent and Roman (1960)) that

$$P \geqq |\gamma_{xy}| \tag{6.52}$$

and that

$$P = \mathrm{Max}\,|\gamma_{xy}|. \tag{6.53}$$

The angle of rotation φ of the system (x, y) to reach this maximum is given by

$$\tan(2\varphi) = \frac{\mathscr{I}_{yy} - \mathscr{I}_{xx}}{\mathscr{I}_{xy} + \mathscr{I}_{yx}}. \tag{6.54}$$

In this new system of coordinates (x', y') it holds that $\mathcal{J}_{x'x'} = \mathcal{J}_{y'y'} = I/2$. If (x, y) is the main system of coordinates in which $\mathcal{J}$ is diagonalized with eigenvalues A_1 and A_2 $(\gamma_{xy} = 0)$, then the system (x', y') makes with (x, y) the angle $45°$.

6.6 Probability interpretation of the eigenvalues of the coherence matrix

In analogy to the case of partial coherence and the probability interpretation of the eigenvalues of the intensity matrix (Sec. 5.5) we can give a similar interpretation to the eigenvalues A_1 and A_2 of the coherence matrix. Writing

$$\frac{1}{A_1 + A_2}\begin{pmatrix} A_1 & 0 \\ 0 & A_2 \end{pmatrix} = \frac{1}{A_1 + A_2}\begin{pmatrix} A_1 & 0 \\ 0 & 0 \end{pmatrix} + \frac{1}{A_1 + A_2}\begin{pmatrix} 0 & 0 \\ 0 & A_2 \end{pmatrix}, \qquad (6.55)$$

we see that the coherence matrix may be written as the incoherent superposition of the coherent matrices of completely polarized light. We may say that the probability p_1 that a photon belongs to the x-polarization is $p_1 = A_1/(A_1 + A_2)$ and the probability that it belongs to the y-polarization is $p_2 = A_2/(A_1 + A_2)$. The entropy will have its minimum if only one of the numbers A_1 and A_2 is not zero, which is the case of completely polarized light; it will have its maximum if $A_1 = A_2$, which is the case of completely unpolarized light. The entropy of partially polarized light will take on values between these two cases.

The eigenvalues A_1 and A_2 may be expressed in terms of P, with respect to (6.46), in the form

$$A_{1,2} = \tfrac{1}{2}I(1 \pm P). \qquad (6.56)$$

6.7 Operational definition of coherence matrix and Stokes parameters

The elements of the coherence matrix may be defined operationally from the interference law (6.31a) (Born and Wolf (1965))

$$\mathcal{J}_{xx} = I(0, 0), \qquad \mathcal{J}_{yy} = I(\pi/2, 0),$$

$$\mathcal{J}_{xy} = \tfrac{1}{2}\{I(\pi/4, 0) - I(3\pi/4, 0)\} + \tfrac{1}{2}i\{I(\pi/4, \pi/2) - I(3\pi/4, \pi/2)\},$$

$$\mathcal{J}_{yx} = \mathcal{J}_{xy}^*. \qquad (6.57)$$

The Stokes parameters S_j may be defined operationally by means of the elements of the coherence matrix, since both these methods are equivalent for the description of the second-order statistical polarization properties of optical fields. They represent a complete description for Gaussian light, for which knowledge of the second-order moment is sufficient. For non-Gaussian light moments of all orders are necessary for a complete description.

6.8 Analogy of coherence matrix and density matrix

The above equations show that in general a measurable quantity M may be represented by the trace relation

$$M = \mathrm{Tr}\,(\hat{M}\mathscr{J}), \tag{6.58}$$

where the coherence matrix $\mathscr{J}$ describes the state of the field and the operator $\hat{M}$ describes the system. This equation is completely analogous to the corresponding quantum mechanical equations with the coherence matrix replaced by the density matrix. The transformation equations for $\mathscr{R}$ [(6.15)] and for $\mathscr{J}$ are analogous to the corresponding equations for the density matrix. However, even if there is a very close analogy between properties of $\mathscr{J}$ and the density matrix $\hat{\varrho}$, this analogy is not complete (Parrent and Roman (1960)). The similarity between $\mathscr{J}$ and $\hat{\varrho}$ has been investigated on the basis of the interference law (Peřina (1963c)). The analogy between the coherence matrix and the density matrix was further extensively investigated by Fano (1957), ter Haar (1961), Parrent and Roman (1960) and Gamo (1964), among others.

A generalization of the Stokes parameters for a quasi-monochromatic but not plane waves was considered by Roman (1959). In this case the coherence matrix possesses (3×3) elements and it may be decomposed in terms of 9 linearly independent (3×3) matrices forming a complete system. The coherence matrix for N beams has been considered by Barakat (1963).

Note that an extensive treatment of partial polarization phenomena in the framework of the second-order theory, including the theory of correlation tensors given in the next chapter, has been provided by Potechin and Tatarinov (1978).

CHAPTER 7

FIELD EQUATIONS AND CONSERVATION LAWS FOR CORRELATION TENSORS

The scalar theory of coherence phenomena developed in Chapters 1 – 5 is applicable to linearly polarized light and it may serve as an approximate description of the coherence properties of light. Chapter 6 was devoted to correlations between the different polarization components of the field where we took into account the vectorial properties of the electromagnetic field, but the coincidence of the space points was assumed there. We now introduce a system of correlation tensors which provide a mathematical as well as physical framework for a unified treatment of coherence and polarization phenomena.

The correlation tensors were introduced into the theory of coherence by Wolf (1954a, 1956) and they have been extensively studied by Roman and Wolf (1960), Roman (1961a, b), Kano (1962), Beran and Parrent (1962, 1964) and Horák (1969a), using a classical non-relativistic formulation, and by Kujawski (1966) and Dialetis (1969b), using a relativistic formulation. Analogous studies of the quantum correlation tensors were carried out by Mehta and Wolf (1967a) and by Horák (1969b, 1971). Some applications of this method were proposed by Germey (1963) and Karczewski (1963a, b) (to problems of interference and diffraction) and by Bourret (1960), Mehta and Wolf (1964, 1967b) and Brevik and Suhonen (1968, 1970) (to the study of the coherence properties of blackbody radiation). Just recently the correlation theory has been employed by Carter (1980) to study properties of radiation from partially correlated current distributions and by Ross and Nieto-Vesperinas (1981) to discuss light scattering by random media.

7.1 Definition of correlation tensors and their properties

We introduce four *correlation tensors* $\mathscr{F}$, $\mathscr{H}$, $\mathscr{M}$ and $\mathscr{N}$ ordered in the matrix $\mathscr{K}$ as follows

$$\mathscr{K}_{jk}(x_1, x_2) = \left\langle \begin{pmatrix} E_j^*(x_1) \\ H_j^*(x_1) \end{pmatrix} \otimes (E_k(x_2), H_k(x_2)) \right\rangle =$$

$$= \begin{pmatrix} \mathscr{E}_{jk}(x_1, x_2) & \mathscr{M}_{jk}(x_1, x_2) \\ \mathscr{N}_{jk}(x_1, x_2) & \mathscr{H}_{jk}(x_1, x_2) \end{pmatrix}, \tag{7.1}$$

where E_j and H_j are the complex analytic signals of the components of the electric and magnetic fields respectively and

$$\mathscr{E}_{jk}(x_1, x_2) = \langle E_j^*(x_1)\, E_k(x_2) \rangle,$$

$$\mathscr{H}_{jk}(x_1, x_2) = \langle H_j^*(x_1)\, H_k(x_2)\rangle,$$
$$\mathscr{M}_{jk}(x_1, x_2) = \langle E_j^*(x_1)\, H_k(x_2)\rangle,$$
$$\mathscr{N}_{jk}(x_1, x_2) = \langle H_j^*(x_1)\, E_k(x_2)\rangle, \qquad j, k = 1, 2, 3. \tag{7.2}$$

Of course, interpreting the brackets here as the average using the density matrix with E_j and H_j represented by their corresponding operators in the normal form, we may define the same system of quantum correlation tensors (cf. Sec. 12.1). Relations of the type (2.26) and (2.28)–(2.31) hold also for the correlation tensors. Higher-order correlation tensors, which may be ordered in a matrix $\mathscr{K}^{(m,n)}$, can also be defined and some of their properties can be studied (Horák (1969b)); but since their structure is rather complicated we restrict ourselves to demonstrate the method of correlation tensors using the second-order tensors.

First we see from (7.1) that

$$\mathscr{K}_{jk}^{+}(x_1, x_2) = \mathscr{K}_{kj}(x_2, x_1) \tag{7.3}$$

and for later convenience we define the tensors

$$U_{jk}(x_1, x_2) = \mathscr{E}_{jk}(x_1, x_2) + \mathscr{H}_{jk}(x_1, x_2) = \operatorname{Tr} \mathscr{K}_{jk}(x_1, x_2), \tag{7.4a}$$

$$S_{jk}(x_1, x_2) = \mathscr{M}_{jk}(x_1, x_2) - \mathscr{N}_{jk}(x_1, x_2) = \operatorname{Tr} \{\hat{\sigma} \mathscr{K}_{jk}(x_1, x_2)\}, \tag{7.4b}$$

where

$$\hat{\sigma} = \begin{pmatrix} 0 & -1 \\ 1 & 0 \end{pmatrix}. \tag{7.5}$$

With the tensor U_{jk} we can associate the scalar U and the vector $\mathbf{U}(U_1, U_2, U_3)$ defined as follows

$$U(x_1, x_2) = \sum_{k=1}^{3} U_{kk}(x_1, x_2) = \langle \mathbf{E}^*(x_1) . \mathbf{E}(x_2)\rangle + \langle \mathbf{H}^*(x_1) . \mathbf{H}(x_2)\rangle, \tag{7.6a}$$

$$U_i(x_1, x_2) = \sum_{j,k=1}^{3} \varepsilon_{ijk} U_{jk}(x_1, x_2) = \\ = \{\langle \mathbf{E}^*(x_1) \times \mathbf{E}(x_2)\rangle + \langle \mathbf{H}^*(x_1) \times \mathbf{H}(x_2)\rangle\}_i, \tag{7.6b}$$

where . and $\times$ denote the scalar and the vector products respectively and ε_{ijk} is the Levi-Civita unit antisymmetric tensor ($\varepsilon_{ijk} = 1\ (-1)$ if (i, j, k) is an even (odd) permutation of (1, 2, 3) and $\varepsilon_{ijk} = 0$ if at least two indices are the same). We may also associate a scalar and a vector with the tensor S_{jk} in the same way

$$S(x_1, x_2) = \sum_{k=1}^{3} S_{kk}(x_1, x_2) = \langle \mathbf{E}^*(x_1) . \mathbf{H}(x_2)\rangle - \langle \mathbf{H}^*(x_1) . \mathbf{E}(x_2)\rangle, \tag{7.7a}$$

$$S_i(x_1, x_2) = \sum_{j,k=1}^{3} \varepsilon_{ijk} S_{jk}(x_1, x_2) = \\ = \{\langle \mathbf{E}^*(x_1) \times \mathbf{H}(x_2)\rangle - \langle \mathbf{H}^*(x_1) \times \mathbf{E}(x_2)\rangle\}_i. \tag{7.7b}$$

One can see that there is a simple physical meaning to the quantities (7.6a) and (7.7b). The first one multiplied by $1/4\pi$ and considered for $x_1 \equiv x_2$ ($\mathbf{x}_1 \equiv \mathbf{x}_2$, $t_1 = t_2$) represents the expectation value of the electromagnetic energy density; the second one multiplied by $c/4\pi$ or $1/4\pi c$ for $x_1 \equiv x_2$ represents the component of the Poynting vector or the component of the field momentum density in vacuo.

Some further tensors may be defined as follows

$$T_{jk}(x_1, x_2) = U_{jk}(x_1, x_2) + U_{kj}(x_1, x_2) - \delta_{jk} \sum_{i=1}^{3} U_{ii}(x_1, x_2), \tag{7.8a}$$

$$Q_{jk}(x_1, x_2) = S_{jk}(x_1, x_2) + S_{kj}(x_1, x_2) - \delta_{jk} \sum_{i=1}^{3} S_{ii}(x_1, x_2). \tag{7.8b}$$

The tensor $(1/4\pi)\, T_{jk}(x_1, x_2)\big|_{x_1 \equiv x_2}$ represents the Maxwell stress tensor of the electromagnetic field. If the quantum tensors are considered, a vacuum contribution occurs as an additive constant to these tensors as a consequence of the commutation rules for the field operators (Mehta and Wolf (1967a)). However the renormalized quantum correlation tensors in which this contribution of the vacuum is omitted are identical with the classical correlation tensors.

From the definition of these correlations and from (7.3) it follows that

$$U_{kj}(x_1, x_2) = U^*_{jk}(x_2, x_1), \tag{7.9a}$$

$$S_{kj}(x_1, x_2) = -S^*_{jk}(x_2, x_1), \tag{7.9b}$$

$$T_{kj}(x_1, x_2) = T_{jk}(x_1, x_2) = T^*_{jk}(x_2, x_1), \tag{7.9c}$$

$$Q_{kj}(x_1, x_2) = Q_{jk}(x_1, x_2) = -Q^*_{jk}(x_2, x_1). \tag{7.9d}$$

Therefore we obtain from (7.6a) and (7.9a)

$$\operatorname{Im} U(x, x) = 0, \tag{7.10a}$$

from (7.7b) and (7.9b)

$$\operatorname{Im} S_i(x, x) = 0, \tag{7.10b}$$

and from (7.8a) and (7.9c)

$$\operatorname{Im} T_{jk}(x, x) = 0. \tag{7.10c}$$

Similarly from (7.7a) and (7.9b)

$$\operatorname{Re} S(x, x) = 0, \tag{7.11a}$$

from (7.6b) and (7.9a)

$$\operatorname{Re} U_i(x, x) = 0, \tag{7.11b}$$

and from (7.8b) and (7.9d)

$$\operatorname{Re} Q_{jk}(x, x) = 0. \tag{7.11c}$$

We can easily modify all these tensors, vectors and scalars for stationary fields and all the quantities will then depend upon the time difference $\tau = t_2 - t_1$ so that the space-time dependence of the quantities is expressed by $(x_1, x_2) \equiv (\mathbf{x}_1, \mathbf{x}_2, \tau)$; if x_1 and x_2 coincide, then $(x, x) \equiv (\mathbf{x}, \mathbf{x}, 0)$. The bracket (x_2, x_1) is equivalent to $(\mathbf{x}_2, \mathbf{x}_1, -\tau)$.

7.2 Dynamical equations for correlation tensors

We begin with the Maxwell equations for the electromagnetic field in vacuo. These equations may be written, using the Levi-Civita tensor ε_{jkl}, in the form

$$\sum_{k,l} \varepsilon_{jkl} \frac{\partial}{\partial x_k} \begin{pmatrix} E_l^*(x) \\ H_l^*(x) \end{pmatrix} - \frac{\hat{\sigma}}{c} \frac{\partial}{\partial t} \begin{pmatrix} E_j^*(x) \\ H_j^*(x) \end{pmatrix} = \hat{0}, \qquad j = 1, 2, 3, \tag{7.12}$$

where the matrix $\hat{\sigma}$ is given by (7.5) and E_j and H_j are understood to be the analytic signals. The divergence conditions may be written as

$$\sum_j \frac{\partial}{\partial x_j} \begin{pmatrix} E_j^*(x) \\ H_j^*(x) \end{pmatrix} = \hat{0}. \tag{7.13}$$

Considering equations (7.12) and (7.13) for $x \equiv x_1$ and multiplying them directly by $(E_m(x_2), H_m(x_2))$ from the right and taking the average, we arrive at

$$\sum_{k,l} \varepsilon_{jkl} \frac{\partial}{\partial x_{1k}} \mathscr{K}_{lm}(x_1, x_2) - \frac{\hat{\sigma}}{c} \frac{\partial}{\partial t_1} \mathscr{K}_{jm}(x_1, x_2) = \hat{0}, \tag{7.14}$$

$$\sum_j \frac{\partial}{\partial x_{1j}} \mathscr{K}_{jm}(x_1, x_2) = \hat{0}, \tag{7.15}$$

which is the first system of dynamical equations for the correlation tensors.

Another set of dynamical equations may be obtained in the following way. We consider the Hermitian conjugate set of equations to (7.12) with $x \equiv x_2$, multiply this matrix equation directly from the left by

$$\begin{pmatrix} E_m^*(x_1) \\ H_m^*(x_1) \end{pmatrix}$$

and take the average. Performing the same with equation (7.13), we arrive at the equations

$$\sum_{k,l} \varepsilon_{jkl} \frac{\partial}{\partial x_{2k}} \mathscr{K}_{ml}(x_1, x_2) + \frac{1}{c} \frac{\partial}{\partial t_2} \mathscr{K}_{mj}(x_1, x_2)\, \hat{\sigma} = \hat{0}, \tag{7.16}$$

$$\sum_j \frac{\partial}{\partial x_{2j}} \mathscr{K}_{mj}(x_1, x_2) = \hat{0}, \tag{7.17}$$

where $\hat{\sigma}^+ = -\hat{\sigma}$ has been used.

Equations (7.14)–(7.17) can be regarded as the basic differential equations of the second-order correlation theory of the electromagnetic field, although the two sets (7.14)–(7.15) and (7.16)–(7.17) of dynamical equations are equivalent in a certain sense. If we use (7.3), we can obtain (7.16) and (7.17) from (7.14) and (7.15) if the first system is Hermitian conjugated and the following substitutions are used: $x_{1k} \to x_{2k}$, $t_1 \to t_2$, $\partial/\partial x_{1k} \to \partial/\partial x_{2k}$, $\partial/\partial t_1 \to \partial/\partial t_2$ ($\hat{\sigma}^+ = -\hat{\sigma}$). In the converse way the first system can be obtained from the second one.

Note that both systems of equations are invariant with respect to the transformations $\mathscr{K} \to \hat{\sigma}\mathscr{K}\hat{\sigma}^{-1} = \hat{\sigma}^{-1}\mathscr{K}\hat{\sigma} = -\hat{\sigma}\mathscr{K}\hat{\sigma}$ ($\hat{\sigma}^{-1} = -\hat{\sigma}$) and also the system (7.14) to (7.15) is invariant with respect to the transformation

$$\mathscr{K} \to \mathscr{K}\begin{pmatrix} 0 & 1 \\ 1 & 0 \end{pmatrix},$$

while the system (7.16)–(7.17) is invariant with respect to the transformation

$$\mathscr{K} \to \begin{pmatrix} 0 & 1 \\ 1 & 0 \end{pmatrix}\mathscr{K}.$$

Considering the first system, applying the operator $\partial/\partial x_{1j}$ to (7.14) and summing over j, we obtain

$$\sum_j \frac{\partial}{\partial x_{1j}} \mathscr{K}_{jm}(x_1, x_2) = \text{constant},$$

since the first term is zero as a consequence of the antisymmetry of ε_{jkl}. Hence, if this constant is equal to zero for $t_1 = t_0$ (i.e. the divergence condition (7.15) holds), then this condition holds for all times.

7.3 Second-order equations – wave equations – for correlation tensors

The first-order dynamical equations derived for the correlation tensors can be used to obtain second-order equations for the correlation tensors – the wave equations in vacuo.

Applying the operator $\partial/\partial(ct_1)$ to (7.14) and using (7.14) again to express the first term in the form

$$\frac{1}{c}\sum_{k,l} \varepsilon_{jkl} \frac{\partial}{\partial x_{1k}} \frac{\partial}{\partial t_1} \mathscr{K}_{lm} = \sum_{k,l,i,n} \hat{\sigma}^{-1} \varepsilon_{jkl}\varepsilon_{lin} \frac{\partial}{\partial x_{1k}} \frac{\partial}{\partial x_{1i}} \mathscr{K}_{nm}, \tag{7.18}$$

we obtain

$$\sum_{k,l,i,n} \varepsilon_{jkl}\varepsilon_{lin} \frac{\partial}{\partial x_{1k}} \frac{\partial}{\partial x_{1i}} \mathscr{K}_{nm} = -\frac{1}{c^2}\frac{\partial^2}{\partial t_1^2} \mathscr{K}_{jm}, \tag{7.19}$$

where $\hat{\sigma}^2 = -\hat{1}$ has been used. Using the identity

$$\sum_l \varepsilon_{jkl}\varepsilon_{lin} = \delta_{ji}\delta_{kn} - \delta_{jn}\delta_{ki}, \tag{7.20}$$

where δ_{jk} is the Kronecker symbol, we finally obtain from (7.19), using also (7.15),

$$\nabla_1^2 \mathscr{K}_{jm} = \frac{1}{c^2}\frac{\partial^2}{\partial t_1^2}\mathscr{K}_{jm}, \tag{7.21a}$$

which is the required wave equation for the correlation tensors. Here $\nabla_1^2 \equiv \equiv \sum_i \partial^2/\partial x_{1i}^2$ is the Laplace operator acting with respect to the coordinates of the point $\mathbf{x}_1$. One can derive the same type of wave equation with respect to the second space-time point in an identical way,

$$\nabla_2^2 \mathscr{K}_{jm} = \frac{1}{c^2}\frac{\partial^2}{\partial t_2^2}\mathscr{K}_{jm}. \tag{7.21b}$$

A further second-order equation can be obtained by applying the operator $\partial/\partial(ct_2)$ to (7.14) and using (7.16),

$$\sum_{k,l,i,n} \varepsilon_{jkl}\varepsilon_{min}\frac{\partial}{\partial x_{1k}}\frac{\partial}{\partial x_{2i}}\mathscr{K}_{ln} = \frac{1}{c^2}\frac{\partial^2}{\partial t_1\,\partial t_2}\hat{\sigma}\mathscr{K}_{jm}\hat{\sigma}^{-1}. \tag{7.22}$$

7.4 Conservation laws for correlation tensors

Taking the trace of (7.14) and (7.15) and making use of (7.4a) and (7.4b), we obtain the conservation laws

$$\sum_{k,l} \varepsilon_{jkl}\frac{\partial}{\partial x_{1k}}U_{lm} - \frac{1}{c}\frac{\partial}{\partial t_1}S_{jm} = 0, \tag{7.23a}$$

$$\sum_j \frac{\partial}{\partial x_{1j}}U_{jm} = 0. \tag{7.23b}$$

Multiplying (7.14) and (7.15) by $\hat{\sigma}$ from the left (or right) and taking the trace, we obtain another system of conservation laws

$$\sum_{k,l} \varepsilon_{jkl}\frac{\partial}{\partial x_{1k}}S_{lm} + \frac{1}{c}\frac{\partial}{\partial t_1}U_{jm} = 0, \tag{7.24a}$$

$$\sum_j \frac{\partial}{\partial x_{1j}}S_{jm} = 0. \tag{7.24b}$$

Both these systems of conservation laws may be written in the matrix form again

$$\sum_{k,l} \varepsilon_{jkl}\frac{\partial}{\partial x_{1k}}\begin{pmatrix}U\\S\end{pmatrix}_{lm} + \frac{\hat{\sigma}}{c}\frac{\partial}{\partial t_1}\begin{pmatrix}U\\S\end{pmatrix}_{jm} = \hat{0}, \tag{7.25a}$$

$$\sum_j \frac{\partial}{\partial x_{1j}}\begin{pmatrix}U\\S\end{pmatrix}_{jm} = \hat{0}. \tag{7.25b}$$

These equations are invariant with respect to the transformation

$$\begin{pmatrix} U \\ S \end{pmatrix} \to \hat{\sigma} \begin{pmatrix} U \\ S \end{pmatrix}. \tag{7.26}$$

The system of conservation laws with respect to the space-time point $(\boldsymbol{x}_2, t_2)$ may be obtained from (7.25) by Hermitian conjugation if the substitutions $x_1 \to x_2$, $\partial/\partial x_{1k} \to \partial/\partial x_{2k}$, $\partial/\partial t_1 \to \partial/\partial t_2$ and the equations (7.9a) and (7.9b) are used (or they may be obtained from (7.23) and (7.24) making use of the substitutions $S_{jm} \to -S_{mj}$, $U_{jm} \to U_{mj}$, $x_1 \to x_2$, $\partial/\partial x_{1k} \to \partial/\partial x_{2k}$ and $\partial/\partial t_1 \to \partial/\partial t_2$).

From (7.21a) and (7.21b) we obtain the wave equations for U_{jm} and S_{jm} taking the trace of (7.21a, b) and the trace of (7.21a, b) multiplied by $\hat{\sigma}$,

$$\nabla_i^2 \begin{pmatrix} U \\ S \end{pmatrix}_{jm} = \frac{1}{c^2} \frac{\partial^2}{\partial t_i^2} \begin{pmatrix} U \\ S \end{pmatrix}_{jm}, \qquad i = 1, 2. \tag{7.27}$$

Equations (7.25a, b) may be called the tensor conservation laws. Putting $j = m$ and summing over j in (7.23a) and (7.24a), we obtain

$$\nabla_1 \cdot \boldsymbol{U} = \frac{1}{c} \frac{\partial}{\partial t_1} S, \tag{7.28}$$

$$\nabla_1 \cdot \boldsymbol{S} = -\frac{1}{c} \frac{\partial}{\partial t_1} U, \tag{7.29}$$

where $U, \boldsymbol{U}$ and $S, \boldsymbol{S}$ are defined in (7.6) and (7.7). These two equations represent the scalar conservation laws.

Multiplying (7.23a) and (7.24a) by ε_{rjm} and summing over j and m, we obtain (with the use of the identity (7.20) and (7.25b))

$$\sum_j \frac{\partial}{\partial x_{1j}} T_{rj} = \frac{1}{c} \frac{\partial}{\partial t_1} S_r, \tag{7.30}$$

$$\sum_j \frac{\partial}{\partial x_{1j}} Q_{rj} = -\frac{1}{c} \frac{\partial}{\partial t_1} U_r, \tag{7.31}$$

where the quantities T_{rj}, S_r, Q_{rj} and U_r are defined by (7.8a), (7.7b), (7.8b) and (7.6b) respectively. Equations (7.30) and (7.31) represent the vectorial conservation laws. The second set of the conservation laws may be obtained from (7.28)–(7.31) by the substitutions $\nabla_1 \to \nabla_2$, $\partial/\partial x_{1j} \to \partial/\partial x_{2j}$ and $\partial/\partial t_1 \to \partial/\partial t_2$.

The conservation laws just derived are linear since they contain the correlation tensors linearly. Some nonlinear conservation laws, which involve the correlation tensors quadratically, have been derived by Roman (1961b) and Horák (1969a).

7.5 Stationary fields

We have noted at the end of Sec. 7.1 that the corresponding equations for stationary fields may be obtained by the substitutions $(x_1, x_2) \to (\mathbf{x}_1, \mathbf{x}_2, \tau)$, $\tau = t_2 - t_1((x_2, x_1) \to (\mathbf{x}_2, \mathbf{x}_1, -\tau)$, $(x, x) \to (\mathbf{x}, \mathbf{x}, 0)$, $-\partial/\partial t_1 = \partial/\partial t_2 = \partial/\partial\tau)$. For example we obtain from (7.14) and (7.16)

$$\sum_{k,l} \varepsilon_{jkl} \frac{\partial}{\partial x_{1k}} \mathscr{K}_{lm}(\mathbf{x}_1, \mathbf{x}_2, \tau) + \frac{\hat{\sigma}}{c} \frac{\partial}{\partial \tau} \mathscr{K}_{jm}(\mathbf{x}_1, \mathbf{x}_2, \tau) = \hat{0}, \tag{7.32}$$

$$\sum_{k,l} \varepsilon_{jkl} \frac{\partial}{\partial x_{2k}} \mathscr{K}_{ml}(\mathbf{x}_1, \mathbf{x}_2, \tau) + \frac{1}{c} \frac{\partial}{\partial \tau} \mathscr{K}_{mj}(\mathbf{x}_1, \mathbf{x}_2, \tau)\, \hat{\sigma} = \hat{0}. \tag{7.33}$$

The wave equations become in this case

$$\nabla_i^2 \mathscr{K}_{jm} = \frac{1}{c^2} \frac{\partial^2}{\partial \tau^2} \mathscr{K}_{jm}, \qquad i = 1, 2 \tag{7.34}$$

and from (7.22) we have

$$\sum_{k,l,i,n} \varepsilon_{jkl} \varepsilon_{min} \frac{\partial}{\partial x_{1k}} \frac{\partial}{\partial x_{2i}} \mathscr{K}_{ln} = -\frac{1}{c^2} \frac{\partial^2}{\partial \tau^2} \hat{\sigma} \mathscr{K}_{jm} \hat{\sigma}^{-1}. \tag{7.35}$$

The conservation laws (7.28)–(7.31) reduce to

$$\nabla_i \cdot \mathbf{U} = \mp \frac{1}{c} \frac{\partial}{\partial \tau} S, \tag{7.36}$$

$$\nabla_i \cdot \mathbf{S} = \pm \frac{1}{c} \frac{\partial}{\partial \tau} U, \tag{7.37}$$

$$\sum_j \frac{\partial}{\partial x_{ij}} T_{mj} = \mp \frac{1}{c} \frac{\partial}{\partial \tau} S_m, \tag{7.38}$$

$$\sum_j \frac{\partial}{\partial x_{ij}} Q_{mj} = \pm \frac{1}{c} \frac{\partial}{\partial \tau} U_m. \tag{7.39}$$

Here the upper and lower signs are taken according as i takes on the value 1 or 2 respectively.

It has been shown by Roman and Wolf (1960) and Mehta and Wolf (1967a) that the standard conservation laws in the averaged forms follow from the present conservation laws as special cases. Thus from (7.37) one can obtain the averaged form of the energy conservation law, and from (7.38) the averaged form of the momentum conservation law of the electromagnetic field is obtained if $\mathbf{x}_1 \to \mathbf{x}_2$, $\tau \to 0$ and the real part of (7.37) and (7.38) is taken. In this sense the conservation laws (7.36) and (7.39) have no classical analogues, since they reduce to the identity $0 = 0$ as a consequence of (7.11a, b, c).

7.6 Cross-spectral tensors

The cross-spectral tensors $\tilde{\mathscr{E}}_{jk}(\mathbf{x}_1, \mathbf{x}_2, \nu)$, $\tilde{\mathscr{H}}_{jk}(\mathbf{x}_1, \mathbf{x}_2, \nu)$, $\tilde{\mathscr{M}}_{jk}(\mathbf{x}_1, \mathbf{x}_2, \nu)$, $\tilde{\mathscr{N}}_{jk}(\mathbf{x}_1, \mathbf{x}_2, \nu)$, as well as $\tilde{U}_{jk}(\mathbf{x}_1, \mathbf{x}_2, \nu)$, $\tilde{S}_{jk}(\mathbf{x}_1, \mathbf{x}_2, \nu)$, arranged in the spectral matrix $\tilde{\mathscr{K}}_{jk}(\mathbf{x}_1, \mathbf{x}_2, \nu)$, may be introduced by the relation

$$\tilde{\mathscr{K}}_{jk}(\mathbf{x}_1, \mathbf{x}_2, \nu) = \int_{-\infty}^{+\infty} \mathscr{K}_{jk}(\mathbf{x}_1, \mathbf{x}_2, \tau) \exp(\mathrm{i}\, 2\pi\nu\tau)\, \mathrm{d}\tau. \tag{7.40}$$

In agreement with the spectral analysis of general correlation functions for stationary fields contained in Sec. 2.2 (equation (2.23)) we can obtain

$$\langle \tilde{E}_j^*(\mathbf{x}_1, \nu)\, \tilde{E}_k(\mathbf{x}_2, \nu') \rangle = \tilde{\mathscr{E}}_{jk}(\mathbf{x}_1, \mathbf{x}_2, \nu)\, \delta(\nu - \nu'), \tag{7.41}$$

where $\tilde{E}_j(\mathbf{x}, \nu)$ is the spectral component of $E_j(\mathbf{x}, t)$. Thus we can conclude that two spectral components of a stationary electric field are correlated if and only if the frequencies coincide. Of course, the same conclusions are valid for the other correlation tensors of stationary fields.

The inverse form to (7.40),

$$\mathscr{K}_{jk}(\mathbf{x}_1, \mathbf{x}_2, \tau) = \int_{0}^{\infty} \tilde{\mathscr{K}}_{jk}(\mathbf{x}_1, \mathbf{x}_2, \nu) \exp(-\mathrm{i}\, 2\pi\nu\tau)\, \mathrm{d}\nu, \tag{7.42}$$

expresses the Wiener-Khintchine theorem (cf. (2.24)). If this theorem is used, arguments like those above lead to the corresponding dynamical equations and the conservation laws for the cross-spectral tensors.

From (7.32) and (7.33) we obtain

$$\sum_{k,l} \varepsilon_{jkl} \frac{\partial}{\partial x_{1k}} \tilde{\mathscr{K}}_{lm}(\mathbf{x}_1, \mathbf{x}_2, \nu) - \frac{\mathrm{i}\, 2\pi\nu\hat{\sigma}}{c} \tilde{\mathscr{K}}_{jm}(\mathbf{x}_1, \mathbf{x}_2, \nu) = \hat{0}, \tag{7.43}$$

$$\sum_{k,l} \varepsilon_{jkl} \frac{\partial}{\partial x_{2k}} \tilde{\mathscr{K}}_{ml}(\mathbf{x}_1, \mathbf{x}_2, \nu) - \frac{\mathrm{i}\, 2\pi\nu}{c} \tilde{\mathscr{K}}_{mj}(\mathbf{x}_1, \mathbf{x}_2, \nu)\, \hat{\sigma} = \hat{0}, \tag{7.44}$$

and the divergence conditions give

$$\sum_{j} \frac{\partial}{\partial x_{1j}} \tilde{\mathscr{K}}_{jm}(\mathbf{x}_1, \mathbf{x}_2, \nu) = \hat{0}, \tag{7.45}$$

$$\sum_{j} \frac{\partial}{\partial x_{2j}} \tilde{\mathscr{K}}_{mj}(\mathbf{x}_1, \mathbf{x}_2, \nu) = \hat{0}. \tag{7.46}$$

The wave equations reduce to the Helmholtz equations

$$\nabla_i^2 \tilde{\mathscr{K}}_{jm}(\mathbf{x}_1, \mathbf{x}_2, \nu) + \frac{4\pi^2\nu^2}{c^2} \tilde{\mathscr{K}}_{jm}(\mathbf{x}_1, \mathbf{x}_2, \nu) = \hat{0}, \qquad i = 1, 2. \tag{7.47}$$

7.7 Non-negative definiteness conditions for correlation tensors and cross-spectral tensors

Just as the non-negative definiteness condition (4.7a) was derived for the correlation function $\Gamma(x_1, x_2)$, analogous non-negative definiteness conditions may be derived for the cross-spectral and correlation tensors (Mehta and Wolf (1967a), Horák (1971)).

Introducing the quantity

$$A = \sum_i \int [f_i(x)\, E_i(x) + g_i(x)\, H_i(x)]\, d^4x, \tag{7.48}$$

we obtain

$$0 \leqq \langle |A|^2 \rangle = \sum_{i,j} \iint [f_i^*(x_1)\, \mathscr{E}_{ij}(x_1, x_2)\, f_j(x_2) + g_i^*(x_1)\, \mathscr{H}_{ij}(x_1, x_2)\, g_j(x_2) + \\ + f_i^*(x_1)\, \mathscr{M}_{ij}(x_1, x_2)\, g_j(x_2) + g_i^*(x_1)\, \mathscr{N}_{ij}(x_1, x_2) f_j(x_2)]\, d^4x_1\, d^4x_2, \tag{7.49}$$

where $f_i(x)$ and $g_i(x)$, $i = 1, 2, 3$ are arbitrary functions of space-time points $x = (\mathbf{x}, t)$ for which the integrals exist. A number of special conditions may be derived by choosing some special functions f and g. Thus if $g \equiv 0$ we obtain the corresponding non-negative definiteness condition for the correlation tensor $\mathscr{E}_{ij}$; if $f \equiv 0$ we obtain such a condition for $\mathscr{H}_{ij}$. Putting

$$f_i(x) = \sum_{m=1}^{M} \sum_{k=1}^{N} \alpha_{mki} \delta(\mathbf{x}_m - \mathbf{x})\, \delta(t_k - t), \tag{7.50}$$

$$g_i(x) = \sum_{m=1}^{M} \sum_{k=1}^{N} \beta_{mki} \delta(\mathbf{x}_m - \mathbf{x})\, \delta(t_k - t), \tag{7.51}$$

$i = 1, 2, 3$, we obtain from (7.49)

$$\sum_{\substack{m,k,n,l,\\ i,j}} [\alpha^*_{mki} \mathscr{E}_{ij}(\mathbf{x}_m, \mathbf{x}_n, t_k, t_l)\, \alpha_{nlj} + \beta^*_{mki} \mathscr{H}_{ij}(\mathbf{x}_m, \mathbf{x}_n, t_k, t_l)\, \beta_{nlj} + \\ + \alpha^*_{mki} \mathscr{M}_{ij}(\mathbf{x}_m, \mathbf{x}_n, t_k, t_l)\, \beta_{nlj} + \beta^*_{mki} \mathscr{N}_{ij}(\mathbf{x}_m, \mathbf{x}_n, t_k, t_l)\, \alpha_{nlj}] \geqq 0. \tag{7.52}$$

Corresponding conditions for stationary fields follow by introducing $\tau = t_2 - t_1$ in (7.49) and (7.52).

Introducing the quantity

$$\tilde{A} = \sum_i \int_{\nu - \Delta\nu/2}^{\nu + \Delta\nu/2} d\nu\, d^3x [f_i(\mathbf{x})\, \tilde{E}_i(\mathbf{x}, \nu) + g_i(\mathbf{x})\, \tilde{H}_i(\mathbf{x}, \nu)], \tag{7.53}$$

where $\tilde{H}$ has a similar significance to $\tilde{E}$ and $(\nu - \Delta\nu/2, \nu + \Delta\nu/2)$ is an arbitrarily small frequency interval, we obtain in a similar way the non-negative definiteness condition for the cross-spectral tensors

$$\sum_{i,j} \iint d^3x_1\, d^3x_2 [f_i^*(\mathbf{x}_1)\, \tilde{\mathscr{E}}_{ij}(\mathbf{x}_1, \mathbf{x}_2, \nu) f_j(\mathbf{x}_2) + g_i^*(\mathbf{x}_1)\, \tilde{\mathscr{H}}_{ij}(\mathbf{x}_1, \mathbf{x}_2, \nu)\, g_j(\mathbf{x}_2) + \\ + f_i^*(\mathbf{x}_1)\, \tilde{\mathscr{M}}_{ij}(\mathbf{x}_1, \mathbf{x}_2, \nu)\, g_j(\mathbf{x}_2) + g_i^*(\mathbf{x}_1)\, \tilde{\mathscr{N}}_{ij}(\mathbf{x}_1, \mathbf{x}_2, \nu) f_j(\mathbf{x}_2)] \geqq 0. \tag{7.54}$$

Finally let us note that the results derived for the correlation tensors are independent of the kind of averaging (they characterize the space-time behaviour of the field) and consequently they are also valid for the quantum correlation tensors, if the contribution of the physical vacuum is subtracted (Mehta and Wolf (1967a)). The present results can be extended to the case of the existence of random sources in the space (Roman (1961a), Beran and Parrent (1962, 1964), Horák (1969a)) and from many further results obtained in this field we mention the following. The general propagation laws, including both the interference laws for partial coherence and partial polarization, have been derived by Horák (1969a) on the basis of the differential equations for the correlation tensors. A theory of the correlation properties of optical fields using the correlation tensors of arbitrary order has been developed by Horák (1969b, 1971). All these formulations of the correlation theory were non-relativistic, which is the typical case in quantum optics applications. A relativistic correlation theory based on the use of the anti-symmetric electromagnetic tensor $F_{\alpha\beta}(x)$ has been developed by Kujawski (1966) and Dialetis (1969b).

The technique of correlation tensors proved to be useful in connection with studies of the properties of turbulent fluids and plasmas.

CHAPTER 8

GENERAL CLASSICAL STATISTICAL DESCRIPTION OF THE FIELD

In Section 2.2 we introduced both the classical and quantum correlation functions of arbitrary order, which are the fundamental quantities in the present treatment of the coherence of light. We now give a more precise classical definition in terms of classical stochastic processes, whereas a more precise specification of the averaging brackets from a quantum-mechanical point of view will be discussed in Chapters 12 and 14 together with the correspondence between these two approaches.

All experimental results reached in optical coherence, interference, diffraction and polarization experiments prior to about 1955 can be described in terms of the second-order coherence theory which we have developed in the preceding chapters. Since that time, a number of unconventional experiments have been carried out or proposed. For their description a systematic broadening of the second-order coherence theory is necessary. From these experiments we mention the following: the experiments of Forrester, Gudmundsen and Johnson (1955) on beats between light from independent sources; the experiments of Hanbury Brown and Twiss (1956a – c, 1957a, b, 1958) on photon correlations and intensity interferometry; the experiments of Magyar and Mandel (1963, 1964) on transient interference effects with light from independent laser sources; experiments measuring photon statistics or alternatively the statistics of emitted photoelectrons by a photodetector (Chapter 10); experiments to measure the sixth-order correlation function by Davidson and Mandel (1968), Davidson (1969) and Corti and Degiorgio (1974, 1976a, b); an experiment on the interference of individual photons and of independent optical beams by Pfleegor and Mandel (1967a, b, 1968) and Radloff (1968, 1971); and experiments by Kimble, Dagenais and Mandel (1977, 1978) and Dagenais and Mandel (1978) on antibunching of photons and by Short and Mandel (1983) and Teich and Saleh (1985) on sub-Poisson light. Such a general formulation of the theory of coherence phenomena based on the classical theory of stochastic functions was first proposed by Wolf (1963a, 1964) and Mandel (1964d), while a theory based on quantum electrodynamics was introduced by Glauber (1963a, b, 1964).

8.1 Stochastic description of light

Consider an optical field represented by the vector analytic wave amplitude $\mathbf{V}(\mathbf{x}_i, t_i) \equiv \mathbf{V}_i$ at a space-time point $(\mathbf{x}_i, t_i)$. The stochastic behaviour of the vectorial field $\mathbf{V}(\mathbf{x}, t)$ can be described by the sequence of probability densities

$$P_1(\mathbf{V}_1), P_2(\mathbf{V}_1, \mathbf{V}_2), \ldots, P_n(\mathbf{V}_1, \mathbf{V}_2, \ldots, \mathbf{V}_n), \ldots \tag{8.1}$$

having the following significance. The joint n-fold probability distribution

$$P_n(\mathbf{V}_1, \mathbf{V}_2, \ldots, \mathbf{V}_n)\, \mathrm{d}^2\mathbf{V}_1\, \mathrm{d}^2\mathbf{V}_2 \ldots \mathrm{d}^2\mathbf{V}_n, \tag{8.2}$$

where $\mathrm{d}^2\mathbf{V}_j = \mathrm{d}(\mathrm{Re}\, \mathbf{V}_j)\, \mathrm{d}(\mathrm{Im}\, \mathbf{V}_j)$, $j = 1, 2, \ldots, n$, represents the joint n-fold probability that at the space-time point $(\mathbf{x}_1, t_1)$ the quantities $\mathrm{Re}\, \mathbf{V}$ and $\mathrm{Im}\, \mathbf{V}$ lie in the intervals $(\mathrm{Re}\, \mathbf{V}_1, \mathrm{Re}\, \mathbf{V}_1 + \mathrm{d}(\mathrm{Re}\, \mathbf{V}_1))$ and $(\mathrm{Im}\, \mathbf{V}_1, \mathrm{Im}\, \mathbf{V}_1 + \mathrm{d}(\mathrm{Im}\, \mathbf{V}_1))$ respectively, etc., and at the space-time point $(\mathbf{x}_n, t_n)$ they lie in the intervals $(\mathrm{Re}\, \mathbf{V}_n, \mathrm{Re}\, \mathbf{V}_n + \mathrm{d}(\mathrm{Re}\, \mathbf{V}_n))$ and $(\mathrm{Im}\, \mathbf{V}_n, \mathrm{Im}\, \mathbf{V}_n + \mathrm{d}(\mathrm{Im}\, \mathbf{V}_n))$ respectively.

If $F(\mathbf{V}_1, \ldots, \mathbf{V}_n)$ represents a function depending on the values of the field $\mathbf{V}$ at the points $(\mathbf{x}_1, t_1), \ldots, (\mathbf{x}_n, t_n)$, we may define the ensemble average of F by the relation†

$$\langle F(\mathbf{V}_1, \ldots, \mathbf{V}_n)\rangle = \int_{-\infty}^{+\infty} \ldots \int_{-\infty}^{+\infty} F(\mathbf{V}_1, \ldots, \mathbf{V}_n)\, P_n(\mathbf{V}_1, \ldots, \mathbf{V}_n)\, \mathrm{d}^2\mathbf{V}_1 \ldots \mathrm{d}^2\mathbf{V}_n. \tag{8.3}$$

Let us note that P_n is a probability distribution of $6n$ variables and the integrals in (8.3) extend over $3n$ complex planes. The complex representation of the field is very convenient for a description of measurements of the statistical properties of optical fields by means of photoelectric detectors. It represents a bridge between the classical and quantum descriptions of the coherence properties of light, even if, since the specification of the complex field $\mathbf{V}$ is equivalent to the simultaneous specifications of the amplitude and the phase of the wave, it might appear that $P_n(\mathbf{V}_1, \ldots, \mathbf{V}_n)$ has no analogue in the description of the quantized field.

In the theory of coherence correlation measurements involve the higher-order moments – correlation functions or correlation tensors – the classical form of which is the particular case of (8.3),

$$\Gamma^{(m,n)}_{\mu_1 \ldots \mu_{m+n}}(x_1, \ldots, x_{m+n}) = \langle \prod_{j=1}^{m} V^*_{\mu_j}(x_j) \prod_{k=m+1}^{m+n} V_{\mu_k}(x_k)\rangle, \tag{8.4}$$

where $x \equiv (\mathbf{x}, t)$, as we have seen in Chapter 2. A special case of stationary and ergodic fields, for which the probability distributions are invariant with respect to translations of the time origin so that the ensemble average is equal to the time average, has also been discussed in Chapter 2.

In the terminology of this chapter some experiments determining the fourth- and sixth-order correlation functions have been carried out, as mentioned above.

---

† For a complete description of the statistics of the field we must put $n \to \infty$ and the ensemble average of F will be given by the functional integral

$$\langle F\rangle = \int_{-\infty}^{+\infty} \ldots \int_{-\infty}^{+\infty} F\{\mathbf{V}(x)\}\, P\{\mathbf{V}(x)\} \prod_x \mathrm{d}^2\mathbf{V}(x),$$

where $x \equiv (\mathbf{x}, t)$ (see e.g. Beran and Parrent (1964); Sec. 13.3).

Some further experiments provide the distribution of photons in the field, and this is related to distributions of the form (8.1); in this way one may also calculate the higher-order correlation functions from experimental data.

8.2 Functional formulation

We will not go into all details of the functional theory here. We only introduce the characteristic functional by the relation

$$C\{y(x)\} = \int \exp\{\int [y(x)\, V^*(x) - y^*(x)\, V(x)]\, \mathrm{d}^4x\}\, P\{V(x)\} \prod_x \mathrm{d}^2 V(x), \quad (8.5)$$

where $P\{V(x)\}$ is the probability functional and x is a concise notation for (x, μ) $(\int \ldots \mathrm{d}^4x \equiv \sum_\mu \int \ldots \mathrm{d}^4x)$. The correlation function (8.4) can be obtained as the functional derivative of the characteristic functional

$$\prod_{j=1}^{m} \frac{\delta}{\delta y(x_j)} \prod_{k=m+1}^{m+n} \frac{\delta}{\delta(-y^*(x_k))} C\{y(x)\}\,\Big|_{y(x)\equiv 0} = \langle \prod_{j=1}^{m} V^*(x_j) \prod_{k=m+1}^{m+n} V(x_k)\rangle, \quad (8.6)$$

where the functional derivative is defined by

$$\frac{\delta F}{\delta y(x_0)} = \lim_{\varepsilon\to 0} \frac{1}{\varepsilon} [F\{y(x) + \varepsilon\delta(x - x_0)\} - F\{y(x)\}]; \quad (8.7)$$

F is a continuous functional and $\delta(x)$ is the four-dimensional Dirac function. A quantum analogue of these equations is given in Sec. 12.3.

A general mathematical treatment of the functional theory can be found in the book by Volterra (1959) (see also Gelfand and Yaglom (1956), Gelfand and Vilenkin (1964), Chapter 3). Some physical applications of the functional theory to the study of the coherence properties of light are available (Beran and Parrent (1964), Peřina (1965a, b), Ingarden (1965)). Numerous applications exist in the field theory (Feynman (1948), Feynman and Hibbs (1965), Bogolyubov and Shirkov (1959), Akhiezer and Berestetsky (1965), Gelfand and Minlos (1954)). A systematic approach to the coherence theory and the photocount statistics based on the theory of functionals has been developed by Zardecki (1969a, 1971, 1974) and its application to laser theory has been suggested by Arecchi, Asdente and Ricca (1976).

CHAPTER 9

SOME APPLICATIONS OF THE THEORY OF COHERENCE TO OPTICAL IMAGING

In this chapter we apply some of the results of earlier chapters to the determination of the relation between an object and its image for optical systems which image extended polychromatic objects. Although only the second-order coherence effects are usually considered in such an analysis of optical imaging processes (Parrent (1961), Beran and Parrent (1964), Hopkins (1953, 1957a, b), Born and Wolf (1965), Maréchal and Francon (1960)), we can, without particular difficulty, perform the analysis including coherence effects of arbitrary order. Such an analysis provides a description of optical imaging with partially coherent non-chaotic (e.g. laser) light (Peřina (1969), Peřina and Peřinová (1969)), since the complete specification of the statistics of such light demands the knowledge of moments of all orders, while for chaotic light the second-order moment is a sufficient specification (Sec. 17.1). Special problems related to the general relation between an object and its image having practical significance are (i) the reconstruction of the object when the image is given (the image may not resemble the object as a consequence of the diffraction properties and of aberrations) and (ii) the similarity between an object and its image (Peřina (1963b), Peřina and Peřinová (1965, 1969)), which characterizes the imaging process with the nonlinearity typical of partially coherent light.

9.1 Spatial Fourier analysis of optical imaging – transfer functions for partially coherent light

An important characteristic of an optical system is the transfer function, the Fourier transform of the diffraction function of the system if coherent light is considered. This describes the ability of the system to transfer spatial frequencies – the number of lines per unit length on a test object. The transfer function may be introduced under the assumption of a linear and stationary system (in space) and in this case the optical system may be treated as a spatial frequency filter, in terms of the theory of electric circuits. Since the relation between an object and its image has the form of a convolution of the object function with the diffraction function assuming coherent light, the frequency spectrum of the image $\tilde{g}(\mu)$ is equal to the product of the transfer function $\tilde{K}(\mu)$ and the spatial frequency function of the object $\tilde{f}(\mu)$. The dimensions of apertures in optical systems are finite so that the transfer function is zero outside a finite region. Consequently the system filters out higher spatial frequencies and these cannot be present in the image, which decreases its quality. Therefore the object cannot be reconstructed uniquely from the relation $\tilde{f}(\mu) = \tilde{g}(\mu)/\tilde{K}(\mu)$ by means of a Fourier transformation, since $\tilde{f}(\mu) = 0/0$ in some regions and consequently the values of $\tilde{f}(\mu)$ for all μ cannot be deduced. In connection with this deconvolution method some possibilities of obtaining super-resolution, using analytic continuation to determine the uncertainty 0/0, have been discussed (Peřina (1971), Frieden (1971), Peřina, Peřinová and Braunerová (1977), Schmidt-Weinmar, Steinle and Baltes (1978/79), Schmidt-Weinmar (1978)).

Calculation of the transfer functions usually involves numerical integration (Hopkins (1957b), O'Neil (1963)) and the analysis was earlier normally limited to strictly coherent or incoherent light. The transfer function for partially coherent objects serves now as a usual characteristic of the quality of optical systems (see e.g. Steel (1957), Möller (1968)). In the first stage all analysis was performed for quasi-monochromatic light (Hopkins (1953, 1957a), Born and Wolf (1965), Maréchal and Francon (1960)). This restriction was removed by Parrent (1961) and Beran and Parrent (1964) who developed an analysis applicable to polychromatic light. However, this analysis does not in general take into account the detection of light in the process of imaging. A general analysis of optical imaging for polychromatic light and for coherence effects of arbitrary order was also developed and applied (Peřina (1969), Peřina and Peřinová (1969)).

An optical system regarded as a spatial filter may be compared with a temporal filter in electronics. If an electronic system is described by the response function $h(t)$ at time t and if the input and output are described by the functions $f(t)$ and $g(t)$ respectively, then

$$g(t') = \hat{L}f(t) = \int_{-\infty}^{+\infty} h(t' - t) f(t)\, \mathrm{d}t, \tag{9.1}$$

where the integral operator $\hat{L}$ is obviously linear and the system is stationary since the response function depends on the time difference $(t' - t)$. The physical condition expressing the causality reads

$$h(t) = 0, \qquad t < 0 \tag{9.2}$$

and the stability condition demands that

$$\int_{-\infty}^{+\infty} |h(t)|\, \mathrm{d}t < \infty, \tag{9.3}$$

since from (9.1) it follows that

$$|g(t')| \leqq \int_{-\infty}^{+\infty} |h(t' - t)|\, |f(t)|\, \mathrm{d}t \leqq \mathrm{Max}\, |f(t)| \int_{-\infty}^{+\infty} |h(t)|\, \mathrm{d}t < \infty,$$

i.e. the response to a finite input is finite.

All these conditions may be realized in the imaging process except the condition (9.2) which has no spatial analogue since the distribution of light exists on both sides of the axis of an optical system (we are dealing with the space variable x instead of the time variable t in this case). The stationary condition holds for optical systems in the isoplanatic regions. A typical relation between the object and its image is two-dimensional and it may be written as

$$g(x', y') = \iint_{-\infty}^{+\infty} K(x' - x, y' - y) f(x, y)\, \mathrm{d}x\, \mathrm{d}y, \tag{9.4}$$

where (x, y) and (x', y') are coordinates in the object and image planes respectively and K is the diffraction function. The condition (9.2) may sometimes be replaced by

$$f(x, y) = 0, \qquad x, y \notin \mathscr{A}, \tag{9.5}$$

where $\mathscr{A}$ is a region of the object plane. Equation (9.5) expresses the fact that objects are usually finite in extension. The transfer function $\tilde{K}(\mu, \nu)$ may be defined by using a Fourier transform as

$$\tilde{g}(\mu, \nu) = \tilde{K}(\mu, \nu) \tilde{f}(\mu, \nu), \tag{9.6}$$

where μ, ν are spatial frequencies and

$$\tilde{K}(\mu, \nu) = \iint_{-\infty}^{+\infty} K(x, y) \exp\left[\mathrm{i}\, 2\pi(\mu x + \nu y)\right] \mathrm{d}x\, \mathrm{d}y = \frac{\tilde{g}(\mu, \nu)}{\tilde{f}(\mu, \nu)}, \tag{9.7}$$

under the assumption that $\tilde{f}(\mu, \nu) \not\equiv 0$ for all μ, ν ($\tilde{K}(\mu, \nu)$ may be zero at isolated points, $\tilde{f} = \tilde{g}/\tilde{K}$ being then defined from the continuity requirement).

Such a formulation is useful for coherent light and, if K is replaced by $|K|^2$ and the functions f, g represent the intensities, for incoherent light. If the light is partially coherent, we must begin with the propagation law (3.21) for the mutual spectral density $G(\boldsymbol{x}_1, \boldsymbol{x}_2, \nu)$. More generally, solving a boundary problem for the wave equations (2.35) (or (2.37) if a stationary field is assumed) we obtain instead of (3.29)

$$G^{(n,n)}(Q_1, \ldots, Q_{2n}, \nu) = \frac{1}{(2\pi)^{2n}} \int_{\mathscr{A}} \cdots \int_{\mathscr{A}} G^{(n,n)}(P_1, \ldots, P_{2n}, \nu) \times$$
$$\times R(P_1, \ldots, P_{2n}, Q_1, \ldots, Q_{2n}, \nu) \prod_{j=1}^{n} \mathrm{d}P_j\, \mathrm{d}P_{n+j}, \tag{9.8}$$

where

$$R(P_1, \ldots, P_{2n}, Q_1, \ldots, Q_{2n}, \nu) = \prod_{j=1}^{n} (1 + \mathrm{i}k_j r_j)(1 - \mathrm{i}k_{n+j} r_{n+j}) \times$$
$$\times \frac{\exp\left[-\mathrm{i}(k_j r_j - k_{n+j} r_{n+j})\right]}{r_j^2 r_{n+j}^2} \cos(\vartheta_j) \cos(\vartheta_{n+j}).$$

Considering this propagation law on the hypersurface (2.21) (i.e. $\nu_1 = -\sum_{j=2}^{2n} \varepsilon_j \nu_j$) with $k \gg 1/r$, we obtain using (2.19)

$$\Gamma^{(n,n)}(Q_1, \ldots, Q_{2n}, \tau) = \frac{1}{(2\pi)^{2n}} \int_{\mathscr{A}} \cdots \int_{\mathscr{A}} \Gamma^{(n,n)}(P_1, \ldots, P_{2n}, \tau_2', \ldots, \tau_{2n}') \times$$
$$\times \prod_{j=1}^{n} \frac{\bar{\Lambda}_j^* \bar{\Lambda}_{n+j}}{r_j r_{n+j}} \mathrm{d}P_j \, \mathrm{d}P_{n+j} = \frac{1}{(2\pi)^{2n}} \int_{\mathscr{A}} \cdots \int_{\mathscr{A}} \gamma^{(n,n)}(P_1, \ldots, P_{2n}, \tau_2', \ldots, \tau_{2n}') \times$$
$$\times \prod_{j=1}^{n} \frac{[I^{(n)}(P_j) \, I^{(n)}(P_{n+j})]^{1/2n}}{r_j r_{n+j}} \bar{\Lambda}_j^* \bar{\Lambda}_{n+j} \, \mathrm{d}P_j \, \mathrm{d}P_{n+j}, \tag{9.9}$$

where

$$\tau_j' = \tau_j + \frac{r_1 - r_j}{c}, \qquad j = 2, 3, \ldots, 2n.$$

Here (2.42a) has been used and $I^{(n)}(P) \equiv \Gamma^{(n,n)}(P, \ldots, P, 0, \ldots, 0)$. Equation (9.9) is a generalization of the propagation law (3.34). Under the usual assumptions analogous to (3.43) the corresponding propagation laws can be obtained for quasi-monochromatic light.

Linear process of imaging

A still more general propagation law, valid in non-homogeneous media, may be written in the form

$$^{i}G^{(m,n)}(\mathbf{x}_1', \ldots, \mathbf{x}_{m+n}', \mathbf{v}) = \int_{\mathscr{A}} \cdots \int_{\mathscr{A}} \prod_{j=1}^{m} K^*(\mathbf{x}_j' - \mathbf{x}_j, \nu_j) \times$$
$$\times \prod_{k=m+1}^{m+n} K(\mathbf{x}_k' - \mathbf{x}_k, \nu_k) \, ^{o}G^{(m,n)}(\mathbf{x}_1, \ldots, \mathbf{x}_{m+n}, \mathbf{v}) \prod_{l=1}^{m+n} \mathrm{d}\mathbf{x}_l, \tag{9.10}$$

where ^{o}G and ^{i}G are the object and image spectral correlation functions respectively; the equation is considered in the isoplanatic region so that K depends on the difference $(\mathbf{x}' - \mathbf{x})$. If the field is assumed to be stationary, (9.10) must be considered on the hypersurface (2.21) again. The image correlation function $^{i}\Gamma^{(m,n)}$ may then be obtained by means of a Fourier transformation.

The transfer functions may also be introduced to describe the quality of the transfer properties of the optical system for partially coherent polychromatic light of arbitrary statistical behaviour. As (9.10), relating the object and image characteristics, is linear we obtain, using the Fourier transformation,

$$^{i}\tilde{G}^{(m,n)}(\boldsymbol{\mu}_1, \ldots, \boldsymbol{\mu}_{m+n}, \mathbf{v}) =$$
$$= \prod_{j=1}^{m} \tilde{K}^*(-\boldsymbol{\mu}_j, \nu_j) \prod_{k=m+1}^{m+n} \tilde{K}(\boldsymbol{\mu}_k, \nu_k) \, ^{o}\tilde{G}^{(m,n)}(\boldsymbol{\mu}_1, \ldots, \boldsymbol{\mu}_{m+n}, \mathbf{v}), \tag{9.11}$$

where $\boldsymbol{\mu}$ is the spatial frequency vector associated with $\mathbf{x}$. The quantities

$$\mathscr{L}^{(m,n)}(\boldsymbol{\mu}_1, \ldots, \boldsymbol{\mu}_{m+n}, \mathbf{v}) = \prod_{j=1}^{m} \tilde{K}^*(-\boldsymbol{\mu}_j, \nu_j) \prod_{k=m+1}^{m+n} \tilde{K}(\boldsymbol{\mu}_k, \nu_k) \tag{9.12}$$

are the transfer functions.

Nonlinear process of imaging

If the field is detected with the use of say n quadratic detectors (placed at points $\mathbf{x}'_j$) whose outputs are correlated so that we are able in principle to measure the quantities ${}^i\Gamma^{(n,n)}(\mathbf{x}'_1, \ldots, \mathbf{x}'_n, \mathbf{x}'_n, \ldots, \mathbf{x}'_1)$†, we must set $m = n$ and $\mathbf{x}'_j = \mathbf{x}'_{n+j}$ in (9.10). Thus we arrive at the nonlinear relation

$$ {}^iG^{(n)}(\mathbf{x}'_1, \ldots, \mathbf{x}'_n, \mathbf{v}) = \int_{\mathscr{A}} \ldots \int_{\mathscr{A}} {}^og^{(n,n)}(\mathbf{x}_1, \ldots, \mathbf{x}_{2n}, \mathbf{v}) \times $$

$$ \times \prod_{j=1}^{n} K^*(\mathbf{x}'_j - \mathbf{x}_j, \nu_j)\, K(\mathbf{x}'_{n+j} - \mathbf{x}_{n+j}, \nu_{n+j})\, {}^ou^*(\mathbf{x}_j)\, {}^ou(\mathbf{x}_{n+j})\, \mathrm{d}\mathbf{x}_j\, \mathrm{d}\mathbf{x}_{n+j}, \tag{9.13} $$

where ${}^og^{(n,n)}$ is the normalized spectral correlation function (in the sense of (2.42a)), i.e. it is a Fourier transform of ${}^o\gamma^{(n,n)}$ with respect to τ_j, ${}^ou(\mathbf{x})$ is the object amplitude and ${}^iG^{(n)}(\mathbf{x}'_1, \ldots, \mathbf{x}'_n, \mathbf{v}) \equiv {}^iG^{(n,n)}(\mathbf{x}'_1, \ldots, \mathbf{x}'_n, \mathbf{x}'_n, \ldots, \mathbf{x}'_1, \mathbf{v})$.

Restricting ${}^og^{(n,n)}$ by the spatial stationary condition

$$ {}^og^{(n,n)}(\mathbf{x}_1, \ldots, \mathbf{x}_{2n}, \mathbf{v}) = {}^og^{(n,n)}(\mathbf{x}_{n+1} - \mathbf{x}_1, \ldots, \mathbf{x}_{2n} - \mathbf{x}_n, \mathbf{v}), \tag{9.14} $$

we obtain from (9.13), by spatial Fourier analysis, the following equations describing the transfer of spatial frequencies

$$ {}^iG^{(n)}(\mathbf{x}'_1, \ldots, \mathbf{x}'_n, \mathbf{v}) = \int_{-\infty}^{+\infty} \ldots \int_{-\infty}^{+\infty} \mathscr{L}^{(n,n)}(\boldsymbol{\eta}_1, \ldots, \boldsymbol{\eta}_{2n}, \mathbf{v}) \times $$

$$ \times \prod_{j=1}^{n} {}^o\tilde{u}^*(\boldsymbol{\eta}_j)\, {}^o\tilde{u}(\boldsymbol{\eta}_{n+j}) \exp\left[\mathrm{i}\, 2\pi \mathbf{x}'_j . (\boldsymbol{\eta}_j - \boldsymbol{\eta}_{n+j})\right] \mathrm{d}\boldsymbol{\eta}_j\, \mathrm{d}\boldsymbol{\eta}_{n+j} \tag{9.15a} $$

and

$$ {}^i\tilde{G}^{(n)}(\boldsymbol{\mu}_1, \ldots, \boldsymbol{\mu}_n, \mathbf{v}) = \int_{-\infty}^{+\infty} \ldots \int_{-\infty}^{+\infty} \mathscr{L}^{(n,n)}(\boldsymbol{\eta}_1, \ldots, \boldsymbol{\eta}_n, \boldsymbol{\eta}_1 + \boldsymbol{\mu}_1, \ldots, \boldsymbol{\eta}_n + \boldsymbol{\mu}_n, \mathbf{v}) \times $$

$$ \times \prod_{j=1}^{n} {}^o\tilde{u}(\boldsymbol{\eta}_j)\, {}^o\tilde{u}(\boldsymbol{\eta}_j + \boldsymbol{\mu}_j)\, \mathrm{d}\boldsymbol{\eta}_j . \tag{9.15b} $$

The generalized transfer functions $\mathscr{L}^{(n,n)}$, generalizing the transmission cross-coefficient of Hopkins (1953, 1957a), are given by

$$ \mathscr{L}^{(n,n)}(\boldsymbol{\eta}_1, \ldots, \boldsymbol{\eta}_{2n}, \mathbf{v}) = \int_{-\infty}^{+\infty} \ldots \int_{-\infty}^{+\infty} {}^o\tilde{g}^{(n,n)}(\boldsymbol{\mu}'_1, \ldots, \boldsymbol{\mu}'_n, \mathbf{v}) \times $$

$$ \times \prod_{j=1}^{n} \tilde{K}^*(\boldsymbol{\mu}'_j + \boldsymbol{\eta}_j, \nu_j)\, \tilde{K}(\boldsymbol{\mu}'_j + \boldsymbol{\eta}_{n+j}, \nu_{n+j})\, \mathrm{d}\boldsymbol{\mu}'_j . \tag{9.16} $$

These generalized transfer functions are completely determined by the diffraction properties of the system (the function K) and the coherence properties of light (the function g). The integration in (9.16) is taken over the common parts of regions in which both the functions $\tilde{K}$ and $\tilde{g}$ are non-zero (as there are finite apertures in the system).

---

† A detailed analysis is given in Sec. 10.6.

If light is detected with the use of n equivalently placed detectors, the process of imaging will be described by equation (9.15a) with $\mathbf{x}'_1 = \mathbf{x}'_2 = \ldots = \mathbf{x}'_n = \mathbf{x}'$. Denoting ${}^{i}G^{(n)}(\mathbf{x}', \ldots, \mathbf{x}', \mathbf{v})$ as ${}^{i}G^{(n)}(\mathbf{x}', \mathbf{v})$ we obtain the following equation in the spatial frequency region

$$ {}^{i}\tilde{G}^{(n)}(\boldsymbol{\mu}, \mathbf{v}) = \int_{-\infty}^{+\infty} \ldots \int_{-\infty}^{+\infty} \mathscr{L}^{(n,n)}(\boldsymbol{\eta}_1, \ldots, \boldsymbol{\eta}_{2n}, \mathbf{v}) \times $$

$$ \times\, \delta\Big(\sum_{j=1}^{2n} \varepsilon_j \boldsymbol{\eta}_j + \boldsymbol{\mu}\Big) \prod_{j=1}^{n} {}^{o}\tilde{u}^*(\boldsymbol{\eta}_j)\, {}^{o}\tilde{u}(\boldsymbol{\eta}_{n+j})\, \mathrm{d}\boldsymbol{\eta}_j\, \mathrm{d}\boldsymbol{\eta}_{n+j}. \tag{9.17} $$

Equation (9.17) shows that the transfer of spatial frequencies is non-trivial if and only if the spatial frequencies $\boldsymbol{\eta}_1, \ldots, \boldsymbol{\eta}_{2n}$ of $2n$ waves going from the object satisfy the relation

$$ \sum_{j=1}^{2n} \varepsilon_j \boldsymbol{\eta}_j + \boldsymbol{\mu} = 0. \tag{9.18} $$

All these results can be specialized for the case of stationary fields if ν_1 is substituted from (2.21).

Quasi-monochromatic analysis

In the quasi-monochromatic approximation described in Sec. 2.4 we obtain the following equation for the transmission of spatial frequencies

$$ {}^{i}\tilde{\Gamma}^{(n)}(\boldsymbol{\mu}_1, \ldots, \boldsymbol{\mu}_n, \boldsymbol{\tau}) = \exp\Big(\mathrm{i}\, 2\pi\bar{\nu} \sum_{j=2}^{2n} \varepsilon_j \tau_j\Big) \times $$

$$ \times \int_{-\infty}^{+\infty} \ldots \int_{-\infty}^{+\infty} \mathscr{L}^{(n,n)}(\boldsymbol{\eta}_1, \ldots, \boldsymbol{\eta}_n, \boldsymbol{\eta}_1 + \boldsymbol{\mu}_1, \ldots, \boldsymbol{\eta}_n + \boldsymbol{\mu}_n) \times $$

$$ \times \prod_{j=1}^{n} {}^{o}\tilde{u}^*(\boldsymbol{\eta}_j)\, {}^{o}\tilde{u}(\boldsymbol{\eta}_j + \boldsymbol{\mu}_j)\, \mathrm{d}\boldsymbol{\eta}_j, \tag{9.19} $$

where ${}^{i}\Gamma^{(n)}$ corresponds to ${}^{i}G^{(n)}$ and

$$ \mathscr{L}^{(n,n)}(\boldsymbol{\eta}_1, \ldots, \boldsymbol{\eta}_{2n}) = \int_{-\infty}^{+\infty} \ldots \int_{-\infty}^{+\infty} \prod_{j=1}^{n} \tilde{K}^*(\boldsymbol{\mu}'_j + \boldsymbol{\eta}_j, \bar{\nu})\, \tilde{K}(\boldsymbol{\mu}'_j + \boldsymbol{\eta}_{n+j}, \bar{\nu}) \times $$

$$ \times\, {}^{o}\tilde{\gamma}^{(n,n)}(\boldsymbol{\mu}'_1, \ldots, \boldsymbol{\mu}'_n) \prod_{k=1}^{n} \mathrm{d}\boldsymbol{\mu}'_k. \tag{9.20} $$

If the sizes of apertures in the optical system and the size of the source are finite (i.e. the *pupil function* $\tilde{K}$ and the degree of coherence ${}^{o}\tilde{\gamma}$ over the source are non-zero only in finite regions), then the generalized transfer functions $\mathscr{L}^{(n,n)}$ (as well as the function ${}^{i}\tilde{\Gamma}^{(n)}$) are also non-zero only in finite regions. Spatial frequencies of the object structure filtered by the system through $\mathscr{L}^{(n,n)}$ do not contribute to the image function ${}^{i}\Gamma^{(n)}$. This fact means that in general the object cannot be reconstructed uniquely from its image. However, considering finite-sized objects and using the so-called Paley-Wiener theorem on entire functions, square integrable objects may be reconstructed uniquely (Sec. 9.2).

The case $n = 1$

Putting $n = m = 1$ in (9.12) we arrive at the transfer function introduced by Parrent (1961) (if it is specialized to stationary fields). Further from (9.15a) we obtain for stationary fields and $n = 1$

$$ {}^{i}I^{(1)}(\mathbf{x}') = \iint_{-\infty}^{+\infty} \mathscr{L}^{(1,1)}(\boldsymbol{\eta}_1, \boldsymbol{\eta}_2)\, {}^{o}\tilde{u}^*(\boldsymbol{\eta}_1)\, {}^{o}\tilde{u}(\boldsymbol{\eta}_2) \exp\left[\mathrm{i}\, 2\pi \mathbf{x}' . (\boldsymbol{\eta}_1 - \boldsymbol{\eta}_2)\right] \mathrm{d}\boldsymbol{\eta}_1\, \mathrm{d}\boldsymbol{\eta}_2, \tag{9.21} $$

where

$$ \mathscr{L}^{(1,1)}(\boldsymbol{\eta}_1, \boldsymbol{\eta}_2) = \int_0^{\infty} \left\{ \int_{-\infty}^{+\infty} \tilde{K}^*(\boldsymbol{\mu}' + \boldsymbol{\eta}_1, \nu)\, \tilde{K}(\boldsymbol{\mu}' + \boldsymbol{\eta}_2, \nu)\, {}^{o}\tilde{g}^{(1,1)}(\boldsymbol{\mu}', \nu)\, \mathrm{d}\boldsymbol{\mu}' \right\} \mathrm{d}\nu \tag{9.22} $$

is the transmission cross-coefficient for imaging with polychromatic light; here second-order coherence effects are included. For quasi-monochromatic light we get the transmission cross-coefficient introduced by Hopkins (1953, 1957a)

$$ \mathscr{L}^{(1,1)}(\boldsymbol{\eta}_1, \boldsymbol{\eta}_2) = \int_{-\infty}^{+\infty} \tilde{K}^*(\boldsymbol{\mu}' + \boldsymbol{\eta}_1, \bar{\nu})\, K(\boldsymbol{\mu}' + \boldsymbol{\eta}_2, \bar{\nu})\, {}^{o}\tilde{\gamma}^{(1,1)}(\boldsymbol{\mu}')\, \mathrm{d}\boldsymbol{\mu}'. \tag{9.23} $$

If the apertures and the source have the form of circles, then the integration in (9.23) is taken over the area common to the circle of ${}^{o}\tilde{\gamma}$ with its centre at the origin and the circles of $\tilde{K}^*$ and $\tilde{K}$ centered on $\boldsymbol{\eta}_1$ and $\boldsymbol{\eta}_2$ respectively.

Putting $n = 1$ in (9.19) and using (9.20), we obtain (Maréchal and Francon (1960))

$$ \begin{aligned} {}^{i}\tilde{I}^{(1)}(\boldsymbol{\mu}) &= \int_{-\infty}^{+\infty} \left\{ \int_{-\infty}^{+\infty} \tilde{K}^*(\boldsymbol{\mu}' + \boldsymbol{\eta}, \bar{\nu})\, \tilde{K}(\boldsymbol{\mu}' + \boldsymbol{\eta} + \boldsymbol{\mu}, \bar{\nu})\, {}^{o}\tilde{\gamma}^{(1,1)}(\boldsymbol{\mu}')\, \mathrm{d}\boldsymbol{\mu}' \right\} \times \\ &\quad \times {}^{o}\tilde{u}^*(\boldsymbol{\eta})\, {}^{o}\tilde{u}(\boldsymbol{\eta} + \boldsymbol{\mu})\, \mathrm{d}\boldsymbol{\eta} \\ &= \int_{-\infty}^{+\infty} \left\{ \int_{-\infty}^{+\infty} \tilde{K}^*\left(\boldsymbol{\mu}' - \frac{\boldsymbol{\mu}}{2}, \bar{\nu}\right) \tilde{K}\left(\boldsymbol{\mu}' + \frac{\boldsymbol{\mu}}{2}, \bar{\nu}\right) {}^{o}\tilde{\gamma}^{(1,1)}(\boldsymbol{\mu}' - \boldsymbol{\eta})\, \mathrm{d}\boldsymbol{\mu}' \right\} \times \\ &\quad \times {}^{o}\tilde{u}^*\left(\boldsymbol{\eta} - \frac{\boldsymbol{\mu}}{2}\right) {}^{o}\tilde{u}\left(\boldsymbol{\eta} + \frac{\boldsymbol{\mu}}{2}\right) \mathrm{d}\boldsymbol{\eta}, \end{aligned} \tag{9.24} $$

where we have used the substitutions $\boldsymbol{\mu}' + \boldsymbol{\eta} + (\boldsymbol{\mu}/2) \to \boldsymbol{\mu}'$ and $\boldsymbol{\eta} + (\boldsymbol{\mu}/2) \to \boldsymbol{\eta}$. These are the well-known relations describing imaging with partially coherent quasi-monochromatic light.

Limiting cases of coherent and incoherent light

By analogy with (4.12) we may write for a fully coherent field

$$\Gamma^{(n,n)}(x_1, \ldots, x_{2n}) = \prod_{j=1}^{n} V^*(x_j)\, V(x_{n+j}) \tag{9.25}$$

for all x_1, ..., x_{2n} and all n, where V is a function of x independent of n (see Sec. 12.2). It is obvious that this condition can be fulfilled by a classical field exhibiting a kind of phase fluctuations at most. An analogous relation for stationary fields reads

$$\Gamma^{(n,n)}(\mathbf{x}_1, \ldots, \mathbf{x}_{2n}, \boldsymbol{\tau}) = \prod_{j=1}^{n} U^*(\mathbf{x}_j)\, U(\mathbf{x}_{n+j}) \exp\left(\mathrm{i}\, 2\pi\nu_0 \sum_{k=2}^{2n} \varepsilon_k \tau_k\right), \tag{9.26}$$

where U is a function of $\mathbf{x}$ only (independent of n) and ν_0 is the frequency of the coherent field. A detailed analysis of coherent fields will be given in Sec. 12.2.

Equation (9.26) leads to the conclusion that (4.20c) holds again, i.e. the transfer function equals $\tilde{K}(\boldsymbol{\mu}, \nu_0)$ in this case. However, from (9.26) it also follows that

$$\gamma^{(n,n)}(\mathbf{x}_1, \ldots, \mathbf{x}_{2n}, \boldsymbol{\tau}) = \exp\left\{-\mathrm{i} \sum_{j=1}^{n} [\beta(\mathbf{x}_j) - \beta(\mathbf{x}_{n+j})] + \mathrm{i}\, 2\pi\nu_0 \sum_{k=2}^{2n} \varepsilon_k \tau_k\right\}, \tag{9.27a}$$

where β is the phase of U and so

$$\gamma^{(n,n)}(\mathbf{x}_1, \ldots, \mathbf{x}_n, \mathbf{x}_n, \ldots, \mathbf{x}_1, \boldsymbol{\tau}) \equiv \gamma^{(n,n)}(\mathbf{x}_1, \ldots, \mathbf{x}_n, \boldsymbol{\tau}) =$$
$$= \exp\left(\mathrm{i}\, 2\pi\nu_0 \sum_{k=2}^{2n} \varepsilon_k \tau_k\right). \tag{9.27b}$$

Thus making use of the Fourier transformation we arrive at

$$\tilde{g}_H^{(n,n)}(\boldsymbol{\mu}'_1, \ldots, \boldsymbol{\mu}'_n, \boldsymbol{\nu}) = \prod_{j=1}^{n} \delta(\boldsymbol{\mu}'_j) \prod_{k=2}^{2n} \delta(\nu_k - \nu_0). \tag{9.28}$$

Hence equation (9.16) considered for stationary light ($\nu_1 = -\sum_{j=2}^{2n} \varepsilon_j \nu_j$) gives us

$$\mathscr{L}_H^{(n,n)}(\boldsymbol{\eta}_1, \ldots, \boldsymbol{\eta}_{2n}, \boldsymbol{\nu}) = \prod_{j=1}^{n} \tilde{K}^*(\boldsymbol{\eta}_j, \nu_0)\, \tilde{K}(\boldsymbol{\eta}_{n+j}, \nu_0) \prod_{k=2}^{2n} \delta(\nu_k - \nu_0). \tag{9.29}$$

For ${}^{i}\tilde{\Gamma}^{(n,n)}$ we obtain

$${}^{i}\tilde{\Gamma}^{(n,n)}(\boldsymbol{\mu}_1, \ldots, \boldsymbol{\mu}_{2n}, \boldsymbol{\tau}) = \exp\left(\mathrm{i}\, 2\pi\nu_0 \sum_{j=2}^{2n} \varepsilon_j \tau_j\right) \times$$
$$\times \prod_{j=1}^{n} \tilde{K}^*(\boldsymbol{\mu}_j, \nu_0)\, \tilde{K}(\boldsymbol{\mu}_{n+j}, \nu_0)\, {}^{o}\tilde{u}^*(\boldsymbol{\mu}_j)\, {}^{o}\tilde{u}(\boldsymbol{\mu}_{n+j}). \tag{9.30}$$

This is in agreement with a spatial Fourier analysis of equation (4.20c) performed in the isoplanatic region. However, it can be seen that the relation between the object and its image is nonlinear if the detection is included in the process of imaging, even if coherent light is used.

We shall see in Sec. 17.1 that chaotic light is completely characterized by the second-order correlation function. Thus assuming spatially incoherent light we may restrict ourselves to the second-order correlation function. As $G^{(1,1)}(\mathbf{x}_1, \mathbf{x}_2, \nu) = G^{(1,1)}(\mathbf{x}_1, \mathbf{x}_1, \nu)\, \delta(\mathbf{x}_1 - \mathbf{x}_2)$ (according to (4.2)), equation (9.10) considered for the stationary field reduces to

$$ {}^iG^{(1,1)}(\mathbf{x}'_1, \mathbf{x}'_2, \nu) = \int_{\mathscr{A}} K^*(\mathbf{x}'_1 - \mathbf{x}, \nu)\, K(\mathbf{x}'_2 - \mathbf{x}, \nu)\, {}^oG^{(1,1)}(\mathbf{x}, \mathbf{x}, \nu)\, \mathrm{d}\mathbf{x}. \tag{9.31} $$

A spatial Fourier analysis of this equation gives

$$ {}^i\tilde{G}^{(1,1)}(\boldsymbol{\mu}_1, \boldsymbol{\mu}_2, \nu) = \tilde{K}^*(-\boldsymbol{\mu}_1, \nu)\, \tilde{K}(\boldsymbol{\mu}_2, \nu)\, {}^o\tilde{G}^{(1,1)}(\boldsymbol{\mu}_1 + \boldsymbol{\mu}_2, \nu), \tag{9.32} $$

which shows that the correlation between incoherent light beams associated with the spatial frequencies $\boldsymbol{\mu}_1$ and $\boldsymbol{\mu}_2$ in the image is determined by the object function ${}^o\tilde{G}^{(1,1)}$ associated with the sum frequency $(\boldsymbol{\mu}_1 + \boldsymbol{\mu}_2)$ in the object.

As $\gamma^{(1,1)}(\mathbf{x}_1, \mathbf{x}_2, \tau) = \exp(-\mathrm{i}2\pi\bar{\nu}\tau)\, \delta(\mathbf{x}_1 - \mathbf{x}_2)\, (|\tau| \ll 1/\Delta\nu)$ holds for spatially incoherent quasi-monochromatic light, equation (9.22) gives the standard transfer function for incoherent light in the form of the autocorrelation of the pupil function $\tilde{K}$,

$$ \begin{aligned} \mathscr{L}^{(1,1)}(\boldsymbol{\eta}_1, \boldsymbol{\eta}_2) &= \int_{-\infty}^{+\infty} \tilde{K}^*(\boldsymbol{\mu}' + \boldsymbol{\eta}_1, \bar{\nu})\, \tilde{K}(\boldsymbol{\mu}' + \boldsymbol{\eta}_2, \bar{\nu})\, \mathrm{d}\boldsymbol{\mu}' = \\ &= \int_{-\infty}^{+\infty} \tilde{K}^*(\boldsymbol{\mu}', \bar{\nu})\, \tilde{K}(\boldsymbol{\mu}' + \boldsymbol{\eta}_2 - \boldsymbol{\eta}_1, \bar{\nu})\, \mathrm{d}\boldsymbol{\mu}'. \end{aligned} \tag{9.33} $$

Equation (9.31) considered for $\mathbf{x}'_1 \equiv \mathbf{x}'_2$ shows that only the process of imaging with incoherent light is linear in intensity if detection is involved.

Some further details together with an analysis of weak visibility objects (for which the optical system works as a normal filter) and the matrix formulation of the present results can be obtained (Peřina (1969)).

9.2 Reconstruction of an object from its image and similarity between object and image

The general formulation of the relation between an object and its image just developed may be used to deduce some consequences about the reconstruction of an object from its image and to study the similarity between an object and its image taking into account coherence effects.

The problem of the reconstruction of the object characteristics from those of the image was considered by Beran and Parrent (1964) for imaging with partially coherent light and by Khurgyn and Yakovlev (1971) for imaging with coherent and incoherent light. It has been shown, in general, that this problem cannot be solved uniquely. Khurgyn and Yakovlev (1971), however, have shown, considering finite-sized objects and using the Paley-Wiener theorem on entire functions, that a unique solution of the problem exists in the space of square integrable functions L_2 (in the sense of Lebesgue). Consequently the reconstruction of the object is unique. Further discussions of the uniqueness of the reconstruction and other procedures of the reconstruction have been provided by Hoenders (1978), Martínez-Herrero (1979, 1980, 1981, 1982), Martínez-Herrero and Durán (1981) and Martínez-Herrero and Mejías (1981, 1982a).

Reconstruction of the object from its image and analytic continuation

Confining ourselves to quasi-monochromatic light for simplicity (this restriction is not essential for the considerations below) we may start with the following propagation law

$$
{}^{i}\Gamma^{(n,n)}(x'_1, \ldots, x'_{2n}) = \int_{\mathscr{A}} \ldots \int_{\mathscr{A}} \prod_{j=1}^{n} K^*(x'_j - x_j)\, K(x'_{n+j} - x_{n+j}) \times
\times\, {}^{o}\Gamma^{(n,n)}(x_1, \ldots, x_{2n}) \prod_{k=1}^{n} \mathrm{d}x_k\, \mathrm{d}x_{n+k}, \tag{9.34}
$$

which follows from (9.10) by Fourier transformation and by putting $\mathbf{t} = \hat{0}$ with the assumption that $K(x) \equiv K(x, \bar{\nu})$; only the one-dimensional case will be considered here, for simplicity.

We are to determine the object quantity ${}^{o}\Gamma^{(n,n)}$ (i.e. the degree of coherence ${}^{o}\gamma^{(n,n)}$ and the nth-order intensity ${}^{o}I^{(n)}(x) \equiv {}^{o}\Gamma^{(n,n)}(x, \ldots, x)$) from the image specified by ${}^{i}\Gamma^{(n,n)}$. If we use the spatial Fourier analysis of (9.34) we may solve this equation in the form

$$
{}^{o}\Gamma^{(n,n)}(x_1, \ldots, x_{2n}) = \int_{\boldsymbol{\Theta}} \ldots \int \frac{{}^{i}\tilde{\Gamma}^{(n,n)}(\mu_1, \ldots, \mu_{2n})}{\prod_{j=1}^{n} \tilde{K}^*(-\mu_j)\, \tilde{K}(\mu_{n+j})} \times
\times \exp\left(-\mathrm{i}\, 2\pi \sum_{j=1}^{2n} \mu_j x_j\right) \prod_{k=1}^{2n} \mathrm{d}\mu_k + f(x_1, \ldots, x_{2n}), \tag{9.35}
$$

where $\boldsymbol{\Theta}$ is a region of spatial frequencies where $\tilde{K}(\mu) \not\equiv 0$ and outside of which $\tilde{K}(\mu) \equiv 0$. The function f is an arbitrary function whose Fourier transform is zero in $\boldsymbol{\Theta}$. We see indeed that the reconstruction is not in general unique. Assuming that the function ${}^{o}\Gamma^{(n,n)}(x_1, \ldots, x_{2n})$ is non-zero in a region $\boldsymbol{\Pi} = (\Pi')^{2n}$ only, where Π' is a one-dimensional finite interval of a finite-sized object, also assumed to be square integrable in any x_j, then the function

$$
{}^{o}\tilde{\Gamma}^{(n,n)}(\mu_1, \ldots, \mu_{2n}) = \frac{{}^{i}\tilde{\Gamma}^{(n,n)}(\mu_1, \ldots, \mu_{2n})}{\prod_{j=1}^{n} \tilde{K}^*(-\mu_j)\, \tilde{K}(\mu_{n+j})} \quad (\text{in } \boldsymbol{\Theta}) \tag{9.36}
$$

is an entire function according to the Paley-Wiener theorem (Paley and Wiener (1934), Khurgyn and Yakovlev (1971)). Thus we can define the function ${}^{o}\tilde{\Gamma}^{(n,n)}$ for all μ_j by means of the Taylor series

$$
{}^{o}\tilde{\Gamma}^{(n,n)}(\mu_1, \ldots, \mu_{2n}) = \sum_{r_1, \ldots, r_{2n}} \prod_{j=1}^{2n} \frac{(\mu - \mu_{0j})^{r_j}}{r_j!} \frac{\partial^{r_j}}{\partial \mu_j^{r_j}} {}^{o}\tilde{\Gamma}^{(n,n)}(\mu_1, \ldots, \mu_{2n}) \Bigg|_{\substack{\mu_1 = \mu_{01} \\ \cdots\cdots \\ \mu_{2n} = \mu_{02n}}}, \tag{9.37}
$$

where $\mu_{01}, \ldots, \mu_{02n} \in \Theta$. Thus the square integrable object function ${}^{o}\Gamma^{(n,n)}$ can be determined uniquely by

$$
\begin{aligned}
&{}^{o}\Gamma^{(n,n)}(x_1, \ldots, x_{2n}) = \\
&= \int_{-\infty}^{+\infty} \cdots \int_{-\infty}^{+\infty} \left\{ \sum_{r_1, \ldots, r_{2n}} \prod_{j=1}^{2n} \frac{(\mu - \mu_{0j})^{r_j}}{r_j!} \frac{\partial^{r_j}}{\partial \mu_j^{r_j}} \left[\frac{{}^{i}\tilde{\Gamma}^{(n,n)}(\mu_1, \ldots, \mu_{2n})}{\prod_{k=1}^{n} \tilde{K}^*(-\mu_k) \tilde{K}(\mu_{n+k})} \right]_{\substack{\mu_1 = \mu_{01} \\ \cdots\cdots \\ \mu_{2n} = \mu_{02n}}} \right\} \times \\
&\times \exp\left(-\mathrm{i}\, 2\pi \sum_{k=1}^{2n} \mu_k x_k\right) \prod_{k=1}^{2n} \mathrm{d}\mu_k .
\end{aligned} \tag{9.38}
$$

This is exactly the method used to obtain super-resolution (Peřina (1971), Lukš (1976), Peřina, Peřinová and Braunerová (1977)). In these papers some further possibilities of analytic continuation using orthogonal polynomials were proposed and used (Peřina and Mišta (1969)) to study the existence of the so-called diagonal representation of the density matrix (Sec. 13.3).

The problem of the correspondence between an object and its image was investigated by Toraldo di Francia (1955) from the point of view of communication theory using the sampling theorem. This enabled him to introduce the notion of "degrees of freedom". So the number of degrees of freedom of the object is infinite while the number of degrees of freedom of the image is finite and consequently some information is lost in the imaging process. This fact expresses the filtering properties of the system and the object cannot be reconstructed uniquely from the image – classes of objects corresponding to a given image exist. A priori knowledge about the object then plays an important role in the reconstruction (e.g. the concept of the two-point resolving power of the optical system). However, using analytic continuation, the number of degrees of freedom of both the object and the image is infinite (cf. a discussion by Khurgyn and Yakovlev (1971), Chap. 8). All the degrees of freedom of the object are contained in the non-filtered part of the spatial spectrum of the image and may be obtained by means of precise measurements or more readily by means of analytic continuation. Thus the reconstruction of the object function from that of the image is unique in this case. Some restrictions caused by the presence of noise have been discussed by Toraldo di Francia (1969). Particularly he has shown that the accuracy of experimental data must strongly increase in order to gain substantial additional information using analytic continuation. The number of degrees of freedom for partially coherent sources was discussed by Starikov (1982).

As a demonstration of the utility of these considerations we employ the concept of the two-point resolving power (Khurgyn and Yakovlev (1971)). We consider the one-dimensional case and coherent light for simplicity, and assume the object structure to be composed of two points distant $2b$ apart so that $f(x) = \delta(x - b) + \delta(x + b)$. Choosing the diffraction function in the form

$$K(x) = \frac{\sin(2\pi x)}{\pi x}, \tag{9.39}$$

which is the diffraction function of a slit, we obtain

$$\tilde{K}(\mu) = \begin{cases} 1, & |\mu| < 1, \\ 0, & |\mu| > 1. \end{cases} \tag{9.40}$$

Thus this system filters all spatial frequencies $|\mu| > 1$ from the image. The image $g(x')$ can be calculated as

$$g(x') = \int_{-\infty}^{+\infty} K(x' - x) f(x)\,\mathrm{d}x = \frac{\sin[2\pi(x' - b)]}{\pi(x' - b)} + \frac{\sin[2\pi(x' + b)]}{\pi(x' + b)}. \tag{9.41}$$

Using the well-known theorem on the Fourier transform of the convolution, we obtain

$$\tilde{f}(\mu) = \frac{\tilde{g}(\mu)}{\tilde{K}(\mu)} = \begin{cases} 2\cos(2\pi\mu b), & |\mu| < 1, \\ \dfrac{0}{0}, & |\mu| > 1. \end{cases} \tag{9.42}$$

Hence the function $\tilde{f}(\mu)$ is given for various separations $2b$ of the points by a set of cosine functions which end at $\mu = \pm 1$.

The classical limit of the resolving power of the system may be defined by the minimal root of the equation

$$\left.\frac{\mathrm{d}^2 g(x')}{\mathrm{d}x'^2}\right|_{x'=0} = 0, \tag{9.43a}$$

which gives

$$\tan(2\pi b_0) = \frac{2(2\pi b_0)}{2 - (2\pi b_0)^2}, \quad \text{i.e.} \quad 2\pi b_0 \approx 2.1. \tag{9.43b}$$

This means that only the curves for $b > b_0$ may be determined in the classical sense and the corresponding points may be distinguished only if the a priori knowledge is used, that two points are to be considered (even in this case the function $\tilde{f}(\mu)$ must be known for all μ if $f(x)$ is to be determined uniquely). When $b < b_0$ one cannot determine, from the classical point of view, the form of the function $\tilde{f}(\mu) = \tilde{g}(\mu)/\tilde{K}(\mu)$ if it is calculated for $|\mu| < 1$ only. This interval is too short and the function is nearly constant over this interval. The separation $2b$ of the points cannot be determined in this case and the points are not distinguishable. However, taking into account that $f(x)$ is a finite-sized object ($f(x) = 0$ for

$|x| > b$), then the function $\tilde{f}(\mu)$ is an entire function which can be defined by the Taylor series for all μ. We are then able to find the form of $\tilde{f}(\mu)$ and we arrive at the unique function $f(x) = \delta(x - b) + \delta(x + b)$ without any restrictions on the separation $2b$. In this way we may, in principle, distinguish arbitrarily near points and there is no limit to the resolving power of the system, provided that the knowledge of a piece of $\tilde{f}(\mu)$ is sufficiently accurate.

Of course, an identical analysis can be performed for the correlation function ${}^o\Gamma^{(n,n)}(x_1, \ldots, x_{2n}) = {}^o\gamma^{(n,n)}(x_1, \ldots, x_{2n})\, {}^ou(x_1) \ldots {}^ou(x_{2n})$ with ${}^ou(x) = \delta(x - b) + \delta(x + b)$, which is also a finite-sized function in all variables. Consequently ${}^o\tilde{\Gamma}^{(n,n)}(\mu_1, \ldots, \mu_{2n})$ is an entire function in all variables μ_j and can be reconstructed in the same way.

A more realistic model has been also considered (Peřina, Peřinová and Braunerová (1977)) representing two luminous points as a sum of two Gaussian functions. The slit function has been employed again. In the measured piece of $\tilde{f}(\mu)$ noise has been simulated up to 20% level and a particular sampling theorem for obtaining derivatives of inaccurate functions, derived for this purpose by Ferwerda and Hoenders (1974), has been adopted. Using the Hermite polynomials, corresponding to the use of analytic continuation with the Taylor power series cut-off by a Gaussian function (Lukš (1976)), the results shown in Fig. 9.1 have been obtained. In this figure only one half of the object and the image is given. The broken curve represents the real object in the form of the Gaussian curve, the dotted curve is the image beyond the classical limit ($2\pi b = 1.8 < 2\pi b_0 \approx 2.1$). Full curves $a - c$ represent the reconstructed values of the object with the levels of noise 0, 1 and 5% respectively.

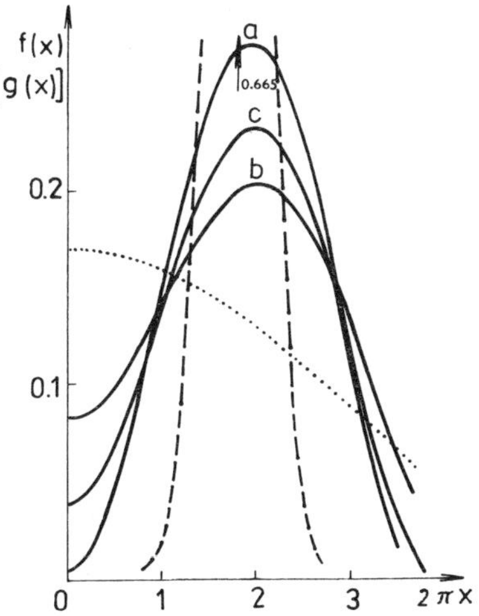

Fig. 9.1 One half of the object $f(x)$ (broken curve) and actual image $g(x)$ (dotted curve) beyond the classical limit; full curves $a - c$ represent the values of the object reconstructed with the use of analytic continuation with the noise levels 0, 1 and 5% respectively (after J. Peřina, V. Peřinová and Z. Braunerová, 1977, *Opt. Appl.* VII/3, 79).

A recent discussion of super-resolution has been presented by Bertero and Pike (1982). The effect of partial coherence on two-point resolution has been investigated by a number of authors (Bhatnagar, Sirohi and Sharma (1971), Kintner and Sillitto (1973), Asakura (1974a, b), Asakura and Mishina (1974), Mehta (1974), Nayyar and Verma (1978), Bertero et al. (1982)) and they have demonstrated that the resolving power is higher with incoherent light than with coherent light and it is intermediate with partially coherent light.

Similarity between an object and its image – algebraic integral equations

The similarity problem, i.e. finding an object structure $f(x)$ which is imaged similarly in the sense that there exists a number λ such that the image function is reproduced, $g(x) = \lambda^{-1} f(x)$, was first formulated by Mandelstam (1947, 1948) for coherent and incoherent light. These results were developed by Khurgyn and Yakovlev (1971) and they have also been generalized to partially coherent light beams (Peřina (1963b, 1965c), Peřina and Peřinová (1965, 1969)). The similarity problem formulated for coherent light is of great importance in the theory of optical resonators for lasers (Toraldo di Francia (1966, 1970), Roncchi (1972)). The role of partial coherence in optical resonators was investigated by Wolf (1963b, 1964), by Streifer (1966) and by Wolf and Agarwal (1984).

If we consider the similarity problem for the correlation functions, expressed by the condition

$$ {}^{i}\Gamma^{(n,n)}(x_1, \dots, x_{2n}) = \frac{1}{|\lambda|^{2n}} {}^{o}\Gamma^{(n,n)}(x_1, \dots, x_{2n}), \tag{9.44} $$

we obtain from (9.34) by using the spatial Fourier transformation

$$ {}^{o}\tilde{\Gamma}^{(n,n)}(\mu_1, \dots, \mu_{2n}) = |\lambda|^{2n} \prod_{j=1}^{n} \tilde{K}^*(-\mu_j)\, \tilde{K}(\mu_{n+j})\, {}^{o}\tilde{\Gamma}^{(n,n)}(\mu_1, \dots, \mu_{2n}). \tag{9.45} $$

As the pupil function $\tilde{K}$ is zero outside a finite region, ${}^{o}\tilde{\Gamma}^{(n,n)}$ must also vanish outside a finite region. Hence, by using the Paley-Wiener theorem, the function ${}^{o}\Gamma^{(n,n)}$ can be continued into the whole complex x-plane by means of the Taylor series and as a consequence of the uniqueness of such a continuation ${}^{o}\Gamma^{(n,n)}$ cannot equal zero in any finite region (if it did it would have to be identically zero). From this result we conclude that for any finite-sized object the similarity problem in the infinite region has no solution, but the similarity problem in a finite region always has a solution.

From (9.45) it follows that one of the following equations

$$ \prod_{j=1}^{n} \tilde{K}^*(-\mu_j)\, \tilde{K}(\mu_{n+j}) = |\lambda|^{-2n}, \tag{9.46} $$

$$^{o}\tilde{\Gamma}^{(n,n)}(\mu_1, \ldots, \mu_{2n}) = 0 \tag{9.47}$$

must be true. Thus, if (9.46) holds in finite regions of the μ-space, then (9.47) must be true outside these regions. If (9.46) holds in the whole space, then K is proportional to the Dirac function and the system is ideal.

When the image is specified by its intensity, and confining ourselves to real functions and real similarity coefficients for simplicity, we arrive at the similarity equation for the real amplitude u

$$u^{2n}(x) = \lambda^{2n} \int_{\mathscr{A}} \ldots \int_{\mathscr{A}} L(x, x_1, \ldots, x_{2n}) \prod_{j=1}^{2n} u(x_j)\, \mathrm{d}x_j, \tag{9.48}$$

where

$$L(x, x_1, \ldots, x_{2n}) = {}^{o}\gamma^{(n,n)}(x_1, \ldots, x_{2n}) \prod_{j=1}^{2n} K(x, x_j). \tag{9.49}$$

One may assume that $K(x, y) = K(y, x)$ and ${}^{o}\gamma^{(n,n)}(x_1, \ldots, x_{2n})$ are continuous functions and that ${}^{o}\gamma^{(n,n)}$ is symmetric with respect to all pairs of variables $x_1, \ldots, x_{2n}$.

Equation (9.48) represents a special case of the general class of nonlinear integral equations which are called algebraic integral equations. They were introduced in mathematics by Schmeidler (1956) and they were further investigated by Peřinová (1969). The following theorems can be shown to be valid (Peřina and Peřinová (1965, 1969)):

If $L(x, x_1, \ldots, x_{2n})$ is a positive continuous function, then there exist a real eigenvalue $\lambda_0 > 0$ and a real normalized eigenfunction $u_0(x) > 0$ (the norm is defined as $\| u_0 \| = \{\int u_0^{2n}(x)\, \mathrm{d}x\}^{1/2n} = 1$) such that (9.48) holds, where

$$\lambda_0 = \{\int_{\mathscr{A}} \ldots \int_{\mathscr{A}} L(x, x_1, \ldots, x_{2n}) \prod_{j=1}^{2n} u_0(x_j)\, \mathrm{d}x_j\, \mathrm{d}x\}^{-1/2n}; \tag{9.50}$$

λ_0 is the smallest eigenvalue and it is simple.

Besides this theorem on the existence of an eigensolution of (9.48) the following theorem on the countable infinity of eigenvalues can be proved.

Let the following assumptions be valid for every eigenvalue λ_0 and the corresponding eigenfunction $u_0(x)$ of equation (9.48):

a) $u_0(x)$ does not equal zero in $\mathscr{A}$,

b) λ_0^{2n} is the simple eigenvalue of the kernel

$$u_0^{-2n+1}(x) \int_{\mathscr{A}} \ldots \int_{\mathscr{A}} L(x, y, x_2, \ldots, x_{2n}) \prod_{j=2}^{2n} u_0(x_j)\, \mathrm{d}x_j,$$

c) the integral $\int_{\mathscr{A}} \varphi(x) \psi^*(x)\, \mathrm{d}x$ is different from zero for the eigenfunction $\varphi(x) \equiv$ $\equiv u_0(x)$ and the associated eigenfunction $\psi(x)$ corresponding to the eigenvalue λ_0^{2n}; then there exists at most a countably infinite set of eigenvalues.

Because of the nonlinearity of equation (9.48) there may in general exist an interval of eigenvalues. As a consequence there may arise branching of a structure

$u_0(x)$ imaged similarly, corresponding to an eigenvalue λ_0, into several structures imaged similarly for eigenvalues belonging to a neighbourhood of λ_0. This branching effect is caused by partial coherence creating typical nonlinearity in imaging.

For coherent and incoherent light we obtain as special cases the linear integral equations for the amplitude and the intensity respectively, studied by Mandelstam (1947, 1948).

In this chapter the space-time behaviour of the correlation functions has again been the main subject of investigation and so the results presented here are correct for both the classical and quantum correlation functions.

CHAPTER 10

FOURTH- AND HIGHER-ORDER COHERENCE PHENOMENA — SEMICLASSICAL TREATMENT

In the preceding chapters of this book, particularly in Chapter 3, we have discussed the possibility of experimental determination of second-order coherence using the interference and diffraction of light. In this chapter we discuss unconventional techniques based on photoelectric correlation measurements and on photon-counting measurements through which the statistical properties of optical fields, including effects of arbitrary order, can be determined. In the first case one measures correlation functions of the type $\Gamma^{(n,n)}$ while in the second case one measures by a photoelectric detector the probability distribution of emitted photoelectrons whose statistics reflect the statistics of the absorbed photons. Thus the higher-order moments – correlation functions – may be calculated using the photocount distribution. Although the higher-order coherence effects will be investigated by quantum methods in the following chapters, we discuss these effects briefly here using the so-called semiclassical treatment which is relatively simple. A typical feature of this semiclassical approach is that the field is treated classically while its interaction with matter in the detection process is described quantum-mechanically.

As we have seen, the second-order coherence effects are characterized by the second-order correlation function which has the physical dimension of intensity (two field amplitudes are multiplied and averaged). The fourth- and higher-order correlation functions may be realized as correlations of intensities of the field at various space-time points. Fourth-order correlation measurements were first performed by Hanbury Brown and Twiss (1956a – c), who used fast photoelectric detectors to measure the correlation of intensity fluctuations. A number of experiments of this type were performed both by correlation and photoelectric coincidence techniques (Hanbury Brown and Twiss (1957a, b), Twiss, Little and Hanbury Brown (1957), Twiss and Little (1959), Rebka and Pound (1957), Brannen, Ferguson and Wehlau (1958), Harwit (1960)). The sixth-order correlation functions were measured by Davidson and Mandel (1968), Davidson (1968) and Corti and Degiorgio (1974, 1976a, b). Another means of gaining information about higher-order moments (correlation functions) is based on measuring the photocount statistics as has been mentioned. This approach provides higher-order moments more readily (see e.g. Arecchi (1965), Arecchi, Berné and Burlamacchi (1966), Arecchi, Berné, Sona and Burlamacchi (1966), Arecchi, Berné and Sona (1966) (for a review see Arecchi (1969), Arecchi and Degiorgio (1972)), Freed and Haus (1965, 1966), Johnson, Jones, McLean and Pike (1966), Johnson, McLean and Pike

(1966), Fray, Johnson, Jones, McLean and Pike (1967) (for reviews see Pike (1969, 1970), Pike and Jakeman (1974)), Martienssen and Spiller (1966a, b), Smith and Armstrong (1966a, b) (for a review see Armstrong and Smith (1967)), Chang, Korenman, Alley and Detenbeck (1969)).

A new branch of interferometry, the so-called *correlation interferometry*, is based on the fourth- and higher-order correlation measurements. By this technique the angular diameters of stars can be obtained, under better conditions than in measurements using the Michelson stellar interferometer (Hanbury Brown and Twiss (1956b, 1958)). The correlation measurements also yield information about the state of polarization of light (Wolf (1960), Mandel and Wolf (1961b), Mandel (1963b)) and about the spectral distribution of light beams (Wolf (1962, 1965)). A theoretical explanation of the fourth-order correlation effect based on the classical or the semiclassical description of light beams was first given by Purcell (1956). The ideas of this paper have been continued by Wolf (1957), Janossy (1957, 1959), Mandel (1958, 1959, 1963a), Kahn (1958), Mandel and Wolf (1961b) and Mandel, Sudarshan and Wolf (1964).

The problem was investigated from a quantum-mechanical point of view by Dicke (1964), Senitzky (1958, 1962), Glauber (1963a–c, 1964, 1965, 1970, 1972), Kelley and Kleiner (1964), Goldberger and Watson (1964, 1965), Holliday and Sage (1964), Lehmberg (1968), Mandel and Meltzer (1969), Peřina (1967a, 1970), Barakat and Blake (1980), Selloni (1980) and Srinivas and Davies (1981), among others.

A basic relation of the semiclassical theory of photodetection expresses the probability $p(t)\,\mathrm{d}t$ of emission (absorption) of a photoelectron (a photon) in the time interval $(t, t + \mathrm{d}t)$ by means of the classical intensity $I(t)$ (it is in general stochastic) of the quasi-monochromatic light which is incident on the photocathode,

$$p(t)\,\mathrm{d}t = \eta I(t)\,\mathrm{d}t = \eta V^*(t)\,V(t)\,\mathrm{d}t, \tag{10.1}$$

where η represents the photo-efficiency of the photoelectric detector and linearly polarized light is assumed. This relation may be derived with the help of the perturbation theory of quantum mechanics (Mandel, Sudarshan and Wolf (1964)) and follows from simple quantum-mechanical consideration (Glauber (1963a–c, 1964, 1965) and Chapter 12).

10.1 Photocount distribution

We determine the probability $p(n, T, t)$ of emission of n photoelectrons (i.e. of n photoelectric counts, or alternatively the probability of detection of n photons if $\eta = 1$, see Chapter 14) by the plane photocathode on which light is normally incident in a time interval $(t, t + T)$, where T is fixed. First we assume that the intensity $I(t)$ has no random fluctuations. Dividing the interval T into sub-intervals with the points $t + \mathrm{i}\,\Delta T \equiv t_i$, $i = 0, 1, \ldots, T/\Delta T$, we can calculate the probability

$p(n, T, t)$ of n counts occurring in the interval $(t, t + T)$. This is the sum over all possible sequences of counts of the product of probabilities of obtaining a count at time t_{r_1}, a count at t_{r_2}, ..., a count at t_{r_n}, multiplied by the probabilities of obtaining no counts in the remaining $T/\Delta T$-n intervals (Mandel (1963a))

$$p(n, T, t) = \lim_{\Delta T \to 0} \sum_{r_1=0}^{T/\Delta T} \dots \sum_{r_n=0}^{T/\Delta T} \frac{1}{n!} \eta^n I(t_{r_1}) \dots I(t_{r_n}) (\Delta T)^n \times$$

$$\times \frac{\prod_{i=0}^{T/\Delta T} [1 - \eta I(t_i) \Delta T]}{\prod_{j=1}^{n} [1 - \eta I(t_{r_j}) \Delta T]} = \frac{1}{n!} [\eta \int_t^{t+T} I(t') \, dt']^n \exp [-\eta \int_t^{t+T} I(t') \, dt'], \tag{10.2}$$

which is a Poisson distribution for the number of counts n with the mean number of counts $\langle n \rangle = \eta \int_t^{t+T} I(t') \, dt'$.

If the intensity $I(t)$ is a stochastic function, (10.2) must be averaged over all possible realizations of $I(t)$ and we arrive at

$$p(n, T, t) = \frac{1}{n!} \langle (\eta W)^n \exp(-\eta W) \rangle = \frac{1}{n!} \int_0^\infty (\eta W)^n \exp(-\eta W) P(W) \, dW, \tag{10.3a}$$

where

$$W = \int_t^{t+T} I(t') \, dt' \tag{10.3b}$$

and $P(W)$ is the probability distribution of the integrated intensity W. The photodetection equation (10.3a) was first derived by Mandel (1958, 1959, 1963a).†

For stationary fields the distribution $p(n, T, t)$ will be independent of t and $p(n, T, t) \equiv p(n, T)$.

In general the distribution of counts will depart from the classical Poisson statistics but an ideal laser source, perfectly stabilized so that the intensity does not fluctuate, leads to the Poisson distribution of counts and will behave like a source of classical particles in this case.

Some properties of the photocount distribution

First we calculate the variance of the number of counts n. We obtain from (10.3a)

$$\langle n \rangle = \sum_{n=0}^{\infty} p(n, T, t) \, n = \left\langle (\eta W) \frac{d}{d(\eta W)} \left[\sum_{n=0}^{\infty} \frac{(\eta W)^n}{n!} \right] \exp(-\eta W) \right\rangle = \eta \langle W \rangle, \tag{10.4a}$$

$$\langle n^2 \rangle = \sum_{n=0}^{\infty} p(n, T, t) \, n^2 = \left\langle \left[(\eta W) \frac{d}{d(\eta W)} \right]^2 \left[\sum_{n=0}^{\infty} \frac{(\eta W)^n}{n!} \right] \exp(-\eta W) \right\rangle =$$

$$= \eta \langle W \rangle + \eta^2 \langle W^2 \rangle. \tag{10.4b}$$

---

† A quantum derivation of the photodetection equation is presented in Chapter 14.

We can obtain in the same way

$$\langle n^3\rangle = \eta\langle W\rangle + 3\eta^2\langle W^2\rangle + \eta^3\langle W^3\rangle, \tag{10.4c}$$

$$\langle n^4\rangle = \eta\langle W\rangle + 7\eta^2\langle W^2\rangle + 6\eta^3\langle W^3\rangle + \eta^4\langle W^4\rangle, \tag{10.4d}$$

etc. The variance of n can be obtained from (10.4a) and (10.4b)

$$\begin{aligned}\langle(\Delta n)^2\rangle &= \langle n^2\rangle - \langle n\rangle^2 = \eta\langle W\rangle + \eta^2\langle(\Delta W)^2\rangle = \\ &= \langle n\rangle + \eta^2\langle(\Delta W)^2\rangle,\end{aligned} \tag{10.5}$$

where

$$\langle(\Delta W)^2\rangle = \langle W^2\rangle - \langle W\rangle^2 \tag{10.6}$$

is the variance of the integrated intensity W. The formula (10.5) has a very simple physical interpretation. It shows that the variance of n, giving fluctuations in the number of ejected photoelectrons, is the sum of fluctuations in the number of classical particles obeying the Poisson distribution ($\langle(\Delta n)^2\rangle = \langle n\rangle$ for the Poisson distribution) and of fluctuations in a classical wave field (the wave interference term $\eta^2\langle(\Delta W)^2\rangle$). This result generalizes the earlier results by Einstein (1909), Bothe (1927) and Fürth (1928a, b) and is correct for any light beams (thermal or non-thermal, stationary or non-stationary). Although the result refers to the fluctuations of photoelectric counts, it can be regarded as reflecting the fluctuation properties of the light itself. For quantum fields having no classical analogues, $\langle(\Delta W)^2\rangle < 0$ and $\langle(\Delta n)^2\rangle < \langle n\rangle$ may occur and then we speak of antibunching of photons and their sub-Poissonian statistics (Chapter 22).

Defining the characteristic function as

$$C^{(n)}(\mathrm{i}x) = \langle\exp(\mathrm{i}xn)\rangle = \sum_{n=0}^{\infty} p(n, T, t)\exp(\mathrm{i}xn), \tag{10.7}$$

we obtain from (10.3a)

$$\begin{aligned}C^{(n)}(\mathrm{i}x) &= \langle\exp[\eta(\mathrm{e}^{\mathrm{i}x} - 1)W]\rangle = \int_0^{\infty} P(W)\exp[\eta(\mathrm{e}^{\mathrm{i}x} - 1)W]\,\mathrm{d}W = \\ &= C^{(W)}(\eta(\mathrm{e}^{\mathrm{i}x} - 1)),\end{aligned} \tag{10.8}$$

where $C^{(W)}(\mathrm{i}x)$ is the characteristic function of the probability distribution $P(W)$, and $\mathrm{i}x$ is a parameter of the characteristic function.† The photocount distribution

† More general photocount distributions have been considered by Teich (1981), Saleh, Tavolacci and Teich (1981), Teich and Saleh (1981), Saleh and Teich (1982) and Saleh, Stoler and Teich (1982). For instance, for the Neyman type-A distribution

$$p(n, W) = \sum_{m=0}^{\infty} \frac{(\alpha m)^n\exp(-\alpha m)}{n!}\frac{W^m\exp(-W)}{m!}, \qquad \alpha > 0,$$

it holds that

$$C^{(n)}(\mathrm{i}x) = \exp\{W[\exp(\alpha(\mathrm{e}^{\mathrm{i}x} - 1)) - 1]\}.$$

This distribution has been used to describe radioluminescence from glass and cathodoluminescence.

$p(n, T, t)$ may be calculated, if the characteristic function $C^{(n)}(\mathrm{i}x)$ is given, by means of a Fourier transform

$$p(n, T, t) = \frac{1}{2\pi} \int_0^{2\pi} \exp(-\mathrm{i}xn)\, C^{(n)}(\mathrm{i}x)\, \mathrm{d}x. \tag{10.9}$$

The moments $\langle n^k \rangle$, $k = 0, 1, \dots$, may be calculated as

$$\langle n^k \rangle = \frac{\mathrm{d}^k}{\mathrm{d}(\mathrm{i}x)^k} C^{(n)}(\mathrm{i}x) \Big|_{\mathrm{i}x=0}. \tag{10.10}$$

Substituting $\exp(\mathrm{i}x) - 1 = \mathrm{i}s$ in (10.8) we obtain for the characteristic function $C^{(W)}(\mathrm{i}\eta s)$

$$C^{(W)}(\mathrm{i}\eta s) = \langle \exp(\mathrm{i}\eta s W) \rangle = \langle (1 + \mathrm{i}s)^n \rangle, \tag{10.11}$$

and hence $p(n, T, t)$ is obtained by means of the derivatives

$$p(n, T, t) = \frac{1}{n!} \frac{\mathrm{d}^n}{\mathrm{d}(\mathrm{i}s)^n} C^{(W)}(\mathrm{i}\eta s) \Big|_{\mathrm{i}s=-1}. \tag{10.12}$$

The factorial moments are equal to

$$\left\langle \frac{n!}{(n-k)!} \right\rangle = \frac{\mathrm{d}^k}{\mathrm{d}(\mathrm{i}s)^k} \langle (1 + \mathrm{i}s)^n \rangle \Big|_{\mathrm{i}s=0} = \frac{\mathrm{d}^k}{\mathrm{d}(\mathrm{i}s)^k} \langle \exp(\mathrm{i}s\eta W) \rangle \Big|_{\mathrm{i}s=0} = \eta^k \langle W^k \rangle. \tag{10.13}$$

One can define the cumulants $\varkappa_j^{(n)}$ as

$$\varkappa_j^{(n)} = \frac{\mathrm{d}^j}{\mathrm{d}(\mathrm{i}x)^j} \ln C^{(n)}(\mathrm{i}x) \Big|_{\mathrm{i}x=0}, \qquad j = 1, 2, \dots, \tag{10.14}$$

so the characteristic function can be written as

$$C^{(n)}(\mathrm{i}x) = \exp\left[\sum_{j=1}^{\infty} \varkappa_j^{(n)} \frac{(\mathrm{i}x)^j}{j!} \right]. \tag{10.15}$$

In the same way the factorial cumulants $\varkappa_j^{(W)}$ are defined by means of $C^{(W)}$ (note that $\varkappa_0^{(n)} = \varkappa_0^{(W)} = 0$). The relation between the cumulants $\varkappa_j^{(n)}$ and $\varkappa_k^{(W)}$ is obtained from (10.8)

$$\sum_{j=1}^{\infty} \varkappa_j^{(n)} \frac{(\mathrm{i}x)^j}{j!} = \sum_{j=1}^{\infty} \varkappa_j^{(W)} \frac{\eta^j (\mathrm{e}^{\mathrm{i}x} - 1)^j}{j!}. \tag{10.16}$$

Comparing coefficients of the powers x^j in the expansion of (10.16) we arrive at the same relations for the cumulants as for the moments (see (10.4)) (Mandel (1959, 1963a))†

† This may be seen by expanding the exponential functions in (10.8); we obtain

$$\sum_{j=1}^{\infty} \langle n^j \rangle \frac{(\mathrm{i}x)^j}{j!} = \sum_{j=1}^{\infty} \langle W^j \rangle \frac{\eta^j (\mathrm{e}^{\mathrm{i}x} - 1)^j}{j!},$$

which is the same relation as (10.16).

$$\varkappa_1^{(n)} = \eta\varkappa_1^{(W)},$$
$$\varkappa_2^{(n)} = \eta\varkappa_1^{(W)} + \eta^2\varkappa_2^{(W)},$$
$$\varkappa_3^{(n)} = \eta\varkappa_1^{(W)} + 3\eta^2\varkappa_2^{(W)} + \eta^3\varkappa_3^{(W)},$$
$$\varkappa_4^{(n)} = \eta\varkappa_1^{(W)} + 7\eta^2\varkappa_2^{(W)} + 6\eta^3\varkappa_3^{(W)} + \eta^4\varkappa_4^{(W)}, \tag{10.17}$$

etc. The significance of the cumulants is as follows

$$\varkappa_1^{(n)} = \frac{1}{\mathrm{i}\langle\exp(\mathrm{i}xn)\rangle}\frac{\mathrm{d}}{\mathrm{d}x}\langle\exp(\mathrm{i}xn)\rangle\bigg|_{\mathrm{i}x=0} = \langle n\rangle, \tag{10.18a}$$

$$\varkappa_2^{(n)} = \frac{1}{\mathrm{i}^2\langle\exp(\mathrm{i}xn)\rangle^2} \times$$
$$\times \left\{\langle\exp(\mathrm{i}xn)\rangle\frac{\mathrm{d}^2}{\mathrm{d}x^2}\langle\exp(\mathrm{i}xn)\rangle - \left[\frac{\mathrm{d}}{\mathrm{d}x}\langle\exp(\mathrm{i}xn)\rangle\right]^2\right\}\bigg|_{\mathrm{i}x=0} =$$
$$= \langle n^2\rangle - \langle n\rangle^2 = \langle(\Delta n)^2\rangle, \tag{10.18b}$$

$$\varkappa_3^{(n)} = \langle n^3\rangle - 3\langle n^2\rangle\langle n\rangle + 2\langle n\rangle^3, \tag{10.18c}$$

etc. The factorial cumulants $\varkappa_j^{(W)}$ have similar significance.

In general we obtain from (10.16)

$$\varkappa_k^{(n)} = \frac{\mathrm{d}^k}{\mathrm{d}(\mathrm{i}x)^k}\sum_{j=1}^{\infty}\varkappa_j^{(W)}\frac{\eta^j}{j!}\sum_{l=0}^{j}\binom{j}{l}(-1)^{j-l}\exp(\mathrm{i}xl)\bigg|_{\mathrm{i}x=0} =$$
$$= \sum_{j=1}^{k}\varkappa_j^{(W)}\eta^j\sum_{l=0}^{j}\frac{(-1)^{j+l}\,l^k}{l!\,(j-l)!}, \tag{10.19}$$

where the identity $\sum_{l=0}^{j}[(-1)^l\,l^k/l!\,(j-l)!] = 0$ for $j > k \geqq 0$ has been used.

It will be shown in Chapter 14 that the relations (10.4) and (10.17) (or (10.19)) are a consequence of the commutation rules for the field operators, which indicates a close formal connection between the classical and quantum descriptions of the statistical properties of light.

When the radiation is very weak, the first term in (10.17) will be dominant and the distribution $p(n, T, t)$ becomes Poissonian ($\varkappa_j^{(n)} = \eta\varkappa_1^{(W)} = \eta\langle W\rangle$ and $C^{(n)}(\mathrm{i}x) = \exp[\eta\langle W\rangle(\mathrm{e}^{\mathrm{i}x} - 1)]$ from (10.15), which is just the generating function for the Poisson distribution). Thus in very weak fields the emitted photoelectrons obey the statistics of classical particles. On the other hand, in strong fields, the last term in (10.17) will be dominant; therefore $\varkappa_j^{(n)} = \eta^j\varkappa_j^{(W)}$ and the distribution of n tends towards the distribution of ηW. The latter distribution is proportional to the probability distribution of the integrated classical intensity and in this case the output of a photoelectric detector can be regarded as a continuous signal.

Further discussions of the cumulants can be found in a paper by Cantrell (1970) and in the monograph by Saleh (1978).

10.2 Determination of the integrated intensity distribution from the photocount distribution

In practice optical fields are usually stationary and the photocount distribution $p(n, T, t) \equiv p(n, T)$ is independent of the time t. If we compute the number of emitted photoelectrons within a time interval of length T in many realizations and compute the number of intervals for the same n, then the normalization determines the photocount distribution $p(n, T)$. The corresponding measurement may be performed using a capacitor which is charged by the emitted photoelectrons so that its voltage is proportional to the number of photoelectrons emitted within the interval T. Or a multichannel analyzer registers automatically the number of intervals with the same channel number n, after a standardization of photoelectron pulses. With knowledge of $p(n, T)$ we can calculate the moments $\langle n^k \rangle$ as well as the factorial moments $\eta^k \langle W^k \rangle = \langle n!/(n-k)! \rangle$ and the statistics of the emitted photoelectrons can be fully specified (Armstrong and Smith (1967), Arecchi and Degiorgio (1972), Pike and Jakeman (1974)). The problem now arises of how information about the statistics of the optical field (that is about the probability distribution $P(W)$) can be obtained from the photocount distribution $p(n, T) \equiv p(n)$; this may be solved, from a mathematical point of view, by inverting the photodetection equation (10.3a).

Denoting $n!\, p(n) = M_n$ and $P(W) \exp(-W) = Q(W)$ in (10.3a) (we also put $\eta = 1$ for simplicity), we find that the inversion problem reduces to the well-known moment problem, that is to determine the function $Q(W)$ if the moment sequence

$$\int_0^\infty Q(W)\, W^k\, \mathrm{d}W = \langle W^k \rangle = M_k, \qquad k = 0, 1, \ldots \tag{10.20}$$

is given; the determination of the integrated intensity distribution $P(W)$ from the photocount distribution $p(n)$ is such a problem.

One can easily see that a formal solution of (10.3a) ($\eta = 1$) may be written in the form

$$P(W) = \exp(W) \sum_{n=0}^{\infty} (-1)^n\, p(n)\, \delta^{(n)}(W), \tag{10.21}$$

where $\delta^{(n)}(W)$ is the nth derivative of the Dirac δ-function.

Although such a mathematical quantity having an infinite number of terms may be meaningless sometimes†, it is well defined here because of the analyticity of the characteristic function

$$C^{(W)}(\mathrm{i}x) = \int_0^\infty P(W) \exp(\mathrm{i}xW)\, \mathrm{d}W; \tag{10.22}$$

† Such a series cannot be considered as defining a normal generalized function either in the space of test functions which are non-zero on a finite interval and continuous in all their derivatives (D-space) or in the space of test functions which decrease like an inverse power of the argument at infinity (S-space); it may be represented as a generalized function in the Z-space of test functions, which is a Fourier transform of the D-space. Such a generalized function is sometimes called an ultradistribution Z'. We shall discuss these questions in connection with the existence of the diagonal representation of the density matrix in Sec. 13.3.

this is analytic in the upper half of the complex x-plane, since $P(W) = 0$ for $W < 0$. Expressing the $\delta^{(n)}$-function in (10.21) by means of the Fourier integral,

$$\delta^{(n)}(W) = \frac{1}{2\pi} \int_{-\infty}^{+\infty} (-\mathrm{i}x)^n \exp(-\mathrm{i}xW)\,\mathrm{d}x, \tag{10.23}$$

we obtain from (10.21) (Wolf and Mehta (1964))

$$P(W) = \frac{1}{2\pi} \exp(W) \int_{-\infty}^{+\infty} C(\mathrm{i}x) \exp(-\mathrm{i}xW)\,\mathrm{d}x, \tag{10.24}$$

where

$$C(\mathrm{i}x) = \sum_{n=0}^{\infty} (\mathrm{i}x)^n\, p(n), \tag{10.25}$$

which is the characteristic function of $P(W)\exp(-W)$. From (10.22) and (10.24) $C(\mathrm{i}x) = C^{(W)}(\mathrm{i}x - 1)$. The analyticity of $C(\mathrm{i}x)$ makes it possible to obtain the values of $C(\mathrm{i}x)$ for all x by analytic continuation if the series (10.25) has a finite radius of convergence. Thus $P(W)$, determined in this way, is unique.

For example, with the Bose-Einstein distribution $p(n) = \langle n \rangle^n/(1 + \langle n \rangle)^{1+n}$ we obtain $C(\mathrm{i}x) = (1 + \langle n \rangle - \mathrm{i}x\langle n \rangle)^{-1}$ and $P(W) = \langle n \rangle^{-1} \exp(-W/\langle n \rangle)$, which is a negative exponential distribution. For the Poisson distribution $p(n) = \langle n \rangle^n \times \times \exp(-\langle n \rangle)/n!$ we obtain $C(\mathrm{i}x) = \exp[-\langle n \rangle (1 - \mathrm{i}x)]$ and $P(W) = \delta(W - \langle n \rangle)$, where δ is the Dirac function. Under certain assumptions, the first case describes chaotic light from traditional sources while the second case describes light from an ideal laser source (see Sec. 10.3).

As every term of (10.21) represents a generalized function with one-point support (roughly speaking only at this point is the function non-zero) we must know an infinite number of members of $p(n)$ to be able to reconstruct $P(W)$ as an ordinary function. This cannot be achieved in experiments and so we give a method based on a decomposition of $P(W)$ in terms of the Laguerre polynomials (Bédard (1967a), Piovoso and Bolgiano (1967), Morse and Feshbach (1953), Klauder and Sudarshan (1968)), in which every term has as its support the whole complex plane. Such a prescription for constructing $P(W)$ also provides an approximate means of determining $P(W)$ given a finite number of $p(n)$ from experiment. This method will be used in Sec. 13.3 in a generalized form to study the existence of the diagonal representation of the density matrix (Peřina and Mišta (1969)).

If $P(W) \in L_2$, we may look for a solution of the inverse problem for the photodetection equation in the form

$$P(W) = \exp[-(\zeta - 1)W] \sum_{j=0}^{\infty} c_j L_j^0(\zeta W), \tag{10.26}$$

where L_j^0 are the Laguerre polynomials defined by Morse and Feshbach (1953), Chap. 6

$$L_j^\mu(x) = [\Gamma(j + \mu + 1)]^2 \sum_{s=0}^{j} \frac{(-x)^s}{s!\,(j - s)!\,\Gamma(s + \mu + 1)}, \tag{10.27}$$

obeying the orthogonality condition

$$\int_0^\infty x^\mu \exp(-x)\, L_j^\mu(x)\, L_k^\mu(x)\, dx = \delta_{jk} \frac{[\Gamma(j+\mu+1)]^3}{\Gamma(j+1)}; \tag{10.28}$$

$\zeta \geqq 1$ is a real number and $\Gamma(\mu) = \int_0^\infty x^{\mu-1} \exp(-x)\, dx$ is the gamma function. Multiplying (10.26) by $L_k^0(\zeta W) \exp(-W)$, integrating over (ζW) and using (10.28) with $\mu = 0$, we obtain

$$c_k = \frac{\zeta}{(k!)^2} \int_0^\infty P(W) \exp(-W)\, L_k^0(\zeta W)\, dW = \zeta \sum_{s=0}^{k} p(s) \frac{(-\zeta)^s}{s!\,(k-s)!}, \tag{10.29}$$

where (10.27) and (10.3a) have also been used. The accuracy of an approximation in which $p(n)$ is obtained for the first $(N+1)$ terms $(n = 0, 1, \ldots, N)$ from measurements is given by (we put $\zeta = 1$ for simplicity)

$$\int_0^\infty [P(W) - P^{(N)}(W)]^2 \exp(-W)\, dW = \sum_{n=N+1}^{\infty} (n!)^2 c_n^2 < \varepsilon, \tag{10.30}$$

where $P^{(N)}$ denotes a function constructed as the Nth partial sum of (10.26) and ε is an arbitrarily small number. As $|L_k^0(x)| \leqq \exp(x/2)$ we have

$$\begin{aligned} |L_k^0(\zeta W) \exp[-(\zeta - 1) W]| &\leqq \exp(\zeta W/2) \exp[-(\zeta - 1) W] = \\ &= \exp(-\zeta W/2 + W) \leqq 1 \quad \text{for} \quad \zeta \geqq 2. \end{aligned}$$

In this case the series (10.26) will be uniformly convergent if $\sum_{j=0}^{\infty} |c_j| < \infty$.

Sometimes it is more suitable for calculations to use a slightly different decomposition showing the explicit dependence on the number of degrees of freedom (modes) M; this may be written in the form

$$P(W) = W^{M-1} \sum_{j=0}^{\infty} c_j L_j^{M-1}(W), \tag{10.31}$$

where

$$c_j = \frac{j!}{\Gamma(j+M)} \sum_{s=0}^{j} \frac{(-1)^s}{(j-s)!\,\Gamma(s+M)}\, p(s). \tag{10.32}$$

If we wish to determine a function $Q(W)$ from its moments $\langle W^k \rangle = \int_0^\infty Q(W)\, W^k\, dW = M_k$, it is sufficient to substitute $P(W) \exp(-W) = Q(W)$ and $p(s)\, s! = M_s$ in (10.31) and (10.32).

10.3 Short-time measurements

We begin with a description of the chaotic field which is the most usual state of the field in nature. Such a field is generated by a thermal source composed of many independent atomic radiators and consists of superpositions of waves of many different frequencies lying within some continuous range. These elementary waves can be regarded as independent waves of indeterminate phases. Using the central limit theorem of mathematical statistics, we can conclude that the field represents a Gaussian random process for the complex amplitude with zero mean value; this was verified by van Cittert (1934, 1939) and Janossy (1957, 1959) by direct calculations and a quantum-mechanical treatment leads to the same conclusion (Sec. 17.1).

A different situation occurs for laser sources (where stimulated emission is dominant) or for Čerenkov radiation; neither can be regarded as Gaussian processes.

Consider first a linearly polarized chaotic field. Since $V^{(r)}(t)$ is a Gaussian variable with zero mean value then, according to (2.7) and (2.31b), $V^{(i)}(t)$ is also a Gaussian variable with zero mean value, uncorrelated with $V^{(r)}(t)$. Further, as the variance of $V^{(r)}$ and the variance of $V^{(i)}$ are equal, say $2\sigma^2$ (compare (2.31a)), it follows that the joint probability distribution of $V^{(r)}$ and $V^{(i)}$ is given by

$$P(V^{(r)}, V^{(i)})\,\mathrm{d}V^{(r)}\,\mathrm{d}V^{(i)} = \frac{1}{2\pi\sigma^2}\exp\left[-\frac{V^{(r)2} + V^{(i)2}}{2\sigma^2}\right]\mathrm{d}V^{(r)}\,\mathrm{d}V^{(i)}. \tag{10.33}$$

However $(V^{(r)2} + V^{(i)2})$ is equal to the intensity I of the field and (10.33) can be rewritten in the form

$$P(I^{1/2}, \arg V)\,\mathrm{d}(I^{1/2})\,\mathrm{d}(\arg V) = \frac{1}{2\pi\sigma^2}\exp\left(-\frac{I}{2\sigma^2}\right) I^{1/2}\,\mathrm{d}(I^{1/2})\,\mathrm{d}(\arg V). \tag{10.34}$$

The probability density $P(I^{1/2}, \arg V)$ is independent of the phase of V, i.e. all phase angles in the range $0 \leqq \arg V \leqq 2\pi$ are equally probable. So, integrating over the phase and taking into account that $\langle I\rangle = 2\sigma^2$, we finally arrive at

$$P(I)\,\mathrm{d}I = \frac{1}{\langle I\rangle}\exp\left(-\frac{I}{\langle I\rangle}\right)\mathrm{d}I, \tag{10.35}$$

which is a negative exponential distribution in the intensity I. The kth moment is

$$\langle I^k\rangle = k!\,\langle I\rangle^k \tag{10.36}$$

and the variance of I is equal to

$$\langle(\Delta I)^2\rangle = \langle I^2\rangle - \langle I\rangle^2 = \langle I\rangle^2. \tag{10.37}$$

Now consider a resolving time T of the detector much smaller than the coherence time. The integrated intensity W is equal to IT since the intensity $I(t)$ may be regarded as practically constant over such a time interval and the photodetection equation becomes

$$p(n, T) \equiv p(n, T, t) = \int_0^\infty \frac{(\eta IT)^n}{n!}\exp(-\eta IT)\,P(I)\,\mathrm{d}I. \tag{10.38}$$

Such short-time measurements provide a deeper insight into the physical problem since the intensity has a simple physical meaning whereas the integrated intensity is a rather more complicated quantity. The distribution $p(n, T)$ of photoelectric counts can be obtained by substituting (10.35) into (10.38) and

$$p(n, T) = \frac{\langle n \rangle^n}{(1 + \langle n \rangle)^{1+n}}, \tag{10.39}$$

where $\langle n \rangle = \eta \langle I \rangle T$. This is the well-known Bose-Einstein distribution for n identical particles in one quantum state. (In one cell of a phase space, this can be understood as follows: in the direction of the beam photons cannot be distinguished in the linear distance cT ($T \approx 1/\Delta\nu$) as a consequence of the uncertainty principle and so they occupy one cell of the phase space; that is they behave as Bose-Einstein particles.) The variance (10.5) can be calculated as

$$\langle (\Delta n)^2 \rangle = \eta \langle I \rangle T + \eta^2 \langle (\Delta I)^2 \rangle T^2 = \langle n \rangle (1 + \langle n \rangle), \tag{10.40}$$

where (10.37) has been used. This variance exceeds $\langle n \rangle$, the value of the variance for the Poisson distribution.

The characteristic function for $P(W)$ is

$$C^{(W)}(\mathrm{i}\eta s) = \int_0^\infty \langle I \rangle^{-1} \exp\left(\mathrm{i}s\eta I T - \frac{I}{\langle I \rangle}\right) \mathrm{d}I = \frac{1}{1 - \mathrm{i}s\langle n \rangle}. \tag{10.41}$$

The results just obtained may be generalized to M degrees of freedom (e.g. $M = T/\tau_c$ if $T \gg \tau_c$, see Sec. 10.5) and to partially polarized light.

If $W = \sum_j W_j$ is a stochastic quantity, where the W_j with the probability distributions $P_j(W_j)$ are statistically independent, then the probability distribution $P(W)$ of W is equal to the convolution of the $P_j(W_j)$, namely

$$P(W) = \int_0^\infty \ldots \int_0^\infty \delta(W - \sum_j W_j) \prod_j P_j(W_j) \, \mathrm{d}W_j. \tag{10.42}$$

Consequently the corresponding characteristic function of $P(W)$ equals the product of the characteristic functions of the $P_j(W_j)$ and we have

$$\langle \exp(\mathrm{i}sW) \rangle = \prod_j \langle \exp(\mathrm{i}sW_j) \rangle, \tag{10.43}$$

or

$$C^{(W)}(\mathrm{i}s) = \prod_j C^{(W_j)}(\mathrm{i}s). \tag{10.44}$$

Applying this result to a system with M independent degrees of freedom with the same mean values $\langle n \rangle / M$ per degree of freedom, we obtain for the resulting characteristic function

$$C^{(W)}(\mathrm{i}\eta s) = \left(1 - \mathrm{i}s \frac{\langle n \rangle}{M}\right)^{-M}. \tag{10.45}$$

This gives (Mandel (1959), Troup (1965))

$$P(W) = \frac{1}{2\pi}\int_{-\infty}^{+\infty}\left(1 - \text{is}\frac{\langle n\rangle}{\eta M}\right)^{-M}\exp(-\text{is}W)\,\text{d}s =$$

$$= \left(\frac{\eta M}{\langle n\rangle}\right)^{M}\frac{W^{M-1}}{\Gamma(M)}\exp\left(-\frac{\eta WM}{\langle n\rangle}\right), \tag{10.46}$$

$$p(n, T) = \frac{1}{n!}\frac{\text{d}^n}{\text{d}(\text{is})^n}\left(1 - \text{is}\frac{\langle n\rangle}{M}\right)^{-M}\Bigg|_{\text{is}=-1} =$$

$$= \frac{\Gamma(n+M)}{n!\,\Gamma(M)}\left(1 + \frac{M}{\langle n\rangle}\right)^{-n}\left(1 + \frac{\langle n\rangle}{M}\right)^{-M}, \tag{10.47}$$

$$\langle W^k\rangle = \frac{1}{\eta^k}\frac{\text{d}^k}{\text{d}(\text{is})^k}\left(1 - \text{is}\frac{\langle n\rangle}{M}\right)^{-M}\Bigg|_{\text{is}=0} = \left(\frac{\langle n\rangle}{\eta}\right)^{k}\frac{\Gamma(k+M)}{\Gamma(M)\,M^k}, \tag{10.48}$$

and

$$\langle(\Delta n)^2\rangle = \langle n\rangle\left(1 + \frac{\langle n\rangle}{M}\right). \tag{10.49}$$

The number $\langle n\rangle/M$ represents the degeneracy parameter.

Consider next partially polarized light. If I_1 and I_2 are the intensities of two linearly independent modes of polarization of Gaussian light, we obtain for the characteristic function

$$C^{(W)}(\text{i}\eta s) = \frac{1}{1 - \text{is}\langle n_1\rangle}\,\frac{1}{1 - \text{is}\langle n_2\rangle}, \tag{10.50}$$

where

$$\langle n_j\rangle = \eta\langle I_j\rangle\, T, \qquad j = 1, 2. \tag{10.51}$$

Using (6.56), we can write

$$\langle n_1\rangle = \tfrac{1}{2}(1 + P)\,\langle n\rangle, \tag{10.52}$$

$$\langle n_2\rangle = \tfrac{1}{2}(1 - P)\,\langle n\rangle \tag{10.53}$$

(assuming $\langle I_1\rangle \geqq \langle I_2\rangle$ without loss of generality), where $\langle n\rangle = \langle n_1\rangle + \langle n_2\rangle$. Writing (10.50) in the form

$$C^{(W)}(\text{i}\eta s) = \frac{1}{\langle n_1\rangle - \langle n_2\rangle}\left[\frac{\langle n_1\rangle}{1 - \text{is}\langle n_1\rangle} - \frac{\langle n_2\rangle}{1 - \text{is}\langle n_2\rangle}\right], \tag{10.54}$$

we obtain (Mandel (1963b))

$$P(I) = \frac{1}{P\langle I\rangle}\left[\exp\left(-\frac{2I}{(1+P)\langle I\rangle}\right) - \exp\left(-\frac{2I}{(1-P)\langle I\rangle}\right)\right] \tag{10.55}$$

and

$$p(n, T) = \frac{1}{P\langle n\rangle}\left[\left(1 + \frac{2}{1+P\langle n\rangle}\right)^{-n-1} - \left(1 + \frac{2}{1-P\langle n\rangle}\right)^{-n-1}\right]. \tag{10.56}$$

The variance of n is equal to

$$\langle(\Delta n)^2\rangle = \langle n^2\rangle - \langle n\rangle^2 = \langle n\rangle\,[1 + \tfrac{1}{2}(1 + P^2)\langle n\rangle], \tag{10.57}$$

so that $M = 2/(1 + P^2)$. If the light is completely polarized $P = 1$ and we obtain (10.40); if it is unpolarized $P = 0$ and we have

$$\langle(\Delta n)^2\rangle = \langle n\rangle(1 + \tfrac{1}{2}\langle n\rangle) \tag{10.58}$$

and $M = 2$.

Another important kind of light is that from an ideal laser whose intensity can be regarded as stabilized (at least approximately) and the intensity probability distribution is therefore a δ-function distribution,

$$P(I) = \delta(I - \langle I\rangle). \tag{10.59}$$

The photodetection equation (10.38) gives in this case

$$p(n, T) = \frac{\langle n\rangle^n}{n!}\exp(-\langle n\rangle), \tag{10.60}$$

which is the Poisson distribution. This result follows simply in the quantum description (Sec. 13.1). The variance $\langle(\Delta n)^2\rangle = \langle n\rangle$ is in agreement with (10.5), since $\langle(\Delta W)^2\rangle = 0$ in this case.

Experimental verification of the applicability of the distributions (10.39) and (10.60) to chaotic and laser light respectively was carried out by Arecchi (1965) (also Arecchi, Berné and Burlamacchi (1966), Arecchi, Berné, Sona and Burlamacchi (1966)) and by others (further details can be found in Chapter 17). A more realistic description of laser light above the threshold of oscillations uses the model of the superposition of coherent and chaotic light, which will be discussed using quantum methods in Sec. 17.3. Chaotic light in Arecchi's experiments was obtained by sending the light of an amplitude-stabilized single-mode He-Ne laser onto a moving ground-glass disk. Such light, called pseudothermal light, was first introduced by Martienssen and Spiller (1964). The photocount distribution $p(n, T)$ for the chaotic field with M degrees of freedom given by (10.47), and for polarized and unpolarized light by (10.56), was experimentally verified by Martienssen and Spiller (1966a) and they have obtained very good agreement with theory. Experimental verification of the validity of the Bose-Einstein distribution for the laser operating below threshold was performed by Freed and Haus (1965).

10.4 Bunching effect of photons

We have noted that the photocount distribution will depart, in general, from the Poisson distribution and that the variance of the number of counts is usually in excess of that given by the Poisson distribution. This must imply that the photoelectrons (and also absorbed photons) do not arrive at random but they have bunching properties.†

In order to describe the bunching effect of photons, as well as the fourth-order correlation effects first observed by Hanbury Brown and Twiss, we need to use the fourth-order correlation function which, for linearly polarized light, is

$$\Gamma^{(2,2)}(\mathbf{x}_1, \mathbf{x}_2, \mathbf{x}_1, \mathbf{x}_2, 0, \tau, 0, \tau) = \langle I_1(t)\, I_2(t+\tau)\rangle = \\ = \langle V_1^*(t)\, V_1(t)\, V_2^*(t+\tau)\, V_2(t+\tau)\rangle; \tag{10.61}$$

this expresses the intensity correlation. For the chaotic field we must calculate the fourth-order correlation function for the Gaussian distribution function of the complex amplitude; it was shown by Wang and Uhlenbeck (1945) and Reed (1962) using the methods of stochastic processes and by Glauber (1963b, 1965) using the methods of quantum theory (cf. Sec. 17.1) that the $(m+n)$th-order correlation function for the chaotic field has the form

$$\Gamma^{(m,n)}(x_1, \ldots, x_{m+n}) = \delta_{mn} \sum_{\pi} \Gamma^{(1,1)}(x_1, x_{n+1}) \ldots \Gamma^{(1,1)}(x_n, x_{2n}), \tag{10.62}$$

where $\sum_{\pi}$ stands for the sum of $n!$ possible permutations of the indices $n+1, \ldots, 2n$ (or $1, \ldots, n$). Equation (10.62) shows that the $2n$th-order correlation function for the chaotic field is determined completely by the second-order correlation function. Putting $x_1 \equiv x_2 \equiv \ldots \equiv x_{2n}$, we obtain

$$\Gamma^{(n,n)}(x, \ldots, x) = \langle I^n(x)\rangle = n!\, \langle I(x)\rangle^n, \tag{10.63}$$

which agrees with (10.36).

Applying (10.62) to (10.61) we obtain (Wolf (1957))

$$\langle I_1(t)\, I_2(t+\tau)\rangle = \Gamma_{11}(0)\,\Gamma_{22}(0) + |\Gamma_{12}(\tau)|^2 = \\ = \langle I_1\rangle \langle I_2\rangle\, [1 + |\gamma_{12}(\tau)|^2] \tag{10.64}$$

and the correlation of the fluctuations $\Delta I_1 = I_1 - \langle I_1\rangle$ and $\Delta I_2 = I_2 - \langle I_2\rangle$ is

$$\langle \Delta I_1(t)\, \Delta I_2(t+\tau)\rangle = \langle I_1(t)\, I_2(t+\tau)\rangle - \langle I_1\rangle \langle I_2\rangle = \langle I_1\rangle \langle I_2\rangle\, |\gamma_{12}(\tau)|^2. \tag{10.65}$$

This shows that the modulus of the degree of coherence may be determined by measuring the correlation of intensity fluctuations and this result is the basis of correlation interferometry and spectroscopy.

If partially polarized light is assumed, we can obtain (in analogy to (10.57))

$$\langle \Delta I_1(t)\, \Delta I_2(t+\tau)\rangle = \tfrac{1}{2}\langle I_1\rangle \langle I_2\rangle (1 + P^2)\, |\gamma_{12}(\tau)|^2, \tag{10.66}$$

which shows that measurements of the correlation of intensity fluctuations may also provide information about the degree of polarization of beams.

---

† If the variance of n is less than that for Poisson distribution, photons exhibit sub-Poisson statistics. However, this is possible only for purely quantum fields (Chapter 22).

The conditional probability $p_c(t \mid t + \tau)\,\Delta\tau$ is the probability that a photoelectric count will be registered in the time interval $(t + \tau, t + \tau + \Delta\tau)$ if a count has been registered at a time t (Mandel (1963a)). The probability of observing a count at both times t and $t + \tau$ within $\mathrm{d}t$ and $\mathrm{d}\tau$ is obviously $\eta^2 I(t)\, I(t + \tau)\, \mathrm{d}t\, \mathrm{d}\tau$ and the probability of two counts separated by the interval τ is $\langle \eta^2 I(t)\, I(t + \tau)\rangle\, \mathrm{d}t\, \mathrm{d}\tau$. If this is divided by the ensemble average of the probability $\eta I(t)\, \mathrm{d}t$ of finding one count at t within $\mathrm{d}t$, we arrive at the conditional probability $p_c(\tau)$ of obtaining a second count τ seconds after the first one within $\mathrm{d}\tau$,

$$p_c(\tau)\,\mathrm{d}\tau = \frac{\eta^2 \langle I(t + \tau)\, I(t)\rangle\, \mathrm{d}t\, \mathrm{d}\tau}{\eta \langle I(t)\rangle\, \mathrm{d}t} = \eta\langle I(t)\rangle \left[1 + \tfrac{1}{2}(1 + P^2)\,|\gamma(\tau)|^2\right] \mathrm{d}\tau, \tag{10.67a}$$

where (10.66) has been used. Assuming completely polarized light we have, putting $P = 1$,

$$p_c(\tau) = \eta\langle I(t)\rangle \left[1 + |\gamma(\tau)|^2\right]. \tag{10.67b}$$

Since $|\gamma(\tau)| \approx 1$ for $\tau \ll \tau_c = 1/\Delta\nu$ and $|\gamma(\tau)| \approx 0$ for $\tau \gg \tau_c$, $p_c(\tau)$ starts at the value $2\eta\langle I(t)\rangle$ for small τ and tends to $\eta\langle I(t)\rangle$ for large τ. This illustrates the bunching properties of a chaotic photon beam. The bunching effect was observed by Arecchi, Gatti and Sona (1966) and Morgan and Mandel (1966). Their results are given in Figs. 10.1 and 10.2 and the experimental results are in very good agreement with the theoretical predictions.

In these measurements, light from a quasi-monochromatic chaotic source is incident on a photomultiplier with a fast response. The output of the photomultiplier

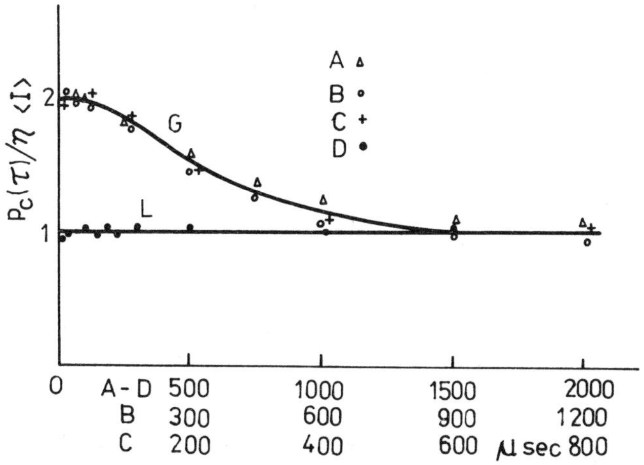

Fig. 10.1 The conditional probability $p_c(\tau)$ of a second count occurring at a time τ after a first has occurred at time $\tau = 0$. Experimental results apply to a laser (L, ●) and to an artificially synthesized chaotic source (G) created by passing laser radiation through ground glass which is rotated at speeds $v = 1.25$ cm/s (△), 2.09 cm/s (○), and 3.14 cm/s (+) (after F. T. Arecchi, E. Gatti and A. Sona, 1966, *Phys. Lett.* **20**, 27).

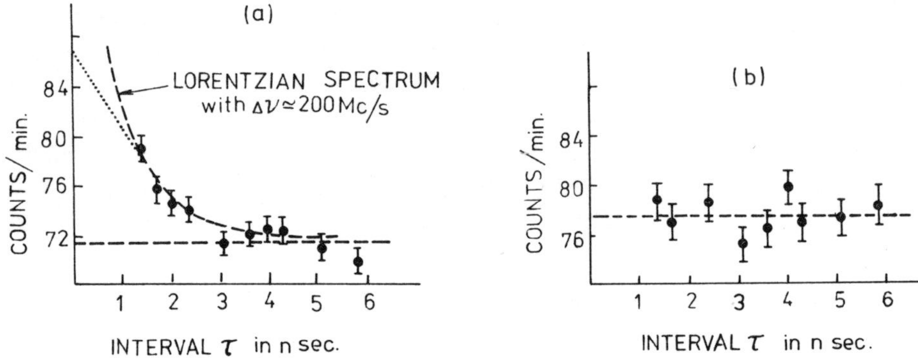

Fig. 10.2 Counting rates illustrating the phenomenon of photon bunching with light from (a) ^{198}Hg source, (b) tungsten lamp; in the latter case the wide frequency spectrum leads to intensity correlations in an immeasurably short time interval (after B. L. Morgan and L. Mandel, 1966, *Phys. Rev. Lett.* **16**, 1012).

reaches a coincidence counter via two different paths, one of which contains a time delay line so that only pulses in the output of the photomultiplier which are separated by the time delay τ can produce an output from the coincidence counter. The conditional probability $p_c(\tau)$ may then be determined from an analysis of the distribution of coincidences as a function of τ. While in the experiment of Morgan and Mandel light from a low pressure ^{198}Hg light source was used, in the experiment of Arecchi, Gatti and Sona pseudothermal light, obtained by randomizing the output of a laser by means of a rotating ground-glass disk, was used. A measurement of time intervals between photoelectrons emitted by a photodetector illuminated by a one-mode laser and by a pseudothermal source (Glauber (1967)) was reported by Bendjaballah (1969). All these experiments are in very good agreement with the theory.

For laser light, which is non-Gaussian light as discussed above, $\langle I_1(t)\, I_2(t+\tau)\rangle = \langle I_1\rangle \langle I_2\rangle$ (cf. (10.59)) and $\langle \Delta I_1(t)\, \Delta I_2(t+\tau)\rangle$ is practically zero. Consequently $p_c(\tau) = \eta\langle I(t)\rangle$ and no bunching effect occurs. This was indeed observed by Arecchi, Gatti and Sona (1966) (Fig. 10.1). It is true that the second term in (10.67b) (typical for chaotic light) is connected with departure from Poisson statistics, and is responsible for the bunching effect of photons. A very illuminating demonstration of the bunching effect of pseudothermal light and of its absence for laser light has been obtained by Bendjaballah (1971).

10.5 Hanbury Brown-Twiss effect – correlation interferometry and spectroscopy of the fourth order

Historically the correlation of photoelectrons in the output from two photodetectors was demonstrated first by Hanbury Brown and Twiss (1956a) in the form of a correlation between two photocurrents treated as continuous signals. The scheme of their apparatus is shown in Fig. 10.3. Light from a thermal source (mercury lamp) S was divided into two beams by a half-silvered mirror M and fell on two photocells P_1, P_2 whose outputs were sent through bandlimited amplifiers A_1, A_2 to a correlater C (one of the photomultipliers is movable to enable changes of the spatial coherence). The averaged value of the product of the intensity fluctuations was recorded by an integrating motor Mo. The experiment has also been repeated by Martienssen and Spiller (1964) with a pseudothermal source.

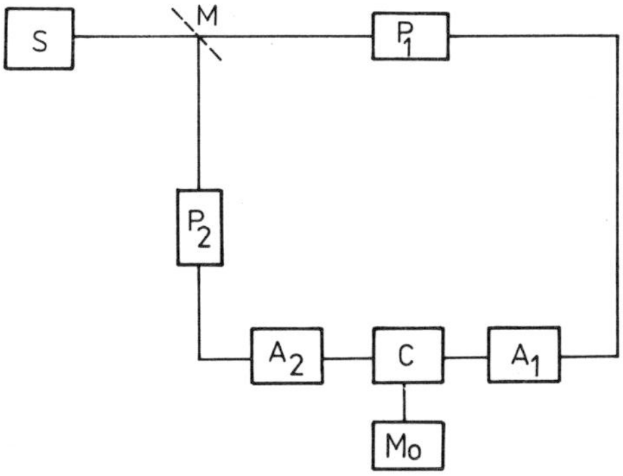

Fig. 10.3 An outline of the apparatus for demonstrating the correlation between intensity fluctuations; S is a thermal source, M is a half-silvered mirror, P_1 and P_2 are photomultipliers, A_1 and A_2 are amplifiers, C is a correlater and Mo is an integrating motor.

To explain this experiment we must use a deeper physical model working with the field intensity rather than with the integrated intensity.

The variance of the number of photoelectrons in one photodetector (assuming linearly polarized light) will be determined as follows: using the photodetection equation (10.3a) we have

$$\langle n^2 \rangle = \langle n \rangle + \eta^2 \iint\limits_t^{t+T} \langle I(t')\, I(t'') \rangle \, dt' \, dt'' = \\ = \langle n \rangle + \eta^2 \iint\limits_0^T \langle I(t + t' - t'')\, I(t) \rangle \, dt' \, dt'', \tag{10.68}$$

where $\langle n \rangle = \eta \langle I \rangle T$ and the stationary condition is assumed. Denoting (Mandel (1958, 1959, 1963a))

$$T\xi(T) = \iint\limits_0^T |\gamma(t' - t'')|^2 \, dt' \, dt'' = 2 \int\limits_0^T (T - t') |\gamma(t')|^2 \, dt', \tag{10.69}$$

we obtain from (10.68)

$$\langle n^2 \rangle = \langle n \rangle + \eta^2 \langle I \rangle^2 T^2 + \eta^2 \langle I \rangle^2 T^2 \frac{\xi(T)}{T} = \\ = \langle n \rangle^2 + \langle n \rangle \left(1 + \langle n \rangle \frac{\xi(T)}{T} \right), \tag{10.70}$$

where (10.64) has also been used. Since $|\gamma(t'-t'')| \leqq 1$ then $\xi(T) \leqq T$, and if $T \ll 1/\Delta\nu$ then $|\gamma(\tau)|$ is practically equal to unity and $\xi(T) \approx T$. Thus the variance following from (10.70) agrees with (10.40). However, if $T \gg 1/\Delta\nu$, then, since $\gamma(\tau)$ is non-vanishing only on an interval of length of a few $1/\Delta\nu$,

$$\xi(T) \approx \xi(\infty) = \int_{-\infty}^{+\infty} |\gamma(t)|^2 \, dt; \tag{10.71}$$

this can be adopted as the coherence time τ_c (cf. (2.53), Mandel (1958, 1959, 1963a)).

From (10.70) we obtain

$$\langle(\Delta n)^2\rangle = \langle n\rangle \left(1 + \langle n\rangle \frac{\tau_c}{T}\right), \tag{10.72}$$

which is the variance of the number of bosons divided into T/τ_c cells of phase space – the length of the cell being $c\tau_c$. The photocount distribution (10.47) becomes for $M = T/\tau_c$

$$p(n, T) = \frac{\Gamma(n + T/\tau_c)}{n!\,\Gamma(T/\tau_c)} \left(1 + \frac{T}{\tau_c\langle n\rangle}\right)^{-n} \left(1 + \frac{\langle n\rangle\,\tau_c}{T}\right)^{-T/\tau_c}. \tag{10.73}$$

It can be shown (Bédard, Chang and Mandel (1967)) that this Mandel-Rice formula (10.47) is valid not only for $T \gg \tau_c$ but represents a good approximation for all T. Similar results have been obtained for the superposition of coherent and chaotic light and are discussed in Sec. 17.3 (Horák, Mišta and Peřina (1971a, b), Peřina, Peřinová and Mišta (1971), Mišta and Peřina (1971)).

Departures from classical Poissonian counting statistics, for which $\langle(\Delta n)^2\rangle = \langle n\rangle$ (valid for ideal laser light), depend on the degeneracy parameter

$$\delta = \frac{\langle n\rangle\,\tau_c}{T} = \eta\langle I\rangle\,\tau_c, \tag{10.74}$$

which gives the mean number of counts per cell of phase space: $\delta \ll 1$ for thermal light and $\delta \gg 1$ for laser light as noted above. The fact that $\delta \ll 1$ for thermal light implies that a direct measurement of the excess photon noise is difficult. The position is substantially better for pseudothermal light.

The correlation between counts from two photodetectors may be obtained in the same manner

$$\langle n_1 n_2\rangle = \eta_1\eta_2 \iint_t^{t+T} \langle I_1(t')\,I_2(t'')\rangle \, dt' \, dt'' =$$
$$= \eta_1\eta_2 \iint_0^T \langle I_1(t + t' - t'')\,I_2(t)\rangle \, dt' \, dt''. \tag{10.75}$$

If we assume the cross-spectral purity condition $\gamma_{12}(\tau) = \gamma_{12}(0)\,\gamma_{11}(\tau)$ to be valid, we obtain

$$\langle n_1 n_2\rangle = \langle n_1\rangle\langle n_2\rangle \left(1 + \frac{\xi(T)}{T} |\gamma_{12}(0)|^2\right) \tag{10.76a}$$

and

$$\langle \Delta n_1 \, \Delta n_2 \rangle = \frac{\xi(T)}{T} \, |\gamma_{12}(0)|^2 \langle n_1 \rangle \langle n_2 \rangle. \tag{10.76b}$$

The normalized correlation is

$$\frac{\langle \Delta n_1 \, \Delta n_2 \rangle}{[\langle (\Delta n_1)^2 \rangle \langle (\Delta n_2)^2 \rangle]^{1/2}} = \frac{\delta}{1 + \delta} \, |\gamma_{12}(0)|^2, \tag{10.77}$$

where the degeneracy parameter δ is given by (10.74) and $\langle n_1 \rangle$ and $\langle n_2 \rangle$ are assumed to be the same. As $\delta \approx 10^{-3}$ or less for a thermal source, the effect is very difficult to be observed. For pseudothermal light the conditions for observation are much better since $\delta \approx 10^{12}$ or more.

If partially polarized light is assumed one obtains, using (10.66),

$$\langle (\Delta n)^2 \rangle = \langle n \rangle \left[1 + \frac{1}{2} (1 + P^2) \langle n \rangle \frac{\xi(T)}{T} \right], \tag{10.78}$$

and

$$\langle \Delta n_1 \, \Delta n_2 \rangle = \frac{1}{2} (1 + P^2) \langle n_1 \rangle \langle n_2 \rangle \, |\gamma_{12}(0)|^2 \frac{\xi(T)}{T}. \tag{10.79}$$

Experiments using the coincidence technique instead of the correlater have been carried out by Twiss, Little and Hanbury Brown (1957), Twiss and Little (1959), Rebka and Pound (1957), Brannen, Ferguson and Wehlau (1958), and Martienssen and Spiller (1964), among others.

If non-Gaussian light is assumed one can write for the correlation of intensities

$$\begin{aligned} \langle I_1(t) \, I_2(t + \tau) \rangle &= \eta^2 [\langle I_1 \rangle \langle I_2 \rangle + \langle \Delta I_1(t) \, \Delta I_2(t + \tau) \rangle] = \\ &= \eta^2 \langle I_1 \rangle \langle I_2 \rangle (1 + |\zeta_{12}(\tau)|^2), \end{aligned} \tag{10.80}$$

where

$$|\zeta_{12}(\tau)|^2 = \frac{\langle \Delta I_1(t) \, \Delta I_2(t + \tau) \rangle}{\langle I_1 \rangle \langle I_2 \rangle}. \tag{10.81}$$

For thermal light $|\zeta_{12}(\tau)| = |\gamma_{12}(\tau)|$. The correlation of counts is given as

$$\langle n_1 n_2 \rangle = \langle n_1 \rangle \langle n_2 \rangle + \eta_1 \eta_2 \iint_0^T \langle \Delta I_1(t') \, \Delta I_2(t'') \rangle \, dt' \, dt''. \tag{10.82}$$

The correlation of fluctuations is therefore equal to

$$\langle \Delta n_1 \, \Delta n_2 \rangle = \langle n_1 \rangle \langle n_2 \rangle \frac{\bar{\xi}(T)}{T}, \tag{10.83}$$

where

$$\begin{aligned} \bar{\xi}(T) &= \frac{1}{T \langle I_1 \rangle \langle I_2 \rangle} \iint_0^T \langle \Delta I_1(t') \, \Delta I_2(t'') \rangle \, dt' \, dt'' = \\ &= \frac{2}{T \langle I_1 \rangle \langle I_2 \rangle} \int_0^T (T - \tau) \langle \Delta I_1(\tau) \, \Delta I_2(0) \rangle \, d\tau, \end{aligned} \tag{10.84}$$

if the stationary condition is assumed. As we have pointed out no Hanbury Brown-Twiss effect will occur for light from an ideal laser since $\Delta I \approx 0$ and consequently $\langle \Delta n_1 \Delta n_2 \rangle \approx 0$. There is no departure from the classical Poisson statistics in this case.

These expressions can serve to define the coherence time for arbitrary statistics of the field (Mandel (1981a)). With the help of (10.81) one can define the coherence time

$$\tau_c = \int_{-\infty}^{+\infty} \left| \frac{\zeta(\tau)}{\zeta(0)} \right|^2 d\tau, \tag{10.85a}$$

which reduces to (10.71) for chaotic fields. Some corrections to the coherence time, coherence area and coherence volume, based on the photocount statistics for the superposition of coherent and chaotic fields, have also been found (Peřina and Mišta (1974)). Similarly we have for the coherence volume in analogy to (3.61)

$$V_c = c \iiint_{-\infty}^{+\infty} \left| \frac{\zeta(x, y, \tau)}{\zeta(0, 0, 0)} \right|^2 dx\, dy\, d\tau, \tag{10.85b}$$

provided that ζ_{12} is stationary in space. For the coherence area for light of arbitrary statistics we have

$$A_c = \iint_{-\infty}^{+\infty} \left| \frac{\zeta(x, y, 0)}{\zeta(0, 0, 0)} \right|^2 dx\, dy. \tag{10.85c}$$

The correlation technique has been applied to measurements of angular diameters of stars by Hanbury Brown and Twiss (1956b, 1958) and forms the basis of correlation interferometry. A scheme of the correlation interferometer is shown in Fig. 10.4. According to (10.79) such correlation measurements allow $|\gamma_{12}(0)|$ to be determined when the separation d of the mirrors is changed (the results obtained will follow the curve shown in Fig. 3.5). The angular radius of a star can be determined in the way described in connection with the Michelson stellar inter-

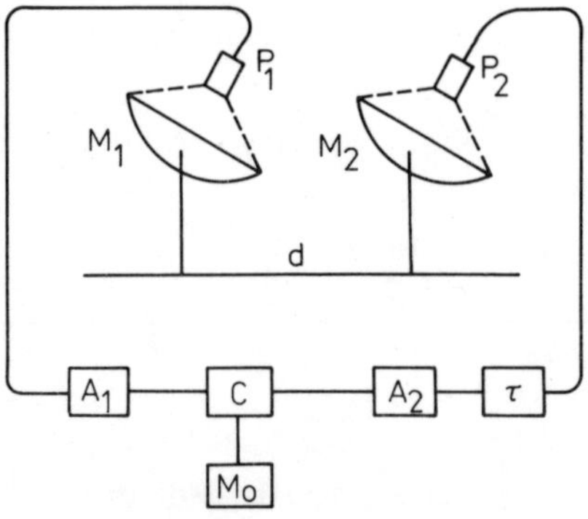

Fig. 10.4 An outline of a stellar correlation interferometer; M_1 and M_2 are mirrors, P_1 and P_2 are photodetectors, A_1 and A_2 are amplifiers, τ is a time delay line, C is a correlater and Mo is an integrating motor.

ferometer (Chapter 3). A large stellar interferometer was built at Narrabri in Australia (Hanbury Brown (1964), see also Wolf (1966)). Further applications of intensity interferometry in physics and astronomy have been reviewed by Twiss (1969).

There are some advantages in the correlation technique for measuring diameters of stars compared with the use of the Michelson stellar interferometer. Correlation measurements involve the intensity $I = |V|^2$ which varies slowly in comparison with the rapidly varying amplitude V, and the phase of V is not involved (the degree of coherence appears in the above equations only as the modulus $|\gamma_{12}(0)|$); consequently phase distortions due to turbulence in the air, changes in the index of refraction of the atmosphere, etc., do not affect the measurements, whereas they may be intolerable for the Michelson stellar interferometer. Also much larger separations d (hundreds of meters) can be used in the correlation interferometer than with the Michelson stellar interferometer, making it possible to increase the resolving power of the interferometer.

On the other hand, the phase information about the degree of coherence is lost in the correlation measurements, even if such measurements (as well as photon bunching measurements) provide information about the degree of polarization and the modulus of the degree of coherence. The problem of recovering the phase of the degree of coherence was discussed in Sec. 4.4. Some information about the spectrum can also be gained from the spectrum $h(\nu)$ of $|\gamma(\tau)|^2$ using measurements of $p_c(\tau)$ (Mandel (1963a, 1967), Sillitto (1963), Wolf (1965)), since

$$h(\nu) = \int_{-\infty}^{+\infty} |\gamma(\tau)|^2 \exp(\mathrm{i}\, 2\pi\nu\tau)\, \mathrm{d}\tau = \int_{0}^{\infty} g(\nu')\, g(\nu' + \nu)\, \mathrm{d}\nu', \tag{10.85d}$$

where g is the normalized power spectrum of light. An experimental recovery of the phase of γ can also be based on the measurements of the sixth-order correlation function (Gamo (1963a, b), Mandel (1963a)). Some other possibilities of determining the phase of γ were discussed in Sec. 4.4.

Finally let us note that an interesting use of the intensity correlation technique in the problem of scattering and of the determination of the phase of the scattering matrix has been discussed by Goldberger, Lewis and Watson (1963) and Goldberger and Watson (1964, 1965). Some techniques for spectroscopic analysis based on using intensity fluctuations have been studied by Goldberger, Lewis and Watson (1966). One of the first intensity correlation linewidth measurements has been performed by Phillips, Kleiman and Davis (1967), while a determination of linewidth using the measurement of photocount statistics has been referred to by Jakeman, Oliver and Pike (1968a). The spectroscopic correlation measurements employ the sensitivity of the photocount statistics to the spectral parameters of radiation, such as the mean frequency and the halfwidth of the spectrum (see Sec. 17.3). These spectral parameters are then determined by the fit of the experimental data and the theoretical model. The resolving power of correlation spectroscopy lies within the range $1 - 10^8$ Hz.

10.6 Higher-order coherence effects

We have shown that light from a chaotic source can be described by a Gaussian probability distribution function, which is fully specified by the second-order correlation function (the first-order moment is zero). For a full description of the statistics of non-Gaussian light (for example laser light) the moments – correlation functions – of all orders are needed. The correlation functions of higher order can in principle be measured by means of multiple coincidence measurements.

Consider a system of N photodetectors illuminated by beams of partially coherent linearly polarized light. If n_j is the number of counts registered by the jth detector in a time interval $(t_j, t_j + T_j)$, then, according to (10.3a),

$$p(n_1, \ldots, n_N, T_1, \ldots, T_N, t_1, \ldots, t_N) = \left\langle \prod_{j=1}^{N} \frac{[\eta_j W_j(T_j, t_j)]^{n_j}}{n_j!} \times \right.$$
$$\left. \times \exp\left[-\eta_j W_j(T_j, t_j)\right] \right\rangle, \tag{10.86}$$

where η_j is the photoefficiency of the jth photodetector and

$$W_j(T_j, t_j) = \int_{t_j}^{t_j+T_j} I_j(\mathbf{x}_j, t'_j)\, \mathrm{d}t'_j. \tag{10.87}$$

The brackets in (10.86) stand for the average over the N-fold probability distribution $P(W_1, \ldots, W_N)$. As with our previous results, we can introduce the characteristic function

$$C^{(\{n_j\})}(\{\mathrm{i}x_j\}) = \left\langle \prod_{j=1}^{N} \exp(\mathrm{i}x_j n_j) \right\rangle = \sum_{\{n_j\}} p(\{n_j\}, \{T_j\}, \{t_j\}) \prod_{j=1}^{N} \exp(\mathrm{i}x_j n_j), \tag{10.88}$$

where $\{n_j\} \equiv (n_1, \ldots, n_N)$, etc., and we obtain from (10.86)

$$C^{(\{n_j\})}(\{\mathrm{i}x_j\}) = C^{(\{W_j\})}(\{\eta_j(\mathrm{e}^{\mathrm{i}x_j} - 1)\}), \tag{10.89}$$

where

$$C^{(\{W_j\})}(\{\mathrm{i}\eta_j x_j\}) = \left\langle \prod_{j=1}^{N} \exp(\mathrm{i}\eta_j x_j W_j) \right\rangle = \sum_{\{n_j\}} p(\{n_j\}, \{T_j\}, \{t_j\}) \prod_{j=1}^{N} (1 + \mathrm{i}x_j)^{n_j}. \tag{10.90}$$

The correlation function of the numbers of counts is

$$\langle n_1 \ldots n_N \rangle = \prod_{j=1}^{N} \frac{\partial}{\partial(\mathrm{i}x_j)} \left\langle \prod_{k=1}^{N} \exp(\mathrm{i}x_k n_k) \right\rangle \Bigg|_{\{\mathrm{i}x_j\}=0} =$$
$$= \eta_1 \ldots \eta_N \int_{t_1}^{t_1+T_1} \cdots \int_{t_N}^{t_N+T_N} \Gamma^{(N,N)}(\mathbf{x}_1, \ldots, \mathbf{x}_N, \mathbf{x}_N, \ldots, \mathbf{x}_1, t'_1, \ldots, t'_N, t'_N, \ldots, t'_1)\, \mathrm{d}t'_1 \ldots \mathrm{d}t'_N, \tag{10.91}$$

where $\Gamma^{(N,N)}$ is the $2N$th-order correlation function of the complex amplitudes. Thus the correlation of photoelectric counts is fully determined by the $2N$th-order correlation function and if the field is assumed to be stationary, then $\int_{t_j}^{t_j+T_j}$ can be replaced by $\int_0^{T_j}$ and the correlation function is given by

$$\Gamma^{(N,N)}(\mathbf{x}_1, \ldots, \mathbf{x}_N, \mathbf{x}_N, \ldots, \mathbf{x}_1, T_1, \ldots, T_N, T_N, \ldots, T_1) = \\ = \frac{\partial^N}{\partial T_1 \ldots \partial T_N} \frac{\langle n_1 \ldots n_N \rangle}{\eta_1 \ldots \eta_N}. \tag{10.92}$$

Writing (10.91) in the form

$$\langle n_1 \ldots n_N \rangle = \eta_1 \ldots \eta_N \langle W_1 \ldots W_N \rangle, \tag{10.93}$$

we can show that

$$\langle \Delta n_1 \ldots \Delta n_N \rangle = \eta_1 \ldots \eta_N \langle \Delta W_1 \ldots \Delta W_N \rangle. \tag{10.94}$$

This correlation of fluctuations will be very small for laser light since $\Delta W \approx 0$.

Some applications of the N-fold formalism to the detection of Gaussian light have been studied by Bédard (1967d), who has shown that information about the spectral profile of the light can be obtained from the three-fold joint photocount distribution. The two-fold photocount distribution for Gaussian light has been experimentally investigated by Arecchi, Berné and Sona (1966). Measurements of higher-order quantities have been made by Davidson and Mandel (1968), Davidson (1969) and Corti and Degiorgio (1974, 1976a, b), who directly measured the sixth-order correlation functions, further Chang, Detenbeck, Korenman, Alley and Hochuli (1967), Chang, Korenman and Detenbeck (1968) and Chang, Korenman, Alley and Detenbeck (1969) experimentally determined the factorial cumulants $\varkappa_2^{(W)}$, $\varkappa_3^{(W)}$ and $\varkappa_4^{(W)}$ of the second, third and fourth order for laser light at the threshold of oscillations; the third reduced factorial moment $H_3 = \langle n(n-1)(n-2)\rangle/\langle n\rangle^3 - 1$ was experimentally obtained by Arecchi (1969).

An exact approach to the study of the coherence properties of optical fields including higher-order correlation effects based on the theory of stochastic processes, and particularly a study of experiments described by the fourth-order correlation fucntions, was given by Picinbono and Boileau (1968) and Boileau and Picinbono (1968). The statistical properties of photoelectrons, including some special models of the field, have been studied in a similar way by Rousseau (1969a).

Multifold photocount statistics have been further investigated by Fillmore (1969), Cantrell (1970), Blake and Barakat (1973, 1976), Aoki (1977), Bendjaballah (1979) and Srinivasan and Gururajan (1981).

We have now completed the classical and semiclassical treatments of the coherence properties of optical fields. The next chapters will be devoted to the quantum description of the statistical properties of fields, based on quantum electrodynamics, as well as to the relation between the classical and quantum descriptions, and to the application of the general quantum coherent-state technique to the study of the quantum statistical properties of radiation in random and nonlinear media.

CHAPTER 11

BASIC IDEAS OF THE QUANTUM THEORY OF THE ELECTROMAGNETIC FIELD

With the improvement of measuring techniques and the development of new techniques (e.g. coincidence circuits) new possibilities of detecting single photons in the visible region occurred, allowing the measurement of the statistics of photons; this was in contrast to earlier investigations where the detecting devices were able to respond only to intensive fluxes of photons. Thus the particle character of the electromagnetic field may be studied directly. Under such circumstances it is obviously advantageous to describe the coherence phenomena in the field by the methods of quantum electrodynamics, where a field operator corresponds to the classical field and the quanta of the field correspond to the classical wave field. As is well-known this correspondence is realized by the second quantization of the field.

The standard treatments of quantum electrodynamics are formulated on the basis of perturbation theory and such formulations lead automatically to discussions of processes involving a few particles. The treatment of many-particle processes is mathematically complex and consequently the classical limit of quantum electrodynamics has never been fully developed using such an approach. In contrast a discussion of cooperative phenomena such as coherence involves a considerable number of photons and their mutual relationships in space and time. (It is only in this way that the operation of a laser source – a typical nonlinear device functioning only at sufficiently high field intensities – can be explained.) Consequently it is necessary to develop quantum electrodynamics in a manner which exhibits clearly the classical limit. To this end we will not use the Fock states of occupation numbers, the basis of the usual formulations, but instead use coherent states expressing the cooperation of photons; these are infinite linear combinations of the Fock states and have the advantage that they retain information about the phase state of the field – information lost when the Fock states are used. With these coherent states it is possible to obtain a close formal correspondence between the quantum and classical descriptions; of course this does not imply the physical equivalence of these descriptions. In general, one can say that the classical description fails for any measurements which are sensitive to fluctuations of the physical vacuum. Such measurements are described by functions of the field operators in orderings other than the normal ordering. In the normal order all creation operators stand to the left of all the annihilation operators and the vacuum expectation value is zero; in other orderings the vacuum expectation value may be non-zero. The principles of the quantum theory of optical coherence employing the coherent states were given by Glauber (1963a – c, 1964, 1965, 1966a, b, 1967, 1969, 1970, 1972).

11.1 Quantum description of the field

We give in this chapter some basic ideas and results of the second quantization of the electromagnetic field in vacuo. For a fuller treatment we refer the reader to the texts (Akhiezer and Berestetsky (1965), Bogolyubov and Shirkov (1959), Dirac (1958), Schweber (1961), Messiah (1961, 1962)) or to the texts with particular attention to quantum optics (Louisell (1964, 1973), Yariv (1975), Klauder and Sudarshan (1968), Loudon (1973)).

In the quantum theory of the electromagnetic field the electric and magnetic strength vectors $\hat{\mathbf{E}}$ and $\hat{\mathbf{H}}$ are regarded as operators in the space of states which describe the field. These quantities satisfying Maxwell's equations can be derived from the vector potential operator $\hat{\mathbf{A}}$ according to the relations

$$\hat{\mathbf{E}} = -\frac{1}{c}\frac{\partial \hat{\mathbf{A}}}{\partial t}, \tag{11.1a}$$

$$\hat{\mathbf{H}} = \nabla \times \hat{\mathbf{A}}. \tag{11.1b}$$

Considering a field in a normalization volume L^3 with periodic boundary conditions we can decompose the vector potential operator $\hat{\mathbf{A}}$ in the form of plane waves as follows

$$\hat{\mathbf{A}}(x) = \frac{(2\pi\hbar c)^{1/2}}{L^{3/2}} \sum_{\mathbf{k}} \sum_{s=1}^{2} k^{-1/2}\{\mathbf{e}^{(s)}(\mathbf{k})\, \hat{a}_{\mathbf{k}s} \exp[\mathrm{i}(\mathbf{k}\,.\,\mathbf{x} - ckt)] + $$
$$+ \mathbf{e}^{(s)*}(\mathbf{k})\, \hat{a}^{+}_{\mathbf{k}s} \exp[-\mathrm{i}(\mathbf{k}\,.\,\mathbf{x} - ckt)]\}, \tag{11.2}$$

where $\hat{a}_{\mathbf{k}s}$ is the annihilation operator for a photon of momentum $\hbar\mathbf{k}$ and polarization s, and $\hat{a}^{+}_{\mathbf{k}s}$ (which is the Hermitian conjugate to $\hat{a}_{\mathbf{k}s}$) is a photon creation operator. These operators obey the following commutation rules

$$[\hat{a}_{\lambda}, \hat{a}^{+}_{\lambda'}] = \hat{1}\delta_{\lambda\lambda'}, \quad [\hat{a}_{\lambda}, \hat{a}_{\lambda'}] = [\hat{a}^{+}_{\lambda}, \hat{a}^{+}_{\lambda'}] = \hat{0}; \tag{11.3}$$

$\lambda \equiv (\mathbf{k}, s)$ is the mode index, $[\hat{A}, \hat{B}] = \hat{A}\hat{B} - \hat{B}\hat{A}$ stands for the commutator of the operators $\hat{A}$, $\hat{B}$, and $\delta_{\lambda\lambda'} = \delta_{\mathbf{k}\mathbf{k}'}\delta_{ss'}$. The unit polarization vector $\mathbf{e}^{(s)}(\mathbf{k})$ satisfies

$$\mathbf{e}^{(s)*}(\mathbf{k})\,.\,\mathbf{e}^{(s')}(\mathbf{k}) = \delta_{ss'}, \tag{11.4a}$$

$$\mathbf{k}\,.\,\mathbf{e}^{(s)}(\mathbf{k}) = 0, \tag{11.4b}$$

that is the vectors $\mathbf{e}^{(1)}$, $\mathbf{e}^{(2)}$ and $\mathbf{k}/k$ $(k = |\mathbf{k}|)$ form an orthogonal system

$$\sum_{s=1}^{2} e^{(s)*}_{\mu}(\mathbf{k})\, e^{(s)}_{\nu}(\mathbf{k}) + \frac{k_\mu k_\nu}{k^2} = \delta_{\mu\nu}, \qquad \mu,\nu = 1, 2, 3. \tag{11.5}$$

In (11.2) only transverse polarizations are present $(s = 1, 2)$ since longitudinal and scalar photons are eliminated by the Lorentz condition $\sum_{\mu=1}^{4} \partial \hat{A}_\mu/\partial x_\mu = 0$ $(x_4 = \mathrm{i}ct)$.

We define the number operator $\hat{n} = \hat{a}^+\hat{a}$ (omitting the mode index λ for simplicity), for which, with the use of the commutation rules (11.3),

$$[\hat{a}, \hat{n}] = \hat{a}, \tag{11.6a}$$

$$[\hat{n}, \hat{a}^+] = \hat{a}^+. \tag{11.6b}$$

Repeated use of these relations gives

$$\hat{n}\hat{a}^m = \hat{a}^m(\hat{n} - m), \tag{11.7a}$$

$$\hat{n}\hat{a}^{+m} = \hat{a}^{+m}(\hat{n} + m) \tag{11.7b}$$

for $m = 0, 1, 2, \ldots$ and also

$$\hat{a}^{+m}\hat{a}^m = \hat{n}(\hat{n} - 1)\ldots(\hat{n} - m + 1), \tag{11.8a}$$

$$\hat{a}^m\hat{a}^{+m} = (\hat{n} + 1)(\hat{n} + 2)\ldots(\hat{n} + m). \tag{11.8b}$$

Assume that there exists an eigenstate $|n\rangle$ of the number operator $\hat{n}$ such that

$$\hat{n}|n\rangle = n|n\rangle, \tag{11.9}$$

where n is a real number. From this equation $\langle n|\hat{n}|n\rangle = n\langle n|n\rangle$, where $\langle n|$ is Hermitian conjugate to $|n\rangle$. Thus $n = \langle n|\hat{a}^+\hat{a}|n\rangle/\langle n|n\rangle \geqq 0$. Further, if $|n\rangle$ is an eigenstate of $\hat{n}$, then $\hat{a}^+\hat{a}(\hat{a}^k|n\rangle) = (n - k)(\hat{a}^k|n\rangle)$ from (11.7a), i.e. the state $(\hat{a}^k|n\rangle)$ is also an eigenstate of $\hat{n}$ with eigenvalues $(n - k)$, $k = 0, 1, 2, \ldots$. As $n \geqq 0$, n must be a positive integer or zero to terminate this sequence and it can be seen from (11.7b) that all integers are eigenvalues of $\hat{n}$ since $\hat{n}(\hat{a}^{+k}|n\rangle) = (n + k)(\hat{a}^{+k}|n\rangle)$, $k = 0, 1, 2, \ldots$. The ground state (the physical vacuum) can be defined as the state with $n = 0$, that is $\hat{n}(\hat{a}^{+k}|0\rangle) = k(\hat{a}^{+k}|0\rangle)$ (making use of (11.7b)), and so the state $\hat{a}^{+k}|0\rangle$ must be proportional to the state $|k\rangle$. From (11.8b) $\langle 0|\hat{a}^k\hat{a}^{+k}|0\rangle = k!$ and we can define the normalized states

$$|n\rangle = \frac{1}{(n!)^{1/2}}\hat{a}^{+n}|0\rangle; \tag{11.10}$$

these are called the Fock states or the occupation number states. It is obvious that these states are orthonormal, $\langle n|m\rangle = \delta_{nm}$, and

$$\hat{a}^+|n\rangle = (n + 1)^{1/2}|n + 1\rangle, \tag{11.11a}$$

$$\hat{a}|n\rangle = n^{1/2}|n - 1\rangle. \tag{11.11b}$$

From here the probabilities of emission and absorption are seen to be proportional to

$$|\langle n + 1|\hat{a}^+|n\rangle|^2 = n + 1, \tag{11.11c}$$

$$|\langle n - 1|\hat{a}|n\rangle|^2 = n, \tag{11.11d}$$

respectively; for $n = 0$ spontaneous emission and for $n \gg 1$ stimulated emission occur. The vacuum stability condition demands that $\hat{a}|0\rangle = 0$. Equations (11.11a, b) show that the operators $\hat{a}^+$ and $\hat{a}$ can indeed be called the creation and annihilation operators respectively of a photon in the field.

If we write the Hamiltonian of the field in the form

$$\hat{H}_F = \frac{1}{8\pi} \int_{L^3} (\hat{\boldsymbol{E}}^2 + \hat{\boldsymbol{H}}^2)\, \mathrm{d}^3 x, \tag{11.12}$$

we obtain by using (11.2) and (11.1)

$$\hat{H}_F = \sum_\lambda (\hat{n}_\lambda + \tfrac{1}{2})\, \hbar\omega_\lambda, \tag{11.13}$$

where $\omega_\lambda = kc$. This operator is, apart from the additive constant $\sum_\lambda (\hbar\omega_\lambda/2)$ (the energy of the physical vacuum), the Hamiltonian of a set of dynamically independent harmonic oscillators with energies $n_\lambda \hbar\omega_\lambda$. Introducing the canonical variables $\hat{p}$ and $\hat{q}$ by the relations

$$\begin{aligned} \hat{a} &= (2\hbar\omega)^{-1/2}(\omega\hat{q} + \mathrm{i}\hat{p}), \\ \hat{a}^+ &= (2\hbar\omega)^{-1/2}(\omega\hat{q} - \mathrm{i}\hat{p}), \end{aligned} \tag{11.14}$$

so that

$$\begin{aligned} \hat{q} &= \left(\frac{\hbar}{2\omega}\right)^{1/2} (\hat{a} + \hat{a}^+), \\ \hat{p} &= -\mathrm{i}\left(\frac{\hbar\omega}{2}\right)^{1/2} (\hat{a} - \hat{a}^+), \end{aligned} \tag{11.15}$$

the commutation rule becomes

$$[\hat{q}, \hat{p}] = \hat{1}\mathrm{i}\hbar \tag{11.16}$$

and (11.13) gives us

$$\hat{H}_F = \sum_\lambda \tfrac{1}{2}(\hat{p}_\lambda^2 + \omega_\lambda^2 \hat{q}_\lambda^2), \tag{11.17}$$

which shows clearly that the free electromagnetic field is equivalent to an infinite set of independent harmonic oscillators.

The states $|n\rangle$ for all n form a complete orthonormal system, which is expressed by the completeness condition (condition of resolution of unity)

$$\sum_{n=0}^{\infty} |n\rangle\langle n| = \hat{1}, \tag{11.18}$$

where $\hat{1}$ is the identity operator. Thus every state $|\rangle$ may be decomposed into states $|n\rangle$ in the form

$$|\rangle = \sum_{k=0}^{\infty} c_k |k\rangle, \tag{11.19}$$

where

$$c_k = \langle k|\rangle. \tag{11.20}$$

The space they span is the familiar Fock space of quantum field theory and nearly all treatments of quantum electrodynamics have been formulated using these states. As we have mentioned such formulations do not retain phase information about the field unless off-diagonal elements are derived. In early applications this was unimportant since such information played a minimal role, but in discussing coherence the phase information must be used. On the other hand, the coherent states retain the phase information and also allow one to consider the classical limit of quantum electrodynamics when the fields are strong and n is large and uncertain.

The vector potential operator $\hat{\boldsymbol{A}}$ is Hermitian ($\hat{\boldsymbol{A}}^+ = \hat{\boldsymbol{A}}$) since the classical field is real and we see from (11.2) that $\hat{\boldsymbol{A}}$ (and also $\hat{\boldsymbol{E}}$ and $\hat{\boldsymbol{H}}$) can be decomposed into the sum of a positive frequency part $\hat{\boldsymbol{A}}^{(+)}$ and a negative frequency part $\hat{\boldsymbol{A}}^{(-)}$ as follows

$$\hat{\boldsymbol{A}} = \hat{\boldsymbol{A}}^{(+)} + \hat{\boldsymbol{A}}^{(-)}, \tag{11.21}$$

where

$$\hat{\boldsymbol{A}}^{(+)}(x) = \frac{(2\pi\hbar c)^{1/2}}{L^{3/2}} \sum_{\boldsymbol{k},s} k^{-1/2}\mathbf{e}^{(s)}(\boldsymbol{k})\, \hat{a}_{\boldsymbol{k}s} \exp\left[\mathrm{i}(\boldsymbol{k}\,.\,\boldsymbol{x} - kct)\right] \tag{11.22}$$

and $\hat{\boldsymbol{A}}^{(-)} = [\hat{\boldsymbol{A}}^{(+)}]^+$. Hence the operators $\hat{\boldsymbol{A}}^{(+)}(x)$ and $\hat{\boldsymbol{A}}^{(-)}(x)$ represent the annihilation and creation operators at x, respectively. They correspond to the classical fields $\boldsymbol{V}$ and $\boldsymbol{V}^*$ and play an important role in the process of detection of fields. For this reason the decomposition of the field operator $\hat{\boldsymbol{A}}$ into two complex quantities has a basic significance in quantum theory, whereas in the classical theory it is largely a matter of mathematical convenience. In the classical theory, where energy quantities with values comparable with the field quantum $\hbar\omega$ are neglected, one cannot distinguish between absorption and emission; only the real part of the complex field $\boldsymbol{V}$ can be measured. In the quantum theory on the other hand, atoms considered as detectors are sensitive to the annihilation operator $\hat{\boldsymbol{A}}^{(+)}$ if they are in unexcited states and to the creation operator $\hat{\boldsymbol{A}}^{(-)}$ if they are in excited states.

To study the detection of the electromagnetic field we also introduce the so-called detection operator (Mandel (1964c))

$$\hat{\mathbf{A}}(x) = L^{-3/2} \sum_{\boldsymbol{k},s} \mathbf{e}^{(s)}(\boldsymbol{k})\, \hat{a}_{\boldsymbol{k}s} \exp\left[\mathrm{i}(\boldsymbol{k}\,.\,\boldsymbol{x} - kct)\right], \tag{11.23}$$

which is closely related to the positive frequency part of the vector potential operator. The number operator $\hat{n}_{Vt}$ of photons localized in a finite volume V (the linear dimensions of which are large compared with the wavelength) is (Mandel (1964c), Cook (1982)).

$$\hat{n}_{Vt} = \int_V \hat{\mathbf{A}}^+(x)\,.\,\hat{\mathbf{A}}(x)\,\mathrm{d}^3x, \tag{11.24a}$$

which gives for $V \equiv L^3$, using (11.4),

$$\hat{n} = \sum_{\lambda} \hat{a}_{\lambda}^{+} \hat{a}_{\lambda}. \tag{11.24b}$$

For quasi-monochromatic fields, the most usual case in practice, all the operators $\hat{\mathbf{E}}^{(+)}$, $\hat{\mathbf{H}}^{(+)}$, $\hat{\mathbf{A}}^{(+)}$ and $\hat{\mathbf{A}}$ are proportional to one another.

11.2 Statistical states

From the quantum-mechanical point of view all information about the statistics of a system is contained in the density matrix $\hat{\varrho}$ (see e.g. Landau and Lifshitz (1964)). It can be decomposed in terms of the Fock states into

$$\hat{\varrho} = \sum_{n,m} \varrho(n, m) \,|\, n\rangle \langle m \,|, \tag{11.25}$$

where $\varrho(n, m) = \langle n \,|\, \hat{\varrho} \,|\, m\rangle$ since $\langle n \,|\, m\rangle = \delta_{nm}$ and the normalization condition reads

$$\operatorname{Tr} \hat{\varrho} = 1, \qquad \text{or} \qquad \sum_{n} \varrho(n, n) = 1 \tag{11.26}$$

and $\hat{\varrho}^{+} = \varrho$ ($\varrho^{*}(n, m) = \varrho(m, n)$).

As an example we can discuss thermal radiation in thermal equilibrium at temperature T. Denoting $\Theta = \hbar\omega/KT$, where K is Boltzmann constant, we have

$$\hat{\varrho} = \sum_{n=0}^{\infty} \varrho(n) \,|\, n\rangle \langle n \,|, \tag{11.27}$$

where

$$\varrho(n) \equiv \varrho(n, n) = [1 - \exp(-\Theta)] \exp(-n\Theta), \tag{11.28}$$

and so

$$\hat{\varrho} = \sum_{n=0}^{\infty} [1 - \exp(-\Theta)] \exp(-\Theta \hat{a}^{+} \hat{a}) \,|\, n\rangle \langle n \,| =$$

$$= [1 - \exp(-\Theta)] \exp\left(-\frac{\Theta}{\hbar\omega} \hat{H}\right) = \frac{\exp\left(-\frac{\Theta}{\hbar\omega} \hat{H}\right)}{\operatorname{Tr}\left\{\exp\left(-\frac{\Theta}{\hbar\omega} \hat{H}\right)\right\}}, \tag{11.29}$$

where (11.18) has been used and $\hat{H}$ is the renormalized free-field Hamiltonian. The expectation value of $\hat{n}$ is equal to

$$\langle \hat{n} \rangle = \operatorname{Tr} \{\varrho \hat{a}^{+} \hat{a}\} = \sum_{n=0}^{\infty} n \exp(-n\Theta) [1 - \exp(-\Theta)] =$$

$$= -[1 - \exp(-\Theta)] \frac{\mathrm{d}}{\mathrm{d}\Theta} \frac{1}{1 - \exp(-\Theta)} = \frac{1}{\exp(\Theta) - 1} = \langle n \rangle, \tag{11.30}$$

so that the mean energy $\langle U \rangle$ is equal to

$$\langle U \rangle = \hbar\omega\langle \hat{n} \rangle = \frac{\hbar\omega}{\exp(\Theta) - 1}. \tag{11.31}$$

From (11.28) one has for the matrix elements of the density matrix

$$\varrho(n) = \frac{\langle n \rangle^n}{(1 + \langle n \rangle)^{1+n}}, \tag{11.32}$$

which is the well-known Bose-Einstein distribution. The quantum characteristic function becomes

$$\mathrm{Tr}\,\{\hat{\varrho} \exp(\mathrm{i}x\hat{n})\} = \sum_{n=0}^{\infty} \frac{1}{1 + \langle n \rangle} \left[\frac{\langle n \rangle \exp(\mathrm{i}x)}{1 + \langle n \rangle} \right]^n =$$

$$= \frac{1}{1 + \langle n \rangle} \frac{1}{1 - \dfrac{\langle n \rangle}{1 + \langle n \rangle} \exp(\mathrm{i}x)} = \frac{1}{1 - \langle n \rangle (\exp(\mathrm{i}x) - 1)}. \tag{11.33}$$

Similarly we find for the Poisson distribution,

$$\varrho(n) = \frac{\langle n \rangle^n}{n!} \exp(-\langle n \rangle), \tag{11.34}$$

that

$$\mathrm{Tr}\,\{\hat{\varrho}\hat{n}\} = \langle n \rangle \tag{11.35a}$$

and

$$\mathrm{Tr}\,\{\hat{\varrho} \exp(\mathrm{i}x\hat{n})\} = \exp[(\mathrm{e}^{\mathrm{i}x} - 1)\langle n \rangle]. \tag{11.35b}$$

11.3 Multimode description

To describe multimode fields we introduce the global Fock states

$$|\{n_\lambda\}\rangle \equiv \prod_\lambda |n_\lambda\rangle \tag{11.36}$$

for which the completeness condition becomes

$$\sum_{\{n_\lambda\}} |\{n_\lambda\}\rangle\langle\{n_\lambda\}| = \hat{1}. \tag{11.37}$$

The density matrix has the form

$$\hat{\varrho} = \sum_{\{n_\lambda\}} \sum_{\{m_\lambda\}} \varrho(\{n_\lambda\}, \{m_\lambda\}) |\{n_\lambda\}\rangle\langle\{m_\lambda\}|, \tag{11.38}$$

where $\varrho(\{n_\lambda\}, \{m_\lambda\}) = \langle\{n_\lambda\}|\hat{\varrho}|\{m_\lambda\}\rangle$. A pure state is described by the density matrix $\hat{\varrho} = |\{n_\lambda\}\rangle\langle\{n_\lambda\}|$.

11.4 Calculation of commutators of the field operators

The correct correspondence between classical fields and the field operators is given by the commutation rules (11.3). However, the commutation rules for space-time dependent field operators, such as $\hat{\mathbf{A}}$, $\hat{\mathbf{E}}$, $\hat{\mathbf{H}}$, etc., are sometimes needed and therefore we derive some of them here, for use in some later calculations. It will be sufficient to derive the commutation rules for $\hat{\mathbf{A}}$ and $\hat{\mathrm{A}}$ since the commutation rules for $\hat{\mathbf{E}}$ and $\hat{\mathbf{H}}$ follow using (11.1).

Making use of (11.2) and (11.3), and also (11.5), we obtain

$$[\hat{A}_\mu(x), \hat{A}_\nu(x')] = \frac{2\pi\hbar c}{L^3}\sum_{\mathbf{k},s} k^{-1} e_\mu^{(s)}(\mathbf{k})\, e_\nu^{(s)*}(\mathbf{k}) \times$$
$$\times \{\exp[ik(x-x')] - \exp[-ik(x-x')]\} =$$
$$= \frac{2\pi\hbar c}{L^3}\sum_{\mathbf{k}} k^{-1}\left(\delta_{\mu\nu} - \frac{k_\mu k_\nu}{k^2}\right)\{\exp[ik(x-x')] - \exp[-ik(x-x')]\}, \tag{11.39}$$

$kx = \mathbf{k}.\mathbf{x} - kct$, and

$$[\hat{A}_\mu^{(+)}(x), \hat{A}_\nu^{(-)}(x')] = \frac{2\pi\hbar c}{L^3}\sum_{\mathbf{k},s} k^{-1} e_\mu^{(s)}(\mathbf{k})\, e_\nu^{(s)*}(\mathbf{k}) \exp[ik(x-x')] =$$
$$= \frac{2\pi\hbar c}{L^3}\sum_{\mathbf{k}} k^{-1}\left(\delta_{\mu\nu} - \frac{k_\mu k_\nu}{k^2}\right)\exp[ik(x-x')]. \tag{11.40}$$

The commutation rule for the detection operator (11.23) becomes

$$[\hat{\mathrm{A}}_\mu(x), \hat{\mathrm{A}}_\nu^+(x')] = L^{-3}\sum_{\mathbf{k}}\left(\delta_{\mu\nu} - \frac{k_\mu k_\nu}{k^2}\right)\exp[ik(x-x')] \tag{11.41}$$

and for $t = t'$ it reduces to

$$[\hat{\mathrm{A}}_\mu(\mathbf{x}, t), \hat{\mathrm{A}}_\nu^+(\mathbf{x}', t)] = \delta_{\mu\nu}\delta(\mathbf{x} - \mathbf{x}') - (2\pi)^{-3}\int \frac{k_\mu k_\nu}{k^2}\exp[i\mathbf{k}.(\mathbf{x}-\mathbf{x}')]\,\mathrm{d}^3k. \tag{11.42}$$

Here we have replaced the sum $\sum_{\mathbf{k}}$ by the integral $(L/2\pi)^3 \int \ldots \mathrm{d}^3k$. On the other hand, it is easily seen that

$$[\hat{A}_\mu^{(+)}(x), \hat{A}_\nu^{(+)}(x')] = [\hat{A}_\mu^{(-)}(x), \hat{A}_\nu^{(-)}(x')] = [\hat{\mathrm{A}}_\mu(x), \hat{\mathrm{A}}_\nu(x')] =$$
$$= [\hat{\mathrm{A}}_\mu^+(x), \hat{\mathrm{A}}_\nu^+(x')] = \hat{0}. \tag{11.43}$$

The functions on the right-hand sides of (11.39)–(11.42) are singular functions which are zero outside the light cone, that is when $(\mathbf{x} - \mathbf{x}')^2 > c^2(t - t')^2$. This is a reflection of the fact that measurements of fields at such pairs of points can be performed with arbitrary accuracy and they cannot influence one another (we remember that if two operators $\hat{A}$, $\hat{B}$ satisfy $[\hat{A}, \hat{B}] = \hat{1}C$, then $\langle(\Delta\hat{A})^2\rangle \times \langle(\Delta\hat{B})^2\rangle \geqq |C|^2/4$, cf. (13.24)). This expresses the principle of causality in quantum electrodynamics (signals cannot propagate more rapidly than with the velocity of light c in vacuo) (Rosenfeld (1958)).

By analogy with the commutation rules (11.6) one can derive the following commutation rules (Mandel (1966a))

$$[\hat{\mathbf{A}}(\mathbf{x}, t), \hat{n}_{Vt'}] = \begin{cases} \hat{\mathbf{A}}(\mathbf{x}, t), & \text{if } (\mathbf{x}, t) \text{ is conjoint with } (V, t'), \\ 0 \quad , & \text{if } (\mathbf{x}, t) \text{ is disjoint with } (V, t') \end{cases} \tag{11.44a}$$

and

$$[\hat{n}_{Vt'}, \hat{\mathbf{A}}^{+}(\mathbf{x}, t)] = \begin{cases} \hat{\mathbf{A}}^{+}(\mathbf{x}, t), & \text{if } (\mathbf{x}, t) \text{ is conjoint with } (V, t'), \\ 0 \quad , & \text{if } (\mathbf{x}, t) \text{ is disjoint with } (V, t'). \end{cases} \tag{11.44b}$$

The terms conjoint and disjoint are defined in the following way: we define the function $U(\mathbf{x}, V) = 1$ if $\mathbf{x} \in V$ and $U = 0$ if $\mathbf{x} \notin V$; then $(\mathbf{x}, t)$ is conjoint with (V, t') if $U(\mathbf{x}', V) = 1$ and disjoint if $U(\mathbf{x}', V) = 0$, where $\mathbf{x}' = \mathbf{x} - c\mathbf{k}(t - t')/|\,\mathbf{k}\,|$.

11.5 Time development in quantum electrodynamics

It is well known that in quantum electrodynamics a state of the field is described by a vector in the Hilbert space of states. The time development of a state can be interpreted either as the time development of the vector in the space if the system of coordinates is fixed (the Schrödinger picture) or as the time development of the system of coordinates if the state vector is fixed (the Heisenberg picture).

In the Schrödinger picture the Schrödinger equation holds for the state vector $|\,\psi(t)\rangle$

$$\mathrm{i}\hbar \frac{\partial}{\partial t} |\,\psi(t)\rangle = \hat{H}\,|\,\psi(t)\rangle, \tag{11.45a}$$

where $\hat{H}$ is the Hamiltonian, independent of time, and for an operator $\hat{F}$

$$-\mathrm{i}\hbar \frac{\mathrm{d}\hat{F}}{\mathrm{d}t} = \hat{0}. \tag{11.45b}$$

Performing the canonical transformations

$$\hat{F}(t) = \exp\left(\mathrm{i}\frac{\hat{H}t}{\hbar}\right) \hat{F} \exp\left(-\mathrm{i}\frac{\hat{H}t}{\hbar}\right), \tag{11.46a}$$

$$|\,\psi(t)\rangle = \exp\left(-\mathrm{i}\frac{\hat{H}t}{\hbar}\right) |\,\psi\rangle, \tag{11.46b}$$

which provide the correspondence between quantities in the Schrödinger and Heisenberg pictures, we obtain in the Heisenberg picture,

$$-\mathrm{i}\hbar \frac{\mathrm{d}}{\mathrm{d}t} \hat{F}(t) = [\hat{H}, \hat{F}(t)], \tag{11.47a}$$

$$\mathrm{i}\hbar \frac{\partial\,|\,\psi\rangle}{\partial t} = 0. \tag{11.47b}$$

For the density matrix $\hat{\varrho} = \{|\rangle\langle|\}_{\text{average}}$ we obtain

$$i\hbar \frac{\partial \hat{\varrho}(t)}{\partial t} = [\hat{H}, \hat{\varrho}(t)] \tag{11.48a}$$

in the Schrödinger picture, whereas

$$i\hbar \frac{\partial \hat{\varrho}}{\partial t} = \hat{0} \tag{11.48b}$$

in the Heisenberg picture and

$$\hat{\varrho}(t) = \exp\left(-i\frac{\hat{H}t}{\hbar}\right)\hat{\varrho}(0)\exp\left(i\frac{\hat{H}t}{\hbar}\right). \tag{11.49}$$

In calculations of the expectation values of an operator $\hat{F}(t)$ it may be convenient to use the following rule

$$\operatorname{Tr}\{\hat{\varrho}\hat{F}(t)\} = \operatorname{Tr}\left\{\hat{\varrho}\exp\left(i\frac{\hat{H}t}{\hbar}\right)\hat{F}\exp\left(-i\frac{\hat{H}t}{\hbar}\right)\right\} = \\ = \operatorname{Tr}\left\{\exp\left(-i\frac{\hat{H}t}{\hbar}\right)\hat{\varrho}\exp\left(i\frac{\hat{H}t}{\hbar}\right)\hat{F}\right\} = \operatorname{Tr}\{\hat{\varrho}(t)\hat{F}\}. \tag{11.50}$$

Sometimes it is convenient to introduce the interaction picture. If $\hat{H} = \hat{H}_0 + \hat{H}_i$, where $\hat{H}_0$ and $\hat{H}_i$ are the free and interaction Hamiltonians, it follows that

$$i\hbar \frac{\partial |\varphi(t)\rangle}{\partial t} = \mathscr{H}_i |\varphi(t)\rangle, \tag{11.51a}$$

$$-i\hbar \frac{\partial \mathscr{F}(t)}{\partial t} = [\hat{H}_0, \mathscr{F}(t)], \tag{11.51b}$$

where

$$|\varphi(t)\rangle = \exp\left(i\frac{\hat{H}_0 t}{\hbar}\right)|\psi(t)\rangle, \tag{11.52a}$$

$$\mathscr{H}_i = \exp\left(i\frac{\hat{H}_0 t}{\hbar}\right)\hat{H}_i\exp\left(-i\frac{\hat{H}_0 t}{\hbar}\right), \tag{11.52b}$$

$$\mathscr{F}(t) = \exp\left(i\frac{\hat{H}_0 t}{\hbar}\right)\hat{F}\exp\left(-i\frac{\hat{H}_0 t}{\hbar}\right). \tag{11.52c}$$

The Schrödinger equation (11.45a) (with $\hat{H}$ generally dependent on t) can be solved in the form

$$|\psi(t)\rangle = \hat{S}(t)|\psi(0)\rangle, \tag{11.53}$$

where $\hat{S}(t)$ is the unitary operator of time development obeying the Schrödinger equation

$$i\hbar \frac{\partial \hat{S}(t)}{\partial t} = \hat{H}(t)\hat{S}(t). \tag{11.54}$$

Solving this equation by the usual perturbation method with the initial condition $\hat{S}(0) = \hat{1}$, we obtain

$$\hat{S}(t) = \mathcal{T} \exp\left[-\frac{\mathrm{i}}{\hbar} \int_0^t \hat{H}(t')\, \mathrm{d}t' \right] =$$

$$= \sum_{n=0}^{\infty} \frac{(-\mathrm{i}/\hbar)^n}{n!} \int_0^t \dots \int_0^t \mathcal{T}\{\hat{H}(t_1) \dots \hat{H}(t_n)\}\, \mathrm{d}t_1 \dots \mathrm{d}t_n, \tag{11.55}$$

where the operator $\mathcal{T}$ is the time-ordering operator. This operator formally reorders the terms so that all operator products are taken with the factors in an increasing time-order sequence from the right to the left.

OPTICAL CORRELATION PHENOMENA

Having prepared the basic quantum formalism we can start a quantum-mechanical treatment of coherence phenomena in the electromagnetic field and show what kind of experiments are described by the quantum correlation functions in various operator orderings. We consider the detection of a field using the interaction of this radiation field with matter. The detection device, composed of atoms, will be assumed to be in its ground state so that photons are registered by absorption. Such detectors of light are photodetectors, photographic plates, etc.

12.1 Quantum correlation functions

Consider an ideal detector which is quite small in size and has a sensitivity independent of the photon frequency. One atom may be considered as such a detector. Suppose the initial state of the field is characterized by the state vector $|i\rangle$ and the atom is initially in its ground state. If $|f\rangle$ is a final state of the field after the absorption of a photon described by the annihilation operator $\hat{A}^{(+)}(x)$ (a linearly polarized field is assumed again), then the transition amplitude of the process is equal to $\langle f|\hat{A}^{(+)}(x)|i\rangle$. The transition probability per unit time from the state $|i\rangle$ to the state $|f\rangle$ of the radiation field due to the absorption of a photon at a space-time point $(\mathbf{x}, t)$ is then proportional to $|\langle f|\hat{A}^{(+)}(x)|i\rangle|^2$. The final state $|f\rangle$ of the field usually remains unobserved so that we must sum over all possible states $|f\rangle$, which gives

$$\sum_f \langle i|\hat{A}^{(-)}(x)|f\rangle\langle f|\hat{A}^{(+)}(x)|i\rangle = \langle i|\hat{A}^{(-)}(x)\hat{A}^{(+)}(x)|i\rangle \tag{12.1}$$

because of the completeness condition $\sum_f |f\rangle\langle f| = \hat{1}$. The initial states $|i\rangle$ of the field depend on some random or unknown parameters and consequently all field measurements must be carried out as averages over the ensemble of ways in which the field can be prepared. As we know the information about the statistics of the field as a statistical dynamic system is contained in the density matrix $\hat{\varrho} = \{|i\rangle\langle i|\}_{\text{average}}$ and we arrive at

$$\{\langle i|\hat{A}^{(-)}(x)\hat{A}^{(+)}(x)|i\rangle\}_{\text{average}} = \operatorname{Tr}\{\hat{\varrho}\hat{A}^{(-)}(x)\hat{A}^{(+)}(x)\}, \tag{12.2}$$

which represents the quantum mean intensity at x. This intensity is a particular value (at $x' = x$) of the correlation function

$$\Gamma_{\mathcal{N}}^{(1,1)}(x, x') = \operatorname{Tr}\{\hat{\varrho}\hat{A}^{(-)}(x)\hat{A}^{(+)}(x')\}, \tag{12.3}$$

which corresponds to the classical correlation function $\Gamma^{(1,1)}(x, x')$. If the vectorial properties of the field are taken into account, vectorial indices have to be added to $\hat{A}^{(+)}$ and $\hat{A}^{(-)}$ and we obtain the second-rank tensor $\Gamma^{(1,1)}_{\mathcal{N},\mu\nu}$ as earlier. For simplicity we use x to denote $(\mathbf{x}, t, \mu)$. Sometimes one considers fields obtained using polarizing filters which select photons of polarization $\mathbf{e}$; then the use of only the scalar field operator $\hat{A} = \mathbf{e} \cdot \hat{\mathbf{A}}$ is sufficient.

More generally, we define the transition probability per unit (time)^n that a photon of a polarization μ_1 is absorbed at $(\mathbf{x}_1, t_1) \equiv x_1$, a photon of a polarization μ_2 at $(\mathbf{x}_2, t_2) \equiv x_2$, etc., and a photon of a polarization μ_n at $(\mathbf{x}_n, t_n) \equiv x_n$. We obtain in the same way that this probability is proportional to

$$\Gamma^{(n,n)}_{\mathcal{N}}(x_1, \ldots, x_n, x_n, \ldots, x_1) = \{\sum_f |\langle f | \hat{A}^{(+)}(x_1) \ldots \hat{A}^{(+)}(x_n) | i\rangle |^2\}_{\text{average}} =$$

$$= \operatorname{Tr} \{\hat{\varrho} \hat{A}^{(-)}(x_1) \ldots \hat{A}^{(-)}(x_n) \hat{A}^{(+)}(x_n) \ldots \hat{A}^{(+)}(x_1)\}. \tag{12.4}$$

This is a particular value of the general correlation function (tensor) $\Gamma^{(m,n)}_{\mathcal{N}}(x_1, \ldots, x_{m+n})$ defined by

$$\Gamma^{(m,n)}_{\mathcal{N}}(x_1, \ldots, x_{m+n}) = \operatorname{Tr} \{\hat{\varrho} \hat{A}^{(-)}(x_1) \ldots \hat{A}^{(-)}(x_m) \hat{A}^{(+)}(x_{m+1}) \ldots \hat{A}^{(+)}(x_{m+n})\} \tag{12.5}$$

if $m = n$ and $x_j = x_{n+j}$, $j = 1, 2, \ldots, n$. Note that all the $\hat{A}^{(-)}$ as well as the $\hat{A}^{(+)}$ can be interchanged since the commutation rules (11.43) hold. These correlation functions agree exactly with the classical correlation functions (8.4) via the correspondence $\hat{A}^{(+)} \leftrightarrow V$, $\hat{A}^{(-)} \leftrightarrow V^*$, $\hat{\varrho} \leftrightarrow P$. Of course the operators $\hat{A}^{(+)}$ and $\hat{A}^{(-)}$ do not commute in general (cf. (11.40)) so that this correspondence of the classical and quantum correlation functions is not unique. The Hanbury Brown-Twiss effect discussed in Chapter 10 is described by the quantum correlation function $\Gamma^{(2,2)}_{\mathcal{N}}(x_1, x_2, x_2, x_1)$.

An important property of the correlation functions (12.5) is that the operators $\hat{A}^{(+)}$ and $\hat{A}^{(-)}$ are in their normal order, i.e. all annihilation operators $\hat{A}^{(+)}$ stand to the right of all creation operators $\hat{A}^{(-)}$. A consequence of this, if the vacuum stability condition $\hat{A}^{(+)}|0\rangle = 0$ is used, is that the vacuum expectation value of these normally ordered products is zero,

$$\Gamma^{(n,n)}_{\mathcal{N},\text{vac}}(x_1, \ldots, x_n, x_n, \ldots, x_1) = \langle 0 | \prod_{j=1}^{n} \hat{A}^{(-)}(x_j) \prod_{k=1}^{n} \hat{A}^{(+)}(x_k) | 0\rangle = 0, \tag{12.6}$$

and the physical vacuum gives no contribution to the quantities characterizing these photoelectronic correlation measurements; however, it does give contribution to measurements which are related to other orderings of field operators. As an example we shall discuss some devices (quantum counters) in Sec. 16.4 whose operation is connected with the antinormally ordered products of field operators ($\hat{A}^{(+)}$ and $\hat{A}^{(-)}$ are interchanged). These devices detect the field by means of stimulated emission rather than by absorption. (In the latter case $\langle 0 | \hat{A}^{(+)}(x) \times \times \hat{A}^{(-)}(x') | 0\rangle = [\hat{A}^{(+)}(x), \hat{A}^{(-)}(x')]$, where the commutator is given in (11.40) and the physical vacuum contributes.) This is responsible for the fact that the probability distributions of absorbed photons and emitted photoelectrons have a similar form whereas the distribution of photons and the distribution of counts of the quantum counter differ from one another (as a consequence of the contribution of the physical vacuum).

From the correlation functions defined in terms of $\hat{A}$ we can obtain the correlation functions (tensors) defined in terms of $\hat{E}$ using (11.1),

$$\operatorname{Tr}\{\hat{\varrho}\prod_{j=1}^{m}\hat{E}^{(-)}(x_j)\prod_{k=m+1}^{m+n}\hat{E}^{(+)}(x_k)\}, \tag{12.7}$$

(or in terms of $\hat{H}$) as well as various mixed forms. The correlation functions (12.7) were used in Glauber's treatment of quantum coherence (Glauber (1963a–c, 1964, 1965, 1969, 1970, 1972)). In practice quasi-monochromatic light is mostly used and then all these quantities are proportional to each other. Further kinds of correlation functions (tensors), based on the electromagnetic antisymmetric tensor $\hat{F}_{\mu\nu}(x) = \partial\hat{A}_\nu/\partial x_\mu - \partial\hat{A}_\mu/\partial x_\nu$, may be introduced (Glauber (1963a)) and some of their properties have been investigated (Kujawski (1966), Dialetis (1969b)).

A more precise treatment of the detection of an electromagnetic field must use the time-dependent perturbation theory. This was done using the electric-dipole approximation by Glauber (1965); the interaction Hamiltonian is

$$\hat{H}_i = -e\sum_{\gamma}\hat{\mathbf{q}}_\gamma\,.\,\hat{\mathbf{E}}(\mathbf{x},t), \tag{12.8a}$$

where $(-e)$ is the charge of an electron, $\hat{\mathbf{q}}_\gamma$ is the spatial coordinate operator of the γth electron of the atom relative to its nucleus, assumed to be located at $\mathbf{x}$. More generally, the interaction Hamiltonian can be written as

$$\hat{H}_i = -\frac{e}{m}\sum_{\gamma}\hat{\mathbf{p}}_\gamma\,.\,\hat{\mathbf{A}}(\mathbf{x},t), \tag{12.8b}$$

where $\hat{\mathbf{p}}_\gamma$ is the momentum operator and m the mass of an electron. A discussion of the interaction Hamiltonians (12.8a, b) from point of view of quantum optics has been given by Mandel (1979a).

Consider N atoms as detectors placed at points $\mathbf{x}_1, \ldots, \mathbf{x}_N$ and apply the formalism for the time development of the quantum system presented in Sec. 11.5. We can solve the Schrödinger equation (11.45a) with $\hat{H} = \hat{H}_i(t)$ (in the interaction picture) in the form (11.53), where the time development operator $\hat{S}$ is given by (11.55). Using the Nth-order approximation of perturbation theory (the Nth term in (11.55)), we obtain for the probability that the first atom has undergone a transition in the interval $(0, T_1)$, the second one in $(0, T_2)$, etc., and Nth one in $(0, T_N)$ (Glauber (1965))

$$p^{(N)}(T_1,\ldots,T_N) = \int_0^{T_1}\cdots\int_0^{T_N}\Gamma_{\mathcal{N}}^{(N,N)}(\mathbf{x}_1,\ldots,\mathbf{x}_N,\mathbf{x}_N,\ldots,\mathbf{x}_1,$$
$$t'_1,\ldots,t'_N,t''_N,\ldots,t''_1)\prod_{j=1}^{N}\mathscr{S}(t'_j - t''_j)\,\mathrm{d}t'_j\,\mathrm{d}t''_j, \tag{12.9}$$

where $\mathscr{S}$ is a weight function (a "response" function of the detectors).

For detectors assumed to be broadband (they do not distinguish between frequencies)

$$\mathscr{S}(t' - t'') = \eta\delta(t' - t''), \tag{12.10}$$

where η is the sensitivity factor and we have, from (12.9),

$$p^{(N)}(T_1, \ldots, T_N) = \eta^N \int_0^{T_1} \ldots \int_0^{T_N} \Gamma_{\mathscr{N}}^{(N,N)}(\mathbf{x}_1, \ldots, \mathbf{x}_N, \mathbf{x}_N, \ldots, \mathbf{x}_1, t'_1, \ldots, t'_N, t'_N, \ldots, t'_1) \prod_{j=1}^{N} \mathrm{d}t'_j. \tag{12.11}$$

This probability must equal the expectation value $\langle \hat{n}_1 \ldots \hat{n}_N \rangle$ which is simply the probability that all the detectors $1, 2, \ldots, N$ have contributed counts in $(0, T_1)$, $(0, T_2), \ldots, (0, T_N)$ respectively; this is a similar result to that obtained in the semiclassical description (equation (10.91)).

The joint counting rate corresponding to times $T_1, \ldots, T_N$ is equal to

$$w(T_1, \ldots, T_N) \doteq \frac{\partial^N}{\partial T_1 \ldots \partial T_N} p^{(N)}(T_1, \ldots, T_N) = \\ = \eta^N \Gamma_{\mathscr{N}}^{(N,N)}(\mathbf{x}_1, \ldots, \mathbf{x}_N, \mathbf{x}_N, \ldots, \mathbf{x}_1, T_1, \ldots, T_N, T_N, \ldots, T_1) \tag{12.12}$$

in analogy to (10.92).

Equation (12.9) can be interpreted as applicable to point detectors which are not sensitive to the instantaneous value of the field but are sensitive rather to an average of the field over a time interval. More generally we can consider detectors which are not sensitive even to the field at points $\mathbf{x}_1, \ldots, \mathbf{x}_N$ but to the averaged field over spatial regions $V_1, \ldots, V_N$ and we obtain

$$p^{(N)} = \int_0^{T_1} \ldots \int_0^{T_N} \int_{V_1} \ldots \int_{V_N} \Gamma_{\mathscr{N}}^{(N,N)}(x'_1, \ldots, x'_N, x''_N, \ldots, x''_1) \times \\ \times \prod_{j=1}^{N} \mathscr{S}(x'_j - x''_j)\, \mathrm{d}^4x'_j\, \mathrm{d}^4x''_j. \tag{12.13}$$

Equation (12.9) is then obtained for $\mathscr{S}(x' - x'') = \delta(\mathbf{x}' - \mathbf{x}'')\,\delta(\mathbf{x}' - \mathbf{x})\,\mathscr{S}(t' - t'')$. In (12.13) x denotes $(\mathbf{x}, t, \mu)$ and the integrals over x imply in addition a summation over the polarization indices μ.

Some properties of the correlation functions were summarized in Chapter 2 and here we briefly outline a derivation of some of them and give some further properties.

Equation (2.39) follows from the inequality

$$\mathrm{Tr}\,\{\hat{\varrho}\hat{B}^+\hat{B}\} \geqq 0, \tag{12.14}$$

valid for an arbitrary operator $\hat{B}$, putting $\hat{B} = \hat{A}^{(+)}(x_1) \ldots \hat{A}^{(+)}(x_n)$. If

$$\hat{B} = \sum_{j=1}^{n} c_j \hat{A}^{(+)}(x_j), \tag{12.15}$$

we obtain the quantum analogue of (4.7a),

$$\sum_{j,k}^{n} c_j^* c_k \Gamma_{\mathcal{N}}^{(1,1)}(x_j, x_k) \geqq 0. \tag{12.16}$$

This condition leads to

$$\text{Det}\,\{\Gamma_{\mathcal{N}}^{(1,1)}(x_j, x_k)\} \geqq 0; \tag{12.17}$$

for $n = 1$ we have $\Gamma_{\mathcal{N}}^{(1,1)}(x, x) \geqq 0$ and for $n = 2$

$$\Gamma_{\mathcal{N}}^{(1,1)}(x_1, x_1)\,\Gamma_{\mathcal{N}}^{(1,1)}(x_2, x_2) \geqq |\Gamma_{\mathcal{N}}^{(1,1)}(x_1, x_2)|^2. \tag{12.18}$$

If

$$\hat{B} = c_1 \hat{A}^{(+)}(x_1) \dots \hat{A}^{(+)}(x_n) + c_2 \hat{A}^{(+)}(x_{n+1}) \dots \hat{A}^{(+)}(x_{2n}), \tag{12.19}$$

(2.41) follows at once.

Another interesting property of $\Gamma_{\mathcal{N}}^{(n,n)}$ is that this function is zero for states with a finite number of photons, say N, in the field when $n > N$, since $\hat{A}^{(+)}$ is the annihilation operator of a photon and the vacuum stability condition holds.

As we have already mentioned the arguments $x_1, \dots, x_m$ as well as $x_{m+1}, \dots, x_{m+n}$ can be interchanged without changing the value of $\Gamma_{\mathcal{N}}^{(m,n)}$ because the operators $\hat{A}^{(+)}$ as well as $\hat{A}^{(-)}$ mutually commute.

12.2 Quantum coherence

Second-order phenomena

Second-order coherence has been discussed in Chapter 3 in connection with the Young experiment and the form of the mutual coherence function for the coherent field was derived in Sec. 4.2. Equation (9.25) has been adopted to define the coherent field in higher orders and we discuss this point more fully in this section.

The description of the Young experiment in terms of quantum correlation functions proceeds exactly as in the classical treatment leading to the results discussed in Sec. 3.1. Thus

$$\begin{aligned} I(x) &= \text{Tr}\,\{\hat{\varrho}[a_1^* \hat{A}^{(-)}(x_1) + a_2^* \hat{A}^{(-)}(x_2)]\,[a_1 \hat{A}^{(+)}(x_1) + a_2 \hat{A}^{(+)}(x_2)]\} = \\ &= |a_1|^2\,\Gamma_{\mathcal{N}}^{(1,1)}(x_1, x_1) + |a_2|^2\,\Gamma_{\mathcal{N}}^{(1,1)}(x_2, x_2) + \\ &\quad + |a_1 a_2|\,\Gamma_{\mathcal{N}}^{(1,1)}(x_1, x_2) + |a_1 a_2|\,\Gamma_{\mathcal{N}}^{(1,1)}(x_2, x_1). \end{aligned} \tag{12.20}$$

The interference terms with $\Gamma_{\mathcal{N}}^{(1,1)}(x_1, x_2)$ and $\Gamma_{\mathcal{N}}^{(1,1)}(x_2, x_1)$ in (12.20) express the fact that it is impossible to distinguish from which pinhole a photon came. We have seen that full coherence gives in second order the maximum visibility of interference fringes and

$$|\Gamma_{\mathcal{N}}^{(1,1)}(x_1, x_2)|^2 = \Gamma_{\mathcal{N}}^{(1,1)}(x_1, x_1)\,\Gamma_{\mathcal{N}}^{(1,1)}(x_2, x_2). \tag{12.21}$$

This is the boundary value of (12.18) corresponding, according to all definitions of the degree of coherence (2.42a – c), to

$$|\gamma_{\mathcal{N}}^{(1,1)}(x_1, x_2)| = 1 \tag{12.22}$$

valid for all x_1 and x_2; in general

$$0 \leqq |\gamma_{\mathcal{N}}^{(1,1)}(x_1, x_2)| \leqq 1. \tag{12.23}$$

Writing (12.21) in the form

$$\Gamma_{\mathcal{N}}^{(1,1)}(x_1, x_2) = A(x_1)\, B(x_2), \tag{12.24}$$

we have from the cross-symmetry condition

$$A(x_1)\, B(x_2) = A^*(x_2)\, B^*(x_1) \tag{12.25}$$

and so

$$\frac{A(x_1)}{B^*(x_1)} = \frac{A^*(x_2)}{B(x_2)} = k, \tag{12.26}$$

where k is a real constant. Therefore $A(x) = kB^*(x)$ and defining the function $V(x) = k^{1/2}B(x)$, we can rewrite (12.24) in the form

$$\Gamma_{\mathcal{N}}^{(1,1)}(x_1, x_2) = V^*(x_1)\, V(x_2), \tag{12.27}$$

in agreement with (4.12).

We can now ask in what manner the higher-order correlation functions are restricted by the factorization condition (12.27), following the method by Titulaer and Glauber (1965). If (12.27) holds, then (12.21) is also fulfilled and this implies that

$$\mathrm{Tr}\,\{\hat{\varrho}\hat{B}^+\hat{B}\} = 0, \tag{12.28}$$

where

$$\hat{B} = \hat{A}^{(+)}(x) - \frac{\Gamma_{\mathcal{N}}^{(1,1)}(x_0, x)}{\Gamma_{\mathcal{N}}^{(1,1)}(x_0, x_0)}\, \hat{A}^{(+)}(x_0); \tag{12.29}$$

here x_0 is an arbitrary space-time point. Therefore $\hat{\varrho}\hat{B}^+ = \hat{B}\hat{\varrho} = \hat{0}$ and we obtain

$$\hat{A}^{(+)}(x)\, \hat{\varrho} = \frac{\Gamma_{\mathcal{N}}^{(1,1)}(x_0, x)}{\Gamma_{\mathcal{N}}^{(1,1)}(x_0, x_0)}\, \hat{A}^{(+)}(x_0)\, \hat{\varrho}, \tag{12.30a}$$

$$\hat{\varrho}\hat{A}^{(-)}(x) = \frac{\Gamma_{\mathcal{N}}^{(1,1)}(x, x_0)}{\Gamma_{\mathcal{N}}^{(1,1)}(x_0, x_0)}\, \hat{\varrho}\hat{A}^{(-)}(x_0). \tag{12.30b}$$

Considering the identities (12.30a) and (12.30b) at x_2 and x_1 respectively, we obtain

$$\mathrm{Tr}\,\{\hat{\varrho}\hat{A}^{(-)}(x_1)\, \hat{A}^{(+)}(x_2)\} = \\ = \frac{\Gamma_{\mathcal{N}}^{(1,1)}(x_1, x_0)\, \Gamma_{\mathcal{N}}^{(1,1)}(x_0, x_2)}{[\Gamma_{\mathcal{N}}^{(1,1)}(x_0, x_0)]^2}\, \mathrm{Tr}\,\{\hat{\varrho}\hat{A}^{(-)}(x_0)\, \hat{A}^{(+)}(x_0)\}, \tag{12.31a}$$

or

$$\Gamma_{\mathcal{N}}^{(1,1)}(x_1, x_2) = \frac{\Gamma_{\mathcal{N}}^{(1,1)}(x_1, x_0)\, \Gamma_{\mathcal{N}}^{(1,1)}(x_0, x_2)}{\Gamma_{\mathcal{N}}^{(1,1)}(x_0, x_0)} = V^*(x_1)\, V(x_2), \tag{12.31b}$$

in agreement with (12.27), where $V(x) = \Gamma_{\mathcal{N}}^{(1,1)}(x_0, x)/[\Gamma_{\mathcal{N}}^{(1,1)}(x_0, x_0)]^{1/2}$.

Higher-order phenomena

For the $(m + n)$th-order correlation function we obtain in an identical manner

$$\Gamma_{\mathcal{N}}^{(m,n)}(x_1, \ldots, x_{m+n}) = \gamma^{(m,n)} \prod_{j=1}^{m} V^*(x_j) \prod_{k=m+1}^{m+n} V(x_k), \tag{12.32}$$

where

$$\gamma^{(m,n)} = \frac{\Gamma_{\mathcal{N}}^{(m,n)}(x_0, \ldots, x_0)}{[\Gamma_{\mathcal{N}}^{(1,1)}(x_0, x_0)]^{(m+n)/2}}\,. \tag{12.33}$$

However, x_0 is an arbitrary point so that (12.33) must be independent of x_0. Consequently the factorization condition (12.27) expressing the second-order coherence leads to the factorization of all correlation functions into the form (12.32).

By analogy with the factorization condition (12.27) for second-order coherence one may construct the following set of factorization conditions,

$$\Gamma_{\mathcal{N}}^{(m,n)}(x_1, \ldots, x_{m+n}) = \prod_{j=1}^{m} V^*(x_j) \prod_{k=m+1}^{m+n} V(x_k), \tag{12.34}$$

in which V is independent of m and n. Thus we can speak of $2N$th-order coherence if (12.34) holds for $m, n \leqq N$. If (12.34) holds for all m, n (in practice for all $m = n$), then the field possesses full coherence. It is clear from a classical point of view that fully coherent fields are noiseless fields whose probability distribution function equals the Dirac function. This fact illustrates the close relation between the noiselessness of fields and coherence. In the quantum theory there also exist quantum states of fields for which (12.34) holds. These are called the *coherent states* (Glauber (1963b)) and their properties will be investigated in the next chapter.

The connection between fluctuations of the field and its coherence properties can be demonstrated as follows. Assume the field to be uniformly fluctuating, $V(x) = cV_{\text{det}}(x)$, c being a random complex variable and V_{det} is a non-fluctuating deterministic field. From the second-order coherence condition $\langle |c|^2 \rangle = 1$ and consequently some phase and amplitude fluctuations are admissible. If additionally fourth-order coherence is requested, $\langle |c|^4 \rangle = 1 = \langle |c|^2 \rangle^2$ and $|c|^2$ must equal unity; therefore only phase fluctuations are admissible in this case, etc.

Comparing (12.34) with (12.32) we see that these two equations differ from one another by the factor $\gamma^{(m,n)}$ and second-order coherence does not lead, in general, to higher-order coherence, for which

$$\gamma^{(m,n)} = 1 \tag{12.35}$$

for all m, n. If a finite number of photons, say M, is present in the field, then $\gamma^{(n,n)} = 0$ for $n > M$ and such a field cannot possess coherence to all orders. On the other hand, for the chaotic field we obtain using (10.63)

$$\gamma^{(m,n)} = \delta_{mn} n!\,, \tag{12.36}$$

and we can conclude that chaotic radiation cannot possess coherence of order higher than two ($m = n = 1$), if the definition (2.42b) of the degree of coherence is adopted. Expression (12.36) as well as some further examples can also be obtained using the results of Chapter 11. Considering one mode only we have from (11.8a)

$$\gamma^{(k,k)} = \frac{\langle \hat{a}^{+k}\hat{a}^{k}\rangle}{\langle \hat{a}^{+}\hat{a}\rangle^{k}} = \frac{\langle \hat{n}(\hat{n}-1)\ldots(\hat{n}-k+1)\rangle}{\langle \hat{n}\rangle^{k}}. \tag{12.37}$$

For a Fock state $|n\rangle$ the density matrix $\hat{\varrho} = |n\rangle\langle n|$ and so

$$\gamma^{(k,k)} = \frac{n!}{(n-k)!\,n^{k}}\,; \tag{12.38}$$

if $k > n$, then $\gamma^{(k,k)} = 0$. For $k = 1$, $|n\rangle$-state is coherent in the 2nd order. An interesting property of the Fock state is that the normal variance is negative, $\langle(\Delta\hat{n})^2\rangle_{\mathcal{N}} = \langle \hat{a}^{+2}\hat{a}^{2}\rangle - \langle \hat{a}^{+}\hat{a}\rangle^2 = n(n-1) - n^2 = -n < 0$ and so the variance of $\hat{n}$ is $\langle(\Delta\hat{n})^2\rangle = \langle\hat{n}\rangle + \langle(\Delta\hat{n})^2\rangle_{\mathcal{N}} = 0$. This expresses the antibunching (sub-Poisson) effect (Chapter 22) for these states and the Fock states therefore have no classical analogue (Miller and Mishkin (1967a), Bertrand and Mishkin (1967)). (Note for comparison that for the coherent states (Chapter 13) $\langle \hat{a}^{+2}\hat{a}^{2}\rangle - \langle \hat{a}^{+}\hat{a}\rangle^2 = 0$.)

For thermal radiation $\hat{\varrho}$ is given by (11.29) and

$$\gamma^{(k,k)} = [\mathrm{Tr}\,\{\exp(-\Theta\hat{n})\}]^{k-1}\,\frac{\mathrm{Tr}\left\{\exp(-\Theta\hat{n})\dfrac{\hat{n}!}{(\hat{n}-k)!}\right\}}{[\mathrm{Tr}\,\{\exp(-\Theta\hat{n})\,\hat{n}\}]^{k}}. \tag{12.39}$$

Further

$$\begin{aligned}\mathrm{Tr}\left\{\exp(-\Theta\hat{n})\frac{\hat{n}!}{(\hat{n}-k)!}\right\} &= \sum_{n=0}^{\infty}\exp(-\Theta n)\frac{n!}{(n-k)!} = \\ &= \exp(-\Theta k)\frac{\mathrm{d}^k}{\mathrm{d}[\exp(-\Theta)]^k}\sum_{n=0}^{\infty}\exp(-\Theta n) = \\ &= \exp(-\Theta k)\frac{\mathrm{d}^k}{\mathrm{d}[\exp(-\Theta)]^k}\frac{1}{1-\exp(-\Theta)} = \\ &= \exp(-\Theta k)\frac{k!}{[1-\exp(-\Theta)]^{k+1}}.\end{aligned} \tag{12.40}$$

Substituting this into (12.39) we obtain

$$\gamma^{(k,k)} = \frac{1}{[1-\exp(-\Theta)]^{k-1}} \frac{k!\exp(-\Theta k)}{[1-\exp(-\Theta)]^{k+1}} \frac{[1-\exp(-\Theta)]^{2k}}{\exp(-\Theta k)} = k! \tag{12.41}$$

in agreement with (12.36).

The conditions (12.35) and (12.36) (i.e. the validity of factorization conditions) for laser and chaotic light were experimentally verified by Jakeman, Oliver and Pike (1968b). Very good agreement was found up to $m = n = 6$.

Taking into account that for the fully coherent field

$$\mathrm{Tr}\,\{\hat{\varrho}\hat{A}^{(-)}(x)\,\hat{A}^{(+)}(x)\} - |\mathrm{Tr}\,\{\hat{\varrho}\hat{A}^{(+)}(x)\}|^2 = 0, \tag{12.42}$$

we arrive at

$$V(x) = \mathrm{Tr}\,\{\hat{\varrho}\hat{A}^{(+)}(x)\} = \Gamma_{\mathcal{N}}^{(0,1)}(x). \tag{12.43}$$

We have seen that the idea of second-order coherence can be derived from the Young experiment and that it has a clear physical significance – the interference fringes have the maximum visibility. We can ask also for the physical significance of higher-order coherence conditions. A natural experiment to investigate this is the photon coincidence counting experiment.

Equation (12.34) implies, if the definition (2.42b) of the degree of coherence is used, that $|{}^{(G)}\gamma_{\mathcal{N}}^{(n,n)}(x_1, \ldots, x_{2n})| \equiv 1$, i.e. ${}^{(G)}\gamma_{\mathcal{N}}^{(n,n)}(x_1, \ldots, x_n, x_n, \ldots, x_1) \equiv 1$ and consequently

$$\Gamma_{\mathcal{N}}^{(n,n)}(x_1, \ldots, x_n, x_n, \ldots, x_1) = \prod_{j=1}^{n} \Gamma_{\mathcal{N}}^{(1,1)}(x_j, x_j); \tag{12.44}$$

this means, according to our earlier results, that the N-fold joint counting rate defined by (12.12) is equal to the product of the counting rates which would be measured by each of the N counters in the absence of all others; the responses of the counters are statistically independent of one another. If the average intensity of the field is independent of time (the field is stationary), the counters detect no tendency toward any sort of correlation in the arrival times of photons. This is in agreement with the semiclassical analysis, contained in Chapter 10, of the bunching effect and the Hanbury Brown-Twiss effect for laser light and in this case no Hanbury Brown-Twiss effect will occur. Fields for which bunching occurs cannot be coherent in the fourth- or higher-order. A typical example is the chaotic field, for which there is a tendency for photons to arrive in pairs (cf. Figs. 10.1 and 10.2).

The full coherence conditions are equivalent to the fact that the photon statistics are Poissonian (cf. Chapter 10). Since $\langle \hat{a}^{+n}\hat{a}^n \rangle = \langle \hat{a}^+\hat{a} \rangle^n$, the normally ordered characteristic function $\langle \exp(\mathrm{i}s\hat{n}) \rangle_{\mathcal{N}}$ is given by

$$\langle \exp(\mathrm{i}s\hat{n}) \rangle_{\mathcal{N}} = \sum_{n=0}^{\infty} \frac{(\mathrm{i}s)^n}{n!} \langle \hat{a}^{+n}\hat{a}^n \rangle = \sum_{n=0}^{\infty} \frac{(\mathrm{i}s)^n}{n!} \langle \hat{a}^+\hat{a} \rangle^n = \exp(\mathrm{i}s\langle \hat{n} \rangle). \tag{12.45}$$

From (10.11) with $\eta = 1$

$$\sum_{n=0}^{\infty} \varrho(n, n)(1 + \mathrm{i}s)^n = \langle \exp(\mathrm{i}s\hat{n}) \rangle_{\mathcal{N}} \tag{12.46}$$

and consequently, in analogy to (10.12), we obtain

$$\varrho(n, n) = \frac{1}{n!} \frac{\mathrm{d}^n}{\mathrm{d}(\mathrm{i}s)^n} \langle \exp(\mathrm{i}s\hat{n}) \rangle_{\mathcal{N}} \bigg|_{\mathrm{i}s=-1} = \frac{\langle n \rangle^n}{n!} \exp(-\langle n \rangle), \tag{12.47}$$

which is the Poissonian distribution; this agrees with (10.60).

While the degrees of coherence $\gamma_{\mathcal{N}}$, ${}^{(G)}\gamma_{\mathcal{N}}$ and ${}^{(S)}\gamma_{\mathcal{N}}$ defined by (2.42a–c) are equivalent if the factorization condition (12.34) holds and $|\gamma_{\mathcal{N}}| = |{}^{(G)}\gamma_{\mathcal{N}}| = |{}^{(S)}\gamma_{\mathcal{N}}| = 1$, they are in general different (except in the case $m = n = 1$, when they are identical). As we have seen the degree of coherence ${}^{(G)}\gamma_{\mathcal{N}}$ is suitable for analyzing photon correlation experiments but it cannot be used as a measure of the partial coherence of higher orders since it has no bound. For this purpose the degrees of coherence $\gamma_{\mathcal{N}}$ and ${}^{(S)}\gamma_{\mathcal{N}}$ can be used, since the inequalities (2.43a, b) hold under the assumption that $\gamma_{\mathcal{N}}$ and ${}^{(S)}\gamma_{\mathcal{N}}$ exist; further (2.43a) holds for fields having classical analogues (Titulaer and Glauber (1965)) for which the weight function of the diagonal representation of the density matrix (Chapter 13) is positive-definite. Adopting for example the definition (2.42a) we can speak of partially coherent fields in higher order if $0 < |\gamma_{\mathcal{N}}^{(n,n)}| < 1$ (in analogy with the case $n = 1$). From the definition of the degree of coherence in quantum terms it follows that this quantity can be defined if the number of photons in the field is larger than n. To define all sequence an infinite number of photons must be present in the field. We can note that the definition (2.42a) is useful in connection with imaging problems (Peřina and Peřinová (1965), Peřina (1969), see Chapter 9), while (2.42c) serves to characterize the visibility in the two-slit experiment if the resulting measured quantity is the n-fold intensity (Klauder and Sudarshan (1968)). It may be concluded that the various definitions of the degree of coherence correspond to various types of experiments.

If $\gamma^{(1,1)} = 1$ for the quantity (12.33), the field is called weakly coherent while if $\gamma^{(n,n)} = 1$ for every n, it is strongly coherent (Picinbono (1967)). It is clear that if for a weakly coherent field $|\gamma_{\mathcal{N}}^{(1,1)}(x_1, x_2)| \equiv 1$, then $|\gamma_{\mathcal{N}}^{(n,n)}(x_1, \dots, x_{2n})| \equiv 1$ in the sense of (2.42a) and vice versa and the same is true in the sense of (2.42c); this can easily be verified by using (12.32). This means that relation (12.22) completely determines a weakly coherent field. Note that for strongly coherent fields considered as classical fields or as quantum fields having classical analogues, a sufficient specification is $\gamma^{(1,1)} = \gamma^{(2,2)} = 1$ (Titulaer and Glauber (1965), Picinbono (1967)).

12.3 Quantum characteristic functional

In analogy to (8.5) we can define the quantum normal characteristic functional

$$C_{\mathcal{N}}\{y(x)\} = \mathrm{Tr}\,\{\hat{\varrho} \exp [\textstyle\int y(x)\, \hat{A}^{(-)}(x)\, \mathrm{d}^4x] \exp [-\textstyle\int y^*(x)\, \hat{A}^{(+)}(x)\, \mathrm{d}^4x]\}, \tag{12.48}$$

from which the complete set of normally ordered correlation functions can be derived as follows

$$\mathrm{Tr}\,\{\hat{\varrho}\hat{A}^{(-)}(x_1) \dots \hat{A}^{(-)}(x_m)\, \hat{A}^{(+)}(x_{m+1}) \dots \hat{A}^{(+)}(x_{m+n})\} =$$

$$= \prod_{j=1}^{m} \frac{\delta}{\delta y(x_j)} \prod_{k=m+1}^{m+n} \frac{\delta}{\delta(-y^*(x_k))} C_{\mathcal{N}}\{y(x)\}\Bigg|_{y(x)\equiv 0}; \tag{12.49}$$

here $x = (\mathbf{x}, t, \mu)$ again so that the integrals over x include the summation over the polarization indices μ.

12.4 Measurements for mixed-order correlation functions

Quadratic detectors only allow the even-order correlation functions of the type $\Gamma_{\mathcal{N}}^{(n,n)}$ to be measured, but some further experiments are possible permitting in principle the measurement of mixed-order correlation functions of the type $\Gamma_{\mathcal{N}}^{(m,n)}$ for $m \neq n$, which contain additional phase information. We follow suggestions by Glauber (1970) based on nonlinear optical processes.

Considering a stationary field and returning to the frequency stationary conditions (2.21), which are necessary if the $\Gamma_{\mathcal{N}}^{(m,n)}$ are to be non-vanishing, we see that they are easily satisfied if $m = n$ and $\nu_j = \nu_{n+j}$. In this case one time dependence cancels the other and it is easy to have a non-vanishing function. A more complicated situation will occur if $m \neq n$, because in this case a very special set of frequencies must be occupied in the field. For example for a non-vanishing $\Gamma_{\mathcal{N}}^{(1,2)}$ it is necessary that $\nu_1 = \nu_2 + \nu_3$, a condition valid in the parametric amplifier or in Raman scattering (Chapter 22). This condition is necessary in order that the correlation function is non-vanishing, but a statistical dependence of modes must also exist, for if modes are independent then the density matrix factorizes into the product of the mode density matrices and $\Gamma_{\mathcal{N}}^{(1,2)} = 0$. In the parametric amplifier and Raman scattering this dependence exists and indeed $\Gamma_{\mathcal{N}}^{(1,2)} \neq 0$. For second harmonic generation in a nonlinear medium we also have $2\nu = \nu + \nu$, since in this case pairs of red photons are joined together to form blue photons. In the outgoing field there are two components; one of them is precisely twice the frequency of the other, that is

$$\hat{A}^{(+)}(t) = \hat{B}^{(+)}(t) + \hat{B}^{(+)2}(t) = \hat{U}^{(+)}(t) \exp(-\mathrm{i}\omega t) + \hat{U}^{(+)2}(t) \exp(-\mathrm{i}\, 2\omega t), \tag{12.50}$$

where $\hat{B}^{(+)}(t) = \hat{U}^{(+)}(t)\exp(-i\omega t)$ is the incident field and $\omega = 2\pi\nu$. There are also phase relations between these two fields and so $\Gamma_{\mathcal{N}}^{(1,2)}$ can be non-zero.

The measurement of the correlation functions with $m \neq n$ implies the measurement of the high frequency field with respect to the low frequency field. Assume we have a photodetector (an atom) with a threshold for photon detection which lies higher in frequency than the red frequency used. A single red photon will not give a photoelectron while two red photons will and so two red photons can be detected simultaneously. The photoelectric effect will be proportional to $\hat{A}_R^{(+)2}$. Of course, a blue photon will also make a photoelectric transition and these two processes interfere with one another giving the correlation function

$$\Gamma_{\mathcal{N}}^{(1,2)} = \mathrm{Tr}\,\{\hat{\varrho}\hat{A}_B^{(-)}\hat{A}_R^{(+)}\hat{A}_R^{(+)}\}. \tag{12.51}$$

Another possibility of measuring $\Gamma_{\mathcal{N}}^{(1,2)}$ (and $\Gamma_{\mathcal{N}}^{(2,1)}$) is to use two frequency converters C_1 and C_2 (Fig. 12.1) formed from nonlinear dielectrics. The first converter C_1 is illuminated by red photons, some of which are converted to blue photons and some of which remain red. Some of the remaining red photons are converted to blue ones by the second converter C_2 and the remaining red photons are filtered by a filter F. The detector D detects the function $\Gamma_{\mathcal{N}}^{(1,1)}$. Denoting the detected field by $\hat{A}_B^{(+)}$, we have

$$\hat{A}_B^{(+)} = \hat{B}_B^{(+)} + \hat{B}_R^{(+)2}. \tag{12.52}$$

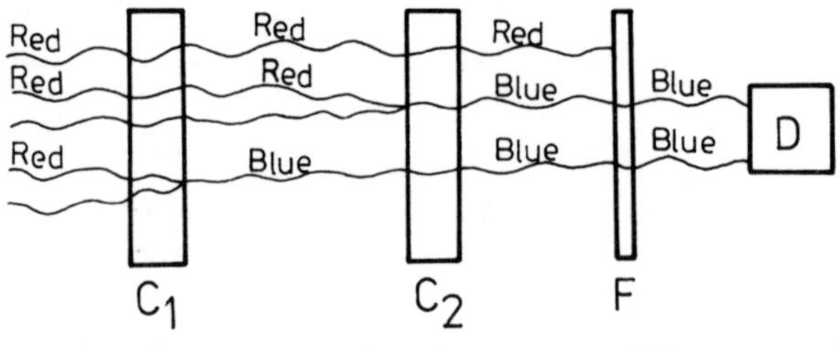

Fig. 12.1 A scheme for measuring $\Gamma_{\mathcal{N}}^{(1,2)}$.

The detector measures the correlation function

$$\begin{aligned}\mathrm{Tr}\,\{\hat{\varrho}\hat{A}_B^{(-)}\hat{A}_B^{(+)}\} &= \mathrm{Tr}\,\{\hat{\varrho}(\hat{B}_B^{(-)} + \hat{B}_R^{(-)2})(\hat{B}_B^{(+)} + \hat{B}_R^{(+)2})\} = \\ &= \mathrm{Tr}\,\{\hat{\varrho}\hat{B}_B^{(-)}\hat{B}_B^{(+)}\} + \mathrm{Tr}\,\{\hat{\varrho}\hat{B}_R^{(-)2}\hat{B}_B^{(+)}\} + \\ &+ \mathrm{Tr}\,\{\hat{\varrho}\hat{B}_B^{(-)}\hat{B}_R^{(+)2}\} + \mathrm{Tr}\,\{\hat{\varrho}\hat{B}_R^{(-)2}\hat{B}_R^{(+)2}\},\end{aligned} \tag{12.53}$$

which also involves the correlation functions $\Gamma_{\mathcal{N}}^{(1,2)}$ and $\Gamma_{\mathcal{N}}^{(2,1)}$.

However, in usual experiments with stationary fields the correlation functions with $m \neq n$ are practically zero (Mandel (1964a), Mehta and Mandel (1967)) and only in the peculiar circumstances of nonlinear optics is $\Gamma_{\mathcal{N}}^{(m,n)} \neq 0$ for $m \neq n$.

An analogue of the Hanbury Brown-Twiss correlation experiment measuring the correlation function $\Gamma_{\mathcal{N}}^{(2,2)}$ can also be performed by using nonlinear dielectrics (instead of two quadratic detectors with the outputs correlated in a correlater).

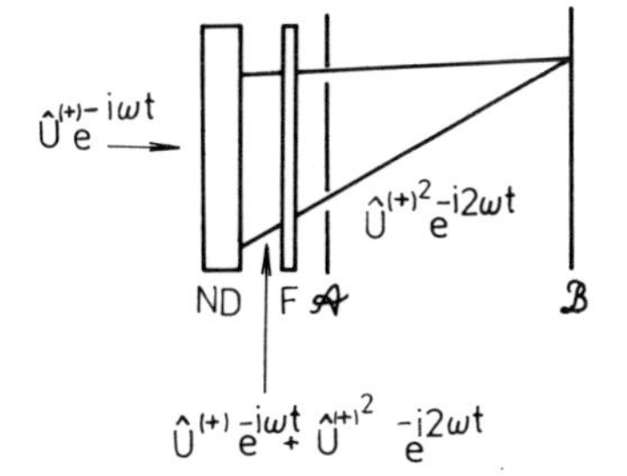

Fig. 12.2 A combination of the nonlinear dielectric ND with the Young's arrangement for measuring $\Gamma_{\mathscr{N}}^{(2,2)}$.

This was suggested by Beran, DeVelis and Parrent (1967). In Fig. 12.2 we see a scheme of the experimental arrangement, where a nonlinear dielectric ND is combined with the Young two-slit arrangement. If the incident field has the complex amplitude $\hat{U}^{(+)}(t)\exp(-\mathrm{i}\omega t)$, then, according to (12.50), the outgoing field has the complex amplitude $\hat{U}^{(+)}(t)\exp(-\mathrm{i}\omega t) + \hat{U}^{(+)2}(t)\exp(-\mathrm{i}\,2\omega t)$. If the first component is filtered by a filter F, then one can observe interference fringes, described by the fourth-order correlation function $\Gamma_{\mathscr{N}}^{(2,2)} = \mathrm{Tr}\,\{\hat{\varrho}\hat{U}^{(-)2}\hat{U}^{(+)2}\}$, on the screen $\mathscr{B}$.

CHAPTER 13

COHERENT-STATE DESCRIPTION OF THE ELECTROMAGNETIC FIELD

In the following we show that in the quantum theory of the electromagnetic field there exist quantum states called coherent states, in which the quantum correlation functions are factorized in the form (12.34), that is, a field in such a state fulfils the full coherence conditions. These states form an overcomplete set and they can be used as a basis for the decomposition of vectors and operators in quantum mechanics. Furthermore, these states make it possible to formulate the quantum theory of optical coherence in a form very close to the classical description developed in earlier chapters. Although the coherent states were first introduced by Schrödinger (1927) and were studied by others (e.g. Klauder (1960)), they were fully utilized in connection with quantum optics by Glauber (1963a, b, 1964, 1965, 1969, 1970, 1972), who introduced them to discuss the quantum coherence of optical fields.

13.1 Coherent states of the electromagnetic field

Definitions

Let us introduce the displaced vacuum states

$$|\alpha\rangle = \hat{D}(\alpha)\,|\,0\rangle, \tag{13.1}$$

where

$$\hat{D}(\alpha) = \exp(\alpha\hat{a}^{+} - \alpha^{*}\hat{a}) \tag{13.2}$$

is a displacement operator. The states $|\alpha\rangle$ are called the *coherent states.* Here $|0\rangle$ is the vacuum state, α is a complex number and $\hat{a}$ and $\hat{a}^{+}$ are the annihilation and creation operators of a photon respectively. Using the Baker-Hausdorff identity (e.g. Louisell (1964))

$$\exp(\hat{A} + \hat{B}) = \exp(\hat{A})\exp(\hat{B})\exp(-\tfrac{1}{2}[\hat{A}, \hat{B}]), \tag{13.3}$$

where $\hat{A}$ and $\hat{B}$ are operators for which the commutator $[\hat{A}, \hat{B}]$ is a *c*-number, i.e. $[[\hat{A}, \hat{B}], \hat{A}] = [[\hat{A}, \hat{B}], \hat{B}] = \hat{0}$, we can rewrite (13.1) in the form

$$\begin{aligned}|\alpha\rangle &= \exp(-\tfrac{1}{2}|\alpha|^2)\exp(\alpha\hat{a}^{+})\exp(-\alpha^{*}\hat{a})\,|\,0\rangle = \\ &= \exp(-\tfrac{1}{2}|\alpha|^2)\sum_{n=0}^{\infty}\frac{\alpha^n}{(n!)^{1/2}}\,|\,n\rangle,\end{aligned} \tag{13.4}$$

where we have expanded both $\exp(\alpha\hat{a}^+)$ and $\exp(-\alpha^*\hat{a})$ in a Taylor series and used the vacuum stability condition $\hat{a}|0\rangle = 0$ together with definition (11.10) of the Fock state $|n\rangle$. Making use of the relation (11.11b) we can easily verify that

$$\hat{a}|\alpha\rangle = \alpha|\alpha\rangle, \tag{13.5a}$$

$$\langle\alpha|\hat{a}^+ = \langle\alpha|\alpha^*, \tag{13.5b}$$

i.e. the coherent state $|\alpha\rangle$ is an eigenstate of the annihilation operator $\hat{a}$ with the eigenvalue α, which is in general a complex number since $\hat{a}$ is not a Hermitian operator. Equation (13.5a) shows that $|\alpha\rangle$ must contain an uncertain and indefinite number of photons since the annihilation operator $\hat{a}$ does not change this state.

One can prove (Miller and Mishkin (1966)) that it is not possible to construct states $\|\alpha\rangle\!\rangle$, with α finite, such that

$$\hat{a}^+\|\alpha\rangle\!\rangle = \alpha\|\alpha\rangle\!\rangle, \tag{13.6a}$$

$$\langle\!\langle\alpha\|\hat{a} = \langle\!\langle\alpha\|\alpha^*, \tag{13.6b}$$

i.e. no eigenstate of the creation operator $\hat{a}^+$ exists with finite eigenvalue. Indeed, assuming that

$$\|\alpha\rangle\!\rangle = \sum_{n=0}^{\infty}|n\rangle\langle n\|\alpha\rangle\!\rangle, \tag{13.7}$$

we obtain the recursion relation for the coefficient $\langle n\|\alpha\rangle\!\rangle$ using (13.6a) and the Hermitian adjoint to (11.11b),

$$\alpha\langle n\|\alpha\rangle\!\rangle = n^{1/2}\langle n-1\|\alpha\rangle\!\rangle, \tag{13.8}$$

which, by repeated use, gives

$$\langle n\|\alpha\rangle\!\rangle = \frac{(n!)^{1/2}}{\alpha^n}\langle 0\|\alpha\rangle\!\rangle. \tag{13.9}$$

Substituting (13.9) into (13.7) we have

$$\|\alpha\rangle\!\rangle = \langle 0\|\alpha\rangle\!\rangle\sum_{n=0}^{\infty}\frac{(n!)^{1/2}}{\alpha^n}|n\rangle \tag{13.10}$$

and consequently we have for the squared norm of $\|\alpha\rangle\!\rangle$

$$\langle\!\langle\alpha\|\alpha\rangle\!\rangle = |\langle 0\|\alpha\rangle\!\rangle|^2\sum_{n=0}^{\infty}\frac{n!}{|\alpha|^{2n}}, \tag{13.11}$$

recalling that $\langle n|m\rangle = \delta_{nm}$. This series is divergent for all finite complex amplitudes α and the states $\|\alpha\rangle\!\rangle$ cannot be regarded as physically admissible states of the radiation field. The states $|n\rangle$ and $|\alpha\rangle$ are the most useful bases for representations of the radiation field.

From (13.4) it is seen that the states $|\alpha\rangle$ are normalized, i.e. $\langle\alpha|\alpha\rangle = 1$ since $\langle n|m\rangle = \delta_{nm}$ and also

$$\langle n|\alpha\rangle = \exp(-\tfrac{1}{2}|\alpha|^2)\frac{\alpha^n}{(n!)^{1/2}}, \tag{13.12a}$$

which gives

$$|\langle n|\alpha\rangle|^2 = \exp(-|\alpha|^2)\frac{|\alpha|^{2n}}{n!}. \tag{13.12b}$$

This is a Poisson distribution with $\langle n\rangle = |\alpha|^2$ and so the probability that a coherent mode of the field contains n quanta is given by the Poisson distribution with $\langle n\rangle = |\alpha|^2$ in agreement with (12.47) and (10.60).

The scalar product of two coherent states is equal to

$$\begin{aligned}\langle\alpha|\beta\rangle &= \exp(-\tfrac{1}{2}|\alpha|^2 - \tfrac{1}{2}|\beta|^2)\sum_{n,m}\frac{\alpha^{*n}\beta^m}{(n!\,m!)^{1/2}}\langle n|m\rangle = \\ &= \exp(-\tfrac{1}{2}|\alpha|^2 - \tfrac{1}{2}|\beta|^2 + \alpha^*\beta)\end{aligned} \tag{13.13a}$$

and so

$$|\langle\alpha|\beta\rangle|^2 = \exp(-|\alpha-\beta|^2), \tag{13.13b}$$

which shows that the coherent states are not orthogonal; they can be regarded as approximately orthogonal if $|\alpha-\beta| \gg 1$.

Expansions in terms of coherent states

Even though the coherent states are not orthogonal they do form an overcomplete system of states, enabling us to expand an arbitrary vector or operator (particularly the density operator) in terms of the coherent states.

We first note that the coherent states are mutually dependent, which is the most characteristic property of an overcomplete set of vectors. On multiplying (13.4) by α^k and integrating over the whole complex α-plane we conclude that

$$\int \alpha^k|\alpha\rangle\,\mathrm{d}^2\alpha = 0, \qquad k = 1, 2, \ldots,$$

where $\alpha = r\exp(\mathrm{i}\varphi)$, r and φ are real and $\mathrm{d}^2\alpha = \mathrm{d}(\mathrm{Re}\,\alpha)\,\mathrm{d}(\mathrm{Im}\,\alpha) = r\,\mathrm{d}r\,\mathrm{d}\varphi$. The Fock state $|n\rangle$ may be expanded in terms of $|\alpha\rangle$ by multiplying (13.4) by $\pi^{-1}(n!)^{-1/2}\alpha^{*n}\exp(-|\alpha|^2/2)$ and integrating over all α giving

$$|n\rangle = \frac{1}{\pi}\int \exp(-\tfrac{1}{2}|\alpha|^2)\frac{\alpha^{*n}}{(n!)^{1/2}}|\alpha\rangle\,\mathrm{d}^2\alpha. \tag{13.14}$$

Substituting (13.14) into (11.18) we obtain

$$\frac{1}{\pi^2}\iint \exp\left(-\frac{|\alpha|^2}{2} - \frac{|\beta|^2}{2} + \alpha\beta^*\right)|\beta\rangle\langle\alpha|\,\mathrm{d}^2\alpha\,\mathrm{d}^2\beta = \hat{1}. \tag{13.15}$$

The integral over β is equal to

$$\frac{1}{\pi}\int |\beta\rangle \exp\left(\frac{|\beta|^2}{2}\right)\exp(-|\beta|^2+\alpha\beta^*)\,d^2\beta = \frac{1}{\pi}\sum_{n=0}^{\infty}\frac{|n\rangle}{(n!)^{1/2}}\times$$

$$\times\sum_{m=0}^{\infty}\frac{\alpha^m}{m!}\int \beta^n\beta^{*m}\exp(-|\beta|^2)\,d^2\beta = \exp\left(\frac{|\alpha|^2}{2}\right)|\alpha\rangle, \tag{13.16}$$

where we have expressed $|\beta\rangle\exp(|\beta|^2/2)$ using (13.4), we have written $\exp(\alpha\beta^*)$ in the form of a series, and we have used the integral

$$\int \beta^n\beta^{*m}\exp(-s|\beta|^2)\,d^2\beta = \tfrac{1}{2}\int_0^{\infty}|\beta|^{n+m}\exp(-s|\beta|^2)\,d|\beta|^2\times$$

$$\times\int_0^{2\pi}\exp[i(n-m)\arg\beta]\,d(\arg\beta) = \pi\int_0^{\infty}|\beta|^{n+m}\exp(-s|\beta|^2)\,d|\beta|^2\delta_{nm} =$$

$$=\frac{\pi n!}{s^{n+1}}\delta_{nm},\qquad \operatorname{Re}s>0 \tag{13.17}$$

for $s = 1$. Thus we arrive at the "resolution of unity" in terms of the coherent states

$$\frac{1}{\pi}\int |\alpha\rangle\langle\alpha|\,d^2\alpha = \hat{1}. \tag{13.18}$$

It can easily be shown that for an entire function $f(\alpha)$, in analogy to (13.16),

$$\frac{1}{\pi}\int f(\beta)\exp(-|\beta|^2+\alpha\beta^*)\,d^2\beta = f(\alpha). \tag{13.19}$$

By using the identity (13.18) we may also express an arbitrary normalized vector $|\rangle$ as

$$|\rangle = \frac{1}{\pi}\int |\alpha\rangle\langle\alpha|\rangle\,d^2\alpha \tag{13.20}$$

and an arbitrary operator $\hat{F}$ as

$$\hat{F} = \frac{1}{\pi^2}\iint |\alpha\rangle\langle\alpha|\hat{F}|\beta\rangle\langle\beta|\,d^2\alpha\,d^2\beta. \tag{13.21}$$

The coefficient $\langle\alpha|\rangle$ of the decomposition (13.20) can be written, expanding the state $|\rangle$ in the Fock states, as

$$\langle\alpha|\sum_n f_n|n\rangle = \exp(-\tfrac{1}{2}|\alpha|^2)\sum_n\frac{\alpha^{*n}}{(n!)^{1/2}}f_n = \exp(-\tfrac{1}{2}|\alpha|^2)f(\alpha^*). \tag{13.22}$$

As the squared norm of $|\rangle$ is $\langle|\rangle = \sum_n |f_n|^2 = 1$, the series $f(\alpha^*) = \sum_n \alpha^{*n}f_n/(n!)^{1/2}$ converges for all values of α in the finite plane and therefore $f(\alpha^*)$ is an entire function.

Minimum-uncertainty wave packets

Defining $\Delta\hat{p}$ and $\Delta\hat{q}$ by

$$\Delta\hat{p} = \hat{p} - \langle\hat{p}\rangle, \qquad \Delta\hat{q} = \hat{q} - \langle\hat{q}\rangle, \tag{13.23}$$

we have from the commutation rule (11.16)

$$\begin{aligned}\tfrac{1}{2}\hbar &= \tfrac{1}{2}|\langle[\hat{q}, \hat{p}]\rangle| = \tfrac{1}{2}|\langle[\Delta\hat{q}, \Delta\hat{p}]\rangle| \leqq \\ &\leqq |\langle\Delta\hat{q}\,\Delta\hat{p}\rangle| \leqq [\langle(\Delta\hat{q})^2\rangle\,\langle(\Delta\hat{p})^2\rangle]^{1/2}.\end{aligned} \tag{13.24}$$

If we use the eigenvalue properties (13.5) of the coherent states we obtain from (11.15)

$$\langle\hat{q}\rangle = \left(\frac{\hbar}{2\omega}\right)^{1/2}(\alpha + \alpha^*), \langle\hat{p}\rangle = -\mathrm{i}\left(\frac{\hbar\omega}{2}\right)^{1/2}(\alpha - \alpha^*), \tag{13.25a}$$

$$\begin{aligned}\langle(\Delta\hat{q})^2\rangle &= \langle\hat{q}^2\rangle - \langle\hat{q}\rangle^2 = \frac{\hbar}{2\omega}, \\ \langle(\Delta\hat{p})^2\rangle &= \langle\hat{p}^2\rangle - \langle\hat{p}\rangle^2 = \frac{\hbar\omega}{2}\end{aligned} \tag{13.25b}$$

and for the coherent states the inequality (13.24) reduces to the equality

$$\langle(\Delta\hat{q})^2\rangle\,\langle(\Delta\hat{p})^2\rangle = \frac{\hbar^2}{4}; \tag{13.26}$$

thus the coherent states are the closest to the classical states that quantum theory allows.

The role of the coherent states in a model of the interaction of radiation and matter applied to the laser was first investigated by Picard and Willis (1965). Using the arithmetic mean–geometric mean inequality and $\langle(\Delta\hat{p})^2\rangle = \omega^2\langle(\Delta\hat{q})^2\rangle$ valid for the coherent states, we have

$$\tfrac{1}{2}[\langle(\Delta\hat{p})^2\rangle + \omega^2\langle(\Delta\hat{q})^2\rangle] \geqq \omega[\langle(\Delta\hat{q})^2\rangle\,\langle(\Delta\hat{p})^2\rangle]^{1/2} = \frac{\hbar\omega}{2}. \tag{13.27}$$

It can easily be seen that the inequality (13.27) reduces to the equality for the coherent states. Thus the quantity

$$\tfrac{1}{2}[\langle(\Delta\hat{p})^2\rangle + \omega^2\langle(\Delta\hat{q})^2\rangle] = \tfrac{1}{2}[\langle\hat{p}^2\rangle + \omega^2\langle\hat{q}^2\rangle] - \tfrac{1}{2}[\langle\hat{p}\rangle^2 + \omega^2\langle\hat{q}\rangle^2] \tag{13.28}$$

represents the difference between the total energy of the field and the coherent energy of the field, so it is the incoherent energy in the field. Therefore it may be said that in the coherent state the incoherent energy takes on its minimum value $\hbar\omega/2$ in every mode arising only from the zero-point fluctuations.

Some properties of displacement operator $\hat{D}(\alpha)$

From the definition of the displacement operator (13.2) it follows that

$$\hat{D}^{+}(\alpha) = \hat{D}^{-1}(\alpha) = \hat{D}(-\alpha). \tag{13.29}$$

Further

$$\hat{a}\hat{D}(\alpha) = \exp(-\tfrac{1}{2}|\alpha|^2)\,\hat{a}\exp(\alpha\hat{a}^{+})\exp(-\alpha^{*}\hat{a}) = \\ = \exp(-\tfrac{1}{2}|\alpha|^2)\sum_{n=0}^{\infty}\frac{\alpha^n}{n!}\hat{a}\hat{a}^{+n}\exp(-\alpha^{*}\hat{a}), \tag{13.30}$$

where we have used (13.3). The commutation rule gives $\hat{a}\hat{a}^{+n} = \hat{a}^{+n}\hat{a} + n\hat{a}^{+n-1}$ and so

$$\hat{a}\hat{D}(\alpha) = \hat{D}(\alpha)\,\hat{a} + \alpha\hat{D}(\alpha). \tag{13.31a}$$

We can prove in the same way for any entire operator $\hat{F}(\hat{a}^{+}, \hat{a})$ that

$$[\hat{a}, \hat{F}] = \frac{\partial\hat{F}}{\partial\hat{a}^{+}}, \qquad [\hat{F}, \hat{a}^{+}] = \frac{\partial\hat{F}}{\partial\hat{a}}, \tag{13.31b}$$

giving the standard commutation rule if $\hat{F} = \hat{a}^{+}$ or $\hat{a}$ respectively. Therefore the operator $\hat{D}(\alpha)$ represents a displacement operator in the sense that

$$\hat{D}^{-1}(\alpha)\,\hat{a}\hat{D}(\alpha) = \hat{a} + \alpha, \\ \hat{D}^{-1}(\alpha)\,\hat{a}^{+}\hat{D}(\alpha) = \hat{a}^{+} + \alpha^{*}. \tag{13.32}$$

More generally for an operator function $\hat{F}(\hat{a}^{+}, \hat{a})$

$$\hat{D}^{-1}(\alpha)\,\hat{F}(\hat{a}^{+}, \hat{a})\,\hat{D}(\alpha) = \hat{F}(\hat{a}^{+} + \alpha^{*}, \hat{a} + \alpha). \tag{13.33}$$

The product of two displacement operators is equal to

$$\hat{D}(\alpha)\,\hat{D}(\beta) = \exp(\alpha\hat{a}^{+} - \alpha^{*}\hat{a})\exp(\beta\hat{a}^{+} - \beta^{*}\hat{a}) = \\ = \hat{D}(\alpha + \beta)\exp[\tfrac{1}{2}(\alpha\beta^{*} - \alpha^{*}\beta)], \tag{13.34}$$

where (13.3) has been used.

Taking the trace of $\hat{D}$ using (13.18), we arrive at

$$\mathrm{Tr}\,\hat{D}(\alpha) = \frac{1}{\pi}\int\langle\beta|\hat{D}(\alpha)|\beta\rangle\,\mathrm{d}^2\beta = \frac{1}{\pi}\int\exp(-\tfrac{1}{2}|\alpha|^2)\langle\beta|\exp(\alpha\hat{a}^{+})\,\times \\ \times\exp(-\alpha^{*}\hat{a})|\beta\rangle\,\mathrm{d}^2\beta = \frac{1}{\pi}\int\exp(-\tfrac{1}{2}|\alpha|^2 + \alpha\beta^{*} - \alpha^{*}\beta)\,\mathrm{d}^2\beta = \\ = \exp(-\tfrac{1}{2}|\alpha|^2)\frac{1}{\pi}\int\exp[2\mathrm{i}\,\mathrm{Im}(\alpha\beta^{*})]\,\mathrm{d}^2\beta = \exp(-\tfrac{1}{2}|\alpha|^2)\,\times \\ \times\frac{1}{\pi}\iint_{-\infty}^{+\infty}\exp[2\mathrm{i}(\mathrm{Im}\,\alpha\,\mathrm{Re}\,\beta - \mathrm{Re}\,\alpha\,\mathrm{Im}\,\beta)]\,\mathrm{d}(\mathrm{Re}\,\beta)\,\mathrm{d}(\mathrm{Im}\,\beta) = \\ = \exp(-\tfrac{1}{2}|\alpha|^2)\,\pi\delta(\mathrm{Re}\,\alpha)\,\delta(\mathrm{Im}\,\alpha) = \pi\delta(\alpha), \tag{13.35}$$

where (13.3), 13.5) and the substitutions $2\,\mathrm{Re}\,\beta = x$ and $2\,\mathrm{Im}\,\beta = y$ have been used for the integration; $\delta(\alpha)$ is the two-dimensional Dirac function.

Equations (13.34) and (13.35) give the completeness condition

$$\mathrm{Tr}\,\{\hat{D}^{+}(\alpha)\,\hat{D}(\beta)\} = \pi\delta(\alpha - \beta). \tag{13.36}$$

From (13.35) it follows that the $\delta(\alpha)$-function can be represented by a Fourier integral in the complex form

$$\delta(\alpha) = \frac{1}{\pi^2}\int \exp\,(\alpha\beta^* - \alpha^*\beta)\,\mathrm{d}^2\beta. \tag{13.37}$$

Expectation values of operators in coherent states

Assume an operator $\hat{F}^{(\mathcal{N})}(\hat{a}^{+}, \hat{a})$ to be in the normal form. Then it holds that

$$\langle\alpha\,|\,\hat{F}^{(\mathcal{N})}(\hat{a}^{+}, \hat{a})\,|\,\alpha'\rangle = F^{(\mathcal{N})}(\alpha^*, \alpha')\,\langle\alpha\,|\,\alpha'\rangle \tag{13.38}$$

and in particular

$$\langle\alpha\,|\,\hat{F}^{(\mathcal{N})}(\hat{a}^{+}, \hat{a})\,|\,\alpha\rangle = F^{(\mathcal{N})}(\alpha^*, \alpha), \tag{13.39}$$

that is, the expectation value of the normally ordered operator $\hat{F}^{(\mathcal{N})}$ is obtained by the substitutions $\hat{a} \to \alpha$, $\hat{a}^{+} \to \alpha^*$ in the operator function. If $\hat{F}^{(\mathcal{N})}(\hat{a}^{+}, \hat{a}) = \hat{a}^{+k}\hat{a}^{l}$ (k, l are integers), then $\langle\alpha\,|\,\hat{a}^{+k}\hat{a}^{l}\,|\,\alpha\rangle = \alpha^{*k}\alpha^{l}$.

Further we can prove, using (13.4), that

$$\hat{a}^{+}\,|\alpha\rangle = \left(\frac{\alpha^*}{2} + \frac{\partial}{\partial\alpha}\right)|\alpha\rangle,$$

$$\langle\alpha\,|\,\hat{a} = \left(\frac{\alpha}{2} + \frac{\partial}{\partial\alpha^*}\right)\langle\alpha\,|, \tag{13.40}$$

and

$$\hat{a}^{+}\,|\alpha\rangle\,\langle\alpha\,| = \left(\alpha^* + \frac{\partial}{\partial\alpha}\right)|\alpha\rangle\,\langle\alpha\,|,$$

$$|\alpha\rangle\,\langle\alpha\,|\,\hat{a} = \left(\alpha + \frac{\partial}{\partial\alpha^*}\right)|\alpha\rangle\,\langle\alpha\,|. \tag{13.41}$$

For an operator $\hat{F}(\hat{a}^{+}, \hat{a})$ we obtain

$$\hat{F}(\hat{a}^{+}, \hat{a})\,|\,\alpha\rangle\,\langle\alpha\,| = F\left(\alpha^* + \frac{\partial}{\partial\alpha}, \alpha\right)|\,\alpha\rangle\,\langle\alpha\,|,$$

$$|\,\alpha\rangle\,\langle\alpha\,|\,\hat{F}(\hat{a}^{+}, \hat{a}) = F\left(\alpha^*, \alpha + \frac{\partial}{\partial\alpha^*}\right)|\,\alpha\rangle\,\langle\alpha\,| \tag{13.42}$$

and multiplying this by $|\alpha\rangle$ from the right and $\langle\alpha|$ from the left we arrive at

$$\langle\alpha|\hat{F}(\hat{a}^+,\hat{a})|\alpha\rangle = F\left(\alpha^* + \frac{\partial}{\partial\alpha},\alpha\right) = F\left(\alpha^*,\alpha + \frac{\partial}{\partial\alpha^*}\right). \tag{13.43}$$

These rules are useful for solving the equations of motion. As an example we consider the Schrödinger equation for an operator $\hat{S}$

$$i\hbar\frac{\partial\hat{S}(\hat{a}^+,\hat{a},t)}{\partial t} = \hat{H}(\hat{a}^+,\hat{a})\,\hat{S}(\hat{a}^+,\hat{a},t). \tag{13.44}$$

If we multiply (13.44) by $|\alpha\rangle$ from the right and by $\langle\alpha|$ from the left and use (13.43) we arrive at

$$i\hbar\frac{\partial S\left(\alpha^* + \frac{\partial}{\partial\alpha},\alpha,t\right)}{\partial t} = H\left(\alpha^* + \frac{\partial}{\partial\alpha},\alpha\right)S\left(\alpha^* + \frac{\partial}{\partial\alpha},\alpha,t\right), \tag{13.45}$$

which is a classical equation. Solving this equation for S we can obtain

$$S\left(\alpha^* + \frac{\partial}{\partial\alpha},\alpha,t\right) = S^{(\mathcal{N})}(\alpha^*,\alpha,t) = \langle\alpha|\hat{S}^{(\mathcal{N})}(\hat{a}^+,\hat{a},t)|\alpha\rangle; \tag{13.46}$$

hence this procedure leads directly to the normal form of $\hat{S}$.

Generalized coherent states

Generalized coherent states have been introduced by Titulaer and Glauber (1966) and may be written as

$$|\alpha,\{\vartheta_n\}\rangle = \exp\left(-\tfrac{1}{2}|\alpha|^2\right)\sum_{n=0}^{\infty}\frac{\alpha^n\exp(i\vartheta_n)}{(n!)^{1/2}}|n\rangle, \tag{13.47}$$

where $\{\vartheta_n\}$ is a sequence of real numbers. These states may be regarded as the most general pure fully coherent states. Some of their properties were studied by Titulaer and Glauber (1966), Crosignani, Di Porto and Solimeno (1968a) and Bialynicka-Birula (1968). Some properties of a subset of the coherent states were investigated by Campagnoli and Zambotti (1968).

Putting $\vartheta_n = n\vartheta$ we see that $\hat{a}|\alpha,\{\vartheta_n\}\rangle = \alpha\exp(i\vartheta)|\alpha,\{\vartheta_n\}\rangle$ and so $\langle\alpha,\{\vartheta_n\}|\hat{a}^{+k}\hat{a}^k|\alpha,\{\vartheta_n\}\rangle = \alpha^{*k}\alpha^k$ independently of ϑ. This shows that the field in the coherent state fulfils the full coherence factorization conditions and also that mixtures of the coherent states which differ from one another by phase shifts also describe coherent fields (for $m = n$). The difference between the coherent states $|\alpha\rangle$ and the generalized coherent states $|\alpha,\{\vartheta_n\}\rangle$ does not manifest itself in experiments described by even-order correlation functions and consequently we may use only the coherent states $|\alpha\rangle$ in the following. However, only the coherent states $|\alpha\rangle$ satisfy the factorization conditions (12.34) for all m and n.

Multimode description

The above coherent-state formulation can be applied to the electromagnetic field in a finite volume having a finite or at most a countably infinite number of degrees of freedom. (A generalization to fields in an infinite volume having a non-denumerable number of degrees of freedom is also possible if functional (continuous) integration is adopted.) For this purpose we introduce the global coherent states

$$|\{\alpha_\lambda\}\rangle \equiv \prod_\lambda |\alpha_\lambda\rangle \tag{13.48}$$

for which

$$|\{\alpha_\lambda\}\rangle = \sum_{\{n_\lambda\}=0} \left\{ \prod_{\lambda'} \frac{\alpha_{\lambda'}^{n_{\lambda'}}}{(n_{\lambda'}!)^{1/2}} \exp\left(-\frac{|\alpha_{\lambda'}|^2}{2}\right) |\{n_\lambda\}\rangle \right\} \tag{13.49}$$

and the completeness condition becomes

$$\int |\{\alpha_\lambda\}\rangle \langle\{\alpha_\lambda\}| \, d\mu(\{\alpha_\lambda\}) = \hat{1}, \tag{13.50}$$

where

$$d\mu(\{\alpha_\lambda\}) \equiv \prod_\lambda \frac{d^2\alpha_\lambda}{\pi} \tag{13.51}$$

and $|\{n_\lambda\}\rangle$ is given by (11.36).

If (11.22) and (13.5a) are used, we obtain the following eigenvalue property for the global coherent states

$$\hat{\mathbf{A}}^{(+)}(x) |\{\alpha_\lambda\}\rangle = \mathbf{V}(x) |\{\alpha_\lambda\}\rangle \tag{13.52a}$$

and its Hermitian adjoint

$$\langle\{\alpha_\lambda\}| \hat{\mathbf{A}}^{(-)}(x) = \langle\{\alpha_\lambda\}| \mathbf{V}^*(x), \tag{13.52b}$$

where

$$\mathbf{V}(x) = \frac{(2\pi\hbar c)^{1/2}}{L^{3/2}} \sum_{\mathbf{k},s} k^{-1/2} \mathbf{e}^{(s)}(\mathbf{k}) \, \alpha_{\mathbf{k}s} \exp\left[i(\mathbf{k}\,.\,\mathbf{x} - kct)\right]. \tag{13.53}$$

Hence, choosing the density matrix in the form $\hat{\varrho} = |\{\alpha_\lambda\}\rangle\langle\{\alpha_\lambda\}|$ the correlation functions (12.5) are factorized in the form (12.34), and the field in the coherent state fulfils the full coherence conditions. One can also see that all frequencies are present in (13.53) so that coherence does not in general restrict the spectral composition of the field, but rather it restricts the statistical behaviour of the field.

Time development of the coherent states

Considering a one-mode field for simplicity with the free-field renormalized Hamiltonian $\hbar\omega\hat{a}^+\hat{a}$, the time dependence of the coherent state $|\alpha(t)\rangle$ is determined by the Schrödinger equation

$$i\hbar \frac{|\alpha(t)\rangle}{\partial t} = \hat{H} |\alpha(t)\rangle, \hat{H} = \hbar\omega\hat{a}^+\hat{a}. \tag{13.54}$$

If the initial condition is $|\alpha(0)\rangle = |\alpha\rangle$, its solution can be written as

$$|\alpha(t)\rangle = \exp\left(-\mathrm{i}\frac{\hat{H}t}{\hbar}\right)|\alpha\rangle = \exp(-\mathrm{i}\omega t\hat{a}^{+}\hat{a} - \tfrac{1}{2}|\alpha|^2) \times$$
$$\times \sum_{n=0}^{\infty} \frac{\alpha^n}{(n!)^{1/2}}|n\rangle = |\alpha\exp(-\mathrm{i}\omega t)\rangle. \tag{13.55}$$

Therefore the coherent state remains coherent at all times for the free-field Hamiltonian. The complex amplitude $\alpha\exp(-\mathrm{i}\omega t)$ moves along circles in the complex plane.

It can be shown more generally (Chapter 20) that a coherent state remains coherent at all times when the Hamiltonian takes the form

$$\hat{H} = \sum_{j,k} f_{jk}(t)\,\hat{a}_j^{+}\hat{a}_k + \sum_k [g_k(t)\,\hat{a}_k^{+} + g_k^{*}(t)\,\hat{a}_k] + h(t), \tag{13.56}$$

where $f_{jk} = f_{kj}^{*}$, $h^{*}(t) = h(t)$ and $g_k(t)$ are arbitrary functions of time (Glauber (1966b), Mehta and Sudarshan (1966), Mehta, Chand, Sudarshan and Vedam (1967)). This Hamiltonian describes free energy, the exchange of energies among modes and the interaction of classical currents with the radiation modes. Therefore a field in the coherent state is generated by non-random prescribed currents (Glauber (1963b, 1965)). A Hamiltonian including third-order nonlinear terms yet preserving special coherent states has been discussed by Mišta (1967). Hamiltonians for quantum harmonic oscillators with time-dependent and random frequencies have been discussed by Crosignani, Di Porto and Solimeno (1968b, 1969) and Solimeno, Di Porto and Crosignani (1969). Further investigations of the evolution of the coherent states, including more general classes of coherent states, have been carried out by Kano (1976), Trifonov and Ivanov (1977), Trias (1977), Chand (1979) and Malkin and Man'ko (1979).

13.2 Glauber-Sudarshan representation of the density matrix

As states of the field are mostly mixed states rather than pure states, we must use the density matrix to describe general fields and therefore it is useful to express the density matrix in terms of the coherent states. Using the operator expansion (13.21) for the density matrix we obtain (Glauber (1963b))

$$\hat{\varrho} = \frac{1}{\pi^2}\iint|\alpha\rangle\langle\alpha|\hat{\varrho}|\beta\rangle\langle\beta|\,\mathrm{d}^2\alpha\,\mathrm{d}^2\beta, \tag{13.57}$$

which is an expansion in terms of dyadic products $|\alpha\rangle\langle\beta|$. By using the Fock form (11.25) of $\hat{\varrho}$ and (13.12a), we obtain for the weight function $\langle\alpha|\hat{\varrho}|\beta\rangle$ in (13.57)

$$\langle\alpha|\hat{\varrho}|\beta\rangle = \sum_{n,m}\varrho(n,m)\frac{\alpha^{*n}\beta^m}{(n!\,m!)^{1/2}}\exp(-\tfrac{1}{2}|\alpha|^2 - \tfrac{1}{2}|\beta|^2). \tag{13.58}$$

Since $\mathrm{Tr}\,\hat{\varrho}^2 = \sum_{n,m} |\varrho(n,m)|^2 \leqq 1$ ($|\varrho(n,m)|^2 \leqq \varrho(n,n)\,\varrho(m,m)$, i.e. $\mathrm{Tr}\,\hat{\varrho}^2 \leqq (\mathrm{Tr}\,\hat{\varrho})^2 = 1$), the series in (13.58) converge for all finite values of α^* and β and so $\langle \alpha | \hat{\varrho} | \beta \rangle$ is an entire function in α and β, i.e. it is a well-behaved function.

The advantage of the expansion (13.57) is that the quantum expectation values of normally ordered operators (q-numbers) reduce to expectation values of eigenvalues in the coherent states (c-numbers) with weight function $\langle \alpha | \hat{\varrho} | \beta \rangle$. These expectation values may be calculated simply by integration over the complex planes of α and β. The expansion (13.57) corresponds to the condition of "resolution of unity" in the form

$$\frac{1}{\pi^2} \iint |\alpha\rangle \langle \alpha | \beta \rangle \langle \beta | \, \mathrm{d}^2\alpha \, \mathrm{d}^2\beta =$$

$$= \frac{1}{\pi^2} \iint \exp\left(\alpha^*\beta - \tfrac{1}{2}|\alpha|^2 - \tfrac{1}{2}|\beta|^2\right) |\alpha\rangle \langle \beta | \, \mathrm{d}^2\alpha \, \mathrm{d}^2\beta = \hat{1}, \tag{13.59}$$

where (13.13a) has been used. This is just equation (13.15) which leads to the simple completeness relation (13.18) using (13.16). We can now ask whether the density matrix $\hat{\varrho}$ can be expanded in the simpler form (Glauber (1963b, c), Sudarshan (1963a))

$$\hat{\varrho} = \int \Phi_{\mathcal{N}}(\alpha) \, |\alpha\rangle \langle \alpha| \, \mathrm{d}^2\alpha \tag{13.60}$$

as a "mixture" of the projection operators $|\alpha\rangle\langle\alpha|$ onto the coherent states $((|\alpha\rangle\langle\alpha|)^2 = |\alpha\rangle\langle\alpha|\alpha\rangle\langle\alpha| = |\alpha\rangle\langle\alpha|)$, where $\Phi_{\mathcal{N}}(\alpha)$ is the weight function. This diagonal representation of the density matrix is called the *Glauber-Sudarshan representation.* Since $\mathrm{Tr}\,\hat{\varrho} = 1$,

$$\int \Phi_{\mathcal{N}}(\alpha) \, \mathrm{d}^2\alpha = 1 \tag{13.61}$$

and from the Hermiticity $\hat{\varrho}^+ = \hat{\varrho}$ it follows that $[\Phi_{\mathcal{N}}(\alpha)]^* = \Phi_{\mathcal{N}}(\alpha)$, i.e. $\Phi_{\mathcal{N}}(\alpha)$ is a real function of the complex variable α.

The function $\langle \alpha | \hat{\varrho} | \beta \rangle$ can be expressed in terms of $\Phi_{\mathcal{N}}(\alpha)$, using (13.60) and (13.13a), in the form

$$\langle \alpha | \hat{\varrho} | \beta \rangle = \int \langle \alpha | \gamma \rangle \langle \gamma | \beta \rangle \, \Phi_{\mathcal{N}}(\gamma) \, \mathrm{d}^2\gamma = \exp\left(-\tfrac{1}{2}|\alpha|^2 - \tfrac{1}{2}|\beta|^2\right) \times$$

$$\times \int \Phi_{\mathcal{N}}(\gamma) \exp\left(-|\gamma|^2 + \alpha^*\gamma + \beta\gamma^*\right) \mathrm{d}^2\gamma \tag{13.62a}$$

and for $\alpha = \beta$ we arrive at

$$\langle \alpha | \hat{\varrho} | \alpha \rangle = \int \Phi_{\mathcal{N}}(\beta) \exp\left(-|\alpha - \beta|^2\right) \mathrm{d}^2\beta. \tag{13.62b}$$

Another relation between the weight function $\Phi_{\mathcal{N}}$ and the Fock matrix elements $\varrho(n, m)$ of the density matrix can be obtained, using (13.60), (Sudarshan (1963a), Glauber (1963b))

$$\langle n | \hat{\varrho} | m \rangle = \varrho(n,m) = \int \Phi_{\mathcal{N}}(\alpha) \langle n | \alpha \rangle \langle \alpha | m \rangle \, \mathrm{d}^2\alpha =$$

$$= \int \Phi_{\mathcal{N}}(\alpha) \frac{\alpha^n \alpha^{*m}}{(n!\,m!)^{1/2}} \exp\left(-|\alpha|^2\right) \mathrm{d}^2\alpha. \tag{13.63}$$

Equation (13.60) can formally be derived from the Fock form (11.25) of the density matrix as follows. Substituting (13.14) into (11.25) and making use of (13.63) we find

$$\hat{\varrho} = \frac{1}{\pi^2} \iiint \exp\left(-\tfrac{1}{2}|\alpha|^2 - \tfrac{1}{2}|\beta|^2 - |\gamma|^2 + \alpha^*\gamma + \beta\gamma^*\right) \Phi_{\mathcal{N}}(\gamma) \times \\ \times |\alpha\rangle\langle\beta| \, \mathrm{d}^2\alpha \, \mathrm{d}^2\beta \, \mathrm{d}^2\gamma \tag{13.64}$$

and using (13.16) we obtain (13.60).

The diagonal form (13.60) of the density matrix is very convenient for calculations of expectation values of normally ordered operators. For example, for the expectation value of the moment operator $\hat{a}^{+k}\hat{a}^l$ (k, l are integers) we obtain

$$\mathrm{Tr}\,\{\hat{\varrho}\hat{a}^{+k}\hat{a}^l\} = \mathrm{Tr}\,\{\textstyle\int \Phi_{\mathcal{N}}(\alpha)\,|\alpha\rangle\langle\alpha|\,\mathrm{d}^2\alpha\hat{a}^{+k}\hat{a}^l\} = \\ = \textstyle\int \Phi_{\mathcal{N}}(\alpha)\langle\alpha|\hat{a}^{+k}\hat{a}^l|\alpha\rangle\,\mathrm{d}^2\alpha = \int \Phi_{\mathcal{N}}(\alpha)\,\alpha^{*k}\alpha^l\,\mathrm{d}^2\alpha = \langle\alpha^{*k}\alpha^l\rangle_{\mathcal{N}}, \tag{13.65}$$

where we have used the eigenvalue properties (13.5a, b) of the coherent states. The suffix $\mathcal{N}$ on Φ expresses the fact that this function is related to the *normally ordered operators*. Consequently, the quantum expectation value of a normally ordered operator can be expressed as a "classical" expectation value in a generalized phase space with the "probability distribution" $\Phi_{\mathcal{N}}(\alpha)$ if the integration is carried out over the whole complex α-plane. As we have seen, the correlation functions suitable for the description of experiments with photodetectors, yielding information about the coherence properties of light, represent just the expectation values of normally ordered products of the field operators and so they may be expressed (using the Glauber-Sudarshan representation of the density matrix) in a form very close to the classical one studied earlier. This representation provides a basis for a general *formal* equivalence between the classical and quantum descriptions of optical coherence.

The field in the coherent state $|\beta\rangle$ possesses the weight function $\Phi_{\mathcal{N}}(\alpha)$

$$\Phi_{\mathcal{N}}(\alpha) = \delta(\alpha - \beta) \tag{13.66}$$

and the density matrix $\hat{\varrho}$ is just the projection operator $|\beta\rangle\langle\beta|$ onto the coherent state. A general density matrix describing a mixed state of the field can be interpreted, according to (13.60), as a superposition of the projection operators $|\alpha\rangle\langle\alpha|$ with the weight function $\Phi_{\mathcal{N}}(\alpha)$.

Although $\Phi_{\mathcal{N}}(\alpha)$ has some of the properties of a probability distribution (it is a real-valued function fulfilling the normalization (13.61)) it cannot generally be interpreted as a classical probability distribution since it is not non-negative and it may have singularities stronger than the δ-function. This property is a consequence of the fact that the quantities $\mathrm{Re}\,\alpha$ and $\mathrm{Im}\,\alpha$, which are mean values of the non-commuting operators $(\hat{a} + \hat{a}^+)/2$ and $(\hat{a} - \hat{a}^+)/2\mathrm{i}$ in the coherent state, are not simultaneously measurable and so $\Phi_{\mathcal{N}}(\alpha)$ cannot be measured directly. We call the $\Phi_{\mathcal{N}}$-function the *quasi-probability function*; it represents one of a whole family of quasi-probability functions, or quasi-distributions. Another quasi-probability function was introduced by Wigner (1932)† and $\pi^{-1}\langle\alpha|\hat{\varrho}|\alpha\rangle$ is in fact a quasi-probability function, too. This latter function has many properties in common with a probability density, including positive-definiteness and regularity, but it cannot be measured directly.

Note that phase operators in terms of the coherent states have been discussed by Paul (1974, 1976).

Some of the properties of $\langle\alpha|\hat{\varrho}|\alpha\rangle$ can be derived as follows. First we see that the function $\pi^{-1}\langle\alpha|\hat{\varrho}|\alpha\rangle$ plays the same role in the averaging of antinormally ordered products of field operators as $\Phi_{\mathcal{N}}(\alpha)$ plays for normally ordered products, since

$$\mathrm{Tr}\,\{\hat{\varrho}\hat{a}^l\hat{a}^{+k}\} = \mathrm{Tr}\,\{\hat{a}^{+k}\hat{\varrho}\hat{a}^l\} = \mathrm{Tr}\left\{\frac{1}{\pi}\int|\alpha\rangle\langle\alpha|\,\mathrm{d}^2\alpha\hat{a}^{+k}\hat{\varrho}\hat{a}^l\right\} = \\ = \frac{1}{\pi}\int\alpha^{*k}\alpha^l\langle\alpha|\hat{\varrho}|\alpha\rangle\,\mathrm{d}^2\alpha. \tag{13.67}$$

We have carried out a cyclic permutation of the operators, which does not change the trace, and introduced the unit operator (13.18). We may denote $\pi^{-1}\langle\alpha|\hat{\varrho}|\alpha\rangle$ as $\Phi_{\mathcal{A}}(\alpha)$, which emphasizes that this function is related to *antinormally ordered operators*. From (13.62b) we have

$$\Phi_{\mathcal{A}}(\alpha) = \frac{1}{\pi}\int\Phi_{\mathcal{N}}(\beta)\exp(-|\alpha-\beta|^2)\,\mathrm{d}^2\beta. \tag{13.68}$$

Putting $k = l = 0$ in (13.67) we arrive at the normalization condition

$$\frac{1}{\pi}\int\langle\alpha|\hat{\varrho}|\alpha\rangle\,\mathrm{d}^2\alpha = 1. \tag{13.69}$$

Introducing the characteristic function (Glauber (1966a))

$$C(\beta) = \mathrm{Tr}\,\{\hat{\varrho}\hat{D}(\beta)\} = \mathrm{Tr}\,\{\hat{\varrho}\exp(\beta\hat{a}^+ - \beta^*\hat{a})\} \tag{13.70}$$

and making use of (13.3) we obtain

$$C_{\mathcal{N}}(\beta) = \exp(\tfrac{1}{2}|\beta|^2)\,C(\beta) = \exp(|\beta|^2)\,C_{\mathcal{A}}(\beta), \tag{13.71}$$

where the normal characteristic function is given by

$$C_{\mathcal{N}}(\beta) = \mathrm{Tr}\,\{\hat{\varrho}\exp(\beta\hat{a}^+)\exp(-\beta^*\hat{a})\} = \int\Phi_{\mathcal{N}}(\alpha)\exp(\alpha^*\beta - \alpha\beta^*)\,\mathrm{d}^2\alpha \tag{13.72}$$

and the antinormal characteristic function is given by

$$C_{\mathcal{A}}(\beta) = \mathrm{Tr}\,\{\hat{\varrho}\exp(-\beta^*\hat{a})\exp(\beta\hat{a}^+)\} = \int\Phi_{\mathcal{A}}(\alpha)\exp(\alpha^*\beta - \alpha\beta^*)\,\mathrm{d}^2\alpha. \tag{13.73}$$

† This quasi-probability function is analogous to the generalized radiance (4.74).

If the complex δ-function (13.37) is employed, the inverse Fourier transforms determine the corresponding quasi-distributions,

$$\Phi_{\mathcal{N}}(\alpha) = \frac{1}{\pi^2} \int C_{\mathcal{N}}(\beta) \exp(\alpha\beta^* - \alpha^*\beta)\, \mathrm{d}^2\beta \tag{13.74}$$

and

$$\Phi_{\mathcal{A}}(\alpha) = \frac{1}{\pi^2} \int C_{\mathcal{A}}(\beta) \exp(\alpha\beta^* - \alpha^*\beta)\, \mathrm{d}^2\beta. \tag{13.75}$$

The Wigner function related to symmetric ordering can be introduced as

$$\Phi_{sym}(\alpha) = \frac{1}{\pi^2} \int C(\beta) \exp(\alpha\beta^* - \alpha^*\beta)\, \mathrm{d}^2\beta, \tag{13.76}$$

where the characteristic function $C(\beta)$ is given by (13.70). The moments of the Wigner function $\Phi_{sym}(\alpha)$ can be calculated as

$$\int \alpha^{*k}\alpha^l \Phi_{sym}(\alpha)\, \mathrm{d}^2\alpha = \frac{\partial^k}{\partial\beta^k} \frac{\partial^l}{\partial(-\beta^*)^l} C(\beta)\bigg|_{\beta=\beta^*=0} = \langle \alpha^{*k}\alpha^l \rangle_{sym}, \tag{13.77}$$

where

$$C(\beta) = \int \Phi_{sym}(\alpha) \exp(\alpha^*\beta - \alpha\beta^*)\, \mathrm{d}^2\alpha. \tag{13.78}$$

Calculating derivatives in (13.77) with the use of (13.70), we can see that they are equal to averages of the *symmetrically* (Weyl) ordered products of k factors of $\hat{a}^+$ and l factors of $\hat{a}$. Thus the Wigner function is the quasi-probability function appropriate to symmetric (Weyl) ordering. If $k = l = 0$ we have the normalization condition

$$\int \Phi_{sym}(\alpha)\, \mathrm{d}^2\alpha = C(0) = 1. \tag{13.79}$$

In the same way the moments of $\Phi_{\mathcal{N}}(\alpha)$ and $\Phi_{\mathcal{A}}(\alpha)$ can be calculated from the characteristic functions $C_{\mathcal{N}}(\beta)$ and $C_{\mathcal{A}}(\beta)$ respectively. From (13.72) and (13.73) we have

$$\frac{\partial^k}{\partial\beta^k} \frac{\partial^l}{\partial(-\beta^*)^l} C_{\mathcal{N}}(\beta)\bigg|_{\beta=\beta^*=0} = \mathrm{Tr}\,\{\hat{\varrho}\hat{a}^{+k}\hat{a}^l\} = \int \Phi_{\mathcal{N}}(\alpha)\, \alpha^{*k}\alpha^l\, \mathrm{d}^2\alpha = \langle \alpha^{*k}\alpha^l \rangle_{\mathcal{N}}, \tag{13.80}$$

$$\frac{\partial^k}{\partial\beta^k} \frac{\partial^l}{\partial(-\beta^*)^l} C_{\mathcal{A}}(\beta)\bigg|_{\beta=\beta^*=0} = \mathrm{Tr}\,\{\hat{\varrho}\hat{a}^l\hat{a}^{+k}\} = \int \Phi_{\mathcal{A}}(\alpha)\, \alpha^{*k}\alpha^l\, \mathrm{d}^2\alpha = \langle \alpha^{*k}\alpha^l \rangle_{\mathcal{A}}. \tag{13.81}$$

If $|\psi_n\rangle$ are normalized eigenstates of $\hat{\varrho}$ with eigenvalues λ_n, then

$$\hat{\varrho} = \sum_n \lambda_n |\psi_n\rangle\langle\psi_n|$$

and since $0 \leqq \lambda_n \leqq 1$ ($\mathrm{Tr}\,\hat{\varrho} = \sum_n \lambda_n = 1$) we obtain

$$\Phi_{\mathscr{A}}(\alpha) = \frac{1}{\pi} \sum_n \lambda_n |\langle \alpha | \psi_n \rangle|^2 \geqq 0. \tag{13.82a}$$

Also

$$\Phi_{\mathscr{A}}(\alpha) \leqq \frac{1}{\pi} \sum_n |\langle \alpha | \psi_n \rangle|^2 = \frac{1}{\pi}, \tag{13.82b}$$

that is $0 \leqq \Phi_{\mathscr{A}}(\alpha) \leqq 1/\pi$, since $|\alpha\rangle = \sum_n |\psi_n\rangle \langle \psi_n | \alpha \rangle$ and $\langle \alpha | \alpha \rangle = \sum_n |\langle \alpha | \psi_n \rangle|^2 = 1$. Further, since $\hat{D}(\alpha)$ is a unitary operator ($\hat{D}^+\hat{D} = \hat{1}$), then $|C(\beta)| \leqq 1$ and from (13.71)

$$|C_{\mathscr{A}}(\beta)| \leqq \exp(-\tfrac{1}{2}|\beta|^2) \tag{13.83a}$$

and

$$|C_{\mathscr{N}}(\beta)| \leqq \exp(\tfrac{1}{2}|\beta|^2). \tag{13.83b}$$

Consequently $|C_{\mathscr{A}}(\beta)|$ always decreases at least as fast as $\exp(-|\beta|^2/2)$ for $|\beta| \to \infty$, while $|C_{\mathscr{N}}(\beta)|$ may diverge as rapidly as $\exp(|\beta|^2/2)$ in that limit.

One can see from (13.58) that $\pi^{-1}\langle \alpha | \hat{\varrho} | \alpha \rangle = \Phi_{\mathscr{A}}(\alpha)$ represents an entire analytic function

$$\Phi_{\mathscr{A}}(\alpha) = \frac{1}{\pi} \sum_{n,m} \varrho(n, m) \frac{\alpha^{*n}\alpha^m}{(n!\,m!)^{1/2}} \exp(-|\alpha|^2). \tag{13.84}$$

The function $\Phi_{\mathscr{A}}(\alpha)$ was first introduced and its properties were studied by Glauber (1965), Kano (1964a, 1965) and Mehta and Sudarshan (1965).

Mutual relations between the functions $\Phi_{\mathscr{N}}(\alpha)$, $\Phi_{\mathscr{A}}(\alpha)$ and $\Phi_{sym}(\alpha)$ can be derived from the relation (13.71) between the characteristic functions using the convolution theorem (cf. also Sec. 16.2)

$$\Phi_{\mathscr{A}}(\alpha) = \frac{2}{\pi} \int \exp(-2|\alpha - \beta|^2)\, \Phi_{sym}(\beta)\, \mathrm{d}^2\beta, \tag{13.85a}$$

$$\Phi_{sym}(\alpha) = \frac{2}{\pi} \int \exp(-2|\alpha - \beta|^2)\, \Phi_{\mathscr{N}}(\beta)\, \mathrm{d}^2\beta; \tag{13.85b}$$

equation (13.68) is the corresponding relation between $\Phi_{\mathscr{A}}(\alpha)$ and $\Phi_{\mathscr{N}}(\alpha)$. In general, the functions $\Phi_{\mathscr{A}}$ and Φ_{sym} on the left-hand side represent averages of the functions Φ_{sym} and $\Phi_{\mathscr{N}}(\alpha)$ on the right-hand side with exponential weight factors. The averaging process which takes us from $\Phi_{\mathscr{N}}$ to Φ_{sym} and $\Phi_{\mathscr{A}}$ tends to smooth out any unruly behaviour of $\Phi_{\mathscr{N}}$ and to transform it into a smooth function; this smoothing process ensures that $\Phi_{\mathscr{A}}$ is non-negative and regular. While the functions $\Phi_{\mathscr{A}}$ and Φ_{sym} exist for all quantum states, the cases in which $\Phi_{\mathscr{N}}$ cannot be defined in the usual sense are precisely those in which the convolution integrals (13.68) and (13.85b) cannot be inverted.

Ordering of the density matrix and quasi-probabilities

An important rule for obtaining $\Phi_{\mathcal{N}}$ and $\Phi_{\mathcal{A}}$ from the density matrix ordered in a certain way can be derived as follows (Lax and Louisell (1967), Lax (1967a), Haken, Risken and Weidlich (1967)). Suppose that an operator $\hat{F}$ can be written in the normal and antinormal forms

$$\hat{F}(\hat{a}^+, \hat{a}) = \hat{F}^{(\mathcal{N})}(\hat{a}^+, \hat{a}) = \hat{F}^{(\mathcal{A})}(\hat{a}^+, \hat{a}) = \sum_{r,s} F_{rs}^{(\mathcal{N})}\hat{a}^{+r}\hat{a}^s = \sum_{r,s} F_{rs}^{(\mathcal{A})}\hat{a}^s\hat{a}^{+r}. \tag{13.86}$$

Assuming similar expansions to be valid for the density matrix

$$\hat{\varrho}(\hat{a}^+, \hat{a}) = \hat{\varrho}^{(\mathcal{N})}(\hat{a}^+, \hat{a}) = \hat{\varrho}^{(\mathcal{A})}(\hat{a}^+, \hat{a}) = \sum_{r,s} G_{rs}^{(\mathcal{N})}\hat{a}^{+r}\hat{a}^s = \sum_{r,s} G_{rs}^{(\mathcal{A})}\hat{a}^s\hat{a}^{+r}, \tag{13.87}$$

we obtain

$$\begin{aligned}
\operatorname{Tr}\{\hat{\varrho}\hat{F}\} &= \operatorname{Tr}\{\hat{\varrho}^{(\mathcal{A})}\hat{F}^{(\mathcal{N})}\} = \\
&= \operatorname{Tr}\left\{\frac{1}{\pi}\int |\alpha\rangle\langle\alpha|\,\mathrm{d}^2\alpha \sum_{r,s,u,v} G_{rs}^{(\mathcal{A})}F_{uv}^{(\mathcal{N})}\,\hat{a}^s\hat{a}^{+r}\hat{a}^{+u}\hat{a}^v\right\} = \\
&= \frac{1}{\pi}\int \sum_{r,s,u,v} G_{rs}^{(\mathcal{A})}F_{uv}^{(\mathcal{N})}\langle\alpha|\,\hat{a}^{+r}\hat{a}^{+u}\hat{a}^v\hat{a}^s\,|\alpha\rangle\,\mathrm{d}^2\alpha = \\
&= \frac{1}{\pi}\int \varrho^{(\mathcal{A})}(\alpha^*, \alpha)\,F^{(\mathcal{N})}(\alpha^*, \alpha)\,\mathrm{d}^2\alpha,
\end{aligned} \tag{13.88}$$

where

$$\frac{1}{\pi}\varrho^{(\mathcal{A})}(\alpha^*, \alpha) \equiv \Phi_{\mathcal{N}}(\alpha) = \sum_{r,s} G_{rs}^{(\mathcal{A})}\alpha^{*r}\alpha^s \tag{13.89a}$$

and

$$F^{(\mathcal{N})}(\alpha^*, \alpha) = \sum_{r,s} F_{rs}^{(\mathcal{N})}\alpha^{*r}\alpha^s. \tag{13.89b}$$

Similarly

$$\operatorname{Tr}\{\hat{\varrho}\hat{F}\} = \operatorname{Tr}\{\hat{\varrho}^{(\mathcal{N})}\hat{F}^{(\mathcal{A})}\} = \frac{1}{\pi}\int \varrho^{(\mathcal{N})}(\alpha^*, \alpha)\,F^{(\mathcal{A})}(\alpha^*, \alpha)\,\mathrm{d}^2\alpha, \tag{13.90}$$

that is

$$\Phi_{\mathcal{A}}(\alpha) \equiv \varrho^{(\mathcal{N})}(\alpha^*, \alpha)/\pi. \tag{13.91}$$

(We also have $\Phi_{\mathcal{A}}(\alpha) = \langle\alpha|\,\hat{\varrho}\,|\alpha\rangle/\pi = \langle\alpha|\,\hat{\varrho}^{(\mathcal{N})}\,|\alpha\rangle/\pi = \varrho^{(\mathcal{N})}(\alpha^*, \alpha)/\pi$.) Further

$$\begin{aligned}
\hat{\varrho} &= \int \Phi_{\mathcal{N}}(\alpha)\,|\alpha\rangle\langle\alpha|\,\mathrm{d}^2\alpha = \sum_{r,s} G_{rs}^{(\mathcal{A})}\hat{a}^s\hat{a}^{+r} = \sum_{r,s} G_{rs}^{(\mathcal{A})}\hat{a}^s\frac{1}{\pi}\int |\alpha\rangle\langle\alpha|\,\mathrm{d}^2\alpha\,\hat{a}^{+r} = \\
&= \frac{1}{\pi}\int \Big(\sum_{r,s} G_{rs}^{(\mathcal{A})}\alpha^s\alpha^{*r}\Big)|\alpha\rangle\langle\alpha|\,\mathrm{d}^2\alpha,
\end{aligned}$$

that is $\Phi_{\mathcal{N}}(\alpha) = \varrho^{(\mathcal{A})}(\alpha^*, \alpha)/\pi$ again. Hence the function $\Phi_{\mathcal{N}}$ ($\Phi_{\mathcal{A}}$) can be calculated from the antinormal (normal) form of the density matrix (obtained using the commutation rules for the operators) with the substitutions $\hat{a} \to \alpha$, $\hat{a}^+ \to \alpha^*$; substituting $\alpha \to \hat{a}$, $\alpha^* \to \hat{a}^+$ in $\Phi_{\mathcal{N}}$ ($\Phi_{\mathcal{A}}$) we obtain the antinormal (normal) form of the density matrix.

An example

As an example, consider the case of Gaussian light (the mean number of photons is $\langle n \rangle$) with a coherent component, which is described by the density matrix

$$\hat{\varrho} = \frac{\langle n \rangle^{\hat{b}^+\hat{b}}}{(1 + \langle n \rangle)^{1+\hat{b}^+\hat{b}}}, \tag{13.92}$$

where $\hat{b} = \hat{a} - \beta$; here β is the complex amplitude of the coherent component. It is obvious that $[\hat{b}, \hat{b}^+] = [\hat{a}, \hat{a}^+] = \hat{1}$ and we can write

$$\hat{\varrho} = \frac{1}{\langle n \rangle}\left(1 + \frac{1}{\langle n \rangle}\right)^{-\hat{b}^+\hat{b}-1} = \frac{1}{\langle n \rangle}\sum_{n=0}^{\infty}\frac{(-1)^n}{n!}\frac{(\hat{b}^+\hat{b}+n)!}{(\hat{b}^+\hat{b})!}\langle n \rangle^{-n}. \tag{13.93}$$

Making use of (11.8b) we obtain

$$\hat{\varrho}^{(\mathcal{A})} = \frac{1}{\langle n \rangle}\sum_{n=0}^{\infty}\frac{(-1)^n}{n!}\frac{\hat{b}^n\hat{b}^{+n}}{\langle n \rangle^n} = \frac{1}{\langle n \rangle}\mathcal{A}\exp\left(-\frac{\hat{b}^+\hat{b}}{\langle n \rangle}\right), \tag{13.94}$$

which gives by the substitution $\hat{b} \to \alpha - \beta$, $\hat{b}^+ \to \alpha^* - \beta^*$,

$$\Phi_{\mathcal{N}}(\alpha) = \frac{1}{\pi\langle n \rangle}\exp\left(-\frac{|\alpha - \beta|^2}{\langle n \rangle}\right). \tag{13.95}$$

In (13.94) $\mathcal{A}$ denotes the antinormal ordering operator.

Similarly

$$\begin{aligned}\hat{\varrho} &= \frac{1}{1 + \langle n \rangle}\left(1 - \frac{1}{1 + \langle n \rangle}\right)^{\hat{b}^+\hat{b}} = \\ &= \frac{1}{1 + \langle n \rangle}\sum_{n=0}^{\infty}(-1)^n\binom{\hat{b}^+\hat{b}}{n}(1 + \langle n \rangle)^{-n} = \\ &= \frac{1}{1 + \langle n \rangle}\sum_{n=0}^{\infty}\frac{(-1)^n}{n!}\frac{(\hat{b}^+\hat{b})!}{(\hat{b}^+\hat{b} - n)!}(1 + \langle n \rangle)^{-n}.\end{aligned} \tag{13.96}$$

Making use of (11.8a) we obtain

$$\begin{aligned}\varrho^{(\mathcal{N})} &= \frac{1}{1 + \langle n \rangle}\sum_{n=0}^{\infty}\frac{(-1)^n}{n!}\frac{\hat{b}^{+n}\hat{b}^n}{(1 + \langle n \rangle)^n} = \\ &= \frac{1}{1 + \langle n \rangle}\mathcal{N}\exp\left(-\frac{\hat{b}^+\hat{b}}{1 + \langle n \rangle}\right)\end{aligned} \tag{13.97}$$

and so

$$\Phi_{\mathcal{A}}(\alpha) = \frac{1}{\pi(1 + \langle n \rangle)}\exp\left(-\frac{|\alpha - \beta|^2}{1 + \langle n \rangle}\right). \tag{13.98}$$

In (13.97) $\mathcal{N}$ denotes the normal ordering operator.

Substituting the expression (11.30) for $\langle n \rangle$ into these equations we can obtain the corresponding results for light in thermal equilibrium at the temperature T, with a superimposed coherent component β. Comparing (13.93) with (13.97) and (13.94) we arrive at

$$\begin{aligned}\hat{\varrho} &= (1 - e^{-\Theta}) \exp[-\Theta(\hat{a}^+ - \beta^*)(\hat{a} - \beta)] = \\ &= (1 - e^{-\Theta}) \mathscr{N}\{\exp[-(1 - e^{-\Theta})(\hat{a}^+ - \beta^*)(\hat{a} - \beta)]\} \equiv \hat{\varrho}^{(\mathscr{N})} \qquad (13.99) \\ &= (e^{\Theta} - 1) \mathscr{A}\{\exp[-(e^{\Theta} - 1)(\hat{a}^+ - \beta^*)(\hat{a} - \beta)]\} \equiv \hat{\varrho}^{(\mathscr{A})}. \qquad (13.100)\end{aligned}$$

An equation of the form (13.99) follows simply in the coherent-state algebra ($\beta = 0$ for simplicity),

$$\begin{aligned}\langle \alpha | \exp(-\Theta \hat{a}^+ \hat{a}) | \alpha \rangle &= \sum_{n=0}^{\infty} \exp(-\Theta n) |\langle \alpha | n \rangle|^2 = \\ &= \sum_{n=0}^{\infty} \frac{(\exp(-\Theta)|\alpha|^2)^n}{n!} \exp(-|\alpha|^2) = \exp(-|\alpha|^2(1 - e^{-\Theta})) = \\ &= \langle \alpha | \mathscr{N}\{\exp[-\hat{a}^+\hat{a}(1 - e^{-\Theta})]\} | \alpha \rangle, \qquad (13.101)\end{aligned}$$

where we have used (11.18) and (13.12b) (cf. (10.8) in the semiclassical treatment).

Superimposition of fields

Sometimes we need to consider superimposed fields. If the complex amplitude α is composed of two components β and γ, i.e. $\alpha = \beta + \gamma$, then the density matrix describing fluctuations of β is

$$\hat{\varrho}_1 = \int \Phi_{\mathscr{N}}^{(1)}(\beta) | \beta + \gamma \rangle \langle \beta + \gamma | \, \mathrm{d}^2\beta, \qquad (13.102)$$

while that describing the fluctuations of γ is

$$\hat{\varrho}_2 = \int \Phi_{\mathscr{N}}^{(2)}(\gamma) | \beta + \gamma \rangle \langle \beta + \gamma | \, \mathrm{d}^2\gamma. \qquad (13.103)$$

The resulting density matrix will be equal to

$$\hat{\varrho} = \iint \Phi_{\mathscr{N}}^{(1)}(\beta) \Phi_{\mathscr{N}}^{(2)}(\gamma) | \beta + \gamma \rangle \langle \beta + \gamma | \, \mathrm{d}^2\beta \, \mathrm{d}^2\gamma = \int \Phi_{\mathscr{N}}(\alpha) | \alpha \rangle \langle \alpha | \, \mathrm{d}^2\alpha, \qquad (13.104)$$

where

$$\Phi_{\mathscr{N}}(\alpha) = \iint \Phi_{\mathscr{N}}^{(1)}(\beta) \Phi_{\mathscr{N}}^{(2)}(\gamma) \, \delta(\alpha - \beta - \gamma) \, \mathrm{d}^2\beta \, \mathrm{d}^2\gamma. \qquad (13.105)$$

More generally, if $\alpha = \sum_j \alpha_j$,

$$\Phi_{\mathscr{N}}(\alpha) = \int \delta(\alpha - \sum_j \alpha_j) \prod_j \Phi_{\mathscr{N}}^{(j)}(\alpha_j) \, \mathrm{d}^2\alpha_j, \qquad (13.106)$$

which is in close analogy to the classical convolution law (10.42) for the probability distributions of statistically independent quantities. This convolution law for the quasi-distributions $\Phi_{\mathcal{N}}^{(j)}$ is a characteristic property of the description of the field in terms of $\Phi_{\mathcal{N}}(\alpha)$, and it is an expression of the general quantum superposition principle

$$\langle\alpha\rangle_{\mathcal{N}} = \int \Phi_{\mathcal{N}}(\alpha)\,\alpha\,\mathrm{d}^2\alpha = \iint \Phi_{\mathcal{N}}^{(1)}(\beta)\,\Phi_{\mathcal{N}}^{(2)}(\gamma)\,(\beta + \gamma)\,\mathrm{d}^2\beta\,\mathrm{d}^2\gamma = \\ = \langle\beta\rangle_{\mathcal{N}} + \langle\gamma\rangle_{\mathcal{N}}, \tag{13.107a}$$

$$\langle|\alpha|^2\rangle_{\mathcal{N}} = \langle|\beta|^2\rangle_{\mathcal{N}} + \langle|\gamma|^2\rangle_{\mathcal{N}} + \langle\beta\gamma^*\rangle_{\mathcal{N}} + \langle\beta^*\gamma\rangle_{\mathcal{N}}. \tag{13.107b}$$

13.3 The existence of the Glauber-Sudarshan quasi-probability

A criterion for $\Phi_{\mathcal{N}}$ to be an L_2-function (a square integrable function) can be derived from the characteristic function $C_{\mathcal{N}}$ (Glauber (1966a, 1972)). It can be seen from (13.74) that if $C_{\mathcal{N}}(\beta) \in L_2$, then $\Phi_{\mathcal{N}}(\alpha) \in L_2$. However, since (13.83b) holds, $C_{\mathcal{N}}(\beta)$ is not in general an L_2-function and consequently $\Phi_{\mathcal{N}}(\alpha)$ need not exist as an ordinary function; nevertheless it may exist as a *generalized function* (*distribution*).

Another L_2-criterion was proposed by Mehta (1967). Using (13.60) and (13.13a) we obtain

$$\langle -\alpha|\hat{\varrho}|\alpha\rangle = \int \Phi_{\mathcal{N}}(\beta)\langle -\alpha|\beta\rangle\langle\beta|\alpha\rangle\,\mathrm{d}^2\beta = \\ = \int \Phi_{\mathcal{N}}(\beta)\exp(-|\alpha|^2 - |\beta|^2 - \alpha^*\beta + \alpha\beta^*)\,\mathrm{d}^2\beta \tag{13.108}$$

and so

$$\Phi_{\mathcal{N}}(\beta) = \frac{\exp(|\beta|^2)}{\pi^2}\int\langle -\alpha|\hat{\varrho}|\alpha\rangle\exp(|\alpha|^2)\exp(\alpha^*\beta - \alpha\beta^*)\,\mathrm{d}^2\alpha. \tag{13.109}$$

Hence, if $\langle -\alpha|\hat{\varrho}|\alpha\rangle\exp(|\alpha|^2)$ is an L_2-function, then $\Phi_{\mathcal{N}}(\beta)\exp(-|\beta|^2)$ is also an L_2-function.

Before going into further details of the existence problems of the Glauber-Sudarshan representation we give some definitions of classes of generalized functions.

Definitions and classes of generalized functions

A generalized function f is a continuous linear functional which maps each *test function* φ of a linear space onto a complex number $(f, \varphi) = \int f(x)\,\varphi(x)\,\mathrm{d}x$ (see e.g. Gelfand and Shilov (1964)). Linearity implies

$$(f, u\varphi_1 + v\varphi_2) = u(f, \varphi_1) + v(f, \varphi_2) \tag{13.110}$$

for any two functions φ_1, φ_2 of the considered space of test functions and any complex numbers u, v; continuity implies

$$\lim_{n\to\infty}(f, \varphi_n) = (f, \lim_{n\to\infty}\varphi_n) = (f, \varphi). \tag{13.111}$$

The set $\{\varphi_n\}$ is a sequence of elements of the space which converges to φ in the space.

The class of infinitely differentiable test functions of bounded support (they are zero outside a finite interval) defines the linear space D of test functions. All the continuous linear functionals defined on D form the space of generalized functions (distributions) D'.

The space S of the test functions is composed of infinitely differentiable test functions which together with their derivatives vanish at infinity more rapidly than any negative power of $|x|$. Then S' is the linear space of continuous linear functionals defined on S; these functionals are called *tempered distributions.* If $f(x)$ is a locally integrable function of polynomial growth, then the linear functional $\int f(x)\varphi(x)\,\mathrm{d}x$ converges for $\varphi \in S$ and hence it defines a tempered distribution. It is obvious that $D \subset S$, i.e. every continuous linear functional defined on S is also a continuous functional on D, i.e. $S' \subset D'$. We see that every tempered distribution is a distribution D'. The space of tempered distributions includes the Dirac function, its finite derivatives, and all their finite combinations.

The entire functions $\psi(z)$ of the complex variable $z = u + \mathrm{i}v$, which satisfy the inequalities

$$|z^n\psi^{(m)}(z)| \leqq c_{n,m} \exp(a|v|), \qquad n, m = 0, 1, \ldots, \tag{13.112}$$

where a and $c_{n,m}$ are constants (which may depend on ψ, n and m), form the linear test function space Z. Then Z' is the space of generalized functions defined on Z; they are sometimes called ultradistributions. Since $\psi(z)$ is an entire function, it cannot vanish along any finite interval of the real axis (or it would have to be identically zero over the whole complex plane). This means that the spaces D and Z have only the zero element in common. But if $\psi(z) \in Z$, then $\psi(u) \in S$ as follows from (13.112), i.e. the space Z consists of all entire analytic functions of the exponential type which decrease rapidly for real values of their argument, and so $Z \subset S$. Therefore $Z' \supset S'$. Relationships between the spaces of test functions and of the corresponding generalized functions are given schematically in Fig. 13.1.

The spaces D and Z as well as D' and Z' are related by means of a Fourier transformation. Let $\varphi(x)$ be an arbitrary test function in D, $\varphi(x) = 0$ for $|x| > a$. Its Fourier transform $\psi(u)$ can be considered as an entire function of $z = u + \mathrm{i}v$ (by using the Paley-Wiener theorem on entire functions, Paley and Wiener (1934), Khurgyn and Yakovlev (1971)), that is

$$\psi(z) = \int_{-a}^{+a} \varphi(x) \exp(\mathrm{i}xz)\,\mathrm{d}x. \tag{13.113}$$

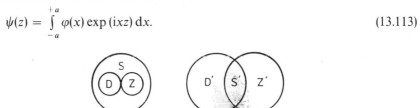

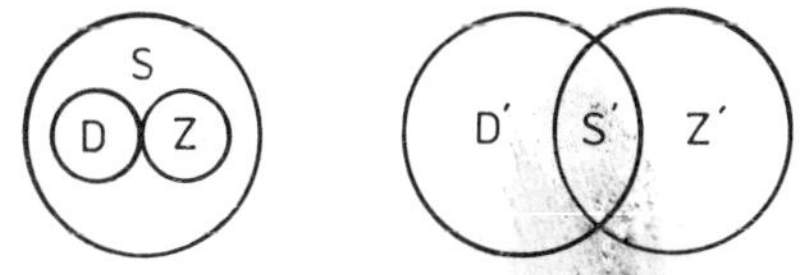

Fig. 13.1 Relationships of the test function spaces and the generalized function spaces.

Following Miller and Mishkin (1967b) we differentiate (13.113) m times with respect to z and integrate it n times by parts with respect to x and we obtain, taking the absolute value,

$$|z^n\psi^{(m)}(z)| \leqq \int_{-a}^{+a}\left|\frac{d^n}{dx^n}[x^m\varphi(x)]\exp(-xv)\right|dx \leqq$$
$$\leqq c_{n,m}\exp(a|v|), \qquad n, m = 0, 1, \ldots, \tag{13.114}$$

where

$$c_{n,m} = \int_{-a}^{+a}\left|\frac{d^n}{dx^n}[x^m\varphi(x)]\right|dx. \tag{13.115}$$

The condition (13.114) is just that required in (13.112). Also every entire function $\psi(z)$ which satisfies (13.114) is the Fourier transform of some infinitely differentiable function $\varphi(x)$ of bounded support $|x| \leqq a$. Thus the Fourier transformation establishes a one-to-one correspondence between the D- and Z-spaces and also between the D'- and Z'-spaces.

An interesting example of an element of Z' is

$$\delta(z + c) = \sum_{n=0}^{\infty}\frac{c^n}{n!}\delta^{(n)}(z), \tag{13.116}$$

where $z = u + iv$ and c is a complex constant. This is an infinite series of derivatives of the δ-function (cf. (10.21)).

An important example connected with the existence of $\Phi_{\mathcal{N}}(\alpha)$ is the series

$$f(u) = \sum_{n=0}^{\infty} c_n\delta^{(n)}(u), \tag{13.117}$$

which is of the form (10.21). Although it has been shown by Cahill (1965) that this series cannot be considered as an element of D' unless $c_n = 0$ for $n > N$ (N finite), it may represent an element of Z' (Miller and Mishkin (1967b)). If $\psi(z) \in Z$ then

$$\int f(u)\,\psi(u)\,du = \sum_{n=0}^{\infty}(-1)^n c_n\psi^{(n)}(0). \tag{13.118}$$

However, from (13.114) it follows that $|\psi^{(n)}(0)| \leqq ca^n$($c, a$ are constants), so that (13.117) is an element of Z' if $\sum_{n=0}^{\infty} c_n z^n = g(z)$ is an entire analytic function.

Further details about generalized functions can be found in the book by Gelfand and Shilov (1964) and about their applications to physics in a review by Güttinger (1966).

Explicit determination of $\Phi_{\mathcal{N}}$

An insight into the existence of $\Phi_{\mathcal{N}}(\alpha)$ can be given by inverting (13.63). This inversion provides information about the quasi-distribution of the complex amplitude of the field expressed by means of the Fock matrix elements $\varrho(n, m)$. This problem was solved by Sudarshan (1963a) in the formal form

$$\Phi_{\mathcal{N}}(\alpha) = \sum_{n=0}^{\infty} \sum_{m=0}^{\infty} \frac{(n!m!)^{1/2} \varrho(n, m)}{(n + m)!\, \pi r} \exp\left[r^2 + \mathrm{i}(m - n)\varphi\right] \left(-\frac{\partial}{\partial r}\right)^{n+m} \delta(r), \tag{13.119}$$

where $\alpha = r\exp(\mathrm{i}\varphi)$. This series is a two-dimensional modification of (10.21) and the solution is expressed by means of a double series of derivatives of the δ-function. As we have noted this series does not lead to a distribution on the D-space of test functions whenever an infinite number of photons are present in the state of the field (Cahill (1965)). From this point of view the prescription (13.119) for $\Phi_{\mathcal{N}}(\alpha)$ was criticized by a number of authors (cf. e.g. Holliday and Sage (1965)). However we have also seen that the series (13.119) may give rise to a generalized function on the Z-space of test functions which is the Fourier transform of D.

We can propose another prescription for constructing the $\Phi_{\mathcal{N}}(\alpha)$ which is based on the decomposition of $\Phi_{\mathcal{N}}(\alpha)$ in terms of the Laguerre polynomials (Peřina and Mišta (1968a, 1969), Lukš (1976)). Consider $\Phi_{\mathcal{N}}(\alpha)$ in the form

$$\Phi_{\mathcal{N}}(\alpha) = \exp\left[-(\zeta - 1)|\alpha|^2\right] \sum_{j=0}^{\infty} \left\{\sum_{\mu=0}^{\infty} c_{j\mu}(\zeta\alpha)^{\mu} L_j^{\mu}(\zeta|\alpha|^2) + \right.$$
$$\left. + \sum_{\mu=1}^{\infty} c_{j\mu}^{*}(\zeta\alpha^{*})^{\mu} L_j^{\mu}(\zeta|\alpha|^2)\right\}, \tag{13.120}$$

where $\zeta \geqq 1$ is a real number and L_j^{μ} are the Laguerre polynomials defined by (10.27) and satisfying the orthogonality condition (10.28). Multiplying (13.120) by $\alpha^{*\mu'} L_k^{\mu'}(\zeta|\alpha|^2)\exp(-|\alpha|^2)$, integrating over $(\zeta\alpha)$ and using (10.28) we obtain for the decomposition coefficients $c_{j\mu}$ in (13.120) the following expression in terms of the Fock matrix elements $\varrho(n, m)$

$$c_{j\mu} = \frac{j!}{\pi[(j + \mu)!]^3} \int \Phi_{\mathcal{N}}(\alpha) \exp(-|\alpha|^2)\, \alpha^{*\mu} L_j^{\mu}(\zeta|\alpha|^2)\, \mathrm{d}^2(\zeta\alpha) =$$
$$= \frac{\zeta}{\pi(j + \mu)!} \sum_{s=0}^{j} (-1)^s \binom{j}{s} \left[\frac{s!}{(s + \mu)!}\right]^{1/2} \zeta^s \varrho(s, s + \mu), \tag{13.121}$$

where we have also used (10.27) and (13.63). This relation can be inverted in the form

$$\varrho(s, s + \mu) = \frac{\pi}{\zeta^{s+1}} \left[\frac{(s + \mu)!}{s!}\right]^{1/2} \sum_{j=0}^{s} (-1)^j \binom{s}{j} (j + \mu)!\, c_{j\mu}. \tag{13.122}$$

One can verify that the substitution of (13.120) and (13.4) into (13.60) with the use of (13.122) leads to the Fock form (11.25) of the density matrix again (Peřina and Mišta (1969)).

The moments $\langle \alpha^{*k}\alpha^l \rangle_{\mathcal{N}}$ of $\Phi_{\mathcal{N}}(\alpha)$ can be calculated using (13.120) (Peřina and Mišta (1969)) or (13.119) (Holliday and Sage (1965)) in the form

$$\langle \alpha^{*k}\alpha^l \rangle_{\mathcal{N}} = \sum_{r=0}^{\infty} \varrho(r + l, r + k) \frac{[(r + l)!\,(r + k)!]^{1/2}}{r!}. \tag{13.123}$$

Equations (13.120) and (13.121) with $\mu = 0$ correspond to (10.26) and (10.29). From (13.120) we obtain (putting $\zeta = 1$ for simplicity)

$$\int \Phi_{\mathcal{N}}^2(\alpha)\, \mathrm{d}\mu(\alpha) = \pi \sum_{j=0}^{\infty} \sum_{\mu=0}^{\infty} |c_{j\mu}|^2 \frac{[(j + \mu)!]^3}{j!} + \pi \sum_{j=0}^{\infty} \sum_{\mu=1}^{\infty} |c_{j\mu}|^2 \frac{[(j + \mu)!]^3}{j!}, \tag{13.124}$$

where $\mathrm{d}\mu(\alpha) = \exp(-|\alpha|^2)\, \mathrm{d}^2\alpha$. Hence, if $\int \Phi_{\mathcal{N}}^2(\alpha)\, \mathrm{d}\mu(\alpha)$ (i.e. $\sum_{j=0}^{\infty} \sum_{\mu=0}^{\infty} |c_{j\mu}|^2 \times$ $\times [(j + \mu)!]^3 (j!)^{-1}$) is finite, then $\Phi_{\mathcal{N}}(\alpha)$ is an L_2-function in the space with the measure $\mathrm{d}\mu$ and the series (13.120) converges to $\Phi_{\mathcal{N}}(\alpha)$ in the norm of L_2, that is

$$\lim_{N\to\infty} \int [\Phi_{\mathcal{N}}(\alpha) - \Phi_{\mathcal{N}}^{(N)}(\alpha)]^2\, \mathrm{d}\mu(\alpha) = 0, \tag{13.125}$$

where $\Phi_{\mathcal{N}}^{(N)}(\alpha)$ is the Nth partial sum of (13.120). In this way a class of physical fields, with $\Phi_{\mathcal{N}}(\alpha)$ as a member of the L_2-space, may be specified by $\varrho(n, m)$ and determined through (13.121).

Examples

We consider first Gaussian light and Gaussian light with a coherent component. In the first case

$$\varrho(n, m) = \delta_{nm} \frac{\langle n \rangle^n}{(1 + \langle n \rangle)^{1+n}}, \tag{13.126}$$

and using (13.121) with $\zeta = 1$ we obtain

$$c_{j0} = \frac{(1 + \langle n \rangle)^{-j-1}}{\pi j!} \tag{13.127}$$

so that

$$\pi^2 \sum_{j=0}^{\infty} c_{j0}^2 (j!)^2 = \sum_{j=0}^{\infty} [(1 + \langle n \rangle)^2]^{-j-1} = \frac{1}{\langle n \rangle^2 + 2\langle n \rangle} < \infty. \tag{13.128}$$

Indeed, the function $\Phi_{\mathcal{N}}(\alpha)$ has the form (13.95) with $\beta = 0$, which is an L_2-function.

In the second case $\varrho(n, m)$ has been given by Mollow and Glauber (1967a) (for a derivation see Sec. 17.3)

$$\varrho(n, m) = \left(\frac{n!}{m!}\right)^{1/2} \exp\left(-\frac{|\beta|^2}{1 + \langle n \rangle}\right) \frac{\langle n \rangle^n}{(1 + \langle n \rangle)^{1+m}} \beta^{*m-n} \times$$
$$\times \frac{1}{m!} L_n^{m-n}\left(-\frac{|\beta|^2}{\langle n \rangle (\langle n \rangle + 1)}\right), \qquad m \geqq n. \tag{13.129}$$

(If $n > m$, then $\varrho(n, m) = \varrho^*(m, n)$.) The coefficients $c_{j\mu}$ are equal to

$$c_{j\mu} = \frac{\zeta}{\pi(1 + \langle n \rangle)^{j+\mu+1}} \exp\left(-\frac{|\beta|^2}{1 + \langle n \rangle}\right) \frac{j!}{[(j + \mu)!]^3} \times$$
$$\times [1 + \langle n \rangle (1 - \zeta)]^j \beta^{*\mu} L_j^\mu\left(\frac{\zeta |\beta|^2}{(1 + \langle n \rangle)[1 + \langle n \rangle (1 - \zeta)]}\right). \tag{13.130}$$

Substituting this result into (13.120) and using the identity (Gradshteyn and Ryzhik (1965), p. 1038)

$$\sum_{n=0}^{\infty} \frac{n!}{[(n + \mu)!]^3} L_n^\mu(x) L_n^\mu(y) z^n = \frac{(xyz)^{-\mu/2}}{1 - z} \exp\left(-z \frac{x + y}{1 - z}\right) \times$$
$$\times I_\mu\left(2 \frac{(xyz)^{1/2}}{1 - z}\right), \tag{13.131}$$

($I_\mu(x)$ is the modified Bessel function), the property $I_{-\mu}(x) = I_\mu(x)$ and the identity

$$\sum_{n=-\infty}^{+\infty} I_n(x) \exp(in\varphi) = \exp(x \cos \varphi), \tag{13.132}$$

we arrive at the diagonal representation function (13.95) again. We can easily verify using (13.131) and (13.132) that the coefficients (13.130) are such that

$$\sum_{j,\mu} |c_{j\mu}|^2 \frac{[(j + \mu)!]^3}{j!} < \infty. \tag{13.133}$$

Note that special cases of completely coherent and chaotic fields can be treated in the limits $\langle n \rangle \to 0$ and $\beta \to 0$, respectively.

One can see that while the support of every term in (13.119) is one point, the support of every term in (13.120) is the whole complex plane. This means that if it is ensured a priori that the function $\Phi_{\mathcal{N}}(\alpha)$ exists as an ordinary function, then all $\varrho(n, m)$ must be known for the reconstruction of this function from $\varrho(n, m)$ using the prescription (13.119). If only a finite number of $\varrho(n, m)$ are different from zero, then $\Phi_{\mathcal{N}}(\alpha)$ has a one-point support and it is reconstructed as a generalized function. On the other hand the prescription (13.120) can be used (terminating the series at a finite term) as an approximation to $\Phi_{\mathcal{N}}(\alpha)$ as an ordinary function if only a finite set of $\varrho(n, m)$ is given. If all $\varrho(n, m)$ are given, then (13.120) represents an ordinary function when this series is uniformly convergent (for all α). If this series is divergent, the sequence of partial sums may converge to a generalized function. Some physical examples of the fields leading to tempered distributions can be given (putting $\zeta = 1$ for simplicity) (Peřina and Mišta (1969)):

a) *The coherent state*

It holds that

$$\Phi_{\mathcal{N}}(\alpha) = \delta(\alpha - \beta) = \lim_{M\to\infty} \frac{1}{\pi} \exp(-|\beta|^2) \sum_{j=0}^{M} \left\{ \sum_{\mu=0}^{M} \frac{j!}{[(j+\mu)!]^3} (\beta\alpha^*)^\mu \times \right.$$
$$\left. \times L_j^\mu(|\alpha|^2) L_j^\mu(|\beta|^2) + \sum_{\mu=1}^{M} (c.\,c.) \right\} \tag{13.134}$$

($c.\,c.$ denotes the complex conjugate of the first expression). This also expresses the completeness condition of the function system $\{\alpha^\mu L_j^\mu\}$ in the L_2-space with the measure $\mathrm{d}\mu(\alpha) = \exp(-|\alpha|^2)\,\mathrm{d}^2\alpha$. The decomposition (13.134) also follows from (13.120) when (13.121) is substituted with $\varrho(n, m) = \beta^n\beta^{*m}\exp(-|\beta|^2)(n!\,m!)^{-1/2}$ (the last expression follows from (13.63) for $\Phi_{\mathcal{N}}(\alpha) = \delta(\alpha - \beta)$).

In experiments the phase information is often lost, so averaging (13.134) over the phase we obtain

$$\Phi_{\mathcal{N}}(\alpha) = \frac{1}{2\pi} \int_0^{2\pi} \delta(\alpha - \beta)\,\mathrm{d}\varphi = \frac{1}{2\pi} \int_0^{2\pi} \frac{\delta(|\alpha| - |\beta|)}{|\alpha|} \delta(\varphi - \psi)\,\mathrm{d}\varphi =$$
$$= \frac{\delta(|\alpha| - |\beta|)}{2\pi|\alpha|} = \frac{\delta(|\alpha| - |\beta|)}{2\pi|\alpha|} + \frac{\delta(|\alpha| + |\beta|)}{2\pi|\alpha|} = \frac{1}{\pi}\delta(|\alpha|^2 - |\beta|^2) =$$
$$= \lim_{M\to\infty} \frac{1}{\pi} \exp(-|\beta|^2) \sum_{j=0}^{M} \frac{1}{(j!)^2} L_j^0(|\alpha|^2) L_j^0(|\beta|^2), \tag{13.135}$$

since $\delta(|\alpha| + |\beta|)/2|\alpha| = 0$ and $\alpha = |\alpha|\exp(\mathrm{i}\varphi)$, $\beta = |\beta|\exp(\mathrm{i}\psi)$.

b) *The Fock state* $|n\rangle$

From (13.119) or from (13.120) it holds for the Fock state $|n\rangle$ that

$$\Phi_{\mathcal{N}}(\alpha) = \frac{1}{\pi} \frac{n!}{(2n)!\,|\alpha|} \exp(|\alpha|^2) \left(\frac{\partial}{\partial|\alpha|}\right)^{2n} \delta(|\alpha|) =$$
$$= \frac{1}{\pi}(-1)^n \exp(|\alpha|^2)\,\delta^{(n)}(|\alpha|^2) = \lim_{M\to\infty} \frac{1}{\pi} \sum_{j=n}^{M} \frac{(-1)^n}{n!\,(j-n)!} L_j^0(|\alpha|^2), \tag{13.136}$$

where we have used the identity

$$\frac{\delta^{(2n)}(|\alpha|)}{|\alpha|} = \frac{\delta^{(2n)}(|\alpha|) + \delta^{(2n)}(-|\alpha|)}{2|\alpha|} = \left(\frac{\partial}{\partial c}\right)^{2n} \times$$
$$\times \left.\frac{\delta(c + |\alpha|) + \delta(c - |\alpha|)}{2|\alpha|}\right|_{c=0} = \left(\frac{\partial}{\partial c}\right)^{2n} \delta(c^2 - |\alpha|^2)\Big|_{c=0} =$$
$$= \left(\frac{\partial}{\partial c}\right)^{2n} \sum_{j=0}^{\infty} \frac{(-1)^j}{j!} c^{2j}\delta^{(j)}(|\alpha|^2)\Big|_{c=0} = (-1)^n \frac{(2n)!}{n!} \delta^{(n)}(|\alpha|^2). \tag{13.137}$$

One can conclude that the state $|n\rangle$ has no classical analogue, since for $n > 0$ the function $\Phi_{\mathcal{N}}(\alpha)$ is more singular than the δ-function (it is proportional to the nth derivative of the δ-function).

General solution of the existence problem

Equation (13.68) can yield a complete solution of the existence problem for $\Phi_{\mathcal{N}}(\alpha)$. This convolution equation has been solved, making the Fourier transformation, by Bonifacio, Narducci and Montaldi (1966) who have shown that if a solution of (13.68) exists it is unique. Taking the Fourier transform of (13.68) we arrive at

$$C_{\mathcal{A}}(\beta) = C_{\mathcal{N}}(\beta)\exp(-|\beta|^2), \tag{13.138}$$

where the characteristic functions $C_{\mathcal{A}}(\beta)$ and $C_{\mathcal{N}}(\beta)$ are defined by (13.72) and (13.73). Hence, if

$$C_{\mathcal{N}}(\beta) = C_{\mathcal{A}}(\beta)\exp(|\beta|^2) \tag{13.139}$$

is a tempered distribution, then the Fourier transform, i.e. $\Phi_{\mathcal{N}}(\alpha)$, is also a tempered distribution. Further, Miller and Mishkin (1967b) have shown that $\Phi_{\mathcal{A}}(\alpha)$ (a function of very regular behaviour) is a member of S' and consequently also of Z'. Then $C_{\mathcal{A}}(\beta) \in D'$ and also $C_{\mathcal{N}}(\beta)$, given by (13.139), lies in D'; therefore the Fourier transform $\Phi_{\mathcal{N}}(\alpha)$ is a member of Z'. It can easily be shown that this solution is unique. Hence, for all physical fields the function $\Phi_{\mathcal{N}}(\alpha)$, the weight function of the Glauber-Sudarshan diagonal representation of the density matrix, exists at least as a distribution Z' (ultradistribution). In fact $\Phi_{\mathcal{N}}(\alpha)$ is an element of a subspace of Z'. In this sense it can be said that the function $\Phi_{\mathcal{N}}(\alpha)$ always exists.

Other approaches to the problem of the existence of the diagonal representation of the density matrix have been suggested. In these approaches, developed by Mehta and Sudarshan (1965), Klauder, McKenna and Currie (1965), Klauder (1966) (see also Klauder and Sudarshan (1968), Sec. 8−4) and Rocca (1967), the weight function $\Phi_{\mathcal{N}}(\alpha)$ is defined as a limit of an infinite sequence of well-behaved weight functions $\{\Phi_{\mathcal{N}}^{(N)}(\alpha)\}$. Thus all density matrices can be expressed as limits of sequences each member of which has the Glauber-Sudarshan diagonal representation with the weight functions $\Phi_{\mathcal{N}}^{(N)}(\alpha)$ as square integrable functions or tempered generalized functions. For the construction of $\Phi_{\mathcal{N}}^{(N)}$ in the first case, the prescription (13.120) terminated after the Nth term in every series can be used, while in the second case the prescription (13.119) terminated after the Nth term in both the series is suitable. However, since in the process of measuring the density matrices fulfilling the condition

$$\mathrm{Tr}\,\{\hat{\varrho}\hat{A} - \hat{\varrho}_N\hat{A}\} \leqq \varepsilon\,\|\hat{A}\| \tag{13.140}$$

(where $\hat{A}$ is a bounded operator with the norm $\|\hat{A}\|$ and ε is an arbitrarily small number) are not distinguished, we can conclude that the density matrix can be expressed in the diagonal form with arbitrary accuracy. This result applies to the physically important operators $\hat{A} = \mathcal{N}\{\exp(\mathrm{i}x\hat{a}^+\hat{a})\}$ and $(1/n!)(\mathrm{d}^n/\mathrm{d}(\mathrm{i}x)^n)\hat{A}|_{\mathrm{i}x=-1} = (1/n!)\mathcal{N}\{(\hat{a}^+\hat{a})^n \exp(-\hat{a}^+\hat{a})\}$, which correspond to the normal characteristic function and to the photocount distribution respectively. The photocount distribution, for example, is given by

$$p(n) \approx \int \Phi_{\mathcal{N}}^{(N)}(\alpha) \frac{|\alpha|^{2n}}{n!} \exp(-|\alpha|^2)\, \mathrm{d}^2\alpha, \tag{13.141}$$

where $p(n) \equiv \varrho(n, n)$ and $\varrho(n, n)$ is given by (13.63).

Note that a regularization procedure for $\Phi_{\mathcal{N}}(\alpha)$ was suggested by Cahill (1969) and a generalized positive complex representation in terms of the coherent states was developed and applied to nonlinear optical processes by Drummond and Gardiner (1980).

13.4 Multimode description

All earlier results and relations can be generalized to fields and systems having a finite or denumerably infinite number of degrees of freedom, if the global states $|\{\alpha_\lambda\}\rangle$ defined by (13.48) are used. The Glauber-Sudarshan representation has the form

$$\hat{\varrho} = \int \Phi_{\mathcal{N}}(\{\alpha_\lambda\})\, |\{\alpha_\lambda\}\rangle\langle\{\alpha_\lambda\}|\, \mathrm{d}^2\{\alpha_\lambda\}, \tag{13.142}$$

where $\Phi_{\mathcal{N}}(\{\alpha_\lambda\}) \equiv \Phi_{\mathcal{N}}(\alpha_1, \alpha_2, \ldots)$ is the weight function (or functional) and $\mathrm{d}^2\{\alpha_\lambda\} \equiv \prod_\lambda \mathrm{d}^2\alpha_\lambda$. For example, we obtain in the multimode case instead of (13.63)

$$\langle\{n_\lambda\}|\hat{\varrho}|\{m_\lambda\}\rangle = \varrho(\{n_\lambda\}, \{m_\lambda\}) = \int \Phi_{\mathcal{N}}(\{\alpha_\lambda\}) \prod_\lambda \frac{\alpha_\lambda^{n_\lambda}\alpha_\lambda^{*m_\lambda}}{(n_\lambda!\, m_\lambda!)^{1/2}} \exp(-|\alpha_\lambda|^2)\, \mathrm{d}^2\alpha_\lambda. \tag{13.143}$$

Further,

$$\Phi_{\mathcal{A}}(\{\alpha_\lambda\}) = \int \Phi_{\mathcal{N}}(\{\beta_\lambda\}) \prod_\lambda \exp(-|\alpha_\lambda - \beta_\lambda|^2) \frac{\mathrm{d}^2\beta_\lambda}{\pi}, \tag{13.144}$$

which corresponds to (13.68) and

$$\Phi_{\mathcal{A}}(\{\alpha_\lambda\}) = \sum_{\{n_\lambda\}} \sum_{\{m_\lambda\}} \varrho(\{n_\lambda\}, \{m_\lambda\}) \prod_\lambda \frac{\alpha_\lambda^{*n_\lambda}\alpha_\lambda^{m_\lambda}}{\pi(n_\lambda!\, m_\lambda!)^{1/2}} \exp(-|\alpha_\lambda|^2), \tag{13.145}$$

which corresponds to (13.84). All other relations can be generalized in an identical way.

Using (13.142) and (13.52) we can write the correlation function (12.5) in the form

$$\Gamma_{\mathcal{N}}^{(m,n)}(x_1, \ldots, x_{m+n}) = \int \Phi_{\mathcal{N}}(\{\alpha_\lambda\}) \prod_{j=1}^{m} V^*(x_j) \prod_{k=m+1}^{m+n} V(x_k)\, \mathrm{d}^2\{\alpha_\lambda\}, \tag{13.146}$$

where $V(x)$ is given by (13.53).

The superposition of two fields described by $\Phi_{\mathcal{N}}^{(1)}(\{\alpha_\lambda\})$ and $\Phi_{\mathcal{N}}^{(2)}(\{\alpha_\lambda\})$ will be characterized, in analogy to (13.105), by

$$\Phi_{\mathcal{N}}(\{\alpha_\lambda\}) = \iint \Phi_{\mathcal{N}}^{(1)}(\{\beta_\lambda\}) \Phi_{\mathcal{N}}^{(2)}(\{\gamma_\lambda\}) \prod_\lambda \delta(\alpha_\lambda - \beta_\lambda - \gamma_\lambda) \, d^2\beta_\lambda \, d^2\gamma_\lambda. \tag{13.147}$$

This convolution law expresses the quantum superposition principle which leads to the multimode interference law identical with the classical interference law

$$\begin{aligned}
&\langle \hat{A}^{(-)}(x) \hat{A}^{(+)}(x) \rangle = \\
&= \int \Phi_{\mathcal{N}}(\{\alpha_\lambda\}) \langle \{\alpha_\lambda\} | \hat{A}^{(-)}(x, \{\hat{a}_\lambda^+\}) \hat{A}^{(+)}(x, \{\hat{a}_\lambda\}) | \{\alpha_\lambda\} \rangle \, d^2\{\alpha_\lambda\} = \\
&= \iint \Phi_{\mathcal{N}}^{(1)}(\{\beta_\lambda\}) \Phi_{\mathcal{N}}^{(2)}(\{\gamma_\lambda\}) [V^*(x, \{\beta_\lambda\}) + V^*(x, \{\gamma_\lambda\})] [V(x, \{\beta_\lambda\}) + \\
&+ V(x, \{\gamma_\lambda\})] \, d^2\{\beta_\lambda\} \, d^2\{\gamma_\lambda\} = \langle | V(x, \{\beta_\lambda\}) |^2 \rangle_{\mathcal{N}} + \langle | V(x, \{\gamma_\lambda\}) |^2 \rangle_{\mathcal{N}} + \\
&+ \langle V^*(x, \{\beta_\lambda\}) V(x, \{\gamma_\lambda\}) \rangle_{\mathcal{N}} + \langle V(x, \{\beta_\lambda\}) V^*(x, \{\gamma_\lambda\}) \rangle_{\mathcal{N}},
\end{aligned} \tag{13.148}$$

since $V(x, \{\alpha_\lambda\})$ is linear in $\{\alpha_\lambda\}$ so that $V(x, \{\beta_\lambda + \gamma_\lambda\}) = V(x, \{\beta_\lambda\}) + V(x, \{\gamma_\lambda\})$. This is in agreement with the rather intuitive formula (12.20).

13.5 More general classes of coherent states

Finally it should be noted that one can introduce more general classes of coherent states for quantum systems described by general potentials if one can use integrals of motion to introduce some generalized boson annihilation operators. These generalized coherent states may then be introduced in one of the above three ways – as eigenstates of the generalized boson operators, or using the displacement operator, or as minimum-uncertainty states; in general, these three procedures need not be equivalent (Perelomov (1977), Malkin and Man'ko (1979), Nieto and Simmons (1978, 1979), Nieto, Simmons and Gutschick (1981)). Such coherent states have proved to be useful also in integrated optics (Krivoshlykov and Sissakian (1979, 1980a, b, 1983)) because the coherent-state matrix elements can describe the propagation of radiation fields in optical waveguides, and they can serve as generating functions for coupling mode coefficients.

In connection with two-photon stimulated emission, provided that atoms of the active medium are considered classically, two-photon coherent states $|\beta\rangle_g$ have been introduced by Yuen (1975, 1976) as eigenstates of the annihilation operator

$$\hat{b} = \mu\hat{a} + \nu\hat{a}^+, \qquad |\mu|^2 - |\nu|^2 = 1, \qquad (\hat{a} = \mu^*\hat{b} - \nu\hat{b}^+), \tag{13.149a}$$

satisfying the commutation rule

$$[\hat{b}, \hat{b}^+] = \hat{1}. \tag{13.149b}$$

Hence

$$\hat{b}|\beta\rangle_g = \beta|\beta\rangle_g, \qquad {}_g\langle\beta|\hat{b}^+ = {}_g\langle\beta|\beta^*. \tag{13.150}$$

Consequently much of the above coherent-state formalism can be transferred to the two-photon coherent-state formalism. Introducing the Hermitian components $\hat{a}_1 = (\hat{a} + \hat{a}^+)/2$ and $\hat{a}_2 = (\hat{a} - \hat{a}^+)/2i$ (proportional to the generalized coordinate and momentum), one can show that the two-photon coherent states may minimize the uncertainty relation (Stoler (1970, 1971, 1972, 1975), Yuen (1975, 1976))

$$\langle(\Delta\hat{a}_1)^2\rangle \langle(\Delta\hat{a}_2)^2\rangle \geqq \frac{1}{16} \tag{13.151}$$

and it holds that

$$\begin{aligned} {}_g\langle\beta\,|\,(\Delta\hat{a}_1)^2\,|\,\beta\rangle_g &= \tfrac{1}{4}\,|\,\mu - \nu\,|^2, \\ {}_g\langle\beta\,|\,(\Delta\hat{a}_2)^2\,|\,\beta\rangle_g &= \tfrac{1}{4}\,|\,\mu + \nu\,|^2. \end{aligned} \tag{13.152}$$

The usual coherent states, related to the harmonic oscillator, are obtained from the two-photon coherent states for $\mu = 1$ and $\nu = 0$. From (13.152) we see that fluctuations in one of the real components of the complex amplitude of the two-photon coherent states may be less than for the coherent states of the harmonic oscillator; thus the two-photon coherent states are squeezed states (for a review, see Walls (1983), and discussions in Mandel and Wolf (1984)). This property makes it possible to reduce quantum noise in optical communication (Yuen and Shapiro (1978), Shapiro, Yuen and Machado Mata (1979), Helstrom (1979a, b), Shapiro (1980)) and in optical interferometers suggested to detect gravitational waves (Caves (1981), Walls and Zoller (1981), Walls and Milburn (1981), Loudon (1981), Milburn and Walls (1981)).

Radcliffe (1971) and Arecchi, Courtens, Gilmore and Thomas (1972, 1973) have introduced another class of coherent states called the *atomic coherent states*. They describe a free system of N two-level atoms and are eigenstates of the angular momentum operator. These correspond to a set of classical dipoles (classical currents) additionally involving physical vacuum fluctuations and are generated by a classical radiation field, in a similar way as the coherent states of a harmonic oscillator are generated by classical currents (for a review, see Peřina (1984)).

Thus we can conclude that under the influence of a classical (non-fluctuating) current an electromagnetic field initially in a coherent state (or in its ground state) will evolve into a coherent state and under the influence of a classical (non-fluctuating) electromagnetic field an atomic system initially in an atomic coherent state (or in its ground state) will evolve into an atomic coherent state.

RELATION BETWEEN QUANTUM AND CLASSICAL DESCRIPTIONS OF COHERENCE

14.1 Quantum and classical correlation functions

Introducing the probability distribution $P_n(V_1, \ldots, V_n)$, $V_j \equiv V(x_j)$, by the relation (Mandel (1963c))

$$P_n(V_1, \ldots, V_n) = \int \Phi_{\mathcal{N}}(\{\alpha'_\lambda\}) \prod_{j=1}^{n} \delta(V_j - V'_j) \, \mathrm{d}^2\{\alpha'_\lambda\} = \langle \prod_{j=1}^{n} \delta(V_j - V'_j) \rangle, \quad (14.1)$$

where the functions V'_j are given by (13.53) with $\alpha_{\mathbf{k}s} \to \alpha'_{\mathbf{k}s}$, we can rewrite (13.146) in the form

$$\Gamma_{\mathcal{N}}^{(m,n)}(x_1, \ldots, x_{m+n}) = \int \ldots \int P_{m+n}(V_1, \ldots, V_{m+n}) \prod_{j=1}^{m} V_j^* \prod_{k=m+1}^{m+n} V_k \prod_{l=1}^{m+n} \mathrm{d}^2 V_l =$$
$$= \langle \prod_{j=1}^{m} V^*(x_j) \prod_{k=m+1}^{m+n} V(x_k) \rangle, \quad (14.2)$$

which is just the form (8.4) of the classical correlation function. Thus there is very close correspondence between the classical and the normally ordered quantum correlation functions. In this correspondence the Glauber-Sudarshan diagonal representation of the density matrix serves as a bridge between the classical and quantum descriptions. More generally, if $\hat{F}\{\hat{A}^{(-)}(x), \hat{A}^{(+)}(x)\}$ is a functional of the field operators $\hat{A}^{(+)}$ and $\hat{A}^{(-)}$ and if we define the displacement operator $\hat{D}(\{\alpha_\lambda\})$ as the product of one-mode displacement operators (13.2) (all operators of different modes commute), we obtain using the property (13.32)

$$\hat{D}^{-1}(\{\alpha_\lambda\}) \hat{A}^{(+)}(x) \hat{D}(\{\alpha_\lambda\}) = \hat{A}^{(+)}(x) + V(x),$$
$$\hat{D}^{-1}(\{\alpha_\lambda\}) \hat{A}^{(-)}(x) \hat{D}(\{\alpha_\lambda\}) = \hat{A}^{(-)}(x) + V^*(x), \quad (14.3)$$

where $V(x)$ is given by (13.53). Hence, one obtains

$$\mathrm{Tr}\,\{\hat{\varrho}\hat{F}\{\hat{A}^{(-)}(x), \hat{A}^{(+)}(x)\}\} =$$
$$= \int \Phi_{\mathcal{N}}(\{\alpha_\lambda\}) \langle 0 | \hat{D}^{-1}(\{\alpha_\lambda\}) \hat{F}\{\hat{A}^{(-)}(x), \hat{A}^{(+)}(x)\} \hat{D}(\{\alpha_\lambda\}) | 0 \rangle \, \mathrm{d}^2\{\alpha_\lambda\} =$$
$$= \int \Phi_{\mathcal{N}}(\{\alpha_\lambda\}) \langle 0 | \hat{F}\{\hat{A}^{(-)}(x) + V^*(x), \hat{A}^{(+)}(x) + V(x)\} | 0 \rangle \, \mathrm{d}^2\{\alpha_\lambda\}, \quad (14.4)$$

since $|\{\alpha_\lambda\}\rangle = \hat{D}(\{\alpha_\lambda\})|0\rangle$. The operators $\hat{A}^{(+)}$ and $\hat{A}^{(-)}$ stand on the right-hand side of (14.4) to give the contribution of the physical vacuum fluctuations. However, if $\hat{F}$ is a normally ordered functional, then the vacuum expectation value of a normally ordered operator is zero since $\hat{A}^{(+)}|0\rangle = \langle 0|\hat{A}^{(-)} = 0$, so the vacuum fluctuations give no contribution, and we have

$$\mathrm{Tr}\,\{\hat{\varrho}\hat{F}\{\hat{A}^{(-)}(x), \hat{A}^{(+)}(x)\}\} = \int \Phi_{\mathcal{N}}(\{\alpha_\lambda\})\, F\{V^*(x), V(x)\}\, \mathrm{d}^2\{\alpha_\lambda\}. \tag{14.5}$$

This result holds no matter how weak the field is, i.e. it holds for arbitrarily low intensities. This result also gives a justification for the procedures used in the calculations of classical optics (Mandel and Wolf (1966)). We can see that there is a formal equivalence between the quantum and classical descriptions of light in this sense; of course, this formal equivalence does not imply a physical equivalence between the classical and quantum theories and we have seen, for example, that the Fock state $|n\rangle$ has no classical analogue, which reflects itself in the strange mathematical properties of $\Phi_{\mathcal{N}}$. However, equation (14.2) shows that all results given in earlier chapters devoted to the classical theory of the space-time-polarization behaviour of the correlation functions are also correct for the quantum correlation functions.

14.2 Photon-number and photocount distributions

The success of the classical or semiclassical theory is a consequence of the fact that the photodetection equation (10.3a), derived by the semiclassical or classical method, follows simply from the properties of the quantized electromagnetic field (Ghielmetti (1964), Jordan and Ghielmetti (1964), Peřina (1965d, 1967a), Glauber (1965), Kelley and Kleiner (1964), Lehmberg (1968), Mandel and Meltzer (1969)).

First consider the probability distribution $p(n)$ of the number of photons n within the normalization volume L^3 at time t. It is obvious that

$$p(m) = \sum_{\{n_\lambda\}} \varrho(\{n_\lambda\}, \{n_\lambda\})\, \delta_{nm}, \tag{14.6}$$

where $\sum_\lambda n_\lambda = n$. The moment $\langle \hat{n}^k \rangle$ ($k \geqq 0$ is integer), where $\hat{n}$ is defined in (11.24b), is, with the help of the Fock and Glauber-Sudarshan representations of the density matrix,

$$\langle \hat{n}^k \rangle = \sum_{\{n_\lambda\}} \varrho(\{n_\lambda\}) \langle \{n_\lambda\} | \hat{n}^k | \{n_\lambda\} \rangle = \int \Phi_{\mathcal{N}}(\{\alpha_\lambda\}) \langle \{\alpha_\lambda\} | \hat{n}^k | \{\alpha_\lambda\} \rangle\, \mathrm{d}^2\{\alpha_\lambda\}, \tag{14.7}$$

where $\varrho(\{n_\lambda\}) \equiv \varrho(\{n_\lambda\}, \{n_\lambda\})$. Substituting the unit operator (11.37) into the right-hand side of (14.7) and using the multimode analogue of (13.12b),

$$|\langle \{n_\lambda\} | \{\alpha_\lambda\} \rangle|^2 = \prod_\lambda \frac{|\alpha_\lambda|^{2n_\lambda}}{n_\lambda!} \exp(-|\alpha_\lambda|^2), \tag{14.8}$$

we obtain

$$\sum_{n=0}^{\infty} p(n)\, n^k = \sum_{\{n_\lambda\}} \left\{ \int \Phi_{\mathcal{N}}(\{\alpha_\lambda\}) \prod_\lambda \frac{|\alpha_\lambda|^{2n_\lambda}}{n_\lambda!} \exp(-|\alpha_\lambda|^2)\, \mathrm{d}^2\{\alpha_\lambda\}\, n^k \right\}, \tag{14.9}$$

where $n = \sum_{\lambda} n_{\lambda}$ is the eigenvalue of the operator $\hat{n}$ in the Fock state $|n\rangle$ and $p(n)$ is given by (14.6). As (14.9) must hold for every k we arrive at (Ghielmetti (1964), Peřina (1967a))

$$p(n) = \int \Phi_{\mathcal{N}}(\{\alpha_{\lambda}\}) \frac{W^n}{n!} \exp(-W) \, d^2\{\alpha_{\lambda}\}; \tag{14.10}$$

here we have used the multinomial theorem

$$\sum_{\{n_\lambda\}}' \prod_{\lambda} \frac{x_{\lambda}^{n_{\lambda}}}{n_{\lambda}!} = \frac{(\sum_{\lambda} x_{\lambda})^n}{n!}, \tag{14.11}$$

where $\sum'$ is taken under the condition $\sum_{\lambda} n_{\lambda} = n$; the quantity W is equal to

$$W = \langle\{\alpha_{\lambda}\} | \hat{n} | \{\alpha_{\lambda}\}\rangle = \int_{L^3} \mathscr{A}^*(x) . \mathscr{A}(x) \, d^3x = \sum_{\lambda} |\alpha_{\lambda}|^2, \tag{14.12}$$

where $\mathscr{A}(x)$ is the eigenvalue of the detection operator $\hat{\mathbf{A}}(x)$ (given by (11.23)) in the coherent state. If we consider a detector with a plane cathode of surface S on which plane waves are normally incident, then $L^3 = ScT$ and (14.12) is proportional to (10.3b). The substitution

$$P_{\mathcal{N}}(W) = \int \Phi_{\mathcal{N}}(\{\alpha'_{\lambda}\}) \, \delta(W - \sum_{\lambda} |\alpha'_{\lambda}|^2) \, d^2\{\alpha'_{\lambda}\} \tag{14.13}$$

finally gives the photodetection equation with the photoefficiency $\eta = 1$,

$$p(n) = \int_0^{\infty} P_{\mathcal{N}}(W) \frac{W^n}{n!} \exp(-W) \, dW. \tag{14.14}$$

Thus we have obtained the same relation for the probability distribution of the number of photons in the volume L^3 at time t as for the probability distribution of emitted photoelectrons within the time interval $(t, t + T)$ and therefore it can be said that in photoelectric detection measurements the statistical properties of absorbed photons are reflected in the statistical properties of emitted photoelectrons. The use of normally ordered operators, ensuring that the contribution of the vacuum is zero, is responsible for this result.

Note that the forms of the photon-number and photocount distributions are rigorously conserved for fields possessing the so-called scaling properties, $p(n, \langle I\rangle, \eta T) = p(n, \eta\langle I\rangle T)$, which means that the characteristic function considered as a function of $is\eta T$ and $\langle I\rangle$ is also a function of is and $\eta\langle I\rangle T$; the kth factorial moment depends then on η^k (Ghielmetti (1976), Jakeman (1981)). In this case the photocount distribution is obtained from the photon-number distribution by the substitution $\langle n\rangle = \langle I\rangle T \to \eta\langle I\rangle T$ (cf. (14.19)). For the Fock states the scaling properties are not satisfied (cf. (14.19)). However, the mostly used fields possess the scaling properties.

We can also show that equation (14.14) holds for an arbitrary volume V (the linear dimensions of which are large compared with the wavelength) which need not coincide with the normalization volume L^3. Considering (V, t') conjoint with $(\mathbf{x}, t)$ and using the commutation rule (11.44a), we obtain (Mandel (1966a))

$$\langle \mathcal{N}\hat{n}^k_{Vt}\rangle \equiv \langle \hat{n}^k_{Vt}\rangle_{\mathcal{N}} = \langle \hat{n}_{Vt}(\hat{n}_{Vt} - 1)\dots(\hat{n}_{Vt} - k + 1)\rangle, \tag{14.15}$$

which is a multimode analogue to (11.8a). Thus the normal characteristic function is equal to

$$\langle \exp(\mathrm{i}x\hat{n}_{Vt})\rangle_{\mathcal{N}} = \sum_{r=0}^{\infty} \frac{(\mathrm{i}x)^r}{r!}\langle \hat{n}^r_{Vt}\rangle_{\mathcal{N}} = \langle (1 + \mathrm{i}x)^{\hat{n}_{Vt}}\rangle, \tag{14.16}$$

which corresponds to (10.11) and substituting $1 + \mathrm{i}x = \exp(\mathrm{i}y)$ we have

$$\langle \exp(\mathrm{i}y\hat{n}_{Vt})\rangle = \langle \exp[\hat{n}_{Vt}(\mathrm{e}^{\mathrm{i}y} - 1)]\rangle_{\mathcal{N}}, \tag{14.17}$$

which corresponds to (10.8) and (13.101); the inverse Fourier transformation leads at once to (14.14) valid in the volume V. As a consequence the coefficients in the moment-expansion $\langle \hat{n}^k_{Vt}\rangle$ in terms of the normal moments $\langle \hat{n}^j_{Vt}\rangle_{\mathcal{N}}$, $j = 1, \dots, k$, are the same as in (10.4), (10.17) or (10.19); this can also be verified directly using the commutation rules (11.42) (Mandel (1964c)). As an example, considering one mode for simplicity and using the commutation rule (11.3), $\langle \hat{n}^2\rangle = \langle \hat{a}^+\hat{a}\hat{a}^+\hat{a}\rangle = \langle \hat{a}^+\hat{a}^+\hat{a}\hat{a}\rangle + \langle \hat{a}^+\hat{a}\rangle = \langle \hat{n}^2\rangle_{\mathcal{N}} + \langle \hat{n}\rangle_{\mathcal{N}}$, in agreement with (10.4b) (or the second equation in (10.17)). These results again confirm the close formal analogy between the quantum and classical descriptions of light beams.

Note that the normalized factorial moments of the photocount distribution,

$$\frac{\langle W^k\rangle_{\mathcal{N}}}{\langle W\rangle^k_{\mathcal{N}}} = \frac{\left\langle \dfrac{n!}{(n-k)!}\right\rangle}{\langle n\rangle^k}, \tag{14.18}$$

which follows from (10.13), are independent of the photoefficiency η so that they are equal to the factorial moments of the photon-number distribution. Thus the photocount distribution can be considered as the photon-number distribution independently of the detector efficiency if the normalized factorial moments are used for the interpretation of experimental data. Denoting the photon-number distribution by $p(n)$ and the photocount distribution by $p_\eta(n)$ for the present, their connection is given by the Bernoulli's distribution

$$p_\eta(n) = \sum_{m=n}^{\infty} \binom{m}{n} \eta^n (1 - \eta)^{m-n}\, p(m). \tag{14.19}$$

For open systems one has $\eta W = \eta\,|\alpha|^2\,T$ in the photodetection equation for the one-mode case, whereas this should be replaced by $(1 - \exp(-\eta T))\,|\alpha|^2$ in closed systems (Mollow (1968a), Scully and Lamb (1969), Arecchi and Degiorgio (1972), Selloni, Schwendimann, Quattropani and Baltes (1978), Shepherd (1981), Srinivas and Davies (1981)), as discussed by Mandel (1981b).

The photocount distribution in the form (10.3a) can be derived with the help of purely quantum methods (Glauber (1965, 1966a, 1972), Kelley and Kleiner

(1964), Lehmberg (1968), Mandel and Meltzer (1969), Klauder and Sudarshan (1968)). Following Glauber (1965, 1966a, 1972) to find the photocount distribution we assume that the sensitive element of a photon counter consists of many atoms, any of which may undergo a photoabsorption process and give rise to a detected photon count.† Let $\hat{n}(T) = \sum_j^N \hat{n}_j(T)$ be interpreted as the number of counts in (0, T), where $\hat{n}_j(T)$ represent the contributions of the individual atoms. The eigenvalues of $\hat{n}_j$ are one or zero; one for final states to which the jth atom has contributed a photon count and zero for states to which it has not contributed. Therefore we can write

$$(1 + \mathrm{i}x)^{\hat{n}(T)} = (1 + \mathrm{i}x)^{\sum_j^N \hat{n}_j(T)} = \prod_j^N (1 + \mathrm{i}x)^{\hat{n}_j(T)} = \prod_j^N [1 + \mathrm{i}x\hat{n}_j(T)], \qquad (14.20)$$

since the eigenvalues of $\hat{n}_j$ are 1 or 0. Expanding (14.20), expressing $\langle \hat{n}_1(T) \dots \hat{n}_m(T) \rangle = p^{(m)}(T, \dots, T)$ from (12.13) and approximating the combination number $N!/(N - m)!\, m!$ appearing in the expansion of (14.20) by $1/m!$ (since N is large and $p^{(m)}$ are non-zero only for $m \ll N$), we obtain

$$\langle (1 + \mathrm{i}x)^{\hat{n}(T)} \rangle = 1 + \sum_{m=1}^{\infty} \frac{(\mathrm{i}x)^m}{m!} F_m, \qquad (14.21)$$

where

$$F_m = \int_0^T \dots \int_0^T \underbrace{\int \dots \int}_{\text{Volume of detector}} \Gamma_{\mathcal{N}}^{(m,m)}(x_1', \dots, x_m', x_m'', \dots, x_1'') \times$$
$$\times \prod_{j=1}^m \mathscr{S}(x_j' - x_j'')\, \mathrm{d}^4 x_j'\, \mathrm{d}^4 x_j''.$$

If we now use the Glauber-Sudarshan representation, we obtain for the characteristic function

$$\langle (1 + \mathrm{i}x)^{\hat{n}(T)} \rangle = \int \Phi_{\mathcal{N}}(\{\alpha_\lambda\}) \exp [\mathrm{i}x\, \bar{W}(\{\alpha_\lambda\})]\, \mathrm{d}^2\{\alpha_\lambda\}, \qquad (14.22)$$

---

† Also two-photon and multi-photon absorptions can be considered. The perturbation theory of quantum mechanics leads to the result (Lambropoulos, Kikuchi and Osborn (1966), Carusotto, Fornaca and Polacco (1967, 1968)) that the probability of a transition for the s-quantum absorption is proportional to $I^s(t)$, i.e. all results concerning the detection of the field given here are correct if $I(t)$ is replaced by $I^s(t)$. The relation between the two-quantum photocount distribution and the intensity fluctuations of the radiation incident on the detector has been studied and compared with the one-quantum results for laser and chaotic radiation by Teich and Diament (1969). Some further investigations of the two- and multi-photon processes were performed (Teich and Wolga (1966), Brunner, Paul and Richter (1965), Lambropoulos (1968), Mollow (1968b), Diament and Teich (1969), Millet and Varnier (1969), Jaiswal and Agarwal (1969), Agarwal (1970), Barashev (1970a, b, 1976), Millet and Usselio-La-Verna (1970), Tunkin and Tchirkin (1970), Peřina, Peřinová and Mišta (1971), Mišta and Peřina (1971), Mišta (1971), Delone and Masalov (1980)). Some experimental results have been referred to by Shiga and Inamura (1967), Teich, Abrams and Gandrud (1970), Clark, Estes and Narducci (1970), Lyons and Troup (1970), Krasiński, Chudzyński and Majewski (1974, 1976) and Glódź (1978). The photon statistics of multiphoton absorption and emission are discussed in Chapter 22.

where

$$\bar{W}(\{\alpha_\lambda\}) = \int_0^T\int_0^T \iint_{\substack{\text{Volume}\\ \text{of detector}}} \mathscr{S}(x' - x'')\, V^*(x')\, V(x'')\, d^4x'\, d^4x''. \tag{14.23}$$

Remember that in (14.23) as well as in (14.21) one must understand the integrals over x as also implying a sum over the polarization indices. If the analogous substitution to (14.13) is used

$$P_{\mathcal{N}}(\bar{W}) = \int \Phi_{\mathcal{N}}(\{\alpha'_\lambda\})\, \delta(\bar{W} - \bar{W}(\{\alpha'_\lambda\}))\, d^2\{\alpha'_\lambda\}, \tag{14.24}$$

we have the final form of the characteristic function

$$\langle (1 + ix)^{\hat{n}(T)} \rangle = \int_0^\infty P_{\mathcal{N}}(\bar{W}) \exp(ix\bar{W})\, d\bar{W}. \tag{14.25}$$

The required photocount distribution is equal to

$$p(n, T) = \frac{1}{n!} \frac{d^n}{d(ix)^n} \langle (1 + ix)^{\hat{n}(T)} \rangle \bigg|_{ix = -1} = \int_0^\infty P_{\mathcal{N}}(\bar{W}) \frac{\bar{W}^n}{n!} \exp(-\bar{W})\, d\bar{W} \tag{14.26}$$

and the factorial moments are from (14.21)

$$\left\langle \frac{\hat{n}!}{(\hat{n} - k)!} \right\rangle = \frac{d^k}{d(ix)^k} \langle (1 + ix)^{\hat{n}(T)} \rangle \bigg|_{ix=0} =$$

$$= \int_0^T \cdots \int_0^T \int \cdots \int_{\substack{\text{Volume}\\ \text{of detector}}} \Gamma^{(k,k)}_{\mathcal{N}}(x'_1, \ldots, x'_k, x''_k, \ldots, x''_1) \prod_{j=1}^k \mathscr{S}(x'_j - x''_j)\, d^4x'_j\, d^4x''_j. \tag{14.27}$$

For a broad-band detector and plane quasi-monochromatic waves normally incident in the z-direction on a plane photocathode of surface S, we can write $\mathscr{S}(x' - x'') = (\eta/ScT)\,\delta(x' - x'')$ and

$$\bar{W} = \frac{\eta}{ScT} \int_0^T dt' \int_0^{cT} dz' \int_S d^2x' I(x') = \eta \int_0^T I(t')\, dt' = \eta W, \tag{14.28}$$

where $I(x) = \sum_\mu V^*_\mu(x)\, V_\mu(x) = \mathbf{V}^*(x) . \mathbf{V}(x)$ is explicitly independent of x, y and z.

This leads, together with (14.26), to the photodetection equation (10.3a) derived by the semiclassical method. Under these simplifications the factorial moments (14.27) are equal to

$$\left\langle \frac{\hat{n}!}{(\hat{n} - k)!} \right\rangle = \eta^k \int_0^T \cdots \int_0^T \Gamma^{(k,k)}_{\mathcal{N}}(t'_1, \ldots, t'_k, t'_k, \ldots, t'_1) \prod_{j=1}^k dt'_j \tag{14.29}$$

(cf. (12.11) and (10.91)). It is a formal matter to generalize these results to multiple detectors in analogy with (10.86)–(10.94).

Finally we can conclude that there is indeed a very close similarity between the quantum and classical descriptions of the coherence properties of the electromagnetic field. If we interpret the Glauber-Sudarshan representation of the density matrix as implying the latter's existence as a Z'-generalized function (ultra-distribution), there is a complete formal equivalence between the two descriptions. This may be regarded as a consequence of the fact that vacuum expectation values of normally ordered operators are zero and thus the physical vacuum gives no contribution. This does not imply physical equivalence, since the classical description is applicable to strong fields where energies comparable with the field quantum $\hbar\omega$ are neglected whereas the quantum description is valid generally including arbitrarily weak fields. The physical inequivalence of the quantum and the classical descriptions reflects itself in the fact that classical distributions are well-behaved non-negative functions while the function $\Phi_{\mathcal{N}}(\{\alpha_\lambda\})$ can take on negative values and can be very singular (e.g. for the Fock state $|n\rangle$ it is proportional to the nth derivative of the δ-function, which has no classical analogue; however, it may be much more singular since it is in general a Z'-distribution). Only in the limit of strong fields where the ordering of the field operators plays no role and $\Phi_{\mathcal{N}} \to \Phi_{\mathcal{A}}$, which is a well-behaved function, may the complete physical equivalence of the classical and quantum descriptions be reached.

STATIONARY CONDITIONS OF THE FIELD

15.1 Time invariance properties of the correlation functions

It is obvious that the quantum stationary condition is expressed by

$$[\hat{H}, \hat{\varrho}] = \hat{0}, \tag{15.1}$$

which follows from (11.48a) since $\partial\hat{\varrho}/\partial t = \hat{0}$ in this case; here $\hat{H}$ is the Hamiltonian of the system. We show that the earlier definition of a stationary field (Chapter 2), i.e. that the $(m + n)$th-order correlation functions are invariant with respect to translations of the time origin or that they depend on $(m + n - 1)$ time differences $(t_j - t_1)$, $j = 2, \ldots, m + n$ only, is a simple consequence of (15.1) (Kano (1964b), Glauber (1965, 1970)).

We can write, in the Heisenberg picture,

$$\begin{aligned} \Gamma_{\mathcal{N}}^{(m,n)}(x_1, \ldots, x_{m+n}) &= \operatorname{Tr}\{\hat{\varrho} \prod_{j=1}^{m} \hat{A}^{(-)}(\mathbf{x}_j, t_j) \prod_{k=m+1}^{m+n} \hat{A}^{(+)}(\mathbf{x}_k, t_k)\} = \\ &= \operatorname{Tr}\left\{\exp\left(\mathrm{i}\frac{\hat{H}\tau}{\hbar}\right)\hat{\varrho}\exp\left(-\mathrm{i}\frac{\hat{H}\tau}{\hbar}\right)\exp\left(\mathrm{i}\frac{\hat{H}\tau}{\hbar}\right)\hat{A}^{(-)}(\mathbf{x}_1, t_1) \times \right. \\ &\quad \left. \times \exp\left(-\mathrm{i}\frac{\hat{H}\tau}{\hbar}\right) \ldots \exp\left(\mathrm{i}\frac{\hat{H}\tau}{\hbar}\right)\hat{A}^{(+)}(\mathbf{x}_{m+n}, t_{m+n})\exp\left(-\mathrm{i}\frac{\hat{H}\tau}{\hbar}\right)\right\}, \end{aligned} \tag{15.2}$$

where we have substituted the unit operator $\exp(-\mathrm{i}\hat{H}\tau/\hbar)\exp(\mathrm{i}\hat{H}\tau/\hbar)$ and used the cyclic property of the trace. Using (11.46a) and assuming (15.1) we arrive at

$$\begin{aligned} \Gamma_{\mathcal{N}}^{(m,n)}(x_1, \ldots, x_{m+n}) &= \operatorname{Tr}\{\hat{\varrho} \prod_{j=1}^{m} \hat{A}^{(-)}(\mathbf{x}_j, t_j + \tau) \prod_{k=m+1}^{m+n} \hat{A}^{(+)}(\mathbf{x}_k, t_k + \tau)\} = \\ &= \Gamma_{\mathcal{N}}^{(m,n)}(\mathbf{x}_1, \ldots, \mathbf{x}_{m+n}, t_1 + \tau, \ldots, t_{m+n} + \tau). \end{aligned} \tag{15.3}$$

Putting $\tau = -t_1$ we have

$$\Gamma_{\mathcal{N}}^{(m,n)}(x_1, \ldots, x_{m+n}) = \Gamma_{\mathcal{N}}^{(m,n)}(\mathbf{x}_1, \ldots, \mathbf{x}_{m+n}, 0, t_2 - t_1, \ldots, t_{m+n} - t_1), \tag{15.4}$$

which is just the above-mentioned stationary condition. From (15.3) or (15.4) the condition (15.1) can also be obtained.

Another stationary condition which makes use of the time derivatives of $\Gamma_{\mathcal{N}}^{(m,n)}$ can be obtained as follows (Horák (1971)). In the Heisenberg picture

$$\mathrm{i}\hbar\frac{\partial \hat{A}^{(+)}(\mathbf{x}_l, t_l)}{\partial t_l} = [\hat{A}^{(+)}(\mathbf{x}_l, t_l), \hat{H}]. \tag{15.5}$$

Multiplying this equation by $\prod_{j=1}^{m} \hat{A}^{(-)}(\mathbf{x}_j, t_j) \prod_{k=m+1}^{l-1} \hat{A}^{(+)}(\mathbf{x}_k, t_k)$ from the left and by $\prod_{k=l+1}^{m+n} \hat{A}^{(+)}(\mathbf{x}_k, t_k)$ from the right and taking the average we arrive at

$$\begin{aligned} i\hbar \langle \prod_{j=1}^{m} \hat{A}^{(-)}(\mathbf{x}_j, t_j) \prod_{k=m+1}^{l-1} \hat{A}^{(+)}(\mathbf{x}_k, t_k) \frac{\partial \hat{A}^{(+)}(\mathbf{x}_l, t_l)}{\partial t_l} \prod_{k=l+1}^{m+n} \hat{A}^{(+)}(\mathbf{x}_k, t_k) \rangle = \\ = \langle \prod_{j=1}^{m} \hat{A}^{(-)}(\mathbf{x}_j, t_j) \prod_{k=m+1}^{l-1} \hat{A}^{(+)}(\mathbf{x}_k, t_k) \hat{A}^{(+)}(\mathbf{x}_l, t_l) \hat{H} \prod_{k=l+1}^{m+n} \hat{A}^{(+)}(\mathbf{x}_k, t_k) \rangle - \\ - \langle \prod_{j=1}^{m} \hat{A}^{(-)}(\mathbf{x}_j, t_j) \prod_{k=m+1}^{l-1} \hat{A}^{(+)}(\mathbf{x}_k, t_k) \hat{H} \hat{A}^{(+)}(\mathbf{x}_l, t_l) \prod_{k=l+1}^{m+n} \hat{A}^{(+)}(\mathbf{x}_k, t_k) \rangle. \quad (15.6) \end{aligned}$$

Summing all these equations for $l = m + 1, \ldots, m + n$ together with the other set of equations obtained with the help of the Hermitian conjugate to (15.5), we obtain the stationary condition (cf. (2.21) in the frequency region)

$$\begin{aligned} i\hbar \sum_{l=1}^{m+n} \frac{\partial}{\partial t_l} \Gamma_{\mathcal{N}}^{(m,n)}(x_1, \ldots, x_{m+n}) = \operatorname{Tr} \{ \hat{\varrho} \prod_{j=1}^{m} \hat{A}^{(-)}(x_j) \prod_{k=m+1}^{m+n} \hat{A}^{(+)}(x_k) \hat{H} \} - \\ - \operatorname{Tr} \{ \hat{\varrho} \hat{H} \prod_{j=1}^{m} \hat{A}^{(-)}(x_j) \prod_{k=m+1}^{m+n} \hat{A}^{(+)}(x_k) \} = \\ = \operatorname{Tr} \{ [\hat{H}, \hat{\varrho}] \prod_{j=1}^{m} \hat{A}^{(-)}(x_j) \prod_{k=m+1}^{m+n} \hat{A}^{(+)}(x_k) \} = 0, \quad (15.7) \end{aligned}$$

provided (15.1) holds. For $m = n = 1$ we have $\partial \Gamma_{\mathcal{N}}^{(1,1)}/\partial t_1 = -\partial \Gamma_{\mathcal{N}}^{(1,1)}/\partial t_2$ as must be since $\Gamma_{\mathcal{N}}^{(1,1)}(t_1, t_2) = \Gamma_{\mathcal{N}}^{(1,1)}(t_2 - t_1)$.

15.2 Stationary conditions in phase space

First let us consider a free one-mode field. Using the renormalized Hamiltonian (11.13), $\hat{H} = \hbar\omega \hat{a}^+ \hat{a}$, and the form (11.25) of the density matrix, we have for (15.1)

$$[\hat{H}, \hat{\varrho}] = \hbar\omega \sum_{n,m} \varrho(n, m)(n - m) | n \rangle \langle m | = \hat{0}, \quad (15.8)$$

that is

$$\langle k | [\hat{H}, \hat{\varrho}] | l \rangle = \hbar\omega \varrho(k, l)(k - l) = 0 \quad (15.9)$$

for all k, l. Consequently $\varrho(k, l)$ must have the form

$$\varrho(k, l) = \varrho(k, k) \delta_{kl}. \quad (15.10)$$

Thus we can see from (13.119) or (13.121) and (13.120) that $\Phi_{\mathcal{N}}(\alpha)$ is independent of the phase φ of α. On the other hand, if $\Phi_{\mathcal{N}}(\alpha)$ is independent of the phase φ, then $\varrho(k, l)$ has the form (15.10) and (15.1) holds. Therefore for a one-mode free field the stationary condition (15.1) is necessary and sufficient for $\Phi_{\mathcal{N}}(\alpha)$ to be independent of the phase φ of the complex amplitude α.

For a multimode free field we again consider the renormalized Hamiltonian (11.13) and, using the multimode form (11.38) of the density matrix, we obtain

$$[\hat{H}, \hat{\varrho}] = \sum_{\{n_\lambda\}} \sum_{\{m_\lambda\}} \varrho(\{n_\lambda\}, \{m_\lambda\}) \left[\sum_\lambda \hbar\omega_\lambda(n_\lambda - m_\lambda)\right] |\{n_\lambda\}\rangle \langle\{m_\lambda\}| = \hat{0}, \tag{15.11}$$

which leads to

$$\varrho(\{n_\lambda\}, \{m_\lambda\}) = \varrho_0(\{n_\lambda\}, \{m_\lambda\})\, \delta_{NM}, \tag{15.12}$$

where

$$N = \sum_\lambda \hbar\omega_\lambda n_\lambda, \qquad M = \sum_\lambda \hbar\omega_\lambda m_\lambda \tag{15.13}$$

and ϱ_0 is equal to ϱ when $N = M$. Substituting (15.12) into the multimode form of (13.119),

$$\Phi_{\mathcal{N}}(\{\alpha_\lambda\}) = \sum_{\{n_\lambda\}} \sum_{\{m_\lambda\}} \varrho(\{n_\lambda\}, \{m_\lambda\}) \prod_\lambda \frac{(n_\lambda!\, m_\lambda!)^{1/2}}{(n_\lambda + m_\lambda)!\, \pi r_\lambda} \times$$
$$\times \exp\left[r_\lambda^2 + \mathrm{i}(m_\lambda - n_\lambda)\varphi_\lambda\right] \left(-\frac{\partial}{\partial r_\lambda}\right)^{n_\lambda + m_\lambda} \delta(r_\lambda), \tag{15.14}$$

or (13.120),

$$\Phi_{\mathcal{N}}(\{\alpha_\lambda\}) = \exp\left[-(\zeta - 1)\sum_\lambda |\alpha_\lambda|^2\right] \times$$
$$\times \sum_{\{j_\lambda\}=0}^{\infty} \left\{ \sum_{\{\mu_\lambda\}=0}^{\infty} c_{\{j_\lambda\},\{\mu_\lambda\}} \prod_\lambda (\zeta\alpha_\lambda)^{\mu_\lambda} L_{j_\lambda}^{\mu_\lambda}(\zeta|\alpha_\lambda|^2) + \sum_{\{\mu_\lambda\}=1}^{\infty} (\text{c. c.})\right\}, \tag{15.15}$$

where

$$c_{\{j_\lambda\},\{\mu_\lambda\}} = \sum_{\{s_\lambda\}=0}^{\{j_\lambda\}} \varrho(\{s_\lambda\}, \{s_\lambda + \mu_\lambda\}) \prod_\lambda \frac{\zeta^{s_\lambda+1}(-1)^{s_\lambda}}{\pi(j_\lambda + \mu_\lambda)!} \binom{j_\lambda}{s_\lambda} \left[\frac{s_\lambda!}{(s_\lambda + \mu_\lambda)!}\right]^{1/2}, \tag{15.16}$$

we can obtain the corresponding function $\Phi_{\mathcal{N}}(\{\alpha_\lambda\})$ for the stationary field. This function will in general depend upon the phases $\{\varphi_\lambda\}$ of $\{\alpha_\lambda\}$.

If the additional condition (Kano (1964b, 1966))

$$[\hat{\varrho}, \hat{a}_\lambda^+ \hat{a}_\lambda] = \hat{0} \tag{15.17}$$

holds for all λ, then

$$\varrho(\{n_\lambda\}, \{m_\lambda\}) = \varrho(\{n_\lambda\}, \{n_\lambda\})\, \delta_{\{n_\lambda\},\{m_\lambda\}} \tag{15.18}$$

and consequently $\Phi_{\mathcal{N}}(\{\alpha_\lambda\})$ is independent of the phases $\{\varphi_\lambda\}$. The Fock form of the density matrix (11.38) is diagonal in this case.

Analogous results for the space dependence of the correlation functions can be obtained from the condition $[\hat{\mathbf{P}}, \hat{\varrho}] = \hat{0}$ (Eberly and Kujawski (1967a)), where $\hat{\mathbf{P}} = \sum_\lambda \hbar\mathbf{k}_\lambda \hat{a}_\lambda^+ \hat{a}_\lambda$ is the total momentum operator of the system. This condition leads to correlation functions which are stationary in space, that is which are homogeneous and isotropic. One can see that if (15.17) holds, then $[\hat{H}, \hat{\varrho}] = [\hat{\mathbf{P}}, \hat{\varrho}] = \hat{0}$ and the field is stationary in time and in space as well. This can also be seen from the form of the correlation functions calculated under the assumption that $\Phi_{\mathcal{N}}(\{\alpha_\lambda\})$ is independent of $\{\varphi_\lambda\}$ (cf. Sec. 17.1, equation (17.26)). In general, if a field is stationary in time it need not be stationary in space, so that for example the function $\Gamma_{\mathcal{N}}^{(1,1)}(\mathbf{x}, \mathbf{x}, t_2 - t_1)$ will generally depend on $\mathbf{x}$.

Hence, the condition (15.17) is a necessary and sufficient condition for $\Phi_{\mathcal{N}}(\{\alpha_\lambda\})$ to be independent of the phases $\{\varphi_\lambda\}$; in this case the phases $\{\varphi_\lambda\}$ are distributed uniformly in the interval $(0, 2\pi)$. The condition (15.1) is a necessary but not a sufficient condition for $\Phi_{\mathcal{N}}(\{\alpha_\lambda\})$ to be independent of the phases $\{\varphi_\lambda\}$.

If $\Phi_{\mathcal{N}}(\{\alpha_\lambda\})$ is independent of $\{\varphi_\lambda\}$ (the field is stationary in time and in space), then $\mathrm{Tr}\,\{\hat{\varrho}\hat{a}_\lambda\} = \mathrm{Tr}\,\{\hat{\varrho}\hat{a}_\lambda^+\} = 0$ and so

$$\mathrm{Tr}\,\{\hat{\varrho}\hat{A}(x)\} = 0; \tag{15.19}$$

and also

$$\mathrm{Tr}\,\{\hat{\varrho}\hat{A}^{(-)}(x_1) \ldots \hat{A}^{(-)}(x_m)\,\hat{A}^{(+)}(x_{m+1}) \ldots \hat{A}^{(+)}(x_{m+n})\} = 0, \tag{15.20}$$

if $m \neq n$, since the integrals over the phases, if the Glauber-Sudarshan representation is used, must equal zero. However, if the field is stationary in time only, then $\Gamma_{\mathcal{N}}^{(m,n)}$ for $m \neq n$ need not be zero, since (15.12) holds in this case and so $\Phi_{\mathcal{N}}(\{\alpha_\lambda\})$ will depend in general on the phases $\{\varphi_\lambda\}$. Also the conditions (2.21) must be fulfilled in the frequency domain for the non-vanishing of $\Gamma_{\mathcal{N}}^{(m,n)}$ for $m \neq n$.

Some possibilities of measuring the correlation functions for $m \neq n$ have been discussed in Sec. 12.4. However, in usual experiments with stationary fields the correlation functions $\Gamma_{\mathcal{N}}^{(m,n)}$ with $m \neq n$ are equal to zero (Mehta and Mandel (1967), Mandel (1964a)).

A class of quasi-stationary fields has been discussed by Picinbono (1969) and Picinbono and Rousseau (1970).

CHAPTER 16

ORDERING OF FIELD OPERATORS IN QUANTUM OPTICS

In the preceding chapters devoted to the quantum theory of coherence we have dealt mainly with normal ordering and its correspondence to c-number functions, although antinormal and symmetric (Weyl) orderings were introduced also. In this chapter we develop a systematic approach to the correspondence between q-number and c-number functions.

The correspondence between functions of q-numbers (operators) and functions of c-numbers (classical functions) was first investigated by Wigner (1932). He introduced a generalized phase space quasi-distribution function in connection with quantum-mechanical corrections to thermodynamic equilibrium. In this formulation the problem of expressing functions of q-numbers according to a prescribed rule of ordering arose together with the closely related problem of the calculation of quantum expectation values as averages with respect to a quasi-distribution function in some generalized phase space. Later Moyal (1949) used the Wigner formulation in a statistical interpretation of quantum mechanics and showed that the Wigner quasi-distribution function is related to a symmetrized (Weyl) ordering of boson field operators. Among other papers devoted to this problem we mention one by Mehta (1964).

The problem of the ordering of field operators in quantum optics plays an important role in connection with generalized phase space descriptions of optical fields (Mandel and Wolf (1965), Klauder and Sudarshan (1968), Peřina (1972), Nussenzveig (1973)), laser theory (Lax (1967a, 1968a, b), Lax and Louisell (1967), Lax and Yuen (1968), Haken (1970a, b), Louisell (1973)), and the photon statistics of nonlinear optical processes (Shen (1967), Schubert and Wilhelmi (1980), Peřina (1980, 1984)). A systematic way of treating problems of the ordering of field operators and the correspondence between functions of q-numbers and functions of c-numbers have been found by Agarwal and Wolf (1968a, 1970) and Wolf and Agarwal (1969). Another general approach to this problem was proposed by Lax (1968b) (see also Louisell (1970)); this theory is valid also for Fermi particles. These methods were used by Agarwal and Wolf (1968b) and Agarwal (1969) to handle quantum dynamical problems in phase space and to obtain an expression for the time-ordered product of Heisenberg operators in terms of products arranged according to a prescribed rule of ordering (Wick's theorem can be deduced as a special case) (Agarwal and Wolf (1968c)). Cahill and Glauber (1969) have given a general treatment of the ordering of operators showing the inter-relationships between the normal, symmetric (Weyl) and antinormal orderings; they also show the correspondence between q-number and c-number functions and how quantum-mechanical expectation values may be calculated using a parametrization making it possible to interpolate between normal and antinormal orderings (this represents a special case of the theory of Agarwal and Wolf). These methods also provide a way of seeing when and how singularities appear in quasi-probability distributions including the weight function of the Glauber-Sudarshan representation. Such a theory for multimode fields has been also developed and demonstrated by an example of the superposition of coherent and chaotic fields (Peřina and Horák (1969a, b, 1970)).

Such results have particular significance for correlation and photon counting experiments; photoelectric detection measurements are related to observables connected with the normally ordered products of field operators and the quantum counter measurements, using stimulated emission rather than absorption, are related to observables connected with the antinormal ordering of field operators (Secs. 12.1 and 16.4). In scattering of light from fluctuations in liquids and turbulent fluids the correlation functions of the field in a symmetrical form are appropriate and these are closely connected to correlation functions of the fluctuating medium (e.g. of the fluctuating density of particles) (Bertolotti, Crosignani, Di Porto and Sette (1967a, b, 1969), Crosignani, Di Porto and Bertolotti (1975)).

16.1 Definition of Ω-ordering and general decompositions of operators

The completeness property (13.36) of the displacement operator $\hat{D}(\alpha)$ enables us to decompose any operator $\hat{A}$ in terms of the displacement operator $\hat{D}(\alpha)$ in the form

$$\hat{A} = \frac{1}{\pi} \int \tilde{a}(\beta)\, \hat{D}^{+}(\beta)\, \mathrm{d}^2\beta, \tag{16.1}$$

where

$$\tilde{a}(\beta) = \mathrm{Tr}\, \{\hat{D}(\beta)\, \hat{A}\}. \tag{16.2}$$

Considering another operator $\hat{B}$ we can write

$$\mathrm{Tr}\, \{\hat{A}\hat{B}\} = \frac{1}{\pi} \int \tilde{a}(-\beta)\, \tilde{b}(\beta)\, \mathrm{d}^2\beta \tag{16.3}$$

and for $\hat{B} \equiv \hat{A}^{+}$

$$\| \hat{A} \|^2 = \mathrm{Tr}\, \{\hat{A}^{+}\hat{A}\} = \frac{1}{\pi} \int |\, \tilde{a}(\beta)\, |^2\, \mathrm{d}^2\beta. \tag{16.4}$$

Thus, if $\tilde{a}(\beta)$ is a square integrable function, then the Hilbert-Schmidt norm $\| \hat{A} \| = [\mathrm{Tr}\, \{\hat{A}^{+}\hat{A}\}]^{1/2}$ is finite and vice versa. In particular for the density matrix $\hat{\varrho} = \hat{\varrho}^{+} \equiv \hat{A}$, $\mathrm{Tr}\, \{\hat{\varrho}^2\} = \int |\, C(\beta)\, |^2\, \mathrm{d}^2\beta/\pi \leqq 1$, where $C(\beta)$ is the characteristic function (13.70); thus we conclude that $0 \leqq |\, C(\beta)\, | \leqq 1$ again.

If $\hat{A}(\hat{a}^+, \hat{a}) \equiv \hat{\varrho}(\hat{a}^+, \hat{a})$, then we have obtained a representation of the density matrix in terms of $\hat{D}(\beta)$. It is obvious that $\tilde{a}(\beta)$ is identical to the Wigner characteristic function $C(\beta)$ given by (13.70) and that $\varrho(\alpha^*, \alpha)/\pi$ obtained with the use of (16.1) is just the Wigner function (13.76). Some further details about the completeness properties of the displacement operators $\hat{D}(\alpha)$ can be found in papers by Cahill and Glauber (1969) and Agarwal and Wolf (1970).

We introduce the operator $\hat{\Delta}(\alpha, \Omega)$ as

$$\hat{\Delta}(\alpha, \Omega) = \frac{1}{\pi}\int \Omega(\beta^*, \beta)\, \hat{D}(\beta) \exp(\alpha\beta^* - \alpha^*\beta) \mathrm{d}^2\beta = \\ = \frac{1}{\pi}\int \Omega(\beta^*, \beta) \exp[\beta(\hat{a}^+ - \alpha^*) - \beta^*(\hat{a} - \alpha)]\, \mathrm{d}^2\beta, \tag{16.5}$$

that is $\hat{\Delta}(\alpha, \Omega)/\pi$ is equal to the Fourier transform of the operator

$$\hat{D}(\beta, \Omega) = \hat{\Omega}\hat{D}(\beta) = \hat{\Omega} \exp(\beta\hat{a}^+ - \beta^*\hat{a}) = \\ = \Omega(\beta^*, \beta) \exp(\beta\hat{a}^+ - \beta^*\hat{a}) \equiv \{\hat{D}(\beta)\}_\Omega, \tag{16.6}$$

where $\hat{\Omega}$ is an operator arranging the expression in an Ω-ordered form denoted by $\{\,\}_\Omega$. The function $\Omega(\beta^*, \beta)$ may be called the *filter function*. Introducing (16.6) into (16.5) we can write

$$\hat{\Delta}(\alpha, \Omega) = \frac{1}{\pi}\int \hat{D}(\beta, \Omega) \exp(\alpha\beta^* - \alpha^*\beta)\, \mathrm{d}^2\beta. \tag{16.7}$$

We assume that the function $\Omega(\alpha^*, \alpha)$ has the following properties: $\Omega(\alpha^*, \alpha)$ is a function obtained for $\beta = \alpha^*$ from a function $\Omega(\beta, \alpha)$, which is an entire analytic function of two complex variables and it has no zeros (minimal filter), $\Omega(0, 0) = 1$ and $\Omega(-\beta, -\alpha) = \Omega(\beta, \alpha)$ (symmetric filter). By using the Weierstrass theorem for entire functions the function $\Omega(\beta, \alpha)$ has the form $\exp[\omega(\beta, \alpha)]$, where $\omega(\beta, \alpha)$ is an entire analytic function with $\omega(0, 0) = 0$. It can be seen by taking the above properties into account that

$$\omega(\beta, \alpha) = \frac{r}{2}\alpha^2 + \frac{t}{2}\beta^2 + \frac{s}{2}\alpha\beta + \text{higher-order terms in } \alpha, \beta, \tag{16.8}$$

where r, t and s are parameters.

As an example we give the cases of normal, symmetric (Weyl) and antinormal orderings. Making use of the Baker-Hausdorff identity (13.3) we obtain for (16.6) identifying Ω with symmetric, normal and antinormal orderings,

$$\hat{D}(\beta, \Omega) \equiv \hat{D}(\beta, s) = \exp\left(\frac{s}{2}|\beta|^2\right)\hat{D}(\beta) = \exp\left(\frac{s}{2}|\beta|^2\right)\exp(\beta\hat{a}^+ - \beta^*\hat{a}) = \\ = \exp\left(\frac{s-1}{2}|\beta|^2\right)\exp(\beta\hat{a}^+)\exp(-\beta^*\hat{a}) = \\ = \exp\left(\frac{s+1}{2}|\beta|^2\right)\exp(-\beta^*\hat{a})\exp(\beta\hat{a}^+), \tag{16.9}$$

that is $r = t = 0$ and other higher-order terms in (16.8) vanish. Hence, $s = 0$ for symmetric ordering, $s = 1$ for normal ordering and $s = -1$ for antinormal ordering.

Obviously $\hat{D}^+(\beta, \Omega) = \hat{D}(-\beta, \Omega^*) = \hat{D}^{-1}(\beta, \Omega^{*-1}) = \hat{D}^{-1}(\beta, \tilde{\Omega}^*)$, where the $\tilde{\Omega}$-ordering is defined as the Ω-ordering with $\Omega(\beta^*, \beta) \to \Omega^{-1}(\beta^*, \beta)$. Therefore it follows that $\hat{\Delta}^+(\alpha, \Omega) = \hat{\Delta}(\alpha, \Omega^*)$. For the s-ordering (as a special case) $\hat{D}^+(\beta, s) = \hat{D}(-\beta, s^*) = \hat{D}^{-1}(\beta, -s^*)$ and $\hat{\Delta}^+(\alpha, s) = \hat{\Delta}(\alpha, s^*)$.

If we express the operators $\hat{a}$ and $\hat{a}^+$ by means of the canonical operators $\hat{p}$, $\hat{q}$ given by (11.15), we can arrange $\hat{D}(\alpha)$ in such a way that all $\hat{q}$ are to the left of all $\hat{p}$ (the standard ordering), or all $\hat{q}$ are to the right of all $\hat{p}$ (the anti-standard ordering). For the first case we obtain $r = 1/2$, $t = -1/2$, $s = 0$ and for the second case $r = -1/2$, $t = 1/2$, $s = 0$ in (16.8).

From (16.7) and (16.6) it follows that $\hat{\Delta}(\alpha, \Omega) = \pi\{\delta(\hat{a} - \alpha)\}_\Omega$ (the δ-function is defined by (13.37)), i.e. the operator $\pi^{-1}\hat{\Delta}(\alpha, \Omega)$ represents the Ω-ordered form of the operator $\delta(\hat{a} - \alpha)$.

Further we obtain

$$\mathrm{Tr}\,\{\hat{D}(\beta, \Omega)\,\hat{D}(\gamma, \tilde{\Omega})\} = \mathrm{Tr}\,\{\hat{D}(\beta, s)\,\hat{D}(\gamma, -s)\} = \mathrm{Tr}\,\{\hat{D}(\beta)\,\hat{D}(\gamma)\} = \pi\delta(\beta + \gamma). \tag{16.10}$$

Thus an operator $\hat{A}$ can be decomposed in the form

$$\hat{A} = \frac{1}{\pi}\int \tilde{a}(\beta, \Omega)\,\hat{D}^{-1}(\beta, \Omega)\,\mathrm{d}^2\beta, \tag{16.11}$$

which is a more general decomposition than (16.1); here

$$\tilde{a}(\beta, \Omega) = \mathrm{Tr}\,\{\hat{D}(\beta, \Omega)\,\hat{A}\}. \tag{16.12}$$

If $\hat{A} \equiv \hat{\varrho}$ and $\hat{D}(\beta, \Omega) \equiv \hat{D}(\beta, s)$ we regain for $s = 1$ and -1 the characteristic functions (13.72), $C_{\mathcal{N}}(\beta) \equiv \tilde{a}(\beta, 1)$, and (13.73), $C_{\mathcal{A}}(\beta) \equiv \tilde{a}(\beta, -1)$. Thus $\varrho(\alpha^*, \alpha)/\pi$ obtained with the help of (16.11) by the substitutions $\hat{a} \to \alpha$ and $\hat{a}^+ \to \alpha^*$ is equal to $\Phi_{\mathcal{N}}(\alpha)$ (see (13.74)) for $s = 1$ and to $\Phi_{\mathcal{A}}(\alpha)$ (see (13.75)) for $s = -1$. For $s = 0$ we have (16.1) and (16.2), which correspond to the symmetric ordering.

By using (16.7) and (16.10) we arrive at the orthogonality relation

$$\mathrm{Tr}\,\{\hat{\Delta}(\beta, \Omega)\,\hat{\Delta}(\gamma, \tilde{\Omega})\} = \pi\delta(\beta - \gamma). \tag{16.13}$$

Other interesting properties of the operator $\hat{\Delta}(\beta, \Omega)$ can also be given using (16.7), such as

$$\int \hat{\Delta}(\beta, \Omega)\,\mathrm{d}^2\beta = \pi\hat{D}(0, \Omega) = \pi \tag{16.14}$$

and

$$\mathrm{Tr}\,\{\hat{\Delta}(\beta, \Omega)\} = 1, \tag{16.15}$$

where (16.5) and (13.35) have been used. Making use of (16.7) and (16.9) for $s = -1$, we arrive at

$$\hat{\Delta}(\beta,\mathscr{A}) \equiv \hat{\Delta}(\beta, -1) =$$
$$= \frac{1}{\pi}\int \exp(-\gamma^*\hat{a})\left(\frac{1}{\pi}\int|\alpha\rangle\langle\alpha|\,\mathrm{d}^2\alpha\right)\exp(\gamma\hat{a}^+)\exp(\gamma^*\beta - \gamma\beta^*)\,\mathrm{d}^2\gamma =$$
$$= \frac{1}{\pi^2}\int|\alpha\rangle\langle\alpha|\,\mathrm{d}^2\alpha\int\exp[\gamma(\alpha^* - \beta^*) - \gamma^*(\alpha - \beta)]\,\mathrm{d}^2\gamma = |\beta\rangle\langle\beta|, \tag{16.16}$$

that is $\hat{\Delta}(\beta, -1)$ is the projection operator onto the coherent state $|\beta\rangle$.

Making use of the completeness property (16.13) of the $\hat{\Delta}$-operator, we can also decompose any operator $\hat{A}$ in terms of $\hat{\Delta}$

$$\hat{A} = \frac{1}{\pi}\int a(\beta, \tilde{\Omega})\,\hat{\Delta}(\beta, \Omega)\,\mathrm{d}^2\beta, \tag{16.17}$$

where

$$a(\beta, \tilde{\Omega}) = \mathrm{Tr}\,\{\hat{\Delta}(\beta, \tilde{\Omega})\,\hat{A}\}. \tag{16.18}$$

As $\hat{\Delta}(\beta, \Omega) = \pi\{\delta(\hat{a} - \beta)\}_{\Omega}$, equation (16.17) provides the Ω-ordered form of the operator $\hat{A}$. Using (16.13) again we have for the trace of the product of two operators $\hat{A}$ and $\hat{B}$

$$\mathrm{Tr}\,\{\hat{A}\hat{B}\} = \frac{1}{\pi^2}\iint a(\beta, \Omega)\,b(\gamma, \tilde{\Omega})\,\mathrm{Tr}\,\{\hat{\Delta}(\beta, \tilde{\Omega})\,\hat{\Delta}(\gamma, \Omega)\}\,\mathrm{d}^2\beta\,\mathrm{d}^2\gamma =$$
$$= \frac{1}{\pi}\int a(\beta, \Omega)\,b(\beta, \tilde{\Omega})\,\mathrm{d}^2\beta. \tag{16.19}$$

If $\hat{B}$ is identified with the density matrix $\hat{\varrho}$ we can write

$$\mathrm{Tr}\,\{\hat{\varrho}\hat{A}\} = \frac{1}{\pi}\int \Phi(\beta, \tilde{\Omega})\,a(\beta, \Omega)\,\mathrm{d}^2\beta, \tag{16.20a}$$

where

$$\Phi(\beta, \tilde{\Omega}) = \mathrm{Tr}\,\{\hat{\varrho}\hat{\Delta}(\beta, \tilde{\Omega})\} = \pi\langle\{\delta(\hat{a} - \beta)\}_{\tilde{\Omega}}\rangle \tag{16.21a}$$

and from (16.17)

$$\hat{\varrho} = \frac{1}{\pi}\int \Phi(\beta, \tilde{\Omega})\,\hat{\Delta}(\beta, \Omega)\,\mathrm{d}^2\beta. \tag{16.22a}$$

Considering the s-ordering as a special case of the Ω-ordering we then have

$$\mathrm{Tr}\,\{\hat{\varrho}\hat{A}\} = \frac{1}{\pi}\int \Phi(\beta, -s)\,a(\beta, s)\,\mathrm{d}^2\beta, \tag{16.20b}$$

where

$$\Phi(\beta, -s) = \mathrm{Tr}\,\{\hat{\varrho}\hat{\Delta}(\beta, -s)\} \tag{16.21b}$$

and

$$\hat{\varrho} = \frac{1}{\pi}\int \Phi(\beta, -s)\,\hat{\Delta}(\beta, s)\,\mathrm{d}^2\beta. \tag{16.22b}$$

Hence, the expectation value of an operator $\hat{A}$ can be calculated as a classical expectation value of the classical function $a(\beta, \Omega)$ corresponding to $\hat{A}$ via $\tilde{\Omega}$-ordering with the quasi-probability $\Phi(\beta, \tilde{\Omega})$ corresponding to $\hat{\varrho}$ via the Ω-ordering as follows from (16.17) and (16.22a). This agrees with the particular cases discussed in Sec. 13.2.

Putting $s = -1$ in (16.22b) and making use of (16.16) we obtain the Glauber-Sudarshan representation of the density matrix

$$\hat{\varrho} = \int \Phi_{\mathcal{N}}(\beta) \,|\, \beta\rangle \langle \beta \,|\, \mathrm{d}^2\beta, \tag{16.23}$$

where $\Phi_{\mathcal{N}}(\beta) \equiv \Phi(\beta, 1)/\pi$.

If $\hat{A} \equiv 1$, then $a(\beta, \Omega) \equiv 1$ from (16.18) with the use of (16.15), and equation (16.20a) provides the normalization condition

$$\operatorname{Tr} \hat{\varrho} = 1 = \frac{1}{\pi} \int \Phi(\beta, \tilde{\Omega})\, \mathrm{d}^2\beta \tag{16.24}$$

and $\hat{\varrho} = \hat{\varrho}^+$ implies $\Phi^*(\beta, \tilde{\Omega}) = \Phi(\beta, \tilde{\Omega}^*)$ $[\Phi^*(\beta, s) = \Phi(\beta, s^*)]$.

Writing (16.21a) in the form

$$\Phi(\alpha, \tilde{\Omega}) = \frac{1}{\pi} \int \operatorname{Tr} \{\hat{\varrho}\hat{D}(\beta, \tilde{\Omega})\} \exp(\alpha\beta^* - \alpha^*\beta)\, \mathrm{d}^2\beta, \tag{16.25}$$

where we have used (16.7), it is obvious that $\operatorname{Tr} \{\hat{\varrho}\hat{D}(\beta, \tilde{\Omega})\}$ is the characteristic function $C(\beta, \tilde{\Omega})$ of the quasi-distribution $\Phi(\alpha, \tilde{\Omega})$ and so

$$\Phi(\alpha, \tilde{\Omega}) = \frac{1}{\pi} \int C(\beta, \tilde{\Omega}) \exp(\alpha\beta^* - \alpha^*\beta)\, \mathrm{d}^2\beta, \tag{16.26a}$$

$$C(\beta, \tilde{\Omega}) = \frac{1}{\pi} \int \Phi(\alpha, \tilde{\Omega}) \exp(-\alpha\beta^* + \alpha^*\beta)\, \mathrm{d}^2\alpha. \tag{16.26b}$$

The characteristic function $C(\beta, \tilde{\Omega})$ is bounded as follows

$$|\, C(\beta, \tilde{\Omega}) \,| = |\, \tilde{\Omega} \,|\; |\operatorname{Tr} \{\hat{\varrho}\hat{D}(\beta)\} \,| \leqq |\, \tilde{\Omega} \,|.$$

The Ω-ordered moments of $\Phi(\alpha, \Omega)$ can be calculated in the following way

$$\left.\frac{\partial^{k+l}}{\partial\beta^k\, \partial(-\beta^*)^l} C(\beta, \Omega)\right|_{\beta=\beta^*=0} = \operatorname{Tr}\left\{\hat{\varrho} \left.\frac{\partial^{k+l}}{\partial\beta^k\, \partial(-\beta^*)^l} \hat{D}(\beta, \Omega)\right\}\right|_{\beta=\beta^*=0} =$$

$$= \operatorname{Tr} \{\hat{\varrho}\{\hat{a}^{+k}\hat{a}^l\}_\Omega\} = \frac{1}{\pi} \int \Phi(\alpha, \Omega)\, \alpha^{*k}\alpha^l\, \mathrm{d}^2\alpha = \langle \alpha^{*k}\alpha^l \rangle_\Omega, \tag{16.27}$$

where we have used (16.6) and (16.26b). If the inverse relation to (16.7) is used we also obtain

$$\{\hat{a}^{+k}\hat{a}^l\}_\Omega = \left.\frac{\partial^{k+l}}{\partial\beta^k\, \partial(-\beta^*)^l} \hat{D}(\beta, \Omega)\right|_{\beta=\beta^*=0} =$$

$$= \left.\frac{\partial^{k+l}}{\partial\beta^k\, \partial(-\beta^*)^l} \frac{1}{\pi} \int \Delta(\alpha, \Omega) \exp(-\beta^*\alpha + \beta\alpha^*)\, \mathrm{d}^2\alpha\right|_{\beta=\beta^*=0} =$$

$$= \frac{1}{\pi} \int \alpha^{*k}\alpha^l \hat{\Delta}(\alpha, \Omega)\, \mathrm{d}^2\alpha. \tag{16.28}$$

Considering these equations for the s-ordering specified by $\Omega(\alpha^*, \alpha) = \exp(s|\alpha|^2/2)$ we obtain from (16.26b) [(16.26a)] equations (13.72), (13.73) and (13.78) [(13.74), (13.75) and (13.76)] and from (16.27) equations (13.80), (13.81) and (13.77) for $s = 1, -1$ and 0 respectively with the replacements $\Phi(\alpha, 1)/\pi \equiv \Phi_{\mathcal{N}}(\alpha)$, $\Phi(\alpha, -1)/\pi \equiv \equiv \Phi_{\mathcal{A}}(\alpha)$, $\Phi(\alpha, 0)/\pi \equiv \Phi_{sym}(\alpha)$, $C(\beta, 1) \equiv C_{\mathcal{N}}(\beta)$, $C(\beta, -1) \equiv C_{\mathcal{A}}(\beta)$, and $C(\beta, 0) \equiv C(\beta)$.

The present general scheme of the correspondence between functions of q-numbers and functions of c-numbers can provide an insight into the problem of the singularities appearing in the quasi-distributions. Consider the s-ordering only in the following. We can decompose the operator $\hat{A}$ in (16.11) into an s-ordered series by expanding the operator $\hat{D}^{-1}$,

$$\hat{A} = \frac{1}{\pi}\int \mathrm{Tr}\,\{\hat{D}(\beta, -s)\,\hat{A}\}\exp\left(\frac{s}{2}|\beta|^2\right)\exp(-\beta\hat{a}^+ + \beta^*\hat{a})\,\mathrm{d}^2\beta = \\ = \sum_n\sum_m \{\hat{a}^{+n}\hat{a}^m\}_s A_{nm}, \tag{16.29}$$

where we have used (16.9), substituted $s \to -s$ and where

$$A_{nm} = \frac{1}{\pi n!\,m!}\int \mathrm{Tr}\,\{\hat{D}(\beta, -s)\,\hat{A}\}(-\beta)^n(\beta^*)^m\,\mathrm{d}^2\beta = \\ = \frac{1}{\pi n!\,m!}\int \mathrm{Tr}\,\{\hat{D}(\beta)\,\hat{A}\}\exp\left(-\frac{s}{2}|\beta|^2\right)(-\beta)^n(\beta^*)^m\,\mathrm{d}^2\beta. \tag{16.30}$$

By applying the Schwarz inequality to (16.30),

$$\left|\frac{1}{\pi}\int \tilde{a}(\beta)\,\tilde{b}(\beta)\,\mathrm{d}^2\beta\right| \leq \left\{\frac{1}{\pi}\int|\tilde{a}(\beta)|^2\,\mathrm{d}^2\beta\right\}^{1/2}\left\{\frac{1}{\pi}\int|\tilde{b}(\beta)|^2\,\mathrm{d}^2\beta\right\}^{1/2}, \tag{16.31}$$

where (16.2) is used and $\tilde{b}(\beta) = \exp(-s|\beta|^2/2)(-\beta)^n(\beta^*)^m$, we see that, as a consequence of the presence of the factor $\exp(-s|\beta|^2/2)$, the coefficients (16.30) are finite for all bounded operators $\hat{A}$ ($\|\hat{A}\|^2 = \mathrm{Tr}\,\{\hat{A}^+\hat{A}\}$, cf. (16.4)) when $\mathrm{Re}\, s > 0$. Therefore it can be said that these coefficients are finite for orderings which are closer to the normal than to the antinormal order. A further investigation of the convergence of the series (16.29) leads to the result (Cahill and Glauber (1969)) that all bounded operators possess convergent s-ordered power series for $\mathrm{Re}\, s > 1/2$, i.e. when the ordering is closer to normal than to symmetric. Note that in the $(\hat{q}, \hat{p})$-representation the exponential factor $\exp(\mathrm{i}\hat{q}\hat{p}/2)$ replaces the factor $\exp(-s|\beta|^2/2)$, as a consequence of the fact that the commutator $[\hat{q}, \hat{p}] = \mathrm{i}\hbar$ is purely imaginary. Thus the integrals for various orderings differ from one another only by unimodular factors in their integrands and a bounded operator that possesses an expansion in one ordering is likely to possess it in another. Such a class of operators is approximately the same as the class for which the symmetrically ordered expansion (16.29) with $s = 0$ is appropriate. For some further details concerning the $(\hat{q}, \hat{p})$-reqresentation in this connection we refer the reader to the original papers by Cahill and Glauber (1969) and by Agarwal and Wolf (1970). In general it is seen from (16.30) that the broadest class of operators possess normally ordered expansions while a relatively smaller class of operators possess antinormally ordered expansions. According to our earlier results this means that, for the density matrix $\hat{\varrho} \equiv \hat{A}$, the class of physical fields which possess the diagonal representation of the density matrix with the weight function $\Phi_{\mathcal{N}}(\alpha) = \pi^{-1}\varrho^{(\mathcal{A})}(\alpha^*, \alpha)$ as an ordinary function is substantially smaller than the class of fields for which the function $\Phi_{\mathcal{A}}(\alpha) = \pi^{-1}\varrho^{(\mathcal{N})}(\alpha^*, \alpha)$ is an ordinary function (in fact for all physical fields $\Phi_{\mathcal{A}}(\alpha)$ is an ordinary function).

A characteristic feature of the expansion (16.22b) is that both the values s and $-s$ appear in it. We have seen that $\hat{\Delta}(\alpha, -1) = |\alpha\rangle\langle\alpha|$, while one can easily verify by a similar calculation such as in (16.16) that the operator $\hat{\Delta}(\alpha, 1)$ (which can be used for decompositions of the density matrix with the weight function $\Phi_{\mathcal{A}}(\alpha)$) is singular. In the first case, when $\hat{\Delta}(\alpha, -1)$ is regular, $\Phi_{\mathcal{N}}(\alpha)$ may have singularities, in the second case, when $\Phi_{\mathcal{A}}(\alpha)$ is regular, $\hat{\Delta}(\alpha, 1)$ is singular. Thus extreme smoothness of one quantity leads to singular behaviour of the other. Only for symmetric ordering for which $s = 0$, are both the quantities $\hat{\Delta}(\alpha, 0)$ and $\Phi_{sym}(\alpha) = \pi^{-1}\Phi(\alpha, 0)$ regular (although the operator $\hat{\Delta}(\alpha, 0)$ is not bounded, it is finite in the sense that $\operatorname{Max}_{\beta} \langle\beta|\hat{\Delta}(\alpha, 0)|\beta\rangle$ is finite for β finite).

16.2 Connecting relations for different orderings

If we consider two orderings Ω_1 and Ω_2 we can write on the basis of (16.5)

$$\hat{\Delta}(\alpha, \Omega_1) = \frac{1}{\pi}\int \Omega_1(\beta^*, \beta)\,\hat{D}(\beta)\exp(\alpha\beta^* - \alpha^*\beta)\,d^2\beta, \tag{16.32a}$$

$$\hat{\Delta}(\alpha, \Omega_2) = \frac{1}{\pi}\int \Omega_2(\beta^*, \beta)\,\hat{D}(\beta)\exp(\alpha\beta^* - \alpha^*\beta)\,d^2\beta. \tag{16.32b}$$

Expressing the operator $\hat{D}(\beta)$ by means of the inverse relation to (16.32a) and substituting it into (16.32b) we arrive at

$$\hat{\Delta}(\alpha, \Omega_2) = \int \hat{\Delta}(\beta, \Omega_1)\,K_{21}(\alpha - \beta)\,d^2\beta, \tag{16.33}$$

where

$$K_{21}(\beta) = \frac{1}{\pi^2}\int \tilde{\Omega}_1(\gamma^*, \gamma)\,\Omega_2(\gamma^*, \gamma)\exp(\beta\gamma^* - \beta^*\gamma)\,d^2\gamma. \tag{16.34}$$

Multiplying (16.33) by the density operator $\hat{\varrho}$, taking the trace and making use of (16.21a) with $\tilde{\Omega} \to \Omega$ we obtain

$$\Phi(\alpha, \Omega_2) = \int \Phi(\beta, \Omega_1)\,K_{21}(\alpha - \beta)\,d^2\beta. \tag{16.35}$$

The corresponding relation between the characteristic functions $C(\alpha, \Omega_1)$ and $C(\alpha, \Omega_2)$ is obviously

$$C(\alpha, \Omega_2) = C(\alpha, \Omega_1)\,\tilde{\Omega}_1(\alpha^*, \alpha)\,\Omega_2(\alpha^*, \alpha). \tag{16.36}$$

Some differential relations between different orderings can also be derived. Writing

$$\hat{\Delta}(\alpha, \Omega_2) = \frac{1}{\pi} \int \tilde{\Omega}_1(\beta^*, \beta)\, \Omega_2(\beta^*, \beta)\, \Omega_1(\beta^*, \beta)\, \hat{D}(\beta) \exp(\alpha\beta^* - \alpha^*\beta)\, d^2\beta =$$

$$= \mathscr{L}_{21}\left(\frac{\partial}{\partial\alpha}, -\frac{\partial}{\partial\alpha^*}\right) \hat{\Delta}(\alpha, \Omega_1), \tag{16.37}$$

where

$$\mathscr{L}_{21}(\gamma^*, \gamma) = \tilde{\Omega}_1(\gamma^*, \gamma)\, \Omega_2(\gamma^*, \gamma), \tag{16.38}$$

we have

$$\Phi(\alpha, \Omega_2) = \mathscr{L}_{21}\left(\frac{\partial}{\partial\alpha}, -\frac{\partial}{\partial\alpha^*}\right) \Phi(\alpha, \Omega_1). \tag{16.39}$$

These general relations can be used for the s-ordering and as a special case relations between normal, symmetric and antinormal orderings can be derived. Since $\Omega(\alpha^*, \alpha) = \exp(s\,|\alpha|^2/2)$ in this case we obtain from (16.36)

$$C(\alpha, s_2) = C(\alpha, s_1) \exp\left(\tfrac{1}{2}(s_2 - s_1)\,|\alpha|^2\right), \tag{16.40a}$$

$$\hat{D}(\alpha, s_2) = \hat{D}(\alpha, s_1) \exp\left(\tfrac{1}{2}(s_2 - s_1)\,|\alpha|^2\right) \tag{16.40b}$$

(recalling that $C(\alpha, s) = \mathrm{Tr}\,\{\hat{\varrho}\hat{D}(\alpha, s)\}$). Calculating $K_{21}(\beta)$ from (16.34),

$$K_{21}(\beta) = \frac{1}{\pi^2} \int \exp\left[\tfrac{1}{2}(s_2 - s_1)\,|\gamma|^2\right] \exp(\beta\gamma^* - \beta^*\gamma)\, d^2\gamma$$

and using the integral

$$\int \exp(-s\,|\gamma|^2 + \alpha\gamma^* + \alpha'\gamma)\, d^2\gamma = \sum_n \sum_m \frac{\alpha^n \alpha'^m}{n!\, m!} \int \exp(-s\,|\gamma|^2)\, \gamma^{*n}\gamma^m\, d^2\gamma =$$

$$= \frac{\pi}{s} \exp\left(\frac{\alpha\alpha'}{s}\right), \qquad \mathrm{Re}\, s > 0 \tag{16.41}$$

(where we have used the integral (13.17) with $s = (s_1 - s_2)/2$, $\alpha = \beta$ and $\alpha' = -\beta^*$), we obtain

$$K_{21}(\beta) = \frac{2}{\pi(s_1 - s_2)} \exp\left(-\frac{2\,|\beta|^2}{s_1 - s_2}\right), \qquad \mathrm{Re}\, s_1 > \mathrm{Re}\, s_2. \tag{16.42}$$

Thus from (16.35)

$$\Phi(\alpha, s_2) = \frac{2}{\pi(s_1 - s_2)} \int \Phi(\beta, s_1) \exp\left(-\frac{2\,|\alpha - \beta|^2}{s_1 - s_2}\right) d^2\beta, \qquad \mathrm{Re}\, s_1 > \mathrm{Re}\, s_2. \tag{16.43}$$

As $\Phi(\alpha, s) = \mathrm{Tr}\,\{\hat{\varrho}\hat{\Delta}(\alpha, s)\}$ the same relation holds between $\hat{\Delta}(\alpha, s_2)$ and $\hat{\Delta}(\alpha, s_1)$ (see (16.33)). Relations (13.85a, b) and (13.68) follow from (16.43) with $s_1 = 0$, $s_2 = -1$; $s_1 = 1$, $s_2 = 0$ and $s_1 = 1$, $s_2 = -1$ respectively. It can be seen from the form of (16.43) that this Gaussian convolution tends to smooth out any unruly behaviour of the function $\Phi(\beta, s_1)$. For example for the coherent state $|\gamma\rangle$ and $s_1 = 1$, $\pi^{-1}\Phi(\beta, 1) \equiv \Phi_{\mathcal{N}}(\beta) = \delta(\beta - \gamma)$ and $\Phi(\alpha, s) = [2/(1-s)] \exp[-2|\alpha - \gamma|^2/ /(1-s)]$, which is a regular function for $s \neq 1$; for $s = -1$, $\pi^{-1}\Phi(\alpha, -1) \equiv \equiv \Phi_{\mathcal{A}}(\alpha) = \pi^{-1} \exp(-|\alpha - \gamma|^2)$, for $s = 0$, $\pi^{-1}\Phi(\alpha, 0) \equiv \Phi_{sym}(\alpha) = 2\pi^{-1} \times \times \exp(-2|\alpha - \gamma|^2)$.

The relation (16.43) can in principle be inverted (in general in a space of generalized functions) as a convolution integral or, in some cases, in terms of a series of the Hermite polynomials H_n (Peřina and Horák (1970)). More generally, making use of (16.39) it formally follows that

$$\Phi(\alpha, s_1) = \exp\left(-\frac{s_1 - s_2}{2} \frac{\partial^2}{\partial\alpha\, \partial\alpha^*}\right) \Phi(\alpha, s_2), \tag{16.44}$$

which is just the operator inversion of (16.43).

We can also derive the corresponding relations between different s-ordered moments $\langle\{\hat{a}^{+k}\hat{a}^l\}_s\rangle \equiv \langle\hat{a}^{+k}\hat{a}^l\rangle_s$ and obtain from (16.43) for $\operatorname{Re} s_1 > \operatorname{Re} s_2$

$$\begin{aligned}
\langle\hat{a}^{+k}\hat{a}^l\rangle_{s_2} &= \langle\alpha^{*k}\alpha^l\rangle_{s_2} = \\
&= \frac{1}{\pi}\int \Phi(\beta, s_1)\, d^2\beta \frac{2}{\pi(s_1 - s_2)} \int \alpha^{*k}\alpha^l \exp\left(-\frac{2|\alpha - \beta|^2}{s_1 - s_2}\right) d^2\alpha = \\
&= \frac{1}{\pi}\int \Phi(\beta, s_1)\, d^2\beta \frac{2}{\pi(s_1 - s_2)} (\gamma^* + \beta^*)^k (\gamma + \beta)^l \exp\left(-\frac{2|\gamma|^2}{s_1 - s_2}\right) d^2\gamma,
\end{aligned} \tag{16.45}$$

where the substitution $\alpha - \beta = \gamma$ has been used. Making use of the binomial theorem and calculating the integral, we obtain

$$\langle\hat{a}^{+k}\hat{a}^l\rangle_{s_2} = \frac{k!}{l!}\left(\frac{s_1 - s_2}{2}\right)^k \left\langle \hat{a}^{l-k} L_k^{l-k}\left(\frac{2\hat{a}^+\hat{a}}{s_2 - s_1}\right)\right\rangle_{s_1}, \quad l \geqq k, \tag{16.46a}$$

$$= \frac{l!}{k!}\left(\frac{s_1 - s_2}{2}\right)^l \left\langle \hat{a}^{+k-l} L_l^{k-l}\left(\frac{2\hat{a}^+\hat{a}}{s_2 - s_1}\right)\right\rangle_{s_1}, \quad l \leqq k; \tag{16.46b}$$

the same relation can be derived with the help of (16.27). First

$$\begin{aligned}
\hat{D}(\beta, s_2) &= \hat{D}(\beta, s_1) \exp\left[\tfrac{1}{2}(s_2 - s_1)|\beta|^2\right] = \\
&= \{\exp(\beta\hat{a}^+ - \beta^*\hat{a} + \tfrac{1}{2}(s_2 - s_1)|\beta|^2)\}_{s_1} = \\
&= \left\{\sum_{m=0}^{\infty} \frac{\beta^{*m}}{m!}(-\hat{a} + \tfrac{1}{2}(s_2 - s_1)\beta)^m \exp(\beta\hat{a}^+)\right\}_{s_1}
\end{aligned} \tag{16.47a}$$

$$= \left\{\sum_{n=0}^{\infty} \frac{\beta^n}{n!}(\hat{a}^+ + \tfrac{1}{2}(s_2 - s_1)\beta^*)^n \exp(-\beta^*\hat{a})\right\}_{s_1}. \tag{16.47b}$$

Making use of the identity (Morse and Feshbach (1953), Chapter 6)

$$(1 + t)^\mu \exp(-xt) = \sum_{n=0}^{\infty} \frac{t^n}{\Gamma(\mu + 1)} L_n^{\mu - n}(x), \tag{16.48}$$

which leads to

$$\hat{D}(\beta, s_2) = \left\{ \sum_{m=0}^{\infty} \sum_{n=0}^{\infty} (-1)^m \frac{\beta^{*m}}{m!} \frac{\hat{a}^{m-n}}{m!} [\tfrac{1}{2}(s_1 - s_2)\beta]^n L_n^{m-n}\left(\frac{2\hat{a}^+\hat{a}}{s_2 - s_1}\right)\right\}_{s_1} \tag{16.49a}$$

$$= \left\{ \sum_{n=0}^{\infty} \sum_{m=0}^{\infty} (-1)^m \frac{\beta^n}{n!} \frac{\hat{a}^{+n-m}}{n!} [\tfrac{1}{2}(s_1 - s_2)\beta^*]^m L_m^{n-m}\left(\frac{2\hat{a}^+\hat{a}}{s_2 - s_1}\right)\right\}_{s_1}, \tag{16.49b}$$

and applying (16.27) we arrive at (16.46a, b) again.

The relations between moments corresponding to normal, symmetric and antinormal orderings can be obtained from (16.46). For example when $s_2 = 1$ and $s_1 = -1$ we have

$$\langle \hat{a}^{+k}\hat{a}^l \rangle_{\mathscr{N}} = \frac{k!}{l!}(-1)^k \langle \hat{a}^{l-k} L_k^{l-k}(\hat{a}^+\hat{a})\rangle_{\mathscr{A}}, \qquad l \geqq k, \tag{16.50}$$

when $s_2 = -1$ and $s_1 = 1$

$$\langle \hat{a}^{+k}\hat{a}^l \rangle_{\mathscr{A}} = \frac{k!}{l!} \langle \hat{a}^{l-k} L_k^{l-k}(-\hat{a}^+\hat{a})\rangle_{\mathscr{N}}, \qquad l \geqq k, \tag{16.51}$$

and when $s_2 = 0$ and $s_1 = 1$

$$\langle \hat{a}^{+k}\hat{a}^l \rangle_{sym} = \frac{k!}{l!}(\tfrac{1}{2})^k \langle \hat{a}^{l-k} L_k^{l-k}(-2\hat{a}^+\hat{a})\rangle_{\mathscr{N}}, \qquad l \geqq k, \tag{16.52}$$

etc.

Some further results concerning the Ω-ordering and s-ordering can be found in the original papers by Agarwal and Wolf (1970) and by Cahill and Glauber (1969).

16.3 Multimode description

The general theory just developed may easily be extended to systems with any number of degrees of freedom (finite or denumerably infinite) in the same way as in Chapter 13. For example a multimode operator $\hat{A}(\{\hat{a}_\lambda\}) \equiv \hat{A}(\{\hat{a}_\lambda^+\}, \{\hat{a}_\lambda\})$ can be decomposed, in analogy to (16.17) and (16.18), as

$$\hat{A}(\{\hat{a}_\lambda\}) = \int a(\{\beta_\lambda\}, \tilde{\Omega})\, \hat{\Delta}(\{\beta_\lambda\}, \Omega) \prod_\lambda \frac{d^2\beta_\lambda}{\pi}, \tag{16.53}$$

where

$$a(\{\beta_\lambda\}, \tilde{\Omega}) = \mathrm{Tr}\,\{\hat{\Delta}(\{\beta_\lambda\}, \tilde{\Omega})\, \hat{A}\} \tag{16.54}$$

and

$$\hat{\Delta}(\{\alpha_\lambda\}, \Omega) = \{\prod_\lambda \pi\delta(\hat{a}_\lambda - \alpha_\lambda)\}_\Omega =$$

$$= \int \Omega(\{\beta_\lambda^*\}, \{\beta_\lambda\}) \prod_\lambda \exp[\beta_\lambda(\hat{a}_\lambda^+ - \alpha_\lambda^*) - \beta_\lambda^*(\hat{a}_\lambda - \alpha_\lambda)] \frac{d^2\beta_\lambda}{\pi} \tag{16.55}$$

and the counterpart to (16.20a) is

$$\operatorname{Tr}\{\hat{\varrho}\hat{A}\} = \int \Phi(\{\beta_\lambda\}, \tilde{\Omega})\, a(\{\beta_\lambda\}, \Omega) \prod_\lambda \frac{d^2\beta_\lambda}{\pi}, \tag{16.56}$$

where Φ is given by (16.54) with $\hat{A}$ replaced by $\hat{\varrho}$. The density operator $\hat{\varrho}$ can be decomposed in the form (16.53) with $a(\{\beta_\lambda\}, \tilde{\Omega})$ replaced by $\Phi(\{\beta_\lambda\}, \tilde{\Omega})$, etc.

Next we shall develop another generalization of the one-mode results (Peřina and Horák (1969b, 1970)) in a manner consistent with the description of the field in terms of the quantities $\hat{\mathbf{A}}(x)$ and $\hat{n}_{Vt}$ defined by (11.23) and (11.24).

First we need to derive the s-ordered form of the operator $\exp(ix\hat{a}^+\hat{a})$. Substituting this operator into (16.30) we have

$$A_{nm} = \frac{1}{\pi n!\, m!} \int \operatorname{Tr}\{\exp(\beta\hat{a}^+ - \beta^*\hat{a}) \exp(ix\hat{a}^+\hat{a})\} \exp\left(-\frac{s}{2}|\beta|^2\right)(-\beta)^n \times$$
$$\times\, \beta^{*m}\, d^2\beta = \frac{1}{\pi^2 n!\, m!} \iint \langle\alpha|\exp(\beta\hat{a}^+)\exp(ix\hat{a}^+\hat{a})\exp(-\beta^*\hat{a})|\alpha\rangle \times$$
$$\times \exp\left(-\frac{s-1}{2}|\beta|^2\right)(-\beta)^n \beta^{*m}\, d^2\beta\, d^2\alpha, \tag{16.57}$$

where we have used (16.9) and the cyclic property of the trace. If we use the normal form of the operator $\exp(ix\hat{a}^+\hat{a})$ given by (14.17), we obtain

$$A_{nm} = \frac{1}{\pi^2 n!\, m!} \iint \exp(\alpha^*\beta) \langle\alpha|\mathcal{N}\{\exp[(e^{ix}-1)\hat{a}^+\hat{a}]\}|\alpha\rangle \exp(-\alpha\beta^*) \times$$
$$\times \exp\left(-\frac{s-1}{2}|\beta|^2\right)(-\beta)^n \beta^{*m}\, d^2\beta\, d^2\alpha = \frac{1}{\pi^2 n!\, m!} \times$$
$$\times \iint \exp\left(\alpha^*\beta - \alpha\beta^* + (e^{ix}-1)|\alpha|^2 - \frac{s-1}{2}|\beta|^2\right)(-\beta)^n \beta^{*m}\, d^2\beta\, d^2\alpha =$$
$$= \frac{(-1)^n}{n!} \delta_{nm} \frac{1}{1-\exp(ix)} \frac{1}{\left(\dfrac{s-1}{2} + \dfrac{1}{1-\exp(ix)}\right)^{n+1}}, \tag{16.58}$$

where the integrals (16.41) and (13.17) have been used. Substituting this result into (16.29) we arrive at

$$\exp(ix\hat{a}^+\hat{a}) = \sum_n \{\hat{a}^{+n}\hat{a}^n\}_s \frac{(-1)^n}{n!} \times$$
$$\times \frac{1}{\left(\dfrac{s-1}{2} + \dfrac{1}{1-\exp(ix)}\right)^n} \frac{1}{\dfrac{s-1}{2}(1-\exp(ix)) + 1} =$$
$$= \frac{2}{1 + e^{ix} + s - s\,e^{ix}} \left\{\exp\left[-\frac{2(1-e^{ix})\hat{a}^+\hat{a}}{1 + e^{ix} + s - s\,e^{ix}}\right]\right\}_s. \tag{16.59}$$

Let us note that, because the operator $\hat{b} = \hat{a} - \beta$ (β is a complex amplitude), describing the superposition of chaotic and coherent fields, obeys with $\hat{b}^+$ the same commutation rule as the operators $\hat{a}$ and $\hat{a}^+$ obey, the relation (16.59) also holds for such operators $\hat{b}$ and $\hat{b}^+$.

First consider the normalization volume of the field; we obtain, introducing the number operator $\hat{n}$ given by (11.24b),

$$\langle \exp(\mathrm{i}x\hat{n})\rangle = \langle \prod_{\lambda}^{M} \exp(\mathrm{i}x\hat{a}_\lambda^+\hat{a}_\lambda)\rangle =$$

$$= \left[\frac{2}{1 + \mathrm{e}^{\mathrm{i}x} + s - s\,\mathrm{e}^{\mathrm{i}x}}\right]^M \left\langle \exp\left[\frac{2\hat{n}(\mathrm{e}^{\mathrm{i}x} - 1)}{1 + \mathrm{e}^{\mathrm{i}x} + s - s\,\mathrm{e}^{\mathrm{i}x}}\right]\right\rangle_s =$$

$$= \left[1 - \frac{1-s}{2}(1 - \mathrm{e}^{\mathrm{i}x})\right]^{-M} \left\langle \exp\left[\frac{-\dfrac{2\hat{n}}{1-s}\dfrac{1-s}{2}(1-\mathrm{e}^{\mathrm{i}x})}{1 - \dfrac{1-s}{2}(1 - \mathrm{e}^{\mathrm{i}x})}\right]\right\rangle_s, \qquad (16.60)$$

where M is the number of modes. This result is clearly valid for an arbitrary volume V (whose linear dimensions are much larger than the wavelength) by the same argument which led to (14.15), (14.16) and (14.17).

The substitution

$$\frac{2(\mathrm{e}^{\mathrm{i}x} - 1)}{1 + \mathrm{e}^{\mathrm{i}x} + s - s\,\mathrm{e}^{\mathrm{i}x}} = \mathrm{i}y \qquad (16.61)$$

in (16.60) gives us the s-ordered characteristic function

$$\langle \exp(\mathrm{i}y\hat{n})\rangle_s \equiv \langle \exp(\mathrm{i}yW)\rangle_s = \left(1 - \frac{1-s}{2}\mathrm{i}y\right)^{-M} \left\langle \left[\frac{1 + \dfrac{1+s}{2}\mathrm{i}y}{1 - \dfrac{1-s}{2}\mathrm{i}y}\right]^n \right\rangle =$$

$$= \left(1 - \frac{1-s}{2}\mathrm{i}y\right)^{-M} \left\langle \left[\frac{2}{(1-s)\left(1 - \dfrac{1-s}{2}\mathrm{i}y\right)} - \frac{1+s}{1-s}\right]^n \right\rangle. \qquad (16.62)$$

Considering two orderings with $s = s_1$ and $s = s_2$ we can write from (16.60)

$$\left[\frac{2}{1 + \mathrm{e}^{\mathrm{i}x} + s_1 - s_1\,\mathrm{e}^{\mathrm{i}x}}\right]^M \left\langle \exp\left[\frac{2\hat{n}(\mathrm{e}^{\mathrm{i}x} - 1)}{1 + \mathrm{e}^{\mathrm{i}x} + s_1 - s_1\,\mathrm{e}^{\mathrm{i}x}}\right]\right\rangle_{s_1} =$$

$$= \left[\frac{2}{1 + \mathrm{e}^{\mathrm{i}x} + s_2 - s_2\,\mathrm{e}^{\mathrm{i}x}}\right]^M \left\langle \exp\left[\frac{2\hat{n}(\mathrm{e}^{\mathrm{i}x} - 1)}{1 + \mathrm{e}^{\mathrm{i}x} + s_2 - s_2\,\mathrm{e}^{\mathrm{i}x}}\right]\right\rangle_{s_2}, \qquad (16.63)$$

from which, using the substitution (16.61) with $s \to s_2$, it follows that

$$\langle \exp(\mathrm{i}y\hat{n})\rangle_{s_2} = \left(1 + \frac{s_2 - s_1}{2}\mathrm{i}y\right)^{-M} \left\langle \exp\left[\frac{\mathrm{i}y\hat{n}}{1 + \dfrac{s_2 - s_1}{2}\mathrm{i}y}\right]\right\rangle_{s_1}, \qquad (16.64)$$

which is the generating function for the Laguerre polynomials L_r^{M-1} (Morse and Feshbach (1953), Chapter 6) so that

$$\langle \exp(\mathrm{i}y\hat{n})\rangle_{s_2} = \sum_{r=0}^{\infty} \frac{(\mathrm{i}y)^r}{\Gamma(r+M)} \left(\frac{s_1 - s_2}{2}\right)^r \left\langle L_r^{M-1}\left(\frac{2\hat{n}}{s_2 - s_1}\right)\right\rangle_{s_1}. \tag{16.65}$$

From (16.64) we can obtain the relation between the distributions $P(W, s_2)$ and $P(W, s_1)$ using the Fourier integral and the residue theorem

$$\begin{aligned} P(W, s_2) &= \frac{1}{2\pi} \int_{-\infty}^{+\infty} \langle \exp(\mathrm{i}y\hat{n})\rangle_{s_2} \exp(-\mathrm{i}yW)\,\mathrm{d}y = \\ &= \frac{1}{2\pi} \int_0^{\infty} P(W', s_1)\,\mathrm{d}W' \int_{-\infty}^{+\infty} \left(1 - \frac{s_1 - s_2}{2}\mathrm{i}y\right)^{-M} \times \\ &\times \exp\left[\frac{\mathrm{i}yW'}{1 - \frac{s_1 - s_2}{2}\mathrm{i}y} - \mathrm{i}yW\right] \mathrm{d}y = \\ &= \frac{2}{s_1 - s_2} \int_0^{\infty} \left(\frac{W}{W'}\right)^{(M-1)/2} \exp\left[-\frac{2(W + W')}{s_1 - s_2}\right] \times \\ &\times I_{M-1}\left(4\frac{(WW')^{1/2}}{s_1 - s_2}\right) P(W', s_1)\,\mathrm{d}W', \qquad \mathrm{Re}\, s_1 > \mathrm{Re}\, s_2, \end{aligned} \tag{16.66}$$

where $I_{M-1}(x)$ is the modified Bessel function.

An inverse relation expressing $P(W, s_1)$ in terms of $P(W, s_2)$ ($\mathrm{Re}\, s_1 > \mathrm{Re}\, s_2$) has a singular character because in this case the above integral (16.66) is non-zero in the complex y-plane only for $W = 0$, which is a one-point support of a generalized function. Such a relation can be written in the form of a series of the derivatives of the δ-function or, in some cases, in the form of a series of the Laguerre polynomials (Peřina and Horák (1969b, 1970)).

From (16.62) one can obtain in the same way the relation between the distribution $P(W, s)$ and the photon-number distribution $p(n)$

$$\begin{aligned} P(W, s) &= \left(\frac{2W}{1-s}\right)^M \frac{\exp\left(-\frac{2W}{1-s}\right)}{W} \times \\ &\times \sum_{n=0}^{\infty} \frac{n!\, p(n)}{[\Gamma(n+M)]^2} \left(\frac{s+1}{s-1}\right)^n L_n^{M-1}\left(\frac{4W}{1-s^2}\right), \end{aligned} \tag{16.67}$$

which can be inverted, using the orthogonality relation (10.28), in the form of the generalized photodetection equation (Peřina and Horák (1969b, 1970), Zardecki (1974))

$$\begin{aligned} p(n) &= \frac{1}{\Gamma(n+M)} \left(\frac{2}{1+s}\right)^M \left(\frac{s-1}{s+1}\right)^n \times \\ &\times \int_0^{\infty} P(W, s)\, L_n^{M-1}\left(\frac{4W}{1-s^2}\right) \exp\left(-\frac{2W}{1+s}\right) \mathrm{d}W. \end{aligned} \tag{16.68}$$

For $s \to 1$, if the asymptotic formula

$$L_n^{M-1}(x) \underset{x\to\infty}{\simeq} \frac{\Gamma(n+M)}{n!}(-x)^n \tag{16.69}$$

is used, equation (16.68) gives the standard photodetection equation (14.14).

The relation between the moments $\langle \hat{n}^k \rangle_{s_2}$ and $\langle \hat{n}^l \rangle_{s_1}$ follows from (16.65)

$$\langle \hat{n}^k \rangle_{s_2} = \frac{d^k}{d(iy)^k} \langle \exp(iy\hat{n}) \rangle_{s_2} \Big|_{iy=0} =$$
$$= \frac{k!}{\Gamma(k+M)} \left(\frac{s_1 - s_2}{2}\right)^k \left\langle L_k^{M-1}\left(\frac{2\hat{n}}{s_2 - s_1}\right)\right\rangle_{s_1}. \tag{16.70}$$

From (16.60), which is also the generating function for the Laguerre polynomials, one obtains

$$\langle \hat{n}^k \rangle = \frac{d^k}{d(ix)^k} \langle \exp(ix\hat{n}) \rangle \Big|_{ix=0} =$$
$$= \frac{d^k}{d(ix)^k} \left\langle \sum_{j=0}^{\infty} \frac{1}{\Gamma(j+M)} \left(\frac{1-s}{2}\right)^j (1 - e^{ix})^j L_j^{M-1}\left(\frac{2\hat{n}}{1-s}\right)\right\rangle_s \Big|_{ix=0} =$$
$$= \sum_{j=0}^{k} \frac{1}{\Gamma(j+M)} \left(\frac{1-s}{2}\right)^j \left[\sum_{r=0}^{j} \binom{j}{r} (-1)^r r^k\right] \left\langle L_j^{M-1}\left(\frac{2\hat{n}}{1-s}\right)\right\rangle_s, \tag{16.71}$$

where the identity used in connection with (10.19) has been used again, and (16.62) provides

$$\langle \hat{n}^k \rangle_s \equiv \langle W^k \rangle_s = \frac{d^k}{d(iy)^k} \langle \exp(iy\hat{n}) \rangle_s \Big|_{iy=0} =$$
$$= \left\langle \sum_{\hat{j}=\hat{0}}^{\hat{n}} \frac{\hat{n}!\,\Gamma(\hat{j}+M+k)}{\hat{j}!\,(\hat{n}-\hat{j})!\,\Gamma(\hat{j}+M)} (-2)^{\hat{j}-k} (s+1)^{\hat{n}-\hat{j}} (s-1)^{k-\hat{n}} \right\rangle. \tag{16.72}$$

A number of special cases for normal, symmetric and antinormal orderings can be obtained in the same way as for the one-mode case by putting $s = 1, 0$ and -1, respectively. For example from (16.66) we have for $s_2 = -1$, $s_1 = 0$,

$$P_{\mathscr{A}}(W) = 2\int_0^\infty \left(\frac{W}{W'}\right)^{(M-1)/2} \exp[-2(W+W')]\, I_{M-1}(4(WW')^{1/2}) \times$$
$$\times P_{sym}(W')\, dW' \tag{16.73a}$$

and for $s_2 = -1$, $s_1 = 1$,

$$P_{\mathscr{A}}(W) = \int_0^\infty \left(\frac{W}{W'}\right)^{(M-1)/2} \exp[-(W+W')]\, I_{M-1}(2(WW')^{1/2})\, P_{\mathscr{N}}(W')\, dW'. \tag{16.73b}$$

The corresponding relation between the moments is obtained from (16.70)

$$\langle \hat{n}^k \rangle_{\mathscr{A}} = \frac{k!}{2^k \Gamma(k+M)} \langle L_k^{M-1}(-2\hat{n}) \rangle_{sym} \tag{16.74a}$$

$$= \frac{k!}{\Gamma(k+M)} \langle L_k^{M-1}(-\hat{n}) \rangle_{\mathscr{N}}, \tag{16.74b}$$

and between the characteristic functions we have from (16.64)

$$\langle \exp(\mathrm{i}y\hat{n}) \rangle_{\mathscr{A}} = \left(1 - \frac{\mathrm{i}y}{2}\right)^{-M} \left\langle \exp\left(\frac{\mathrm{i}y\hat{n}}{1 - \frac{\mathrm{i}y}{2}}\right) \right\rangle_{sym} \tag{16.75a}$$

$$= (1 - \mathrm{i}y)^{-M} \left\langle \exp\left(\frac{\mathrm{i}y\hat{n}}{1 - \mathrm{i}y}\right) \right\rangle_{\mathscr{N}}. \tag{16.75b}$$

Similarly we have from (16.71) putting $s = 1$ and using the asymptotic formula (16.69)†

$$\langle \hat{n}^k \rangle = \sum_{j=0}^{k} \langle \hat{n}^j \rangle_{\mathscr{N}} \sum_{r=0}^{j} \frac{(-1)^{r+j} r^k}{r!\,(j-r)!} = \sum_{j=0}^{k} \langle \hat{n}^j \rangle_{\mathscr{N}} \sum_{m=0}^{j} \frac{(-1)^m (j-m)^k}{m!\,(j-m)!}, \tag{16.76}$$

which corresponds to the semiclassical relations (10.4) and (10.19). For $s = 0$ we can obtain the corresponding relation between the moments $\langle \hat{n}^k \rangle$ and $\langle \hat{n}^j \rangle_{sym}$ and for $s = -1$ we have

$$\langle \hat{n}^k \rangle = \sum_{j=0}^{k} \langle \hat{n}^j \rangle_{\mathscr{A}} \sum_{r=0}^{j} \frac{(-1)^{k+r} (M+r)^k}{r!\,(j-r)!}. \tag{16.77}$$

From (16.72) there follows for $s = -1$ the multimode analogue of (11.8b)

$$\langle \hat{n}^k \rangle_{\mathscr{A}} = \langle (\hat{n} + M)(\hat{n} + M + 1) \ldots (\hat{n} + M + k - 1) \rangle =$$

$$= \left\langle \frac{\Gamma(\hat{n} + M + k)}{\Gamma(\hat{n} + M)} \right\rangle, \tag{16.78}$$

---

† More generally for any function $\hat{F}(\hat{a}^+\hat{a})$ of the number operator $\hat{a}^+\hat{a}$, the coherent-state algebra provides

$$\langle \alpha | \hat{F}(\hat{a}^+\hat{a}) | \alpha \rangle = \sum_n |\langle \alpha | n \rangle|^2 F(n) = \sum_n \frac{|\alpha|^{2n}}{n!} \exp(-|\alpha|^2) F(n) =$$

$$= \sum_{n=0}^{\infty} \sum_{m=0}^{\infty} \frac{|\alpha|^{2(n+m)} (-1)^m}{n!\,m!} F(n) = \sum_{m=0}^{\infty} \sum_{j=m}^{\infty} \frac{(-1)^m |\alpha|^{2j}}{(j-m)!\,m!} F(j-m) =$$

$$= \langle \alpha | \mathscr{N} \left\{ \sum_{j=0}^{\infty} (\hat{a}^+\hat{a})^j \sum_{m=0}^{j} \frac{(-1)^m F(j-m)}{(j-m)!\,m!} \right\} | \alpha \rangle,$$

that is

$$\hat{F}(\hat{a}^+\hat{a}) = \sum_{j=0}^{\infty} \{(\hat{a}^+\hat{a})^j\}_{\mathscr{N}} \sum_{m=0}^{j} \frac{(-1)^m F(j-m)}{m!\,(j-m)!}.$$

which can also be derived directly with the help of the commutation rules (11.44) using the relation $\int_V \hat{\mathbf{A}}(\mathbf{x}, t) \cdot \hat{\mathbf{A}}^+(\mathbf{x}, t)\, \mathrm{d}^3x = \hat{n}_{Vt} + VL^{-3} \sum_{\mathbf{k},s} \mathbf{e}^{(s)}(\boldsymbol{k}) \cdot \mathbf{e}^{(s)*}(\boldsymbol{k}) = \hat{n}_{Vt} + VL^{-3}\mu = \hat{n}_{Vt} + M$, where (11.4a) has been used; μ is the number of modes in the volume L^3 and so M is the number of modes in the volume V. Note that successive factors in (14.15) are decreased by unity, whereas those in (16.78) are increased by unity. The difference may be regarded as a reflection of the fact that normally ordered correlations correspond to photon absorption, whereas anti-normally ordered correlations correspond to photon emission (this point will be discussed in greater detail in the next section).

Equation (16.62) gives us the following characteristic functions

$$\langle \exp(\mathrm{i}y\hat{n}) \rangle_{\mathcal{N}} = \langle (1 + \mathrm{i}y)^{\hat{n}} \rangle, \tag{16.79}$$

$$\langle \exp(\mathrm{i}y\hat{n}) \rangle_{sym} = \left(1 - \frac{\mathrm{i}y}{2}\right)^{-M} \left\langle \left[\frac{1 + \frac{\mathrm{i}y}{2}}{1 - \frac{\mathrm{i}y}{2}}\right]^{\hat{n}} \right\rangle \tag{16.80}$$

and

$$\langle \exp(\mathrm{i}y\hat{n}) \rangle_{\mathscr{A}} = \langle (1 - \mathrm{i}y)^{-\hat{n} - M} \rangle \tag{16.81}$$

for $s = 1, 0$ and -1 respectively and from (16.60) we have

$$\langle \exp(\mathrm{i}x\hat{n}) \rangle = \langle \exp[\hat{n}(\mathrm{e}^{\mathrm{i}x} - 1)] \rangle_{\mathcal{N}} \tag{16.82a}$$

$$= \left(\frac{2}{1 + \mathrm{e}^{\mathrm{i}x}}\right)^M \left\langle \exp\left[\frac{2\hat{n}(\mathrm{e}^{\mathrm{i}x} - 1)}{1 + \mathrm{e}^{\mathrm{i}x}}\right] \right\rangle_{sym} \tag{16.82b}$$

$$= \langle \exp[-\mathrm{i}xM + \hat{n}(1 - \mathrm{e}^{-\mathrm{i}x})] \rangle_{\mathscr{A}}. \tag{16.82c}$$

Equation (16.82a) is in agreement with (14.17) and (16.82c) for $M = 1$ with (13.100).

Finally from (16.67), for $s = -1$,

$$P_{\mathscr{A}}(W) = \exp(-W) \sum_{n=0}^{\infty} \frac{W^{n+M-1}}{\Gamma(n + M)}\, p(n), \tag{16.83}$$

so that

$$p(n) = \frac{\mathrm{d}^{n+M-1}}{\mathrm{d}W^{n+M-1}} [P_{\mathscr{A}}(W) \exp(W)] \Big|_{W=0}. \tag{16.84}$$

In the classical limit, when the average photon number per mode becomes large, the commutator of the field operators $[\hat{a}, \hat{a}^+] = \hat{a}\hat{a}^+ - \hat{a}^+\hat{a}$ is practically zero since $\hat{a}\hat{a}^+ = \hat{a}^+\hat{a} + \hat{1} \approx \hat{a}^+\hat{a}$ (this is also true for the operators $\hat{\mathbf{A}}(x)$ and $\hat{\mathbf{A}}(x)$) and the distinction between different orderings of the field operators vanishes as a consequence of the correspondence principle. Thus all distributions $P(W, s)$ for various s and $p(n)$ are equal and consequently all their moments are also equal. This can be seen for example as follows: for the photodetection equation (14.14) (or alternatively for (16.83)) we can conclude that the function under the integral (summation) sign gives its main contribution to the integral (sum) in the neighbourhood of the point $W = n$ where the function $W^n \exp(-W)/n!$ has its maximum; however this function tends to the function $\delta(W - n)$ for large W. Thus $p(n) \approx P_{\mathcal{N}}(n)$ for n large ($n \approx W$) and $\langle \hat{n}^k \rangle \approx \langle \hat{n}^k \rangle_{\mathcal{N}}$, etc.

Finally we mention a general formulation of the ordering of field operators suitable for boson as well as fermion fields; this was suggested by Lax (1968b) (see also Louisell (1970)).

Consider a complete set of operators $\{\hat{a}_\lambda\}$ in the Schrödinger picture obeying some set of commutation or anti-commutation relations (the anticommutator of two operators $\hat{A}$ and $\hat{B}$ is defined as $[\hat{A}, \hat{B}]_+ = \hat{A}\hat{B} + \hat{B}\hat{A}$) with an associated set of c-numbers $\{\alpha_\lambda\}$. An operator $\hat{A}^{(Q)}(\{\hat{a}_\lambda\})$ in a Q-ordered form can be expressed as

$$\hat{A}^{(Q)}(\{\hat{a}_\lambda\}) = \int A^{(Q)}(\{\alpha_\lambda\}) \prod_\lambda \delta(\hat{a}_\lambda - \alpha_\lambda)\, \mathrm{d}^2\alpha_\lambda, \tag{16.85}$$

where the δ-functions are in the chosen order. If the $\{\hat{a}_\lambda\}$ are not Hermitian and $\{\hat{a}_\lambda^+\}$ are also present, the δ-functions are defined, for example, by

$$\delta(\hat{a} - \alpha) = \frac{1}{\pi^2} \int \exp[-\beta^*(\hat{a} - \alpha)] \exp[\beta(\hat{a}^+ - \alpha^*)]\, \mathrm{d}^2\beta \tag{16.86}$$

if the chosen order is $\hat{a}, \hat{a}^+$. The expectation value of (16.85) then becomes

$$\begin{aligned} \mathrm{Tr}\,\{\hat{\varrho}\hat{A}^{(Q)}\} &= \int A^{(Q)}(\{\alpha_\lambda\}) \langle \prod_\lambda \delta(\hat{a}_\lambda - \alpha_\lambda) \rangle\, \mathrm{d}^2\{\alpha_\lambda\} = \\ &= \int A^{(Q)}(\{\alpha_\lambda\})\, \Phi(\{\alpha_\lambda\})\, \mathrm{d}^2\{\alpha_\lambda\}, \end{aligned} \tag{16.87}$$

where

$$\Phi(\{\alpha_\lambda\}) = \langle \prod_\lambda \delta(\hat{a}_\lambda - \alpha_\lambda) \rangle = \int C(\{\beta_\lambda\}) \prod_\lambda \exp(\alpha_\lambda\beta_\lambda^* - \alpha_\lambda^*\beta_\lambda) \frac{\mathrm{d}^2\beta_\lambda}{\pi^2} \tag{16.88}$$

and the characteristic function is equal to

$$C(\{\beta_\lambda\}) = \mathrm{Tr}\,\{\hat{\varrho} \prod_\lambda \exp(-\beta_\lambda^*\hat{a}_\lambda) \exp(\beta_\lambda\hat{a}_\lambda^+)\}. \tag{16.89}$$

The advantage of this formulation is that the Q-order of operators explicitly occurs in equation (16.85) while in the Agarwal-Wolf formulation the various orders of operators are specified by the function $\Omega(\alpha^*, \alpha)$ and it is not always clear which order of operators corresponds explicitly to a given $\Omega(\alpha^*, \alpha)$.

16.4 Measurements corresponding to antinormally ordered correlation functions – quantum counters

In this section we show that in principle there is a possibility of measuring the statistics of optical fields when the photoelectric detectors normally used are replaced by atomic counting devices, called *quantum counters* (Mandel (1966b)). Such detectors operate by stimulated emission rather than by absorption of photons in an external field, and consequently correlation measurements by means of quantum counters correspond to antinormally ordered products of field operators. The quantum counter is in principle a device useful for measurements in high degeneracy fields only (it is insensitive to fields with low degeneracy, i.e. with a low mean number of photons per mode) and its practical realization is not a simple matter. However, it is interesting from the theoretical point of view to compare this device with the photoelectric detector and to compare the way the two devices measure fields.

A schematic energy level diagram for a quantum counter is given in Fig. 16.1, as suggested by Mandel (1966b). In this figure a represents a terminal energy level and c is a metastable level which is radiatively coupled to a broad energy band d, corresponding to a very short-lived state. The system will make spontaneous radiative transitions from d to a. We shall assume that this system is prepared in the state c by optical pumping from a to the broad energy level b from which it will make non-radiative transitions to c. Further let the interval $(E_c - E_d)/\hbar c$, defined by the energy levels E_c and E_d, be of the same order as the wave number of a typical mode of the external field, and let $(E_d - E_a) > (E_c - E_d)$.

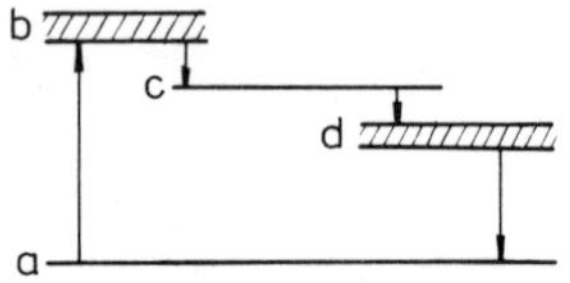

Fig. 16.1 Energy levels of the quantum counter.

Under the interaction of the system placed at a space point $\boldsymbol{x}$ at time t with the external field, the system can make a stimulated transition from c to d with the emission of a photon. As level d is very short lived, it will decay spontaneously from d to a with the further emission of a photon. The latter photon is distinguishable from the former since $(E_d - E_a) > (E_c - E_d)$ and a photodetector placed in a neighbourhood of $\boldsymbol{x}$ with sufficiently high photoelectric threshold will register the second photon alone. The combination of the photodetector with a large number of such atomic systems acts as a quantum counter of the external field functioning by stimulated emission of radiation. It should be noted that two photons are emitted into the field by the system but only one is absorbed by the photodetector; the photodetector plays an auxiliary role only, the field is actually measured by means of the first induced transition.

Making use of arguments similar to those used in connection with the measurement of the field by means of photoelectric detectors (Sec. 12.1) we find that the probability that one count is registered at $(\boldsymbol{x}, t)$ is proportional to

$$\{\sum_f |\langle f|\hat{A}^{(-)}(x)|i\rangle|^2\}_{\text{average}} = \{\sum_f \langle i|\hat{A}^{(+)}(x)|f\rangle\langle f|\hat{A}^{(-)}(x)|i\rangle\}_{\text{average}} =$$

$$= \text{Tr}\,\{\hat{\varrho}\hat{A}^{(+)}(x)\hat{A}^{(-)}(x)\} = \Gamma_{\mathscr{A}}^{(1,1)}(x, x), \tag{16.90}$$

i.e. this probability is proportional to the antinormally ordered correlation function $\Gamma_{\mathscr{A}}^{(1,1)}(x, x)$. Similarly the joint probability that counts will be registered by N quantum counters at space-time points $x_1, \ldots, x_N$ is proportional to

$$\Gamma_{\mathscr{A}}^{(N,N)}(x_1, \ldots, x_N, x_N, \ldots, x_1) = \\ = \mathrm{Tr}\,\{\hat{\varrho}\hat{A}^{(+)}(x_1) \ldots \hat{A}^{(+)}(x_N)\,\hat{A}^{(-)}(x_N) \ldots \hat{A}^{(-)}(x_1)\}, \tag{16.91}$$

which is the antinormal correlation function of the $2N$th order.

From (16.78) it is clear that measurements by quantum counters will be sensitive to the field only if the mean number of photons $\langle n \rangle$ is much larger than the number of modes, i.e. if the degeneracy parameter $\langle n \rangle / M = \langle n \rangle L^3/\mu V = \delta$ is much greater than unity. This is the case for laser fields but for non-degenerate fields, such as thermal fields, $\delta \ll 1$ and the quantum counter will not be a useful measuring device. However for sufficiently strong fields (large δ) the difference between normally and antinormally ordered correlations vanishes and $\langle \hat{n}^k \rangle_{\mathscr{A}} \approx \langle \hat{n}^k \rangle_{\mathscr{N}} \approx \langle \hat{n}^k \rangle$, because the field approaches the classical limit; consequently both photoelectric detectors and quantum counters will give practically the same results in this case.

Mandel (1966b) calculated the probability $p'(n, T, t)$ that n counts will be registered in the time interval $(t, t + T)$ when the quantum counter is exposed to a plane beam of quasi-monochromatic light to which the quantum counter can respond. By an argument identical to that used in the derivation of the photodetection equation (10.3a), he obtained

$$p'(n, T, t) = \int_0^\infty P_{\mathscr{A}}(\bar{W}) \frac{\bar{W}^n}{n!} \exp(-\bar{W})\, \mathrm{d}\bar{W}, \tag{16.92}$$

which corresponds to (10.3a) or to (14.26) with $\bar{W}$ given by (14.28). Consequently all the properties of this relation are the same as those of (10.3a) and they are summarized in Chapter 10. However in this case $\langle (\Delta\hat{n})^2 \rangle_{\mathscr{A}} \geqq 0$ always holds (we have seen that this is not the case for the photoelectric detector, e.g. for the $|n\rangle$-state, $\langle (\Delta\hat{n})^2 \rangle_{\mathscr{N}} = -n$ and $\langle (\Delta\hat{n})^2 \rangle = 0$); thus the variance of n will always be greater than or equal to the variance corresponding to a Poisson distribution. This fact is a consequence of the non-negativeness of the quasi-distribution $\Phi_{\mathscr{A}}(\{\alpha_\lambda\})$ (equation (13.82a)). Hence the antinormal correlations are always positive semi-definite. From (13.68) it can be seen that for strong fields, for which the quasi-probability $\Phi_{\mathscr{N}}(\beta)$ is non-zero for large $|\beta|$ only and the function $\exp(-|\alpha - \beta|^2)$ is sharply peaked, the principal contribution to $\Phi_{\mathscr{A}}(\alpha)$ will come from values of β in a neighbourhood of α. Then if $\Phi_{\mathscr{N}}(\alpha)$ is sufficiently smooth $\Phi_{\mathscr{A}}(\alpha) \approx \Phi_{\mathscr{N}}(\alpha)$, and for multimode field $\Phi_{\mathscr{A}}(\{\alpha_\lambda\}) \approx \Phi_{\mathscr{N}}(\{\alpha_\lambda\})$. Thus we conclude again that the normal and antinormal correlations are equal in this classical limit.

We have discussed in Chapter 14 that the photon-number distribution and the photocount distribution are mostly of the same form (they differ from one another by a scale change given by the photoefficiency η of the detector). This can be regarded as a consequence of the fact that photoelectric detection measurements are related to normally ordered products of field operators for which the vacuum expectation values are zero, so that the vacuum fluctuations give no contribution to the normal correlation functions. A different situation occurs for other orderings; for example for antinormally ordered products of field operators connected with quantum counter measurements the photocount distribution is given by (16.92) while the photon-number distribution is given by (16.84), so they differ substantially from one another. This is caused by the contribution of the physical vacuum through the commutator $\hat{a}\hat{a}^+ - \hat{a}^+\hat{a} = \hat{1}$, i.e. $\hat{a}\hat{a}^+ = \hat{a}^+\hat{a} + \hat{1}$. Note that the last relation generates the convolution law (13.68) by the correspondence $\hat{a}\hat{a}^+ \rightleftarrows \Phi_{\mathscr{A}}(\alpha)$, $\hat{a}^+\hat{a} \rightleftarrows \Phi_{\mathscr{N}}(\alpha)$ and $\hat{1} \rightleftarrows \exp(-|\alpha|^2) = \langle\alpha|0\rangle\langle 0|\alpha\rangle = \langle\alpha|\mathscr{N}\{\exp(-\hat{a}^+\hat{a})\}|\alpha\rangle$ (i.e. $|0\rangle\langle 0| = \mathscr{N}\{\exp(-\hat{a}^+\hat{a})\}$). For M-mode fields the vacuum fluctuations contribute through the number of modes M since $\int_V \hat{\mathbf{A}}(x) . \hat{\mathbf{A}}^+(x)\, d^3x = \int_V \hat{\mathbf{A}}^+(x) . \hat{\mathbf{A}}(x)\, d^3x + M$.

The connection between $p'(n, T, t)$ and $p(n)$ is determined by substitution of (16.83) into (16.92), which gives (Peřina (1968b))

$$p'(n, T, t) = \sum_{m=0}^{\infty} p(m)\, 2^{-m-n-M} \binom{m + n + M - 1}{n}. \tag{16.93}$$

Some alternating (symmetrical) ordering is characteristic for scattering processes (Loudon (1973)). In such a process a photon in an incident beam 1 is absorbed, while a photon in a scattered beam 2 is produced. Therefore the transition probability is proportional to the following correlation function with the alternating order,

$$\mathrm{Tr}\,\{\hat{\varrho}\hat{A}_1^{(-)}\hat{A}_2^{(+)}\hat{A}_2^{(-)}\hat{A}_1^{(+)}\} \sim \langle\hat{n}_1\rangle(\langle\hat{n}_2\rangle + 1); \tag{16.94}$$

spontaneous scattering takes place if $\langle\hat{n}_2\rangle = 0$, stimulated scattering if $\langle\hat{n}_2\rangle \gg 1$.

We can conclude that different orderings of field operators correspond to various types of experiments and they are characteristic for them.

CHAPTER 17

SPECIAL STATES OF THE ELECTROMAGNETIC FIELD

The general methods developed in preceding chapters will be demonstrated now by applying them to the most typical state in nature – the chaotic (Gaussian) radiation (and as a special case to the thermal field). In addition we apply them to laser light and to the superposition of chaotic and coherent fields. In this chapter we consider free optical fields, whereas optical fields in interaction with matter will be considered in Chapters 20–22.

17.1 Chaotic (Gaussian) radiation

Distributions and characteristic functions

It is well known that the state for which the entropy

$$H = -\mathrm{Tr}\,\{\hat{\varrho}\ln\hat{\varrho}\} \tag{17.1}$$

has its maximum under the assumptions that $\mathrm{Tr}\,\hat{\varrho} = 1$ and $\mathrm{Tr}\,\{\hat{\varrho}\hat{a}^+\hat{a}\} = \langle n\rangle$ is the chaotic state for which the density matrix possesses the diagonal Fock form (11.27) with $\varrho(n)$ as members of the Bose-Einstein distribution (11.32), i.e.

$$\hat{\varrho} = \frac{\langle n\rangle^{\hat{a}^+\hat{a}}}{(1+\langle n\rangle)^{\hat{a}^+\hat{a}+1}}. \tag{17.2}$$

For thermal (blackbody) radiation (11.30) is appropriate for $\langle n\rangle$ and we obtain (11.29) for $\hat{\varrho}$.

The density matrix for a multimode field is given, in analogy to (11.29), by

$$\hat{\varrho} = \frac{\prod_\lambda \exp(-\Theta_\lambda \hat{a}_\lambda^+\hat{a}_\lambda)}{\mathrm{Tr}\,\{\prod_\lambda \exp(-\Theta_\lambda \hat{a}_\lambda^+\hat{a}_\lambda)\}} = \frac{\exp\left(-\frac{1}{KT}\hat{H}\right)}{\mathrm{Tr}\left\{\exp\left(-\frac{1}{KT}\hat{H}\right)\right\}}, \tag{17.3}$$

where the Hamiltonian $\hat{H}$ is given by (11.13) with the vacuum energy term $\sum_\lambda \hbar\omega_\lambda/2$ subtracted.

We must still determine what kind of function $\Phi_{\mathcal{N}}(\{\alpha_\lambda\})$ corresponds to (17.2) and (17.3). Because of the statistical independence of different modes the multimode form of (17.2) reads

$$\hat{\varrho} = \prod_\lambda \frac{\langle n_\lambda\rangle^{\hat{a}_\lambda^+\hat{a}_\lambda}}{(1+\langle n_\lambda\rangle)^{\hat{a}_\lambda^+\hat{a}_\lambda+1}} = \sum_{\{n_\lambda\}}\prod_{\lambda'} \frac{\langle n_{\lambda'}\rangle^{n_{\lambda'}}}{(1+\langle n_{\lambda'}\rangle)^{n_{\lambda'}+1}}\,|\{n_\lambda\}\rangle\langle\{n_\lambda\}|. \tag{17.4}$$

Making use of (13.50) and (13.49) we can write

$$\hat{\varrho} = \int |\{\alpha_\lambda\}\rangle\langle\{\alpha_\lambda\}| \prod_\lambda \frac{d^2\alpha_\lambda}{\pi} \prod_{\lambda'} \frac{\langle n_{\lambda'}\rangle^{\hat{a}^+_{\lambda'}\hat{a}_{\lambda'}}}{(1+\langle n_{\lambda'}\rangle)^{\hat{a}^+_{\lambda'}\hat{a}_{\lambda'}+1}} =$$

$$= \int \sum_{\{n_\lambda\}} \sum_{\{m_\lambda\}} |\{n_\lambda\}\rangle\langle\{m_\lambda\}| \prod_\lambda \exp(-|\alpha_\lambda|^2) \frac{\alpha_\lambda^{n_\lambda}\alpha_\lambda^{*m_\lambda}}{(n_\lambda! m_\lambda!)^{1/2}} \times$$

$$\times \prod_{\lambda'} \frac{\langle n_{\lambda'}\rangle^{\hat{a}^+_{\lambda'}\hat{a}_{\lambda'}}}{(1+\langle n_{\lambda'}\rangle)^{\hat{a}^+_{\lambda'}\hat{a}_{\lambda'}+1}} \frac{d^2\alpha_{\lambda'}}{\pi} = \int \sum_{\{n_\lambda\}} \sum_{\{m_\lambda\}} |\{n_\lambda\}\rangle\langle\{m_\lambda\}| \times$$

$$\times \prod_\lambda \frac{\exp(-|\alpha_\lambda|^2)}{1+\langle n_\lambda\rangle} \frac{\alpha_\lambda^{n_\lambda}\alpha_\lambda^{*m_\lambda}}{(n_\lambda! m_\lambda!)^{1/2}} \left(\frac{\langle n_\lambda\rangle}{1+\langle n_\lambda\rangle}\right)^{m_\lambda} \frac{d^2\alpha_\lambda}{\pi}. \tag{17.5}$$

As only those terms of the series with $n_\lambda = m_\lambda$ do not vanish (by integration over the phases), we arrive at

$$\hat{\varrho} = \int \prod_\lambda \frac{\exp\left(-|\alpha_\lambda|^2 + \dfrac{\langle n_\lambda\rangle}{1+\langle n_\lambda\rangle}|\alpha_\lambda|^2\right)}{\pi(1+\langle n_\lambda\rangle)} |\beta_\lambda\rangle\langle\beta_\lambda| \, d^2\alpha_\lambda =$$

$$= \int \prod_{\lambda'} \frac{\exp\left(-\dfrac{|\beta_{\lambda'}|^2}{\langle n_{\lambda'}\rangle}\right)}{\pi\langle n_{\lambda'}\rangle} |\{\beta_\lambda\}\rangle\langle\{\beta_\lambda\}| \, d^2\{\beta_\lambda\}, \tag{17.6}$$

where the substitution $\alpha_\lambda[\langle n_\lambda\rangle/(1+\langle n_\lambda\rangle)]^{1/2} = \beta_\lambda$ has been used. If we write α_λ instead of β_λ again, we obtain the final result

$$\Phi_{\mathcal{N}}(\{\alpha_\lambda\}) = \prod_\lambda \frac{\exp\left(-\dfrac{|\alpha_\lambda|^2}{\langle n_\lambda\rangle}\right)}{\pi\langle n_\lambda\rangle}. \tag{17.7}$$

This is a Gaussian distribution in the complex amplitude α_λ and this distribution is independent of the phases of α_λ; thus the chaotic field is stationary in time and in space. It is easy to verify that the substitution of (17.7) and (13.49) into (13.142) leads to (17.4) again.

This result can also be derived using the quantum analogue of the central limit theorem (Glauber (1963b), Klauder and Sudarshan (1968)). The probability distribution (17.7) for the chaotic field is consistent with the classical probability distribution (10.35). This follows, taking into account the fact that the complex amplitude of the field is a linear superposition of the type (13.53) and that the convolution of two Gaussian distributions is again a Gaussian distribution. Thus we obtain (10.35) with $\langle I\rangle = \langle \mathbf{V}^*(x) . \mathbf{V}(x)\rangle = 2\pi\hbar c L^{-3} \sum_{\mathbf{k},s} k^{-1}\langle n_{\mathbf{k}s}\rangle$, since $\langle \alpha^*_{\mathbf{k}s}\alpha_{\mathbf{k}'s'}\rangle = \delta_{\mathbf{k}\mathbf{k}'}\delta_{ss'}\langle|\alpha_{\mathbf{k}s}|^2\rangle = \delta_{\mathbf{k}\mathbf{k}'}\delta_{ss'}\langle n_{\mathbf{k}s}\rangle$; for blackbody radiation $\langle n_{\mathbf{k}s}\rangle$ is independent of the polarization s and $\langle n_{\mathbf{k}s}\rangle \equiv \langle n_{\mathbf{k}}\rangle \equiv \langle n_k\rangle$ is given by (11.30).

A calculation of the characteristic functional (12.48) for chaotic radiation will enable us to determine the correlation functions according to (12.49) and also the multifold probability distribution (14.1). Writing

$$\varepsilon_\mu(x, \lambda) = \frac{(2\pi\hbar c)^{1/2}}{L^{3/2}} \frac{1}{k^{1/2}} e_\mu^{(s)}(\mathbf{k}) \exp[i(\mathbf{k} . \mathbf{x} - kct)], \tag{17.8}$$

the μ-component of (11.22) can be written in the form

$$\hat{A}_\mu^{(+)}(x) = \sum_\lambda \varepsilon_\mu(x,\lambda)\,\hat{a}_\lambda \equiv \hat{A}^{(+)}(x). \tag{17.9}$$

Substituting (17.9) and (17.7) into (12.48) we obtain

$$C_{\mathcal{N}}\{y(x)\} = \int \prod_\lambda \frac{1}{\pi\langle n_\lambda\rangle} \exp\left(-\frac{|\alpha_\lambda|^2}{\langle n_\lambda\rangle}\right) \times$$
$$\times \exp\left\{\int\left[y(x)\sum_{\lambda'} \varepsilon^*(x,\lambda')\,\alpha^*_{\lambda'} - y^*(x)\sum_{\lambda'}\varepsilon(x,\lambda')\,\alpha_{\lambda'}\right]\mathrm{d}^4x\right\}\mathrm{d}^2\{\alpha_\lambda\} =$$
$$= \int \prod_\lambda \frac{1}{\pi\langle n_\lambda\rangle}\exp\left(-\frac{|\alpha_\lambda|^2}{\langle n_\lambda\rangle} + \alpha^*_\lambda y_\lambda - \alpha_\lambda y^*_\lambda\right)\mathrm{d}^2\alpha_\lambda, \tag{17.10}$$

where

$$y_\lambda = \int y(x)\,\varepsilon^*(x,\lambda)\,\mathrm{d}^4x. \tag{17.11}$$

(The integral over x also includes a sum over μ.) Therefore by using (16.41)

$$C_{\mathcal{N}}\{y(x)\} = \prod_\lambda \exp\left(-\langle n_\lambda\rangle\,|y_\lambda|^2\right) =$$
$$= \exp\left[-\iint y^*(x')\sum_\lambda \langle n_\lambda\rangle\,\varepsilon(x',\lambda)\,\varepsilon^*(x,\lambda)\,y(x)\,\mathrm{d}^4x\,\mathrm{d}^4x'\right]. \tag{17.12}$$

However the second-order correlation function is

$$\Gamma_{\mathcal{N}}^{(1,1)}(x,x') = \mathrm{Tr}\,\{\hat{\varrho}\hat{A}^{(+)}(x)\,\hat{A}^{(-)}(x')\} =$$
$$= \int\prod_\lambda \frac{1}{\pi\langle n_\lambda\rangle}\exp\left(-\frac{|\alpha_\lambda|^2}{\langle n_\lambda\rangle}\right)\sum_\lambda\sum_{\lambda'}\varepsilon^*(x,\lambda)\,\varepsilon(x',\lambda')\,\alpha^*_\lambda\alpha_{\lambda'}\,\mathrm{d}^2\{\alpha_\lambda\} =$$
$$= \sum_\lambda \langle n_\lambda\rangle\,\varepsilon^*(x,\lambda)\,\varepsilon(x',\lambda) \tag{17.13}$$

and we finally arrive at

$$C_{\mathcal{N}}\{y(x)\} = \exp\left[-\iint y(x)\,\Gamma_{\mathcal{N}}^{(1,1)}(x,x')\,y^*(x')\,\mathrm{d}^4x\,\mathrm{d}^4x'\right]. \tag{17.14}$$

The correlation function $\Gamma_{\mathcal{N}}^{(n,n)}$ is, according to (12.49),

$$\Gamma_{\mathcal{N}}^{(n,n)}(x_1,\ldots,x_{2n}) =$$
$$= (-1)^n \prod_{k=n+1}^{2n}\frac{\delta}{\delta(-y^*(x_k))}\left\{\prod_{j=1}^{n}\left[\int\Gamma_{\mathcal{N}}^{(1,1)}(x_j,x')\,y^*(x')\,\mathrm{d}^4x'\right]C_{\mathcal{N}}\{y(x)\}\right\}\Bigg|_{y(x)=0} =$$
$$= \sum_\pi\prod_{j=1}^{n}\Gamma_{\mathcal{N}}^{(1,1)}(x_j,x_{n+j}), \tag{17.15}$$

where $\sum_\pi$ stands for the sum of $n!$ possible permutations of the indices $n+1,\ldots,2n$ (or $1,\ldots,n$). It is obvious that $\Gamma_{\mathcal{N}}^{(m,n)} = 0$ for $m \neq n$. This is just the relation (10.62).

Making use of the complex representation (13.37) of the δ-function, we can write (14.1) in the form

$$P_n(V_1, \ldots, V_n) = \int \Phi_{\mathcal{N}}(\{\alpha'_\lambda\}) \int \prod_{j=1}^{n} \exp(V_j z_j^* - V_j^* z_j) \times$$
$$\times \exp(-V'_j z_j^* + V'^*_j z_j) \frac{d^2 z_j}{\pi^2} d^2\{\alpha'_\lambda\} =$$
$$= \int C_{\mathcal{N}}(\{z_j\}) \prod_{j=1}^{n} \exp(V_j z_j^* - V_j^* z_j) \frac{d^2 z_j}{\pi^2}, \tag{17.16}$$

where the characteristic function $C_{\mathcal{N}}(\{z_j\})$ is given by (17.14) with

$$y(x) = \sum_{j=1}^{n} z_j \delta(x - x_j); \tag{17.17}$$

that is

$$C_{\mathcal{N}}(\{z_j\}) = \exp\left[-\sum_{j,k}^{n} z_j \Gamma_{\mathcal{N}}^{(1,1)}(x_j, x_k) z_k^*\right], \tag{17.18}$$

which, if we introduce the vectors

$$\hat{Z} = \begin{pmatrix} z_1 \\ \vdots \\ z_n \end{pmatrix}, \qquad \hat{V} = \begin{pmatrix} V_1 \\ \vdots \\ V_n \end{pmatrix}, \qquad \hat{V}' = \begin{pmatrix} V'_1 \\ \vdots \\ V'_n \end{pmatrix} \tag{17.19}$$

and the covariance matrix

$$\mathcal{R} = \langle \hat{V}' \otimes \hat{V}'^+ \rangle, \tag{17.20}$$

can be written in the form

$$C_{\mathcal{N}}(\hat{Z}) = \exp(-\hat{Z}^+ \mathcal{R} \hat{Z}). \tag{17.21}$$

Thus we have from (17.16)

$$P_n(V_1, \ldots, V_n) = \frac{1}{\pi^{2n}} \int \exp(-\hat{Z}^+ \mathcal{R} \hat{Z}) \exp(-\hat{V}^+ \hat{Z} + \hat{Z}^+ \hat{V}) \prod_{j=1}^{n} d^2 z_j, \tag{17.22}$$

which, after applying (16.41), finally gives

$$P_n(V_1, \ldots, V_n) = \frac{1}{\pi^n \operatorname{Det} \mathcal{R}} \exp(-\hat{V}^+ \mathcal{R}^{-1} \hat{V}). \tag{17.23}$$

This is the multivariate Gaussian distribution, so that the quantized electromagnetic chaotic field is described as a Gaussian random process.

It is obvious that very similar results are obtainable for the s-ordering. Applying the multimode analogue of (16.43) (with $s_2 = s$ and $s_1 = 1$) to (17.7) we obtain

$$\Phi(\{\alpha_\lambda\}, s) = \prod_\lambda \left(\langle n_\lambda \rangle + \frac{1-s}{2}\right)^{-1} \exp\left(-\frac{|\alpha_\lambda|^2}{\langle n_\lambda \rangle + \frac{1-s}{2}}\right), \tag{17.24}$$

which clearly shows that all the results just derived for chaotic light are also valid for the case of the s-ordering if $\langle n_\lambda \rangle$ is replaced by $\langle n_\lambda \rangle + (1 - s)/2$.

Hence, both the space-time quasi-probabilities as well as the phase space quasi-distributions are multivariate Gaussian distributions with positive-definite covariance matrices for $-1 \leqq s \leqq 1$ (this includes the normal, symmetric and antinormal orderings).

The second-order correlation function

The second-order correlation functions for blackbody radiation have been obtained by a number of authors. Bourret (1960) has used a technique employed in the theory of the isotropic turbulence of an incompressible fluid to derive expressions for the second-order electric correlation tensor. Later Sarfatt (1963) rederived the main results by explicit quantum mechanical calculations.

Correlation functions suitable for the description of the temporal coherence of blackbody radiation were derived by Kano and Wolf (1962) who showed that for this case the corresponding degree of coherence has no zeros in the lower half of the complex τ-plane. Mehta (1963) studied the coherence time and effective bandwidth of blackbody radiation, and a more complete treatment of the electric, magnetic and mixed correlation tensors of the second order for blackbody radiation (as well as of the spectral correlation tensors) was given by Mehta and Wolf (1964, 1967a, b) in a series of papers. In these the explicit behaviour of the second-order correlation tensors and spectral correlation tensors was investigated and the authors showed by direct calculations that the classical and quantum correlation tensors are identical for blackbody radiation. The statistical properties of blackbody radiation in an unbounded domain were investigated by Holliday (1964), Holliday and Sage (1964) and Keller (1965) using the theory of functionals. Glauber (1963b, 1965) derived some general properties of the correlation tensors for blackbody radiation, Eberly and Kujawski (1967b) and Kujawski (1969) studied the properties of correlation tensors in uniformly moving coordinate systems, and Kujawski (1968) showed that the coherence properties of blackbody radiation can be characterized by a single scalar correlation function and discussed a resemblance between commutators of the quantized electromagnetic field and the correlation tensors. A relativistic coherence theory of blackbody radiation has been developed by Brevik and Suhonen (1968, 1970) and Eberly and Kujawski (1972).

We will discuss only some simple properties of the electric correlation tensor here, particularly in connection with the temporal coherence of blackbody radiation; a more detailed treatment of this subject can be found in the review article by Mandel and Wolf (1965).

The quantum analogue of the electric correlation tensor $\mathscr{E}_{ij}(x_1, x_2)$ defined in (7.2) can clearly be written in the form

$$\mathscr{E}_{ij}(x_1, x_2) = \mathrm{Tr}\,\{\hat{\varrho}\hat{E}_i^{(-)}(x_1)\,\hat{E}_j^{(+)}(x_2)\}. \tag{17.25}$$

The use of (17.13) and (11.1a) provides

$$\mathscr{E}_{ij}(x_1, x_2) = \\ = \frac{2\pi\hbar c}{L^3} \sum_{\mathbf{k},s} \langle n_{\mathbf{k}s} \rangle \, k \, e_i^{(s)*}(\mathbf{k}) \, e_j^{(s)}(\mathbf{k}) \exp\left[\mathrm{i}\mathbf{k} \cdot (\mathbf{x}_2 - \mathbf{x}_1) - \mathrm{i}kc(t_2 - t_1)\right]. \quad (17.26)$$

As $\langle n_{\mathbf{k}s} \rangle \equiv \langle n_k \rangle$ (given by (11.30)) we can sum over s with the help of (11.5), and the substitution $\sum_{\mathbf{k}} \to (L/2\pi)^3 \int \ldots \mathrm{d}^3k$ leads to

$$\mathscr{E}_{ij}(x_1, x_2) = \mathscr{E}_{ij}(x_2 - x_1) = \mathscr{E}_{ij}(x) = \\ = \frac{\hbar c}{(2\pi)^2} \int \frac{k}{\exp(\Theta) - 1} \left(\delta_{ij} - \frac{k_i k_j}{k^2} \right) \exp(\mathrm{i}kx) \, \mathrm{d}^3k = \\ = \frac{\hbar c}{(2\pi)^2} \left(-\delta_{ij} \nabla^2 + \frac{\partial}{\partial x_i} \frac{\partial}{\partial x_j} \right) \int \frac{1}{k(\exp(\Theta) - 1)} \exp(\mathrm{i}kx) \, \mathrm{d}^3k, \quad (17.27)$$

where $x \equiv (\mathbf{x}, \tau) = x_2 - x_1$ and the x_i, $i = 1, 2, 3$ are cartesian components of $\mathbf{x}$. Introducing spherical polar coordinates for $\mathbf{k}$ with the polar axis along the direction $\mathbf{x}$ we obtain

$$\mathscr{E}_{ij}(x) = \frac{\hbar c}{(2\pi)^2} \left(\frac{\partial}{\partial x_i} \frac{\partial}{\partial x_j} - \delta_{ij} \nabla^2 \right) \int_0^\infty \frac{k \, \mathrm{d}k}{\exp\left(\dfrac{\hbar ck}{KT}\right) - 1} \times \\ \times \exp(-\mathrm{i}kc\tau) \int_0^\pi \exp(\mathrm{i}kr \cos\vartheta) \sin(\vartheta) \, \mathrm{d}\vartheta \int_0^{2\pi} \mathrm{d}\varphi = \\ = \frac{4\hbar c}{\pi} \sum_{n=1}^\infty \left\{ \frac{\delta_{ij}}{\left[\left(n \dfrac{\hbar c}{KT} + \mathrm{i}c\tau \right)^2 + r^2 \right]^2} + \frac{2(x_i x_j - r^2 \delta_{ij})}{\left[\left(n \dfrac{\hbar c}{KT} + \mathrm{i}c\tau \right)^2 + r^2 \right]^3} \right\}, \quad (17.28)$$

where $r = |\mathbf{x}|$. The degree of coherence is obtained in the form

$$\gamma_{ij}(x) = \frac{\mathscr{E}_{ij}(x)}{[\mathscr{E}_{ii}(0) \mathscr{E}_{jj}(0)]^{1/2}} = \frac{\mathscr{E}_{ij}(x)}{\mathscr{E}_{ii}(0)} = \\ = \frac{90}{\pi^4} \left(\frac{\hbar c}{KT} \right)^4 \sum_{n=1}^\infty \left\{ \frac{\delta_{ij}}{\left[\left(n \dfrac{\hbar c}{KT} + \mathrm{i}c\tau \right)^2 + r^2 \right]^2} + \frac{2(x_i x_j - r^2 \delta_{ij})}{\left[\left(n \dfrac{\hbar c}{KT} + \mathrm{i}c\tau \right)^2 + r^2 \right]^3} \right\}, \quad (17.29)$$

where

$$\mathscr{E}_{ii}(0) = \frac{4\hbar c}{\pi} \left(\frac{KT}{\hbar c} \right)^4 \sum_{n=0}^\infty \frac{1}{(n+1)^4} = \frac{4\hbar c}{\pi} \left(\frac{KT}{\hbar c} \right)^4 \frac{\pi^4}{90}; \quad (17.30)$$

here the sum is equal to the generalized Riemann ζ-function $\zeta(4, 1) = \pi^4/90$ defined by (Whittaker and Watson (1940))

$$\zeta(s, a) = \sum_{n=0}^\infty (n + a)^{-s}. \quad (17.31)$$

As a special case we can consider temporal coherence by putting $\mathbf{x} = \mathbf{0}$ in (17.29) and we obtain

$$\gamma_{ij}(\mathbf{0}, \tau) = \delta_{ij} \frac{90}{\pi^4} \zeta\left(4, 1 + \mathrm{i}\frac{KT}{\hbar}\tau\right). \tag{17.32}$$

Hence any two orthogonal components $E_i(\mathbf{x}, t_1)$ and $E_j(\mathbf{x}, t_2)$ of blackbody radiation are completely uncorrelated. In this connection it should be mentioned that the function (17.32) has no zeros in the lower half of the complex τ-plane (Kano and Wolf (1962)) and therefore the phase of this function can be reconstructed uniquely from its known modulus with the help of the dispersion relations discussed in Sec. 4.4. Thus in this case the radiation spectrum can be determined from the visibility of the interference fringes.

It is evident that for the higher-order correlation tensors a relation of the same form as (17.15) holds, i.e. they are fully determined by the second-order correlation tensor.

Photocount statistics

Photocount statistics for Gaussian light were first considered by Glauber (1963c) in connection with the Hanbury Brown-Twiss experiment; in this paper Glauber introduced, for the first time, the diagonal representation of the density matrix in terms of coherent states.

Basic formulae for the description of the photocount statistics of Gaussian light have already been given in Chapter 10 (e.g. equations (10.45)–(10.49)) but some more general results can also be derived.

Returning to the characteristic generating function in the form (14.22) we obtain for Gaussian light

$$\langle \exp(\mathrm{i}x\bar{W})\rangle_{\mathcal{N}} = \int \exp\left(-\sum_\lambda \frac{|\alpha_\lambda|^2}{\langle n_\lambda\rangle} + \mathrm{i}x\sum_{\lambda,\lambda'} \alpha_\lambda^* \bar{W}_{\lambda\lambda'}\alpha_{\lambda'}\right)\prod_\lambda \frac{\mathrm{d}^2\alpha_\lambda}{\pi\langle n_\lambda\rangle}, \tag{17.33}$$

where

$$\bar{W}_{\lambda\lambda'} = \iint_0^T \iint_V \varepsilon^*(x', \lambda)\, \mathcal{S}(x' - x'')\, \varepsilon(x'', \lambda')\, \mathrm{d}^4x'\, \mathrm{d}^4x'' \tag{17.34}$$

and $V(x)$ corresponding to (17.9) has been used. Making use of the substitutions

$$\beta_\lambda = \frac{\alpha_\lambda}{\langle n_\lambda\rangle^{1/2}}, \qquad \langle n_\lambda\rangle^{1/2}\, \bar{W}_{\lambda\lambda'}\langle n_{\lambda'}\rangle^{1/2} = U_{\lambda\lambda'}, \tag{17.35}$$

equation (17.33) becomes

$$\langle \exp(\mathrm{i}x\bar{W})\rangle_{\mathcal{N}} = \int \exp(-\sum_\lambda |\beta_\lambda|^2 + \mathrm{i}x\sum_{\lambda,\lambda'} \beta_\lambda^* U_{\lambda\lambda'}\beta_{\lambda'})\prod_\lambda \frac{\mathrm{d}^2\beta_\lambda}{\pi} =$$

$$= \int \exp(-\sum_\lambda |\beta_\lambda|^2 + \mathrm{i}x\sum_\lambda \mathcal{U}_\lambda |\beta_\lambda|^2)\prod_\lambda \frac{\mathrm{d}^2\beta_\lambda}{\pi} = \prod_\lambda \frac{1}{(1 - \mathrm{i}x\mathcal{U}_\lambda)}, \tag{17.36}$$

where the $\mathcal{U}_\lambda$ are eigenvalues of the matrix $\hat{U} \equiv (U_{\lambda\lambda'})$. For a field in the normalization volume the quantity W ($\bar{W} = \eta W$) given by (14.12) is suitable for the description of the field and $\mathscr{U}_\lambda = \langle n_\lambda \rangle$. If moreover all the $\langle n_\lambda \rangle$ are assumed to be equal so that $\mathscr{U}_\lambda = \langle n \rangle / M$, where M is the number of modes in the field, we arrive at (10.45).

The characteristic function (17.36) can be written in the form

$$\begin{aligned}\langle \exp(\mathrm{i}x\bar{W}) \rangle_{\mathcal{N}} &= \frac{1}{\operatorname{Det}(\hat{1} - \mathrm{i}x\hat{U})} = \exp\Big[-\sum_\lambda \ln(1 - \mathrm{i}x\mathscr{U}_\lambda)\Big] = \\ &= \exp\left[\sum_\lambda \sum_{n=1}^{\infty} \frac{(\mathrm{i}x\mathscr{U}_\lambda)^n}{n}\right] = \exp\left[\sum_{n=1}^{\infty} \frac{(\mathrm{i}x)^n}{n} \operatorname{Tr} \hat{U}^n\right] = \\ &= \exp\big[-\operatorname{Tr} \ln(\hat{1} - \mathrm{i}x\hat{U})\big] \end{aligned} \tag{17.37}$$

and comparing this result with the relation defining the cumulants $\varkappa_j^{(W)}$ (cf. (10.15)) we can write for the factorial cumulants

$$\begin{aligned}\varkappa_1^{(W)} &= \operatorname{Tr} \hat{U} = \iint_0^T \iint_V \sum_\lambda \varepsilon^*(x', \lambda)\, \mathscr{S}(x' - x'')\, \varepsilon(x'', \lambda) \langle n_\lambda \rangle\, \mathrm{d}^4x'\, \mathrm{d}^4x'' = \\ &= \iint_0^T \iint_V \mathscr{S}(x' - x'')\, \Gamma_{\mathcal{N}}^{(1,1)}(x', x'')\, \mathrm{d}^4x'\, \mathrm{d}^4x'', \\ \varkappa_j^{(W)} &= (j-1)! \operatorname{Tr} \hat{U}^j = (j-1)! \iint \prod_{r=1}^{j} \mathscr{S}(x'_r - x''_r)\, \Gamma_{\mathcal{N}}^{(1,1)}(x'_r, x''_{r+1}) \times \\ &\quad \times \mathrm{d}^4x'_r\, \mathrm{d}^4x''_r, \qquad x'_{j+1} = x'_1, \qquad j \geqq 2. \end{aligned} \tag{17.38a}$$

Note that the first term in the series in (17.37) (i.e. $\mathrm{i}x \operatorname{Tr} \hat{U}$) corresponds to the fully coherent field. The presence of the other terms violates the full coherence of the field. If $\mathscr{S}(x' - x'') = \eta\delta(x'' - x')\,\delta(\mathbf{x}' - \mathbf{x})$ where $\mathbf{x}$ is a fixed point, and if we assume a stationary field, we obtain the following simplified expressions for factorial cumulants

$$\begin{aligned}\varkappa_1^{(W)} &= \eta T \langle I \rangle_{\mathcal{N}}, \\ \varkappa_j^{(W)} &= (j-1)!\, (\eta \langle I \rangle_{\mathcal{N}})^j \int_0^T \ldots \int_0^T \gamma_{\mathcal{N}}(t_1 - t_2)\, \gamma_{\mathcal{N}}(t_2 - t_3) \ldots \gamma_{\mathcal{N}}(t_j - t_1) \times \\ &\quad \times \mathrm{d}t_1\, \mathrm{d}t_2 \ldots \mathrm{d}t_j, \qquad j \geqq 2. \end{aligned} \tag{17.38b}$$

In order to calculate the characteristic function (17.36) we must find the eigenvalues $\mathscr{U}_\lambda$ of the matrix $\hat{U}$ with elements given in (17.35), where $\bar{W}_{\lambda\lambda'}$ is given by (17.34). This problem was considered by Bédard (1966a), Jakeman and Pike (1968), Dialetis (1969a) and Rousseau (1969b) for Gaussian Lorentzian light, by Jaiswal and Mehta (1969) (see also Mehta (1970)) for partially polarized Gaussian light, and by Jakeman and Pike (1969) and Peřina and Peřinová (1971) for the superposition of coherent and Gaussian Lorentzian light. Considering a complete system of orthonormal functions $\{\varphi_\lambda(\mathbf{x}, t)\}$ over the interval $(0, T)$ and the volume V, we can write

$$\hat{A}^{(+)}(x) = \sum_\lambda \hat{b}_\lambda \varphi_\lambda(x), \tag{17.39a}$$

where

$$\hat{b}_\lambda = \int_0^T \int_V \hat{A}^{(+)}(x)\, \varphi_\lambda^*(x)\, d^4x. \tag{17.39b}$$

Thus for a field stationary in time and in space

$$\begin{aligned}\mathrm{Tr}\,\{\hat{\varrho}\hat{b}_\lambda^+ \hat{b}_{\lambda'}\} &= \mathscr{U}_\lambda \delta_{\lambda\lambda'} = \\ &= \iint_0^T \iint_V \mathrm{Tr}\,\{\hat{\varrho}\hat{A}^{(-)}(x)\,\hat{A}^{(+)}(x')\}\, \varphi_\lambda(x)\, \varphi_{\lambda'}^*(x')\, d^4x\, d^4x'.\end{aligned} \tag{17.40}$$

Multiplying this equation by $\varphi_{\lambda'}(x'')$ and summing over λ', we finally obtain the integral equation determining $\mathscr{U}_\lambda$ and $\varphi_\lambda(x)$

$$\mathscr{U}_\lambda \varphi_\lambda(x'') = \int_0^T \int_V \Gamma_{\mathscr{N}}^{(1,1)*}(x'', x)\, \varphi_\lambda(x)\, d^4x, \tag{17.41}$$

since $\sum_\lambda \varphi_\lambda^*(x')\, \varphi_\lambda(x'') = \delta(x' - x'')$. The eigenvalues $\mathscr{U}_\lambda$ will depend on the spectral properties of the radiation since (17.41) involves the correlation function $\Gamma_{\mathscr{N}}^{(1,1)*}(x'', x) = \Gamma_{\mathscr{N}}^{(1,1)}(x - x'')$ as the kernel. The homogeneous integral equation, in a more restricted form (in the time domain only), was solved for a spectral line with a Lorentzian profile in the previously quoted papers.

It is obvious that an exact calculation of the characteristic function is a very difficult problem in general, although all the mathematics is relatively simple if $T \ll \tau_c \approx 1/\Delta\nu$ or $T \gg \tau_c$. In the first case the correlation function is practically constant over the time interval $(0, T)$ ($\mathscr{S}(x' - x'') = \eta\delta(x'' - x')\,\delta(\mathbf{x}' - \mathbf{x})$ is assumed for simplicity) so that $\varkappa_j^{(W)} = (j - 1)!\,(\eta\langle I\rangle_{\mathscr{N}}\, T)^j$ and

$$\begin{aligned}\langle \exp(\mathrm{i}x\hat{W})\rangle_{\mathscr{N}} &= \exp\left[\sum_{n=1}^{\infty} \frac{(\mathrm{i}x)^n}{n} (\eta\langle I\rangle_{\mathscr{N}}\, T)^n\right] = \\ &= \exp\left[-\ln(1 - \mathrm{i}x\eta\langle I\rangle_{\mathscr{N}}\, T)\right] = \frac{1}{1 - \mathrm{i}x\eta\langle I\rangle_{\mathscr{N}}\, T};\end{aligned} \tag{17.42}$$

this corresponds to (10.41) appropriate for the Bose-Einstein distribution (10.39). In the second case we obtain (10.45) with $p(n, T)$ given by (10.47) or (10.73) where $M = T/\tau_c$. Hence formula (10.47) holds for $T \gg \tau_c$ and also for $T \ll \tau_c$ (if $M = 1$ in this case); this suggests that perhaps it represents a good approximation for all T. Indeed it was shown by Bédard, Chang and Mandel (1967), using the exact recursion relation for $p(n, T)$ derived by Bédard (1966a), that it represents a good approximation to $p(n, T)$ for all T if

$$M = \frac{T}{\xi(T)} = \frac{T^2}{2\int_0^T (T - t')\,|\gamma_{\mathscr{N}}(t')|^2\, dt'}, \tag{17.43}$$

where (10.69) has been used and the second exact and approximate moments have been compared. This was explicitly verified for the Lorentzian, Gaussian and rectangular spectral profiles. An interesting feature of the Mandel-Rice formula (10.47) is that it is explicitly independent of the form of the spectrum, which enters through M only. Similar results have been obtained for the superposition of coherent and chaotic fields and are discussed in Sec. 17.3 (Horák, Mišta and Peřina (1971a, b), Mišta and Peřina (1971), Peřina, Peřinová and Mišta (1971)).

Another approximate expression for $p(n, T)$ for a Lorentzian line shape, when

$$\langle n_\lambda \rangle \hbar\omega_\lambda = \frac{\text{constant}}{(\omega_k - \omega_0)^2 + \Gamma^2} \tag{17.44}$$

(ω_0 is the mean frequency and Γ is the halfwidth of the line) and $T \gg 1/\Gamma$, was derived by Glauber (1963c, 1965, 1966a) (see also Klauder and Sudarshan (1968), p. 225). He found that

$$p(n, T) = \frac{1}{n!}\left(\frac{2\Omega T}{\pi}\right)^{1/2}\left(\frac{\Gamma\eta\langle I\rangle_{\mathcal{N}}\, T}{\Omega}\right)^n K_{n-1/2}(\Omega T)\exp(\Gamma T), \tag{17.45}$$

where $\Omega = (\Gamma^2 + 2\eta\langle I\rangle_{\mathcal{N}}\,\Gamma)^{1/2}$ and $K_{n-1/2}$ is a modified Hankel function of half-integral order. The asymptotic form of (17.45) is

$$p(n, T) = \frac{\langle n\rangle}{(2\pi\mu n^3)^{1/2}}\exp\left[-\frac{1}{2\mu}\left(n^{1/2} - \frac{\langle n\rangle}{n^{1/2}}\right)^2\right], \tag{17.46}$$

where $\mu = \eta\langle I\rangle_{\mathcal{N}}/\Gamma$ and $\langle n\rangle = \eta\langle I\rangle_{\mathcal{N}}\,T$. The formula (17.45) was also compared with the exact solution (Bédard, Chang and Mandel (1967)) and was found to be in good agreement for $T/\tau_c \geqq 10$. Yet another asymptotic expression for $p(n, T)$ has been suggested by McLean and Pike (1965).

As we have noted in Chapter 10 the experimental verification of the validity of (10.39) for chaotic light was given by Arecchi (1965), Arecchi, Berné and Burlamacchi (1966), Arecchi, Berné, Sona and Burlamacchi (1966), Freed and Haus (1965), Arecchi (1969), Pike (1969, 1970) and Bendjaballah (1971). The multimode expression (10.47) has been verified for Gaussian light with M degrees of freedom by Martienssen and Spiller (1966a), who also verified the validity of (10.56) for polarized ($P = 1$) and unpolarized ($P = 0$) light. A measurement of the time evolution of a stationary Gaussian field in terms of joint photocount distributions was carried out by Arecchi, Berné and Sona (1966). Further investigations of the photocount statistics of Gaussian light have been performed by Bures, Delisle and Zardecki (1971, 1972a–c) and Cantrell and Fields (1973).

In measurements of the photocount statistics, the so-called dead-time effect occurs; this means that for the counter there exists a time interval τ_D after each registration, during which no photoemission can be registered. Consequently the number of events and the variance registered during the counting interval T will be smaller (a kind of antibunching of pulses) than the actual number of events, so the measured photocount distribution must be corrected. These questions have been studied by Johnson, Jones, McLean and Pike (1966), Bédard (1967c), Cantor and Teich (1975), Srinivasan (1974, 1978), Teich and McGill (1976), Saleh (1978) and Vannucci and Teich (1981).

Finally we give the s-ordered form of some of the foregoing equations. It is clear from (17.24) that the equations related to a particular s-ordering can be obtained by substituting $\langle n_\lambda \rangle \to \langle n_\lambda \rangle + (1 - s)/2$ in the corresponding equations. Thus we obtain from (16.64) (considered for $s_1 = 1$ and $s_2 = s$) and (10.45)

$$\langle \exp(\mathrm{i}xW) \rangle_s = \left[1 - \mathrm{i}x\left(\frac{\langle n \rangle}{\eta M} + \frac{1-s}{2}\right)\right]^{-M}, \tag{17.47}$$

where $\langle n \rangle$ is the mean number of counts. This leads to

$$P(W, s) = \frac{1}{\left(\dfrac{\langle n \rangle}{\eta M} + \dfrac{1-s}{2}\right)^M} \frac{W^{M-1}}{\Gamma(M)} \exp\left\{-\frac{W}{\dfrac{\langle n \rangle}{\eta M} + \dfrac{1-s}{2}}\right\} \tag{17.48}$$

and

$$\langle W^k \rangle_s = \left(\frac{\langle n \rangle}{\eta M} + \frac{1-s}{2}\right)^k \frac{\Gamma(k+M)}{\Gamma(M)}. \tag{17.49}$$

Finally let us mention some results for Fermi particles. Corresponding to the bunching effect for photons, occurring as a consequence of the Bose-Einstein statistics with the requirement of maximum entropy, an antibunching effect for chaotic fermions can occur even though no coherent states exist for them. This point was discussed by Glauber (1970) on the basis of wave packets constructed for chaotic fields and by Bénard (1969, 1970a, b, 1975) using the theory of Goldberger and Watson (1965) which is valid for bosons as well as fermions. For a discussion of the coherence properties of fermions we refer the reader to papers by Ledinegg (1967) and Ledinegg and Schachinger (1983).

17.2 Laser radiation

The statistical properties of laser radiation (above the threshold of oscillations) are qualitatively different from those of Gaussian radiation; this fact was first recognized by Golay (1961). A complete investigation of the statistical and coherence properties of laser radiation and of the laser itself requires the solution of the equations of motion for a system consisting of the radiation in interaction with atoms of the lasing active matter and the corresponding reservoirs (heat baths) describing pumping light, cavity losses, lattice vibrations, etc. The main difference between laser radiation and thermal radiation lies in the fact that laser radiation is produced mainly by stimulated emission rather than by spontaneous emission and that very strong coupling of modes exists, associated with the non-linearity of the equations of motion. This nonlinearity of the laser theory is a characteristic one and only nonlinear theories are able to explain successfully all properties of the laser. Moreover the nonlinearity plays an important role in the stability of the laser. In the literature the statistical properties of laser radiation are treated in three ways, using

a) the Langevin equations for the complex amplitudes (these are the usual equations of motion with stochastic forces, describing a Markoff process, added),

b) the classical Fokker-Planck equation for the probability distribution and

c) the master equation for the density matrix;

the last method is a completely quantal one, and by using the coherent-state technique, the generalized Fokker-Planck equation and the Heisenberg-Langevin equations can be derived, including all the quantum properties of the system (Chapter 20). However, we will demonstrate these methods by examining non-linear optical processes rather than the laser, which has been treated in many works (Haken (1967, 1970a, b, 1972, 1978), Lax (1967a, b, 1968a, b), Risken (1968, 1970), Scully and Lamb (1967, 1968), Willis (1966), Lax and Louisell (1967), Lax and Yuen (1968), Fleck (1966a, b), Paul (1966, 1969), Lax and Zwanziger (1973), Nussenzveig (1973), Loudon (1973), Louisell (1973), Sargent and Scully (1972), Sargent, Scully and Lamb (1974), Yariv (1967, 1975)). An interesting approach to the problem based on an intensive use of the quantum electrodynamical formalism of Schwinger, using generalized Green functions and functional derivatives, was suggested by Korenman (1966, 1967). In this section we treat only the case of the ideal laser and summarize the main results of the nonlinear theory (leading to the Risken distribution) describing the laser statistics over the whole region of operation (below, at and above the threshold of oscillations). A powerful model of the superposition of coherent and chaotic fields, describing the statistical properties of the laser radiation above threshold, is applicable when a linear approximation to the theory can be used. This model will be given in the next section in greater detail. Experimental verifications of the laser theory, which are also mentioned here, were reviewed by Armstrong and Smith (1967), Pike (1970), Arecchi and Degiorgio (1972) and Pike and Jakeman (1974).

Ideal laser model

The ideal laser model is an example of the connection of the coherent state and the atomic coherent state, as discussed at the end of Sec. 13.5. The radiation field is connected with the dipole transitions of atoms in the active medium of the laser. The polarization of these atoms oscillates with the field and they radiate energy into the field. The whole active medium has an oscillating polarization density and, since the time derivative of the polarization density gives the prescribed current distribution, the radiation field can be regarded as a product of oscillating classical (c-number) non-random currents, provided that the laser is stabilized. Therefore the radiation field of the laser is in a coherent state $|\beta\rangle$. The density matrix describing this field is

$$\hat{\varrho} = |\beta\rangle\langle\beta| \tag{17.50}$$

and

$$\Phi_{\mathcal{N}}(\alpha) = \delta(\alpha - \beta) = \frac{\delta(|\alpha| - |\beta|)}{|\alpha|}\,\delta(\varphi - \psi), \tag{17.51}$$

where $\alpha = |\alpha| \exp(i\varphi)$ and $\beta = |\beta| \exp(i\psi)$; the factor $1/|\alpha|$ appears here since $d(\mathrm{Re}\,\alpha)\,d(\mathrm{Im}\,\alpha) = |\alpha|\,d|\alpha|\,d\varphi$. In the Schrödinger picture $\hat{\varrho} = |\beta \exp(-i\omega t)\rangle \times \langle\beta \exp(-i\omega t)|$ so that $\langle\hat{a}\rangle = \beta \exp(-i\omega t)$ is not a stationary state since it varies with time t. The distribution (17.51) expresses the fact that both the intensity $|\alpha|^2$ and the phase φ of the complex amplitude are perfectly stabilized.

Unfortunately the phase φ of oscillations at high optical frequencies is not usually under control and we have to assume that the phase is uniformly distributed in the interval $(0, 2\pi)$. Thus we obtain

$$\Phi_{\mathcal{N}}(\alpha) = \frac{1}{2\pi}\int_0^{2\pi} \frac{\delta(|\alpha| - |\beta|)}{|\alpha|}\,\delta(\varphi - \psi)\,d\varphi = \\ = \frac{\delta(|\alpha| - |\beta|)}{2\pi|\alpha|} = \frac{1}{\pi}\,\delta(|\alpha|^2 - |\beta|^2). \tag{17.52}$$

The multimode function $\Phi_{\mathcal{N}}(\{\alpha_\lambda\})$ will have, for such a single-mode ideal laser at mode λ, the form

$$\Phi_{\mathcal{N}}(\{\alpha_\lambda\}) = \frac{\delta(|\alpha_\lambda| - |\beta_\lambda|)}{2\pi|\alpha_\lambda|} \prod_{\lambda' \neq \lambda} \frac{\delta(|\alpha_{\lambda'}|)}{2\pi|\alpha_{\lambda'}|}, \tag{17.53}$$

the other modes being in the vacuum state. Mixed states formed as a superposition of coherent states with various phases are also coherent and the field described by (17.53) fulfils the full coherence conditions (12.34) for $m = n$. The distribution (17.53) is independent of the phases $\{\varphi_\lambda\}$ and therefore it describes a stationary field. This distribution corresponds to the intensity of a perfectly stabilized laser where the phase is random (this is the case already described in classical terms by equation (10.59)); the corresponding photocount distribution is Poissonian and is given by (10.60). For the probability distribution $P_{\mathcal{N}}(I)$ we have equation (10.59) again from (17.52) (or (17.53)) since $\langle I\rangle_{\mathcal{N}} = (2\pi\hbar c/L^3)\,k^{-1} \times |e_\mu^{(s)}(\mathbf{k})|^2\,|\alpha_{\mathbf{k}s}|^2$.

The corresponding density matrix elements $\varrho(n, m)$ for the distribution (17.51) are obtained from (13.63) as

$$\varrho(n, m) = \frac{\beta^n \beta^{*m}}{(n!\,m!)^{1/2}} \exp(-|\beta|^2) \tag{17.54}$$

and the normally ordered moments are equal to

$$\langle \hat{a}^{+k}\hat{a}^l\rangle_{\mathcal{N}} = \alpha^{*k}\alpha^l. \tag{17.55}$$

Putting $n = m$ in (17.54) we obtain the Poissonian probability $p(n) \equiv \varrho(n, n)$ that n photons are present in the field. The distribution $\Phi(\alpha, s)$ and the corresponding moments follow from (16.43) and (16.46) with $s_1 = 1$ and $s_2 = s$,

$$\Phi(\alpha, s) = \frac{2}{1-s} \exp\left(-\frac{2|\alpha - \beta|^2}{1-s}\right) \tag{17.56}$$

and

$$\langle \hat{a}^{+k}\hat{a}^l \rangle_s = \frac{l!}{k!}\left(\frac{1-s}{2}\right)^l \beta^{*k-l} L_l^{k-l}\left(\frac{2|\beta|^2}{s-1}\right), \qquad k \geqq l. \tag{17.57}$$

The corresponding relations for the symmetric and antinormal orderings follow with $s = 0$ and $s = -1$, respectively.

To obtain the quantities $p(n)\,[p(n, T)]$, $\langle \hat{a}^{+k}\hat{a}^l \rangle_{\mathcal{N}} = \delta_{kl}\langle \hat{a}^{+k}\hat{a}^k \rangle_{\mathcal{N}}$, $\Phi(\alpha, s)$ and $\langle \hat{a}^{+k}\hat{a}^l \rangle_s = \delta_{kl}\langle \hat{a}^{+k}\hat{a}^k \rangle_s$ corresponding to the phase averaged distribution (17.52) or (17.53), we may consider a more general case specified by

$$\Phi_{\mathcal{N}}(\{\alpha_\lambda\}) = \prod_{\lambda=1}^{M} \frac{\delta(|\alpha_\lambda|^2 - |\beta_\lambda|^2)}{\pi} \prod_{\lambda=M+1}^{\infty} \frac{\delta(|\alpha_\lambda|)}{2\pi|\alpha_\lambda|}. \tag{17.58}$$

The statistical properties of this model for $M = 2$ and for general M were investigated both theoretically and experimentally by Bertolotti, Crosignani, Di Porto and Sette (1966, 1967b), particularly from the point of view of interference measurements of the Young type.

The distribution $P(W, s)$ in the normalization volume can be calculated as follows

$$\begin{aligned} P(W, s) &= \int \Phi(\{\alpha'_\lambda\}, s)\, \delta\Big(W - \sum_{\lambda=1}^{M} |\alpha'_\lambda|^2\Big) \prod_{\lambda=1}^{M} \frac{d^2\alpha'_\lambda}{\pi} = \\ &= \frac{1}{2\pi} \int_{-\infty}^{+\infty} \exp(-ixW) \Big\langle \exp\Big(ix \sum_{\lambda=1}^{M} |\alpha'_\lambda|^2\Big)\Big\rangle_s dx, \end{aligned} \tag{17.59}$$

where the characteristic function is, considering M modes only in (17.58):

$$\begin{aligned} \Big\langle \exp\Big(ix \sum_{\lambda=1}^{M} |\alpha'_\lambda|^2\Big)\Big\rangle_s &= \left(\frac{2}{1-s}\right)^M \int \prod_{\lambda=1}^{M} \times \\ &\times \exp\left(-\frac{2|\alpha'_\lambda - \beta_\lambda|^2}{1-s} + ix|\alpha'_\lambda|^2\right) \frac{d^2\alpha'_\lambda}{\pi} \end{aligned} \tag{17.60}$$

(the vacuum terms $\delta(|\alpha_\lambda|)/2\pi|\alpha_\lambda|$ in (17.58) can be obtained by putting $\beta_\lambda = 0$ in some modes). Writing $|\alpha'_\lambda - \beta_\lambda|^2 = |\alpha'_\lambda|^2 + |\beta_\lambda|^2 - 2|\alpha'_\lambda||\beta_\lambda| \cos(\varphi'_\lambda - \psi_\lambda)$, we obtain

$$\begin{aligned} \Big\langle \exp\Big(ix \sum_{\lambda=1}^{M} |\alpha'_\lambda|^2\Big)\Big\rangle_s &= \int_0^\infty \prod_{\lambda=1}^{M} \frac{2}{1-s} \exp\left(-\frac{2|\beta_\lambda|^2}{1-s}\right) \times \\ &\times \exp\left(-\frac{2|\alpha'_\lambda|^2}{1-s} + ix|\alpha'_\lambda|^2\right) I_0\left(4\frac{|\alpha'_\lambda||\beta_\lambda|}{1-s}\right) d|\alpha'_\lambda|^2, \end{aligned} \tag{17.61}$$

where I_0 is the modified Bessel function of zero order. Expressing I_0 in a power series and integrating over $|\alpha'_\lambda|^2$, we arrive at

$$\langle \exp(ixW) \rangle_s = \left(1 - ix\frac{1-s}{2}\right)^{-M} \exp\left(\frac{ix\langle n_c \rangle}{1 - ix\dfrac{1-s}{2}}\right), \tag{17.62}$$

where $\langle n_c \rangle = \sum_\lambda |\beta_\lambda|^2$ is the mean number of photons in the coherent field. This expression is the generating function for the Laguerre polynomials $L_n^{M-1}(x)$ since (Morse and Feshbach (1953))

$$(1 - \mathrm{i}xB)^{-M} \exp\left(\frac{\mathrm{i}xA}{1 - \mathrm{i}xB}\right) = \sum_{n=0}^{\infty} \frac{(\mathrm{i}xB)^n}{\Gamma(n + M)} L_n^{M-1}\left(-\frac{A}{B}\right) \tag{17.63a}$$

$$= (1 + B)^{-M} \exp\left(-\frac{A}{1 + B}\right) \sum_{n=0}^{\infty} \frac{1}{\Gamma(n + M)} \times$$
$$\times \left(1 + \frac{1}{B}\right)^{-n} (1 + \mathrm{i}x)^n L_n^{M-1}\left(-\frac{A}{B(1 + B)}\right). \tag{17.63b}$$

Equation (17.62) also follows from (16.64) putting $s_2 = s$, $s_1 = 1$ and taking into account the fact that

$$P_{\mathcal{N}}(W) = \int \prod_\lambda \delta(|\alpha'_\lambda|^2 - |\beta_\lambda|^2)\, \delta(W - \sum_{\lambda'} |\alpha'_{\lambda'}|^2) \prod_{\lambda'} \mathrm{d}|\alpha'_{\lambda'}|^2 =$$
$$= \delta(W - \langle n_c \rangle). \tag{17.64}$$

This corresponds to

$$p(n) = \sum_{\{n_\lambda\}}' \prod_\lambda \frac{|\beta_\lambda|^{2n_\lambda}}{n_\lambda!} \exp(-|\beta_\lambda|^2) = \frac{(\sum_\lambda |\beta_\lambda|^2)^n}{n!} \exp(-\sum_\lambda |\beta_\lambda|^2), \tag{17.65}$$

where the summation $\sum'$ is restricted by the condition $\sum_\lambda n_\lambda = n$ and the multinomial theorem has been used. Substituting (17.62) into (17.59) and using the residue theorem (or substituting (17.64) into (16.66) with $s_1 = 1$ and $s_2 = s$), we obtain

$$P(W, s) = \frac{2}{1 - s}\left(\frac{W}{\langle n_c \rangle}\right)^{(M-1)/2} \exp\left[-\frac{2(W + \langle n_c \rangle)}{1 - s}\right] \times$$
$$\times I_{M-1}\left(4\frac{(W\langle n_c \rangle)^{1/2}}{1 - s}\right); \tag{17.66}$$

the moments $\langle W^k \rangle_s$ are determined from (17.62) by using (17.63a)

$$\langle W^k \rangle_s = \frac{\mathrm{d}^k}{\mathrm{d}(\mathrm{i}x)^k} \langle \exp(\mathrm{i}xW) \rangle_s \bigg|_{\mathrm{i}x=0} =$$
$$= \frac{k!}{\Gamma(k + M)} \left(\frac{1 - s}{2}\right)^k L_k^{M-1}\left(\frac{2\langle n_c \rangle}{s - 1}\right). \tag{17.67}$$

For the normal ordering $s = 1$ and the characteristic function (17.62) reduces to

$$\langle \exp(\mathrm{i}xW) \rangle_{\mathcal{N}} = \exp(\mathrm{i}x\langle n_c \rangle), \tag{17.68}$$

that is the photon statistics are Poissonian and so

$$p(n) = \frac{\langle n_c \rangle^n}{n!} \exp(-\langle n_c \rangle) \tag{17.69}$$

independently of M; the normal moments are

$$\langle W^k \rangle_{\mathcal{N}} = \langle n_c \rangle^k. \tag{17.70}$$

Equations (17.66) and (17.67) tend to (17.64) and (17.70) in the limit $s \to 1$.

The results corresponding to the single-mode distribution (17.52) can be obtained for $M = 1$.

Realistic laser model

A model more realistic than that of the ideal laser was investigated by Glauber (1965) and also by Klauder and Sudarshan (1968) (Sec. 9.2). Such a model based upon the phase diffusion of the complex amplitude leads to a Lorentzian spectrum for laser radiation (see also Louisell (1973) and Peřina (1984)).

Now we can briefly discuss the main results of the nonlinear laser theory based on the Van der Pol oscillator, as developed by Risken (1965, 1966), Hempstead and Lax (1967), Fleck (1966a, b), Scully and Lamb (1966), Weidlich, Risken and Haken (1967), Lax (1967a) and Carmichael and Walls (1974).

When equations of motion are derived for the system of radiation field, atoms of the lasing medium, and the reservoir variables describing pumping and losses, one obtains, after eliminating the atomic and reservoir variables (see e.g. Louisell (1973)), the Langevin equation for the complex amplitude of the radiation from a single-mode laser in the form

$$\frac{\mathrm{d}\alpha}{\mathrm{d}t} = (w - |\alpha|^2)\,\alpha + L, \tag{17.71a}$$

where w is the pumping parameter (the gain coefficient minus the damping constant), L is a Markoffian δ-correlated Langevin force, $\langle L^*(t)\, L(t') \rangle = 4\delta(t - t')$, describing noise from pumping, spontaneous emission, etc., and we have eliminated the $\exp(-\mathrm{i}\omega t)$-dependence. The nonlinear term in (17.71a) is responsible for the saturation of the laser field. Equation (17.71a) is a natural nonlinear generalization of the equation of motion for a harmonic oscillator and the corresponding Fokker-Planck equation (see Chapter 20) can be written in the form

$$\frac{\partial \Phi_{\mathcal{N}}}{\partial t} + \frac{1}{r}\frac{\partial}{\partial r}\left[(w - r^2)\, r^2 \Phi_{\mathcal{N}}\right] = \frac{1}{r}\frac{\partial}{\partial r}\left(r\frac{\partial \Phi_{\mathcal{N}}}{\partial r}\right) + \frac{1}{r^2}\frac{\partial^2 \Phi_{\mathcal{N}}}{\partial \varphi^2} \tag{17.71b}$$

(r and φ are the modulus and the phase of the complex field amplitude and equation (17.71b) is written in the normalized quantities). The pumping parameter $w < 0$ below threshold, $w = 0$ at threshold and $w > 0$ above the threshold of oscillations. The steady-state solution becomes

$$\begin{aligned}\Phi_{\mathcal{N}}(r) &= \frac{N}{2\pi}\exp\left(-\frac{r^4}{4} + w\frac{r^2}{2}\right) = \\ &= \frac{N}{2\pi}\exp\left(\frac{w^2}{4}\right)\exp\left[-\frac{1}{4}(r^2 - w)^2\right],\end{aligned} \tag{17.72}$$

where

$$\frac{1}{N} = \int_0^\infty \exp\left(-\frac{r^4}{4} + w\frac{r^2}{2}\right) r\,dr. \tag{17.73}$$

Returning to the unnormalized intensity I of the field and denoting by I_0 the average intensity at threshold, we can rewrite (17.72) in the form

$$\begin{aligned} P_{\mathcal{N}}(I) &= \frac{2}{\pi I_0}\frac{\exp(-w^2)}{1+\operatorname{erf} w}\exp\left(-\frac{I^2}{\pi I_0^2} + 2\frac{wI}{\pi^{1/2}I_0}\right) = \\ &= \frac{2}{\pi I_0}\frac{1}{1+\operatorname{erf} w}\exp\left[-\frac{1}{\pi I_0^2}(I - \pi^{1/2}wI_0)^2\right], \qquad I \geqq 0, \\ P_{\mathcal{N}}(I) &= 0, \qquad I < 0, \end{aligned} \tag{17.74}$$

where $\operatorname{erf} w = (2/\pi^{1/2})\int_0^w \exp(-x^2)\,dx$. For the mean intensity we obtain

$$\langle I \rangle_{\mathcal{N}} = I_0\left(\pi^{1/2}w + \frac{\exp(-w^2)}{1+\operatorname{erf} w}\right). \tag{17.75}$$

The photocount distribution has been calculated by Smith and Armstrong (1966a) (see also Armstrong and Smith (1967)) in the form

$$\begin{aligned} p(n, T) &= \frac{1}{\pi^{1/2}}\frac{D^n}{n!}\frac{\exp(-wD + \frac{1}{4}D^2)}{1+\operatorname{erf} w}\sum_{m=0}^{n}\binom{n}{m}c^m \times \\ &\times\left[\Gamma\left(\frac{n-m+1}{2}\right) + (-1)^{n-m}\Gamma\left(\frac{n-m+1}{2}, c^2\right)\right], \qquad c \geqq 0, \\ &= \frac{1}{\pi^{1/2}}\frac{D^n}{n!}\frac{\exp(-wD + \frac{1}{4}D^2)}{1+\operatorname{erf} w}\sum_{m=0}^{n}\binom{n}{m}(-1)^m|c|^m \times \\ &\times\left[\Gamma\left(\frac{n-m+1}{2}\right) - \Gamma\left(\frac{n-m+1}{2}, |c|^2\right)\right], \qquad c < 0, \end{aligned} \tag{17.76}$$

where $D = \pi^{1/2}\eta I_0 T$, $c = w - D/2$, η is the detector efficiency, T is the counting time, $\Gamma(a, x) = \int_x^\infty \exp(-t)\,t^{a-1}\,dt$ is the incomplete gamma function. Another form of $p(n, T)$ was obtained by Bédard (1967b)

$$p(n, T) = \frac{N\mu^n}{(2\pi)^{1/2}}\exp\left[\tfrac{1}{4}(\mu^2 - 2J - 2w^2)\right]D_{-n-1}(\mu - 2^{1/2}w), \tag{17.77}$$

where $\mu = (\pi/2)^{1/2}\eta I_0 T$, $J = \pi^{1/2}\eta I_0 T w$, $N = 2/(1 + \operatorname{erf} w)$ and $D_n(z)$ is the parabolic cylinder function (Gradshteyn and Ryzhin (1965), p. 1064). The factorial moments are (Bédard (1967b), Arecchi, Rodari and Sona (1967))

$$\langle \bar{W}^k \rangle_{\mathcal{N}} = \frac{N\mu^k}{(2\pi)^{1/2}}k!\exp(-\tfrac{1}{2}w^2)\,D_{-k-1}(-2^{1/2}w). \tag{17.78}$$

There are three typical regions of operation for the laser. Below threshold $w < 0$ and if $|w|$ is sufficiently large, then

$$P_{\mathcal{N}}(I) \approx \text{constant} \exp\left(-\frac{2|w|I}{\pi^{1/2}I_0}\right), \tag{17.79}$$

which is a negative exponential distribution; thus the laser radiation below threshold is Gaussian. Near and at threshold the correct distribution $P_{\mathcal{N}}(I)$ is given by (17.74) corresponding to the nonlinear oscillator. Well above threshold, where the linearized theory is appropriate and $w > 0$, we may write $I^{1/2} = \pi^{1/4}(wI_0)^{1/2}(1 + \varepsilon)$, where ε is small since the function (17.74) is sharply peaked and so we have from (17.74)

$$\begin{aligned} P_{\mathcal{N}}(I) &= \frac{2}{\pi I_0}\frac{1}{1 + \operatorname{erf} w}\exp\left[-w^2((1+\varepsilon)^2 - 1)^2\right] \approx \\ &\approx \text{constant} \exp\left[-4w^2\frac{[I^{1/2} - \pi^{1/4}(wI_0)^{1/2}]^2}{\pi^{1/2}wI_0}\right] \approx \\ &\approx \text{constant} \exp\left[-\frac{4w}{\pi^{1/2}I_0}[I^{1/2} - (\pi^{1/2}wI_0)^{1/2}]^2\right]. \end{aligned} \tag{17.80}$$

The amplitude probability distribution $P_{\mathcal{N}}(I^{1/2})$ is a Gaussian function centered at $I^{1/2} = (\pi^{1/2}wI_0)^{1/2}$ and it tends to the distribution $\delta(I - \langle I \rangle_{\mathcal{N}})$ in the limit $w \to \infty$, valid for an ideal laser. The distribution (17.74) can be regarded as the smooth δ-function $\delta(I - \pi^{1/2}wI_0)$ in the form of a Gaussian distribution; such a model of laser statistics was also suggested by Bédard (1966b). Some further laser models were discussed by Sillitto (1968), Mandel (1965) and Troup (1968). The evolution of the quantum statistics of stimulated and spontaneous emission has been discussed by Rockower, Abraham and Smith (1978), and questions of saturation have been discussed by Chung, Huang and Abraham (1980).

Now we mention some experimental studies of laser statistics, and those verifying the present model of the laser photon statistics can also be divided into three groups – a) below threshold, b) near and at threshold and c) above threshold.

a) *Below threshold*

As we have seen the photon statistics in this region are Bose-Einstein statistics. The primary work on laser light below threshold was done by Freed and Haus (1965) and Smith and Armstrong (1966b) who verified the validity of the Bose-Einstein distribution (10.39) for $T < \tau_c$. Freed and Haus (1965) also measured sub-threshold counting statistics for $T > \tau_c$, which leads to the verification of (17.46) and (10.47) with M given by (17.43). A review of these experimental results was given by Armstrong and Smith (1967). Similar results were obtained by Arecchi (1965), Arecchi, Berné and Burlamacchi (1966), Arecchi, Berné, Sona and Burlamacchi (1966), who started with a stabilized laser field from a laser far above threshold and then scattered this beam from a moving scattering plate.

b) *Near and at threshold*

Near and at threshold the nonlinear theory must be used for the description of the photon statistics and equations (17.74), (17.76), (17.77) and (17.78) are appropriate. Equation (17.76) was shown to be valid by Smith and Armstrong (1966a) (see Fig. 17.1), who also determined the reduced factorial moment $H_2 = \langle \hat{n}(\hat{n}-1)\rangle/\langle \hat{n}\rangle^2 - 1 = \langle W^2\rangle_{\mathcal{N}}/\langle W\rangle^2_{\mathcal{N}} - 1 = \varkappa_2^{(W)}/\varkappa_1^{(W)2}$, which varies from 1 for Gaussian light well below threshold to 0 for an ideal amplitude stabilized laser field well above threshold. The same results were obtained by Arecchi,

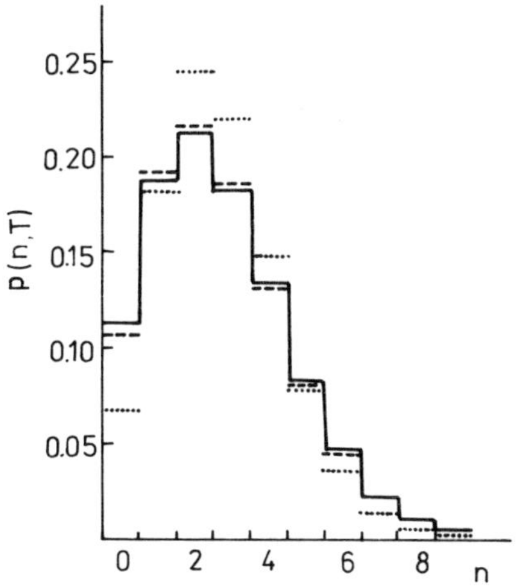

Fig. 17.1 Photocount distribution observed just above threshold (solid line). The Poisson distribution (dotted lines) and the nonlinear oscillator distribution giving the best fit (dashed line) are also shown. For $n = 7, 8$ and 9 the nonlinear oscillator and observed distributions coincide; $\langle I\rangle_{\mathcal{N}}/I_0 \approx 3.25$, $\langle n\rangle = 2.693$, $H_2 = 0.133$ (after A. W. Smith and J. A. Armstrong, 1966, *Phys. Rev. Lett.* **16**, 1169).

Rodari and Sona (1967) and Pike (1969, 1970). Also measurements of the third reduced factorial moment $H_3 = \langle \hat{n}(\hat{n}-1)(\hat{n}-2)\rangle/\langle \hat{n}\rangle^3 - 1 = \langle W^3\rangle_{\mathcal{N}}/\langle W\rangle^3_{\mathcal{N}} - 1$, which varies from 5 for Gaussian light to 0 for laser light, showed very good agreement with the theory (Arecchi (1969)). Further verifications of the validity of the nonlinear theory of the laser were performed by Chang, Detenbeck, Korenman, Alley and Hochuli (1967), Chang, Korenman and Detenbeck (1968) and Chang, Korenman, Alley and Detenbeck (1969), who determined experimentally the second, third and fourth normalized factorial cumulants; and by Davidson and Mandel (1967, 1968) and Davidson (1969), who measured the sixth-order correlation function using the intensity correlation technique. These correlation measurements have been continued by a number of authors (Arecchi, Giglio and Sona (1967), Meltzer and Mandel (1970, 1971), Meltzer, Davis and Mandel (1970), Jakeman, Oliver, Pike, Lax and Zwanziger (1970), Arecchi and Degiorgio (1971), Corti and Degiorgio (1974, 1976a, b), Corti, Degiorgio and Arecchi (1973)), and all the experimental results are in very good agreement with

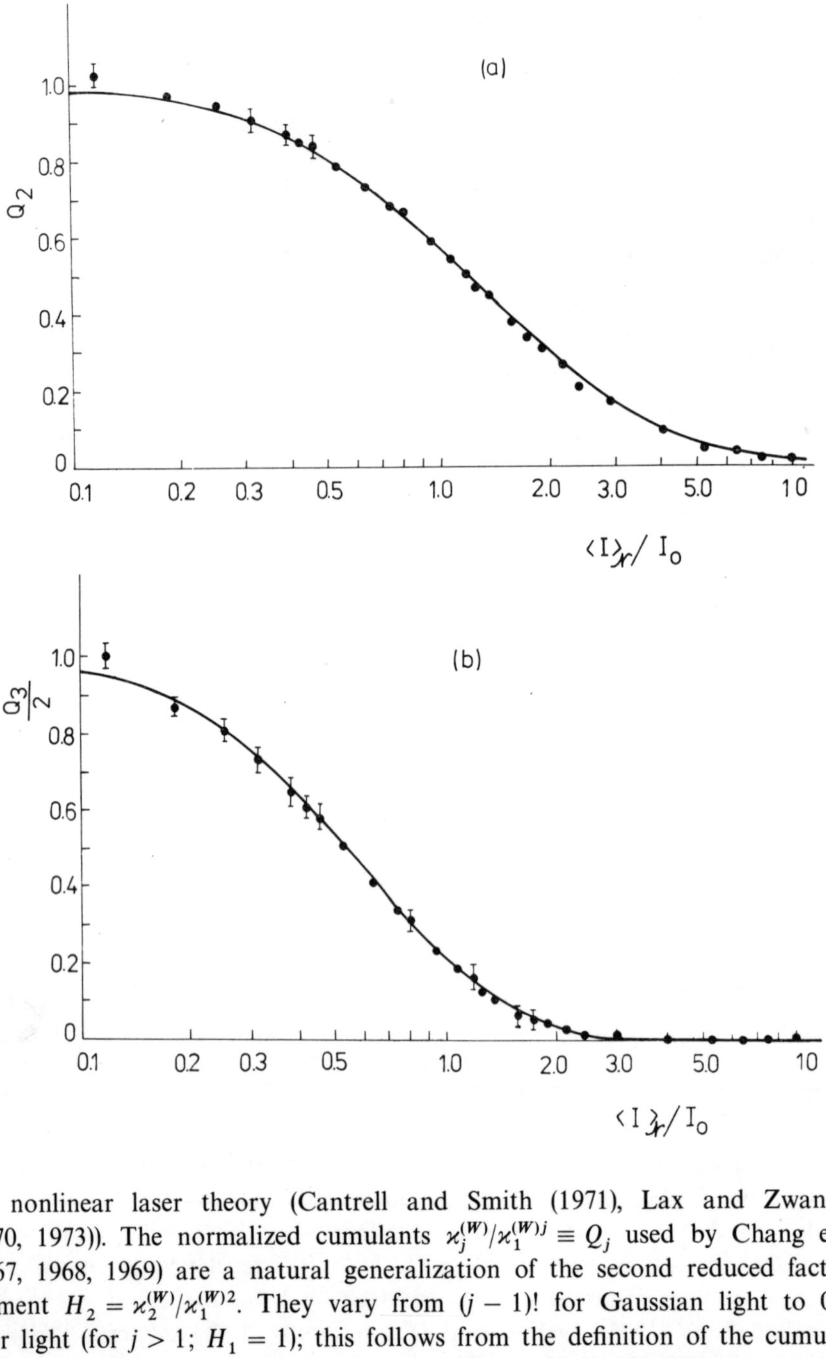

the nonlinear laser theory (Cantrell and Smith (1971), Lax and Zwanziger (1970, 1973)). The normalized cumulants $\varkappa_j^{(W)}/\varkappa_1^{(W)j} \equiv Q_j$ used by Chang et al. (1967, 1968, 1969) are a natural generalization of the second reduced factorial moment $H_2 = \varkappa_2^{(W)}/\varkappa_1^{(W)2}$. They vary from $(j-1)!$ for Gaussian light to 0 for laser light (for $j > 1$; $H_1 = 1$); this follows from the definition of the cumulants (Sec. 10.1) and the characteristic functions (10.41) and (17.68). The measurements of Chang et al. (1967, 1968, 1969) verify the theory in the threshold region from 1/10 to 10 times the threshold intensity. Their results are shown in Fig. 17.2.

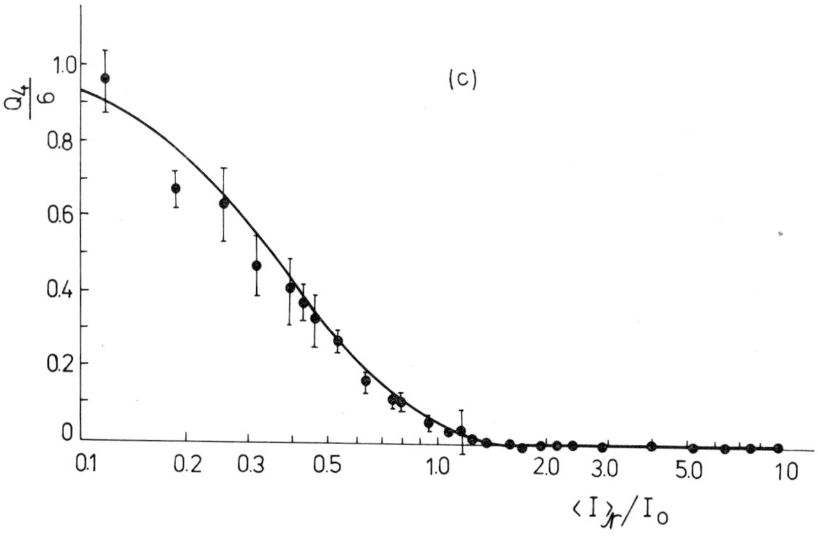

Fig. 17.2 Normalized (a) second, (b) third, and (c) fourth factorial cumulants of laser light plotted as functions of the normalized intensity $\langle I \rangle_{\mathcal{N}}/I_0$ in the threshold region. The curves are the theoretical predictions, the dots are the experimental data (after R. F. Chang, V. Korenman, C. O. Alley and R. W. Detenbeck, 1969, *Phys. Rev.* **178**, 612).

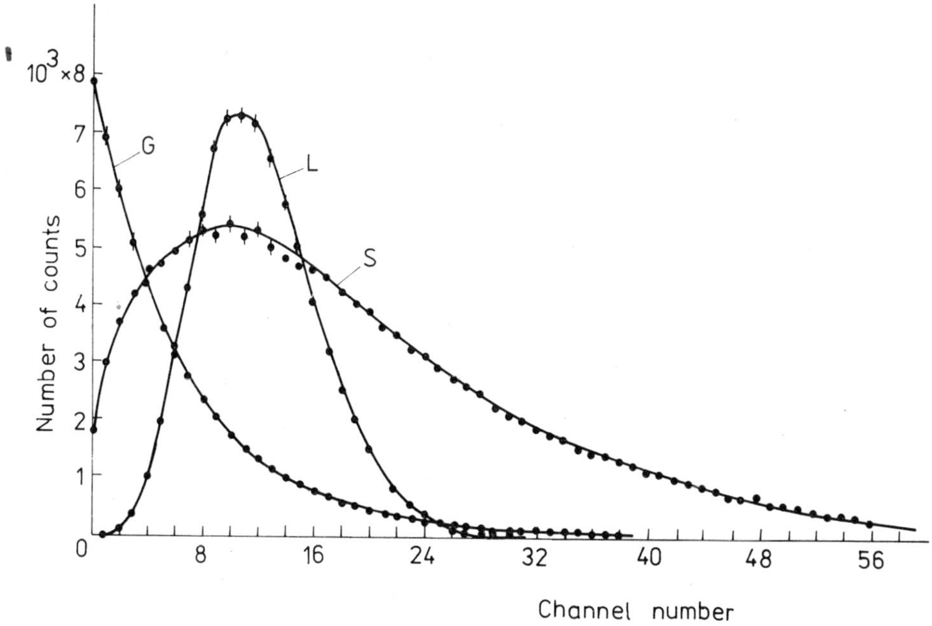

Fig. 17.3 The experimental and theoretical photocount distributions for Gaussian (G), laser (L) and the superposed fields (S) (after F. T. Arecchi, A. Berné, A. Sona and P. Burlamacchi, 1966, *IEEE J. Quant. Electr.* **QE-2**, 341; copyright © 1966 IEEE).

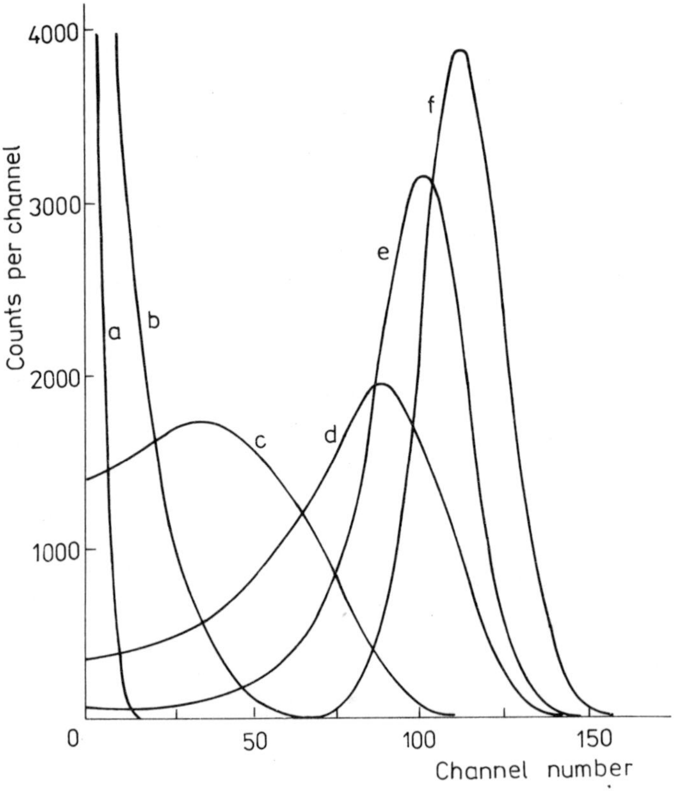

Fig. 17.4 Observed photocount distributions with various time delays for a single-mode gas laser in transient operating conditions for (a) 2.6 μs, (b) 3.7 μs, (c) 4.3 μs, (d) 5 μs, (e) 5.6 μs, and (f) 8.8 μs (after F. T. Arecchi, V. Degiorgio and B. Querzola, 1967, *Phys. Rev. Lett.* **19**, 1168).

Some measurements concerning the dynamics of laser radiation at threshold have been carried out by Arrecchi, Giglio and Sona (1967).

c) *Above threshold*

In the region above threshold with $I/I_0 > 5$ the model of the superposition of fully coherent and Gaussian radiation described by the distribution (17.80) (which is studied in greater detail in the next section) is appropriate for the description of laser statistics. Important experimental work has been done in this region by Arecchi, Berné and Burlamacchi (1966), Arecchi, Berné, Sona and Burlamacchi (1966), Smith and Armstrong (1966a), Freed and Haus (1966), Martienssen and Spiller (1966b) and Magill and Soni (1966). Photocount distributions observed by Arecchi, Berné, Sona and Burlamacchi (1966) are shown in Fig. 17.3. These results verify the validity of the superposition model of coherent and Gaussian light for laser light above threshold, where the linear approximation to the nonlinear laser theory is correct ($I/I_0 > 5$). The corresponding equations for the superposition of signal and noise, which have been verified by experiment, are derived in the next section. The curve L in Fig. 17.3 confirms that well above threshold $p(n, T)$ is Poissonian so that in this region the laser field is in a phase averaged coherent state. The ideal laser model previously given is then appropriate.

The photocount distribution of modulated laser beams was calculated and experimentally determined by Pearl and Troup (1968), Bendjaballah and Perrot (1970), Teich and Vannucci (1978) and Prucnal and Teich (1979).

The transient solutions of the laser Fokker-Planck equation (17.71b) were investigated by Risken and Vollmer (1967), Gnutzmann (1969, 1970), Lax and Zwanziger (1970, 1973) and Zardecki, Bures and Delisle (1972), and transient photocount statistics were measured by Arecchi, Degiorgio and Querzola (1967), Arecchi and Degiorgio (1971) and Jakeman, Oliver, Pike, Lax and Zwanziger (1970). The transient photocount distributions obtained by Arecchi, Degiorgio and Querzola (1967) are shown in Fig. 17.4; they vary from the Bose-Einstein distributions a, b corresponding to operation below threshold, to the stabilized Poisson distribution f well above threshold. A new approach to the investigation of the fluctuations of unstable states, based on a hierarchy of equations for moments following from the Fokker-Planck equation (Chapter 20) has been suggested and used by Arecchi and Politi (1980) and Arecchi, Politi and Ulivi (1982). Note that distributions such as those shown in Fig. 17.4 can be generated in the framework of urn models, reflecting also the dynamics of the system (Blažek (1979)).

17.3 Superposition of coherent and chaotic fields

Because of their importance in describing laser-light statistics above the threshold of operation as well as scattered laser light, the statistics of a superposition of coherent and chaotic fields have been investigated by a number of authors; this work is also relevant to other branches of physics, particularly in optical communication, as a model of the superposition of signal and noise, and in high energy physics (Biyajima (1984), Blažek (1984)). The first papers dealing quantum-mechanically with this subject were published by Lachs (1965), Troup (1966) and Glauber (1966a); in these papers the photocount distribution and its factorial moments for the superposition of one-mode coherent and narrow-band chaotic fields with the same mean frequencies were derived. Another approach to this problem, based on calculation of the correlation functions, was developed by Morawitz (1965, 1966). These results were generalized to multimode fields by Peřina (1967b, 1968a, b), McGill (1967), Lachs (1967, 1971), Peřina and Mišta (1968b) and Fillmore (1969). An extension with respect to the s-ordering was given by Peřina and Horák (1969a, b, 1970). Jakeman and Pike (1969) pointed out that additional spectral information can be obtained using heterodyne detection, the chaotic light being superimposed, before detection, on a known coherent component (Teich (1977)). In this way the central frequency and the spectral halfwidth of the chaotic field can be determined. However, the general heterodyne detection problem for chaotic (thermal) light includes the case when both the central frequency of the chaotic field and the frequency of the coherent field are arbitrary and differ from one another. Single-mode results obtained by these authors were completed and generalized to multimode fields (including s-ordering) by Peřina and Horák (1969a). Experimental results have been reported, and compared with the theory, by Jakeman, Oliver and Pike (1971). These investigations have been continued by a number of authors (Korenman (1967), Aldridge (1969), Fillmore and Lachs (1969), Jaiswal and Mehta (1970), Mehta and Jaiswal (1970), Solimeno, Corti and Nicoletti (1970), Cantrell (1971), Mišta, Peřina and Peřinová (1971), Horák, Mišta and Peřina (1971a, b), Diament and Teich (1971), Mišta (1971), Laxpati and Lachs (1972), Ruggieri, Cummings and Lachs (1972), Saleh (1975a, b, 1978)), including partial polarization (Peřina, Peřinová and Mišta (1971, 1972)), multiphoton processes (Mišta and Peřina (1971)) and propagation of radiation through random media (Diament and Teich (1971), Peřina and Peřinová (1972)).

The quasi-probability function $\Phi_{\mathcal{N}}(\{\alpha_\lambda\})$ for the superposition of coherent and chaotic fields can be derived as the convolution of the Gaussian function (17.7) and the δ-function distribution $\delta(\{\alpha_\lambda\} - \{\beta_\lambda\}) = \prod_\lambda^M \delta(\alpha_\lambda - \beta_\lambda)$ according to the quantum superposition law (13.147), that is

$$\Phi_{\mathcal{N}}(\{\alpha_\lambda\}) = \iint \prod_\lambda^M \delta(\alpha_\lambda - \alpha'_\lambda - \alpha''_\lambda)\, \delta(\alpha'_\lambda - \beta_\lambda) \frac{\exp\left(-\dfrac{|\alpha''_\lambda|^2}{\langle n_{ch\lambda}\rangle}\right)}{\pi\langle n_{ch\lambda}\rangle} \mathrm{d}^2\alpha'_\lambda\, \mathrm{d}^2\alpha''_\lambda =$$

$$= \prod_\lambda^M \frac{1}{\pi\langle n_{ch\lambda}\rangle} \exp\left(-\frac{|\alpha_\lambda - \beta_\lambda|^2}{\langle n_{ch\lambda}\rangle}\right); \tag{17.81}$$

this is the multimode form of (13.95) (we write $\langle n_{ch\lambda}\rangle$ instead of $\langle n_\lambda\rangle$ for the mean number of photons in the mode λ of the chaotic field to distinguish this from the mean number $\langle n_{c\lambda}\rangle = |\beta_\lambda|^2$ of photons in the mode λ of the coherent field). The multimode form of (16.43) for $s_2 = s$ and $s_1 = 1$ gives

$$\Phi(\{\alpha_\lambda\}, s) = \prod_\lambda^M \frac{1}{\langle n_{ch\lambda}\rangle + \dfrac{1-s}{2}} \exp\left(-\frac{|\alpha_\lambda - \beta_\lambda|^2}{\langle n_{ch\lambda}\rangle + \dfrac{1-s}{2}}\right). \tag{17.82}$$

This equation reduces to (17.24) if $\{\beta_\lambda\} = 0$. For $s = -1$ we can obtain the function $\Phi_{\mathcal{A}}(\{\alpha_\lambda\})$ for the superposition of coherent and chaotic fields (cf. (13.98)).

Single-mode field

First we assume a single-mode field and the mode index λ will be omitted for simplicity. Substituting $\Phi_{\mathcal{N}}(\alpha)$ into (13.63) and performing the integration we find $\varrho(n, m)$ in the form (13.129), which led to the correct $\Phi_{\mathcal{N}}(\alpha)$ in the form (13.95). Another way of calculating $\varrho(n, m)$ is to introduce the generating function $R(z^*, z, \lambda)$ as

$$R(z^*, z, \lambda) = \int \frac{\exp\left(-\dfrac{|\alpha - \beta|^2}{\langle n_{ch}\rangle}\right)}{\pi\langle n_{ch}\rangle} \exp(-\lambda|\alpha|^2 + z^*\alpha + z\alpha^*)\,\mathrm{d}^2\alpha =$$

$$= \frac{1}{\lambda\langle n_{ch}\rangle + 1} \exp\left(-\frac{\lambda|\beta|^2}{\lambda\langle n_{ch}\rangle + 1}\right) \exp\left(\frac{\langle n_{ch}\rangle}{\lambda\langle n_{ch}\rangle + 1}|z|^2 + \frac{\beta^* z + \beta z^*}{\lambda\langle n_{ch}\rangle + 1}\right), \tag{17.83}$$

where (16.41) has been used and $\lambda \geqq 0$ is a real parameter. Using relations of the form (16.47) and (16.49) with $(s_2 - s_1)/2$ replaced by $\langle n_{ch}\rangle/(\lambda\langle n_{ch}\rangle + 1)$, $\hat{a}^+$ by $\beta^*/(\lambda\langle n_{ch}\rangle + 1)$ and $\hat{a}$ by $-\beta/(\lambda\langle n_{ch}\rangle + 1)$, we obtain

$$R(z^*, z, \lambda) = \frac{1}{\lambda\langle n_{ch}\rangle + 1} \exp\left(-\frac{\lambda|\beta|^2}{\lambda\langle n_{ch}\rangle + 1}\right) \sum_{j=0}^{\infty} \sum_{l=0}^{\infty} \frac{z^j}{j!\,j!} \times$$

$$\times \left(\frac{\beta^*}{\lambda\langle n_{ch}\rangle + 1}\right)^{j-l} \left(\frac{\langle n_{ch}\rangle}{\lambda\langle n_{ch}\rangle + 1}\right)^{l} z^{*l} L_l^{j-l}\left(-\frac{|\beta|^2}{\langle n_{ch}\rangle(\lambda\langle n_{ch}\rangle + 1)}\right). \tag{17.84}$$

For $\varrho(n, m)$ we have (Mollow and Glauber (1967a))

$$\varrho(n, m) = \frac{1}{(n!\,m!)^{1/2}} \frac{\partial^n}{\partial z^{*n}} \frac{\partial^m}{\partial z^m} R(z^*, z, 1)\Big|_{z = z^* = 0} =$$

$$= \frac{1}{m!}\left(\frac{n!}{m!}\right)^{1/2} \frac{1}{\langle n_{ch}\rangle + 1} \exp\left(-\frac{|\beta|^2}{\langle n_{ch}\rangle + 1}\right) \frac{\langle n_{ch}\rangle^n}{(\langle n_{ch}\rangle + 1)^m} \beta^{*m-n} \times$$

$$\times L_n^{m-n}\left(-\frac{|\beta|^2}{\langle n_{ch}\rangle(\langle n_{ch}\rangle + 1)}\right), \qquad m \geqq n, \tag{17.85}$$

which is just (13.129). If $n > m$, the condition $\varrho(n, m) = \varrho^*(m, n)$ can be used. The normal moments $\langle \hat{a}^{+k}\hat{a}^l\rangle_{\mathcal{N}}$ can be calculated from (17.84) as follows (Mollow and Glauber (1967a), Peřina and Mišta (1968b))

$$\langle \hat{a}^{+k}\hat{a}^l\rangle_{\mathcal{N}} = \frac{\partial^k}{\partial z^k} \frac{\partial^l}{\partial z^{*l}} R(z^*, z, 0)\Big|_{z = z^* = 0} =$$

$$= \frac{l!}{k!} \langle n_{ch}\rangle^l \beta^{*k-l} L_l^{k-l}\left(-\frac{|\beta|^2}{\langle n_{ch}\rangle}\right), \qquad k \geqq l; \tag{17.86}$$

if $k < l$, then $\langle \hat{a}^{+k}\hat{a}^l\rangle_{\mathcal{N}} = \langle \hat{a}^{+l}\hat{a}^k\rangle^*_{\mathcal{N}}$. This is in agreement with (17.57) with $(1 - s)/2$ replaced by $\langle n_{ch}\rangle$.

All these equations can easily be extended to the s-ordering by the substitution $\langle n_{ch}\rangle \to \langle n_{ch}\rangle + (1-s)/2$, as follows from (17.82).

The special cases of coherent and chaotic fields can be obtained in the limits $\langle n_{ch}\rangle \to 0$ and $\beta \to 0$ respectively.

Equation (17.85) for $m = n$ gives the probability that n photons are in the field. The moments (17.86) for $k = l$ are measurable by means of photodetectors.

Multimode field

Employing calculations (17.60)–(17.62) for the distribution function (17.82) we obtain

$$\langle \exp(\mathrm{i}xW)\rangle_s =$$

$$= \prod_\lambda \left[1 - \mathrm{i}x\left(\langle n_{ch\lambda}\rangle + \frac{1-s}{2}\right)\right]^{-1} \exp\left[\frac{\mathrm{i}x\langle n_{c\lambda}\rangle}{1 - \mathrm{i}x\left(\langle n_{ch\lambda}\rangle + \dfrac{1-s}{2}\right)}\right] \tag{17.87a}$$

$$= \sum_{n=0}^{\infty} (\mathrm{i}x)^n \sum_{\sum_\lambda n_\lambda = n} \prod_\lambda \frac{\left(\langle n_{ch\lambda}\rangle + \dfrac{1-s}{2}\right)^{n_\lambda}}{n_\lambda!} L^0_{n_\lambda}\left(-\frac{\langle n_{c\lambda}\rangle}{\langle n_{ch\lambda}\rangle + \dfrac{1-s}{2}}\right) \tag{17.87b}$$

$$= \sum_{n=0}^{\infty} (1 + \mathrm{i}x)^n \sum_{\sum_\lambda n_\lambda = n} \prod_\lambda \left(1 + \langle n_{ch\lambda}\rangle + \frac{1-s}{2}\right)^{-1} \times$$

$$\times \exp\left(-\frac{\langle n_{c\lambda}\rangle}{\langle n_{ch\lambda}\rangle + 1 + \dfrac{1-s}{2}}\right) \frac{1}{n_\lambda!}\left(1 + \frac{1}{\langle n_{ch\lambda}\rangle + \dfrac{1-s}{2}}\right)^{-n_\lambda} \times$$

$$\times L^0_{n_\lambda}\left(-\frac{\langle n_{c\lambda}\rangle}{\left(\langle n_{ch\lambda}\rangle + \dfrac{1-s}{2}\right)\left(\langle n_{ch\lambda}\rangle + 1 + \dfrac{1-s}{2}\right)}\right), \tag{17.87c}$$

where we have also used (17.63a, b) and $\langle n_{ch\lambda}\rangle$ and $\langle n_{c\lambda}\rangle$ represent the mean numbers of photons in the mode λ of the chaotic and coherent fields respectively.

Thus we arrive at (Peřina and Mišta (1968b))

$$\langle W^k\rangle_s = \left.\frac{\mathrm{d}^k}{\mathrm{d}(\mathrm{i}x)^k}\langle \exp(\mathrm{i}xW)\rangle_s\right|_{\mathrm{i}x=0} =$$

$$= k! \sum_{\sum_\lambda n_\lambda = k} \prod_\lambda \frac{1}{n_\lambda!}\left(\langle n_{ch\lambda}\rangle + \frac{1-s}{2}\right)^{n_\lambda} L^0_{n_\lambda}\left(-\frac{\langle n_{c\lambda}\rangle}{\langle n_{ch\lambda}\rangle + \dfrac{1-s}{2}}\right) \tag{17.88}$$

and

$$p(n) = \frac{1}{n!} \frac{d^n}{d(ix)^n} \langle \exp(ixW) \rangle_{\mathcal{N}} \bigg|_{ix=-1} =$$

$$= \sum_{\substack{\Sigma n_\lambda = n \\ \lambda}} \prod_\lambda (1 + \langle n_{ch\lambda} \rangle)^{-1} \exp\left(- \frac{\langle n_{c\lambda} \rangle}{\langle n_{ch\lambda} \rangle + 1} \right) \frac{1}{n_\lambda!} \times$$

$$\times \left(1 + \frac{1}{\langle n_{ch\lambda} \rangle} \right)^{-n_\lambda} L^0_{n_\lambda} \left(- \frac{\langle n_{c\lambda} \rangle}{\langle n_{ch\lambda} \rangle (1 + \langle n_{ch\lambda} \rangle)} \right). \tag{17.89}$$

The distribution $P(W, s)$ may in principle be obtained from

$$P(W, s) = W^{M-1} \exp(-W) \sum_{n=0}^{\infty} \frac{n!\, L_n^{M-1}(W)}{\Gamma(n + M)} \sum_{j=0}^{n} \frac{(-1)^j \langle W^j \rangle_s}{j!\,(n - j)!\, \Gamma(j + M)}, \tag{17.90}$$

which is analogous to (10.31), under the assumption that

$$\int_0^\infty P^2(W, s)\, W^{1-M} \exp(W)\, dW = \sum_{n=0}^{\infty} \frac{[\Gamma(n + M)]^3}{n!} c_n^2 < \infty, \tag{17.91a}$$

where

$$c_n = \frac{n!}{\Gamma(n + M)} \sum_{j=0}^{n} \frac{(-1)^j \langle W^j \rangle_s}{j!\,(n - j)!\, \Gamma(j + M)}. \tag{17.91b}$$

These formulae can be made simpler if all mean occupation numbers per mode in the chaotic field are equal, i.e. if $\langle n_{ch\lambda} \rangle = \langle n_{ch} \rangle / M$, where $\langle n_{ch} \rangle$ is the mean total number of photons in the chaotic field and M is the number of modes. Equations (17.87a), (17.88) and (17.89) give in this case (Peřina and Horák (1969b, 1970))

$$\langle \exp(ixW) \rangle_s = \left[1 - ix \left(\frac{\langle n_{ch} \rangle}{M} + \frac{1 - s}{2} \right) \right]^{-M} \times$$

$$\times \exp \left[\frac{ix \langle n_c \rangle}{1 - ix \left(\frac{\langle n_{ch} \rangle}{M} + \frac{1 - s}{2} \right)} \right], \tag{17.92}$$

$$\langle W^k \rangle_s = \frac{k!}{\Gamma(k + M)} \left(\frac{\langle n_{ch} \rangle}{M} + \frac{1 - s}{2} \right)^k L_k^{M-1} \left(- \frac{\langle n_c \rangle}{\frac{\langle n_{ch} \rangle}{M} + \frac{1 - s}{2}} \right) \tag{17.93}$$

and (Peřina (1967b, 1968a, b))

$$p(n) = \frac{1}{\Gamma(n + M)} \left(1 + \frac{M}{\langle n_{ch} \rangle} \right)^{-n} \left(1 + \frac{\langle n_{ch} \rangle}{M} \right)^{-M} \times$$

$$\times \exp \left(- \frac{\langle n_c \rangle M}{\langle n_{ch} \rangle + M} \right) L_n^{M-1} \left(- \frac{\langle n_c \rangle M^2}{\langle n_{ch} \rangle (\langle n_{ch} \rangle + M)} \right), \tag{17.94}$$

where $\langle n_c \rangle = \sum_\lambda |\beta_\lambda|^2$, and the following identity for the Laguerre polynomials (Gradshteyn and Ryzhik (1965), Sec. 8.97)

$$\sum_{\sum_\lambda^M n_\lambda = j} \prod_\lambda^M \frac{1}{\Gamma(n_\lambda + \mu_\lambda + 1)} L_{n_\lambda}^{\mu_\lambda}(x_\lambda) = \frac{1}{\Gamma(j + \sum_\lambda^M \mu_\lambda + M)} L_j^{\mathscr{P}}(\sum_\lambda^M x_\lambda) \qquad (17.95)$$

with

$$\mathscr{P} = \sum_\lambda^M \mu_\lambda + M - 1,$$

has been used.

The photocount distribution can be obtained using the substitutions $\langle n_{ch} \rangle \rightarrow \rightarrow \eta \langle n_{ch} \rangle$ and $\langle n_c \rangle \rightarrow \eta \langle n_c \rangle$, where η is the photoefficiency.

The Fourier transform of (17.92) provides the distribution $P(W, s)$ (Peřina and Horák (1969b, 1970))

$$P(W, s) = \left(\frac{\langle n_{ch} \rangle}{M} + \frac{1-s}{2} \right)^{-1} \left(\frac{W}{\langle n_c \rangle} \right)^{(M-1)/2} \times$$
$$\times \exp\left(- \frac{W + \langle n_c \rangle}{\frac{\langle n_{ch} \rangle}{M} + \frac{1-s}{2}} \right) I_{M-1}\left(2 \frac{(\langle n_c \rangle W)^{1/2}}{\frac{\langle n_{ch} \rangle}{M} + \frac{1-s}{2}} \right), \qquad (17.96)$$

in analogy to (17.66).

All families of distributions $P(W, s)$ tend to $P(W, 1) = P_{\mathscr{N}}(W)$ in the strong field limit and the special cases $\langle n_{ch} \rangle \rightarrow 0$ and $\langle n_c \rangle \rightarrow 0$ lead to the corresponding equations for coherent and chaotic fields respectively.

For the heterodyne detection of chaotic light, formulae describing the superposition of coherent and chaotic fields with different mean frequencies are needed. These can be derived from equations (17.87), (17.88) and (17.89) and we give them for the normal ordering only.

Consider the characteristic function (17.87a) for $2M$ modes with mean occupation numbers $\langle n_{ch\lambda} \rangle = \langle n_{ch} \rangle / M$ per mode for $\lambda = 1, 2, \ldots, M$ and $\langle n_{ch\lambda} \rangle = 0$ for $\lambda = M + 1, \ldots, 2M$; $\langle n_{ch} \rangle$ is the mean occupation photon number in the whole chaotic field. Let $\omega_{\lambda\mu}$ characterize the frequency shifts between the λ-mode of the chaotic field and the μ-mode of the coherent field and $\langle n_{c\lambda\mu} \rangle$ be the mean number in the coherent mode μ which can be superimposed on the mode λ of the chaotic field. We introduce the shift parameters $\omega_\lambda^2 = \sum_\mu \langle n_{c\lambda\mu} \rangle \omega_{\lambda\mu}^2 / \langle n_{c\lambda} \rangle$, where $\langle n_{c\lambda} \rangle = \sum_\mu \langle n_{c\lambda\mu} \rangle$, and make the replacements $\langle n_{c\lambda} \rangle \rightarrow \langle n_{c\lambda} \rangle \omega_\lambda^2$ and $\langle n_{c,\lambda+M} \rangle \rightarrow \langle n_{c\lambda} \rangle (1 - \omega_\lambda^2)$ $(\lambda = 1, \ldots, M)$ in (17.87a) with $s = 1$. We then arrive at the following characteristic function for the superposition of M-mode coherent and chaotic fields with frequency shifts described by ω_λ

$$\langle \exp(\mathrm{i}xW)\rangle_{\mathcal{N}} =$$
$$= \prod_{\lambda=1}^{M}\left(1 - \mathrm{i}x\frac{\langle n_{ch}\rangle}{M}\right)^{-1}\exp\left[\frac{\mathrm{i}x\omega_\lambda^2\langle n_{c\lambda}\rangle}{1 - \mathrm{i}x\dfrac{\langle n_{ch}\rangle}{M}} + \mathrm{i}x\langle n_{c\lambda}\rangle(1 - \omega_\lambda^2)\right] =$$
$$= \left(1 - \mathrm{i}x\frac{\langle n_{ch}\rangle}{M}\right)^{-M}\exp\left[\frac{\mathrm{i}x\omega^2\langle n_c\rangle}{1 - \mathrm{i}x\dfrac{\langle n_{ch}\rangle}{M}} + \mathrm{i}x\langle n_c\rangle(1 - \omega^2)\right] =$$
$$= \left(1 - \mathrm{i}x\frac{\langle n_{ch}\rangle}{M}\right)^{-M}\exp\left[\frac{(\mathrm{i}x)^2\,\omega^2\langle n_c\rangle\langle n_{ch}\rangle}{M\left(1 - \mathrm{i}x\dfrac{\langle n_{ch}\rangle}{M}\right)} + \mathrm{i}x\langle n_c\rangle\right], \qquad (17.97)$$

where $\omega^2 = \sum_\lambda^M \omega_\lambda^2\langle n_{c\lambda}\rangle/\langle n_c\rangle$ with $\langle n_c\rangle = \sum_\lambda^M \langle n_{c\lambda}\rangle$ (i.e. $\omega^2 = \sum_{\lambda,\mu}\omega_{\lambda\mu}^2\langle n_{c\lambda\mu}\rangle/\sum_{\lambda,\mu}\langle n_{c\lambda\mu}\rangle$). This characteristic function was first derived for $M = 1$ by Jakeman and Pike (1969), Jaiswal and Mehta (1970) and Mehta and Jaïswal (1970), who showed for chaotic light with a Lorentzian spectral profile that $\omega_{\lambda\mu} = 2\sin(\Omega_{\lambda\mu}/2)/\Omega_{\lambda\mu}$; $\Omega_{\lambda\mu}$ is the difference between the mean frequency of the chaotic (Lorentzian) mode λ and the frequency of the coherent mode μ, multiplied by the time interval T of the observation. Thus the characteristic function (17.97) can be regarded as the product of the characteristic function describing the superposition of chaotic and coherent fields, with mean occupation numbers $\langle n_{ch}\rangle$ and $\langle n_c\rangle\omega^2$, and the characteristic function describing a purely coherent field with the mean occupation number $\langle n_c\rangle(1 - \omega^2)$ (the total mean occupation number in the coherent field is $\langle n_c\rangle\omega^2 + \langle n_c\rangle(1 - \omega^2) = \langle n_c\rangle$). The parameter ω distributes the mean occupation number $\langle n_c\rangle$ between the coherent part of the superposition and the purely coherent field. For $\omega = 1$, $\langle n_c\rangle$ belongs fully to the superposition and we obtain the characteristic function for the superposition with the same frequencies; for $\omega = 0$, $\langle n_c\rangle$ belongs fully to the purely coherent field and we obtain the product of the characteristic functions for the purely chaotic and purely coherent fields.

The distribution $P_{\mathcal{N}}(W)$ can be calculated from (17.97) by means of the Fourier transform, and

$$P_{\mathcal{N}}(W) =$$
$$= \frac{M}{\langle n_{ch}\rangle}\left[\frac{W - \langle n_c\rangle(1 - \omega^2)}{\langle n_c\rangle\omega^2}\right]^{(M-1)/2}\exp\left[-\frac{W + \langle n_c\rangle(2\omega^2 - 1)}{\langle n_{ch}\rangle}M\right] \times$$
$$\times I_{M-1}\left(2|\omega|M\frac{[\langle n_c\rangle(W - \langle n_c\rangle(1 - \omega^2))]^{1/2}}{\langle n_{ch}\rangle}\right), \quad W \geqq \langle n_c\rangle(1 - \omega^2),$$
$$P_{\mathcal{N}}(W) = 0, \qquad W < \langle n_c\rangle(1 - \omega^2). \qquad (17.98)$$

More general expressions for the moments $\langle W^k\rangle_{\mathcal{N}}$ and the photon-number distribution $p(n)$ can be obtained using (17.98), the substitution $W = W' + \langle n_c\rangle(1 - \omega^2)$, (17.93) for $s = 1$, and (17.94) (Peřina and Horák (1969a)), or simply by using the above substitutions in (17.88) for $s = 1$ and (17.89) considered for $2M$ modes. In this way we obtain

$$\langle W^k \rangle_{\mathcal{N}} = k! \sum_{\sum_{\lambda}^{2M} n_\lambda = k} \left(\frac{\langle n_{ch} \rangle}{M} \right)^{\sum_{\lambda}^{M} n_\lambda} \prod_{\lambda}^{M} \frac{1}{n_\lambda!} L_{n_\lambda}^{0} \left(- \frac{\langle n_{c\lambda} \rangle M}{\langle n_{ch} \rangle} \right) \times$$

$$\times \prod_{\lambda' = M+1}^{2M} \frac{\langle n_{c\lambda'} \rangle^{n_{\lambda'}}}{n_{\lambda'}!}. \tag{17.99}$$

Writing

$$\sum_{\sum_{\lambda}^{2M} n_\lambda = k} = \sum_{j=0}^{k} \sum_{\sum_{\lambda}^{M} n_\lambda = j} \sum_{\sum_{\lambda = M+1}^{2M} n_\lambda = k-j}$$

and using the identity (17.95) and the polynomial theorem, we arrive at

$$\langle W^k \rangle_{\mathcal{N}} = \left\langle \frac{\hat{n}!}{(\hat{n} - k)!} \right\rangle = \sum_{j=0}^{k} \frac{k!}{(k-j)!\, \Gamma(j + M)} [\langle n_c \rangle (1 - \omega^2)]^{k-j} \times$$

$$\times \left(\frac{\langle n_{ch} \rangle}{M} \right)^j L_j^{M-1} \left(- \frac{\langle n_c \rangle \omega^2 M}{\langle n_{ch} \rangle} \right). \tag{17.100}$$

Similarly one obtains

$$p(n) = \left(1 + \frac{\langle n_{ch} \rangle}{M} \right)^{-M} \exp\left[- \frac{\langle n_c \rangle [M + \langle n_{ch} \rangle (1 - \omega^2)]}{M + \langle n_{ch} \rangle} \right] \times$$

$$\times \sum_{j=0}^{n} \frac{1}{(n-j)!\, \Gamma(j + M)} [\langle n_c \rangle (1 - \omega^2)]^{n-j} \left(1 + \frac{M}{\langle n_{ch} \rangle} \right)^{-j} \times$$

$$\times L_j^{M-1} \left(- \frac{\langle n_c \rangle \omega^2 M^2}{\langle n_{ch} \rangle (\langle n_{ch} \rangle + M)} \right). \tag{17.101}$$

For $M \to \infty$ the distribution $p(n)$ tends to the Poisson distribution $p(n) = \langle n \rangle^n \exp(-\langle n \rangle)/n!$ with $\langle n \rangle = \langle n_{ch} \rangle + \langle n_c \rangle$ since from (17.97) $\lim_{M \to \infty} \langle \exp(\mathrm{i}xW) \rangle_{\mathcal{N}} = \exp[\mathrm{i}x(\langle n_{ch} \rangle + \langle n_c \rangle)]$.

The photocount distribution based on (17.101) is shown in Fig. 17.5 for (a) $M = 1$, $\Omega = 0$, (b) $M = 5$, $\Omega = 0$, (c) $M = 1$, $\Omega = 100$, and (d) $M = 5$, $\Omega = 100$. The curves a, b, c and d correspond to $\langle n_c \rangle / \langle n_{ch} \rangle = 0/20$, 10/10, 16/4 and 20/0, respectively ($\langle n_c \rangle + \langle n_{ch} \rangle = 20$). In cases (c) and (d) the Poisson distribution is not shown. The dotted curves correspond to measurements with a photon counter (Horák, Mišta and Peřina (1971b)). We observe that the photocount curves are sharper with increasing M, Ω and $\langle n_c \rangle / \langle n_{ch} \rangle$ (and aslo with increasing $M = 2/(1 + P^2)$, i.e. with decreasing P). From 1966 a number of measurements have been performed to verify the validity of this model for laser light above threshold (Armstrong and Smith (1967), Pike (1970), Arecchi and Degiorgio (1972), Ruggieri, Cummings and Lachs (1972), Pike and Jakeman (1974)).

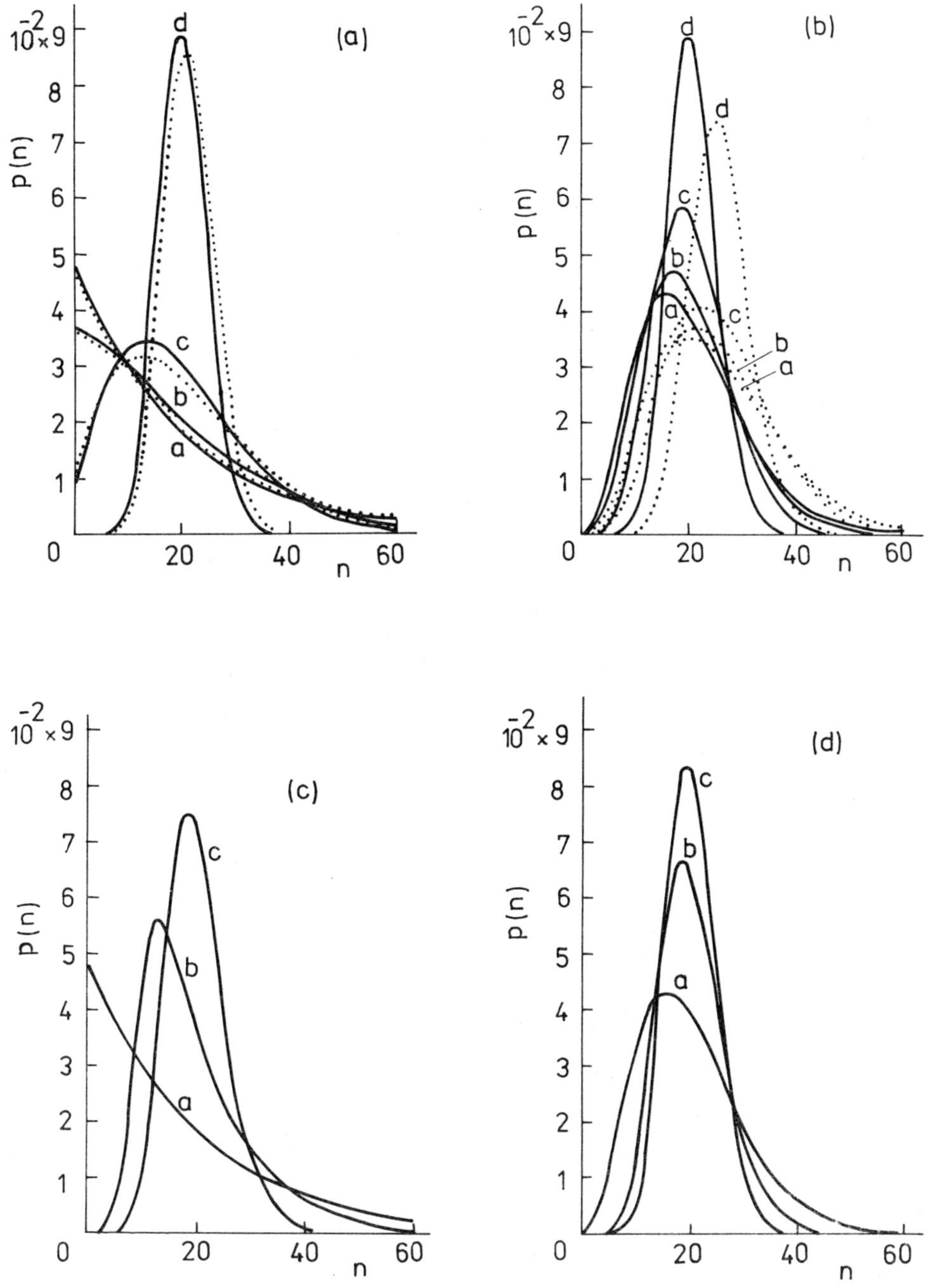

Fig. 17.5 The photocount distribution for (a) $M = 1$, $\Omega = 0$, (b) $M = 5$, $\Omega = 0$, (c) $M = 1$, $\Omega = 100$, and (d) $M = 5$, $\Omega = 100$. The curves *a*, *b*, *c* and *d* are shown for $\langle n_c\rangle/\langle n_{ch}\rangle = 0/20$, 10/10, 16/4 and 20/0 respectively ($\langle n_c\rangle + \langle n_{ch}\rangle = 20$). In figures (c) and (d) the Poisson distribution is not shown. The dotted curves in figures (a) and (b) correspond to measurements with photon counters (after R. Horák, L. Mišta and J. Peřina, 1971, *Czech. J. Phys.* **B 21**, 614).

One can easily write down the corresponding equations when the $\langle n_{ch\lambda}\rangle$ are different from one another (Peřina and Horák (1969a), Peřina, Peřinová and Mišta (1971)). However, they are of a rather complex form.

For the second moment $\langle W^2\rangle_{\mathcal{N}}$ we obtain from (17.100)

$$\langle W^2\rangle_{\mathcal{N}} = \langle n_{ch}\rangle^2\left(1 + \frac{1}{M}\right) + 2\langle n_{ch}\rangle\langle n_c\rangle\left(1 + \frac{\omega^2}{M}\right) + \langle n_c\rangle^2, \quad (17.102a)$$

that is

$$\langle(\Delta W)^2\rangle_{\mathcal{N}} = \frac{\langle n_{ch}\rangle^2 + 2\omega^2\langle n_{ch}\rangle\langle n_c\rangle}{M} \quad (17.102b)$$

and

$$\langle\hat{n}^2\rangle = \langle W\rangle_{\mathcal{N}} + \langle W^2\rangle_{\mathcal{N}} = \langle n_{ch}\rangle + \langle n_c\rangle + \langle n_{ch}\rangle^2\left(1 + \frac{1}{M}\right) + $$
$$+ 2\langle n_{ch}\rangle\langle n_c\rangle\left(1 + \frac{\omega^2}{M}\right) + \langle n_c\rangle^2, \quad (17.103a)$$

giving

$$(\langle\Delta\hat{n})^2\rangle = \langle W\rangle_{\mathcal{N}} + \langle(\Delta W)^2\rangle_{\mathcal{N}} = \langle n_{ch}\rangle + \langle n_c\rangle + $$
$$+ \frac{\langle n_{ch}\rangle^2}{M} + \frac{2\omega^2\langle n_{ch}\rangle\langle n_c\rangle}{M}. \quad (17.103b)$$

The first and the second terms in (17.103b) correspond to the Poisson distribution (with $\langle n\rangle = \langle n_{ch}\rangle + \langle n_c\rangle$), the third term represents the photon bunching effect of the chaotic light and the fourth term represents the photon bunching effect due to interference between the coherent and chaotic components. This term is zero when $\omega = 0$ while the last two terms vanish when $M \to \infty$. The latter case gives the Poissonian variance $\langle(\Delta\hat{n})^2\rangle = \langle n_{ch}\rangle + \langle n_c\rangle$ ($\langle(\Delta W)^2\rangle_{\mathcal{N}} = 0$).

As a special case we obtain for $\omega = 1$ equations (17.92), (17.93), (17.94) and (17.96) (for $s = 1$), and for $\langle n_{ch}\rangle \to 0$ and $\langle n_c\rangle \to 0$ we arrive at the previously given formulae for coherent and chaotic fields, respectively.

The present formulae for the superpositon of coherent and chaotic fields describe generally the superposition of an M-mode chaotic field with an N-mode coherent field ($M \geqq N$). The coherent N-mode field can be generated by a laser operating on N modes well above threshold or by N single-mode lasers operating in this region with their fields superimposed. The statistical properties of light from an M-mode laser operating in the region above threshold where the linearized laser theory is correct (as well as some scattering experiments using an M-mode laser) are also described by the present formulae. In principle the formulae with different mean frequencies are suitable for the description of heterodyne detection of a chaotic M-mode field, when the chaotic field is superimposed on a coherent N-mode field ($M \geqq N$) with frequency shifts characterized by ω_λ. However, the characteristic function (17.97) is of the same form regardless of the number N ($\leqq M$) of coherent modes. Hence the most important case in practice, the detection of an M-mode chaotic field superimposed on a single-mode coherent field, is described by the same characteristic function and the corresponding formulae are obtained by putting $\langle n_{c\lambda} \rangle = 0$ for all modes except the considered mode.

An alternative method of treating the superposition of coherent and chaotic fields uses the correlation function technique. Denoting the s-ordered correlation function as $\Gamma_s^{(n,n)}$, we can write

$$\Gamma_s^{(n,n)}(x_1, \ldots, x_n, x_n, \ldots, x_1) = \int \prod_{\lambda}^{M} \left[\pi \left(\langle n_{ch\lambda} \rangle + \frac{1-s}{2} \right) \right]^{-1} \times$$
$$\times \exp \left(- \frac{|\alpha_\lambda - \beta_\lambda|^2}{\langle n_{ch\lambda} \rangle + \frac{1-s}{2}} \right) V^*(x_1) \ldots V^*(x_n) V(x_n) \ldots V(x_1) \, d^2\{\alpha_\lambda\}. \tag{17.104}$$

The substitution $\alpha_\lambda - \beta_\lambda = \gamma_\lambda$ leads to

$$\Gamma_s^{(n,n)}(x_1, \ldots, x_n, x_n, \ldots, x_1) = \int \prod_{\lambda}^{M} \left[\pi \left(\langle n_{ch\lambda} \rangle + \frac{1-s}{2} \right) \right]^{-1} \times$$
$$\times \exp \left(- \frac{|\gamma_\lambda|^2}{\langle n_{ch\lambda} \rangle + \frac{1-s}{2}} \right) [V^*(x_1) + B^*(x_1)] \ldots [V^*(x_n) + B^*(x_n)] \times$$
$$\times [V(x_n) + B(x_n)] \ldots [V(x_1) + B(x_1)] \, d^2\{\gamma_\lambda\}, \tag{17.105}$$

where $V(x) \equiv V(x, \{\gamma_\lambda\})$ is a Gaussian variable and $B(x) \equiv B(x, \{\beta_\lambda\})$ is a coherent field. Applying the factorization formula (17.15) modified to the s-ordering ($\langle n_\lambda \rangle$ are replaced by $\langle n_{ch\lambda} \rangle + (1-s)/2$) we can in principle calculate all $\Gamma_s^{(n,n)}$. A graphical method for this has been developed by Peřina and Mišta (1968b) and Mišta (1981). Denoting the second-order correlation function of the chaotic field as ${}^{ch}\Gamma_s^{(1,1)}$ we obtain successively

$$\Gamma_s^{(1,1)}(x_1, x_1) = {}^{ch}\Gamma_s^{(1,1)}(x_1, x_1) + |B(x_1)|^2, \tag{17.106a}$$

$$\Gamma_s^{(2,2)}(x_1, x_2, x_2, x_1) = ({}^{ch}\Gamma_s^{(1,1)}(x_1, x_1) + |B(x_1)|^2) \times$$
$$\times ({}^{ch}\Gamma_s^{(1,1)}(x_2, x_2) + |B(x_2)|^2) + |{}^{ch}\Gamma_s^{(1,1)}(x_1, x_2)|^2 +$$
$$+ 2 \operatorname{Re}\{{}^{ch}\Gamma_s^{(1,1)}(x_1, x_2) B^*(x_2) B(x_1)\}, \tag{17.106b}$$

etc. The variance of the number of counts by a detector is (Morawitz (1965), Mandel and Wolf (1966))

$$\langle (\Delta \hat{n})^2 \rangle = \langle n \rangle + \frac{\langle n_{ch} \rangle^2}{T^2} \iint_0^T |{}^{ch}\gamma_{\mathcal{N}}^{(1,1)}(t_1 - t_2)|^2 \, dt_1 \, dt_2 +$$
$$+ 2 \frac{\langle n_{ch} \rangle \langle n_c \rangle}{T^2} \operatorname{Re} \{ \iint_0^T {}^{ch}\gamma_{\mathcal{N}}^{(1,1)}(t_1 - t_2) \, {}^{c}\gamma_{\mathcal{N}}^{(1,1)*}(t_1 - t_2) \, dt_1 \, dt_2 \}, \tag{17.107}$$

where $\langle n \rangle = \langle n_{ch} \rangle + \langle n_c \rangle$, $\langle n_{ch} \rangle = \eta \langle I_{ch} \rangle T$, $\langle n_c \rangle = \eta I_c T$, and ${}^{ch}\gamma_{\mathcal{N}}^{(1,1)}$ and ${}^{c}\gamma_{\mathcal{N}}^{(1,1)}$ are the degrees of coherence for chaotic and coherent fields respectively. In (17.107) the first term is Poissonian, the second describes the Hanbury Brown-Twiss effect of chaotic light, and the third is the interference term.

The comparison of (17.103b) and (17.107) provides the expression for the number of degrees of freedom M,

$$M = \frac{1 + 2\omega^2 \langle n_c \rangle / \langle n_{ch} \rangle}{\mathscr{J}_1 + 2(\langle n_c \rangle / \langle n_{ch} \rangle) \bar{\mathscr{J}}_1} \geqq 1, \tag{17.108}$$

where

$$\mathscr{J}_1 = \frac{1}{T^2} \iint_0^T | {}^{ch}\gamma_{\mathscr{N}}^{(1,1)}(t_1 - t_2) |^2 \, dt_1 \, dt_2,$$

$$\bar{\mathscr{J}}_1 = \frac{1}{T^2} \operatorname{Re} \iint_0^T {}^{ch}\gamma_{\mathscr{N}}^{(1,1)}(t_1 - t_2) \, {}^{c}\gamma_{\mathscr{N}}^{(1,1)}(t_2 - t_1) \, dt_1 \, dt_2. \tag{17.109}$$

For $\langle n_c \rangle = 0$, we have (17.43) from (17.108). The substitution of this M in the formulae (17.98), (17.100) and (17.101) makes it possible to use them as approximations for arbitrary radiation spectra and arbitrary counting time intervals (Horák, Mišta and Peřina (1971a, b)). This approach also permits the inclusion of spatial coherence (Bures, Delisle and Zardecki (1972a), Zardecki, Delisle and Bures (1972, 1973), Peřina and Mišta (1974), Peřina (1977)), partial polarization (Peřina, Peřinová and Mišta (1971, 1972)) and multiphoton absorption (Mišta and Peřina (1971)).

The accuracy of these approximate formulae is better than 1% provided that the signal-to-noise ratio $\langle n_c \rangle / \langle n_{ch} \rangle$ is greater than about 4 and in general the approximate formulae give better agreement for rectangular and Gaussian spectra than for the Lorentzian spectrum (Peřina, Peřinová, Lachs and Braunerová (1973)). A decrease of the degree of polarization P (increase of $M = 2/(1 + P^2)$) and detuning of frequencies lead to an increase of the accuracy (Horák, Mišta and Peřina (1971a, b), Peřina, Peřinová and Mišta (1972), Mišta, Peřina and Braunerová (1973)).

A comparison of the exact (full curves) and approximate (dotted curves) values of the third reduced factorial moment $\langle W^3 \rangle_{\mathscr{N}} / \langle W \rangle_{\mathscr{N}}^3 - 1$ is shown in Fig. 17.6 with $\gamma = \Gamma T$ (Γ being the spectral halfwidth) as independent variable, for (a) $\Omega = 0$, (b) $\Omega = 100$ and for $\langle n_c \rangle / \langle n_{ch} \rangle = 18/2$ (A), $16/4$ (B), $10/10$ (C) and $0/20$ (D) ($\langle n_c \rangle + \langle n_{ch} \rangle = 20$) (Horák, Mišta and Peřina (1971a)).

An interesting behaviour of the photocount distribution occurs for modulated fields (Fray, Johnson, Jones, McLean and Pike (1967), Pearl and Troup (1968), Diament and Teich (1970a), Bendjaballah and Perrot (1971, 1973), Picinbono (1971), Mišta (1973), Kitazima (1974), Teich and Vannucci (1978), Prucnal and Teich (1979), Koňák, Štěpánek, Dvořák, Kupka, Křepelka and Peřina (1982)). Fig. 17.7 represents the comparison of the theoretical photocount distribution (solid curves) with experimental data for triangularly modulated coherent radiation in the absence of the dead-time effect and with the dead-time τ_D (Teich and

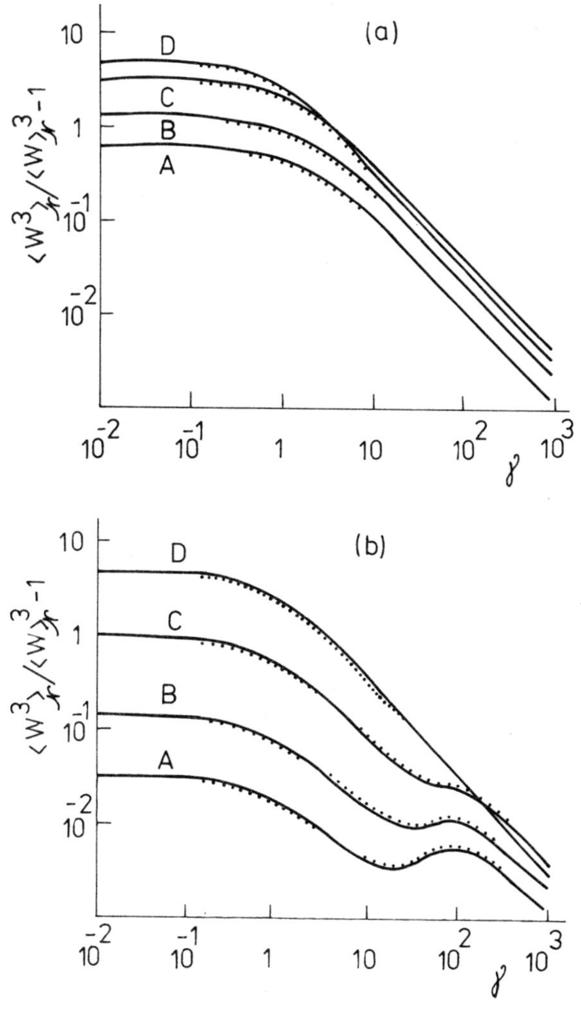

Fig. 17.6 Dependence of the third reduced factorial moment on γ for $\langle n_c\rangle/\langle n_{ch}\rangle = 18/2$ (*A*), 16/4 (*B*), 10/10 (*C*), and 0/20 (*D*) ($\langle n_c\rangle + \langle n_{ch}\rangle = 20$) and for (a) $\Omega = 0$, (b) $\Omega = 100$; the exact values are represented by solid curves and the approximate ones are represented by dotted curves (after R. Horák, L. Mišta and J. Peřina, 1971, *J. Phys.* **A 4**, 231).

Vannucci (1978)). We observe that the photocount distribution can be extremely flat, and that the modulation leads to broadening of the curves and the bunching phenomenon is accentuated. The dead-time decreases both the mean and the variance, leading to a reduction in the number of counts and a kind of "anti-bunching" (Teich and Saleh (1982)). The photocount distributions of extremely weak exponentially modulated luminescence radiation, having a typical two-peak bistable behaviour up to a threshold modulation frequency, have been employed by Koňák et al. (1982) to determine decay times. The comparison of the

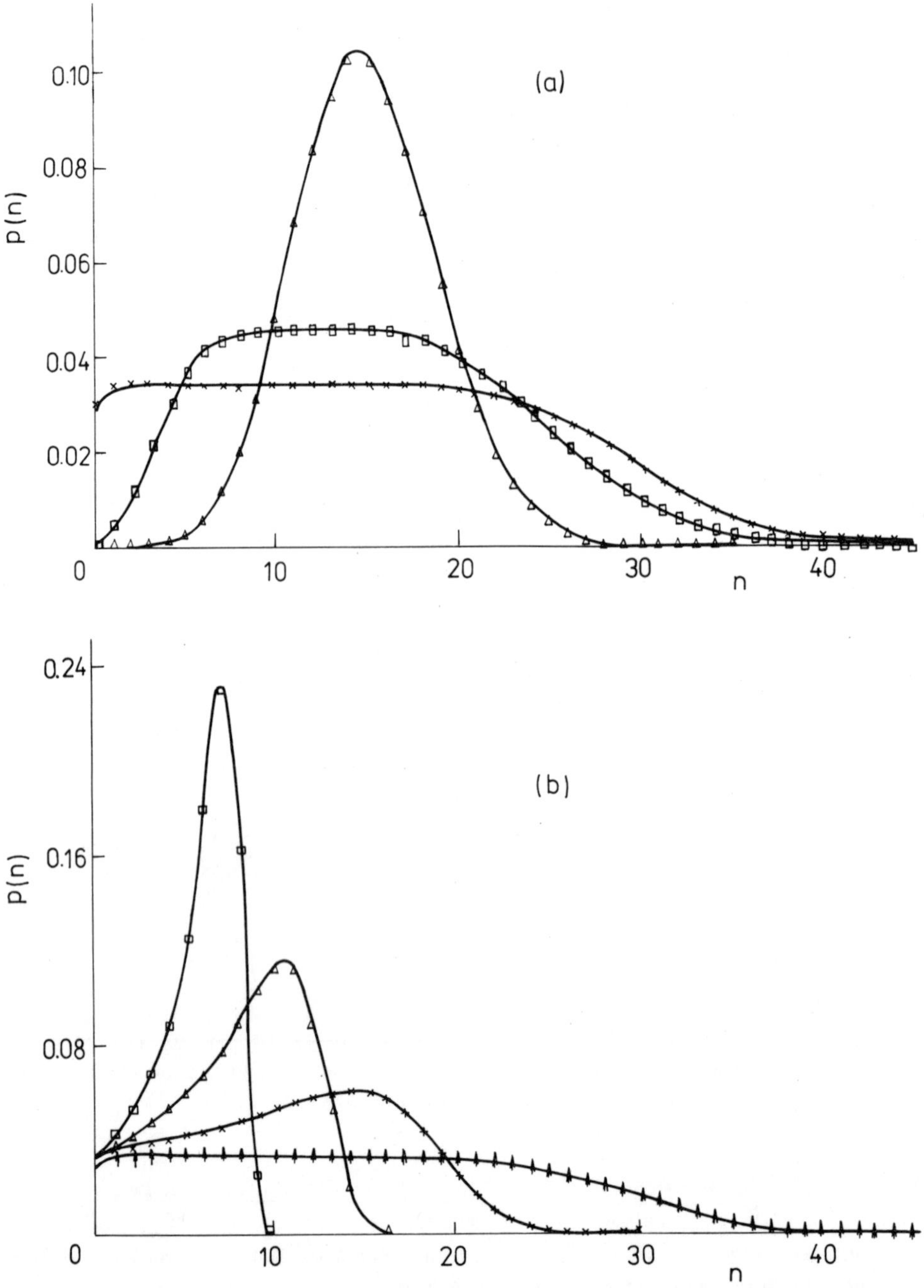

Fig. 17.7 A comparison of experimental and theoretical photocount distributions for triangularly modulated coherent light with $\langle n_c \rangle \approx 15$ for (a) $m = 0$ (△), $m = 0.74$ (□), and $m = 0.99$ (×), m being the depth of modulation, in the absence of the dead-time effect, and (b) with the dead-time effect if $m \approx 1$ and $\tau_D/T = 0$ (↑), 0.02 (×), 0.05 (△), 0.1 (□) (after M. C. Teich and G. Vannucci, 1978, *J. Opt. Soc. Am.* **68**, 1338).

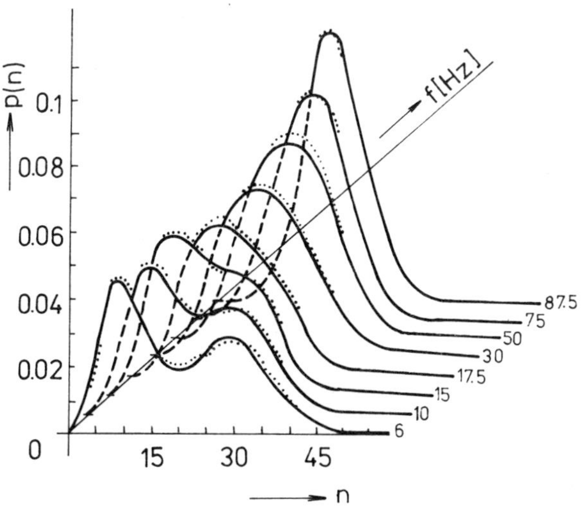

Fig. 17.8 The modulation-frequency dependence of the photocount distribution with an exponential modulation of light; the theoretical distributions are shown by full curves and the experimental distributions by dotted curves (after Č. Koňák, P. Štěpánek, L. Dvořák, Z. Kupka, J. Křepelka and J. Peřina, 1982, *Opt. Acta* **29**, 1105).

theoretical (full curves) and experimental (dotted curves) photocount distributions for various modulation frequencies for a slow luminophore is shown in Fig. 17.8. The theoretical photocount distribution is described by the Mandel-Rice formula with an exponential modulation reflecting transient states of the luminophore.

The superposition of single-mode coherent light and of chaotic light consisting of two Lorentzian spectral lines has been discussed by Tornau and Echtermeyer (1973), Mehta and Gupta (1975) and Mišta and Peřina (1977a).

A substantially extended review of the statistical properties of the superposition of coherent and chaotic fields can be found in the recent monograph by this author (Peřina (1984), see also Peřina (1980)).

CHAPTER 18

INTERFERENCE OF INDEPENDENT LIGHT BEAMS

An important topic related to the coherence of optical fields is the interference of independent light beams. Although such interference effects have been observed in the region of radiowaves it was considerably more difficult to observe them in the optical region. However, it is clear that in principle no limitations exist apart from the difficulty, even in the optical region. An outline of an interference experiment with independent sources is given in Fig. 18.1. Two sources S_1 and S_2 produce slightly convergent light beams which are projected onto the screen $\mathscr{A}$.

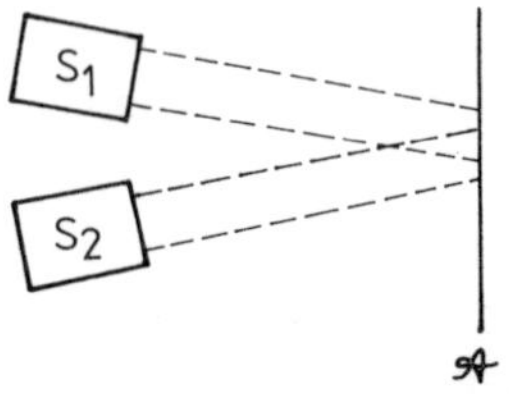

Fig. 18.1 A scheme of the interference experiment with independent light sources.

In the part of the screen common to both beams one observes the resulting field by means of a photodetector system. In general, interference fringes will be present (since both the fields are linear and must interfere) but they will vary with time. If the resolving time of the detector is long, the fringes will be washed out and no interference effect is observed; if the resolving time is sufficiently short, fringes are visible. In this connection the question arises of whether this result contradicts the well-known remark of Dirac (1958) that "each photon interferes only with itself and interference between different photons never occurs".

The first demonstration of beats resulting from the superposition of incoherent light beams was given by Forrester, Gudmundsen and Johnson (1955), who used the two spectral components of a Zeemann doublet. It is interesting to note that this experiment was performed with non-degenerate light and the mean number of photons received on a coherence area, during the time in which a steady beat was observed, was much less than one. With the development of the laser, producing light beams with the degeneracy parameter $\delta \gg 1$, such beat experiments became easier to perform (Javan, Ballik and Bond (1962), Lipsett and Mandel (1963, 1964)). Magyar and Mandel (1963, 1964) have shown that interference fringes may also be observed by superposing two independent laser light beams.

The significance of a large value of the degeneracy parameter δ for these experiments can easily be seen. The number of photons defining the interference pattern in the receiving plane in a time less than the coherence time is limited by δ. When $\delta \ll 1$ it is difficult to think of interference fringes.

All these experiments can be explained classically (see Mandel and Wolf (1965)) and the characteristic feature of the description of these effects is that, since the effects are transient, an averaging process must be avoided. A quantum-mechanical description, based on the coherent state formalism, was given by Paul, Brunner and Richter (1963), Paul (1964, 1966, 1967), Mandel (1964b), Richter, Brunner and Paul (1964), Korenman (1965), Reynolds, Spartalian and Scarl (1969), Teich (1969), Walls (1977), Bertolotti and Sibilia (1980) and Mandel (1983).

Interesting experiments on the interference of independent photon beams were performed by Pfleegor and Mandel (1967a, b, 1968) and by Radloff (1968, 1971).

In the experiment of Pfleegor and Mandel interference fringes were measured under conditions where the light intensity was so low that the mean time interval between photons was large compared with their transit time through the measuring apparatus, i.e. there was a high probability that one photon was absorbed before the next one was emitted by one or other of the laser sources. Since the intensities were very low, a photon correlation technique was required to observe the interference fringes. The interference pattern was received on a stack of thin glass plates, each of which had a thickness corresponding to about a half fringe width. The plates were cut and arranged so that any light falling on the 1st, 3rd, 5th, etc. plate was fed to one photomultiplier, while light falling on the 2nd, 4th, 6th, etc. plate was fed to the other. When the half fringe spacing coincides with the plate thickness, and for example the fringe maxima fall on the odd-numbered plates, one phototube will register nearly all the photons and the other almost none. The position of the fringe maxima is unpredictable and random, but if the number of photons registered by one phototube increases, the number registered by the other must decrease, provided the fringe spacing is right for the plates. Thus there must be a negative correlation between the numbers of counts from the two phototubes and such a negative correlation was indeed observed. This result shows (since the experimental conditions were arranged so that one photon was absorbed before the next one was emitted) that the above mentioned Dirac statement is applicable to these experiments. In general experiments with independent light beams are in agreement with the Dirac remark since any "localization" of a photon in space-time automatically rules out the possibility of knowing its momentum, as a consequence of the uncertainty principle. Thus one cannot say to which beam a given photon belongs – each photon is to be considered as being partly in both beams and interfering only with itself.

Radloff (1968, 1971) performed an analogous experiment in the temporal region under such conditions that in one detection time interval only one photon was present with high probability. After many detections interference fringes developed.

In principle, interference experiments of this sort are also possible, although more difficult, with chaotic sources. Interference experiments with two independent thermal sources were performed by Haig and Sillitto (1968) and McMillan, R. M. Sillitto and W. Sillitto (1979), empoying a photon coincidence measurement for 144 hours.

Let the field generated by the source S_1 be described by the function $\Phi_{\mathcal{N}}^{(1)}(\{\alpha_{\lambda 1}\})$ and the field generated by the source S_2 be described by the function $\Phi_{\mathcal{N}}^{(2)}(\{\alpha_{\lambda 2}\})$. The superposition of both these fields is then described by the convolution (13.147). The expectation value of the intensity at a point $x = (\mathbf{x}, t)$ on the screen $\mathscr{A}$ is given by (13.148), that is

$$I(x) = \langle \hat{A}^{(-)}(x)\hat{A}^{(+)}(x)\rangle = I^{(1)}(x) + I^{(2)}(x) +$$
$$+ 2\,\mathrm{Re}\left\{\int \Phi_{\mathcal{N}}^{(1)}(\{\alpha_{\lambda 1}\})\, V^*(x, \{\alpha_{\lambda 1}\})\, \mathrm{d}^2\{\alpha_{\lambda 1}\} \int \Phi_{\mathcal{N}}^{(2)}(\{\alpha_{\lambda 2}\})\, V(x, \{\alpha_{\lambda 2}\})\, \mathrm{d}^2\{\alpha_{\lambda 2}\}\right\}, \tag{18.1}$$

where $I^{(j)}(x) \equiv \langle \hat{A}^{(-)}(x, \{\alpha_{\lambda j}\})\,\hat{A}^{(+)}(x, \{\alpha_{\lambda j}\})\rangle$, $j = 1, 2$ are averaged intensities produced by each source if the other is absent. We restrict ourselves here and in the following to linearly polarized light for simplicity. The third term in (18.1) represents an interference term and if this term does not vanish one can observe the interference effect. In general this term will not vanish if the fields produced by both the sources contain non-zero coherent components. As an example we give the case when the fields of both the sources are in coherent states, that is

$$\Phi_{\mathcal{N}}^{(j)}(\{\alpha_{\lambda j}\}) = \prod_{\lambda} \delta(\alpha_{\lambda j} - \beta_{\lambda j}), \qquad j = 1, 2. \tag{18.2}$$

In this case the interference term is equal to

$$2\,\mathrm{Re}\left\{V^*(x, \{\beta_{\lambda 1}\})\, V(x, \{\beta_{\lambda 2}\})\right\}, \tag{18.3}$$

and so it is non-vanishing. However this assumes a knowledge of the phases of $\{\beta_\lambda\}$ which is not usually available in practice, since the phase information about optical vibrations is almost always lost. Then the distributions (18.2) must be averaged over the phases of $\{\alpha_{\lambda j}\}$ giving distributions of the form (17.58) independent of the phases describing the stationary fields. The interference term in (18.1) is zero now and no interference effect can be observed in the intensity pattern. This conclusion is in agreement with our earlier results that such an interference effect can occur only when there is at least partial coherence between the beams.

In this case the interference pattern can be detected by two quadratic detectors (photodetectors). We have to examine the correlation of intensities at two space-time points, $\langle I(x_1)\, I(x_2)\rangle_{\mathcal{N}}$, as in describing the Hanbury Brown-Twiss effect. Assuming that the beams are described by $\Phi_{\mathcal{N}}^{(1)}$ and $\Phi_{\mathcal{N}}^{(2)}$ and that the phases of the complex amplitudes are uniformly distributed in the interval $(0, 2\pi)$, we obtain (Mandel (1964b))

$$\begin{aligned}
&\langle I(x_1)\, I(x_2)\rangle_{\mathcal{N}} = \\
&= \langle [V^*(x_1, \{\alpha_{\lambda 1}\}) + V^*(x_1, \{\alpha_{\lambda 2}\})]\, [V(x_1, \{\alpha_{\lambda 1}\}) + V(x_1, \{\alpha_{\lambda 2}\})] \times \\
&\times [V^*(x_2, \{\alpha_{\lambda 1}\}) + V^*(x_2, \{\alpha_{\lambda 2}\})]\, [V(x_2, \{\alpha_{\lambda 1}\}) + V(x_2, \{\alpha_{\lambda 2}\})]\rangle_{\mathcal{N}} = \\
&= \langle |V(x_1, \{\alpha_{\lambda 1}\})|^2\, |V(x_2, \{\alpha_{\lambda 1}\})|^2\rangle_{\mathcal{N}} + \langle |V(x_1, \{\alpha_{\lambda 1}\})|^2\, |V(x_2, \{\alpha_{\lambda 2}\})|^2\rangle_{\mathcal{N}} + \\
&+ \langle |V(x_1, \{\alpha_{\lambda 2}\})|^2\, |V(x_2, \{\alpha_{\lambda 1}\})|^2\rangle_{\mathcal{N}} + \langle |V(x_1, \{\alpha_{\lambda 2}\})|^2\, |V(x_2, \{\alpha_{\lambda 2}\})|^2\rangle_{\mathcal{N}} + \\
&+ \langle V^*(x_1, \{\alpha_{\lambda 1}\})\, V(x_1, \{\alpha_{\lambda 2}\})\, V^*(x_2, \{\alpha_{\lambda 2}\})\, V(x_2, \{\alpha_{\lambda 1}\})\rangle_{\mathcal{N}} + \\
&+ \langle V^*(x_1, \{\alpha_{\lambda 2}\})\, V(x_1, \{\alpha_{\lambda 1}\})\, V^*(x_2, \{\alpha_{\lambda 1}\})\, V(x_2, \{\alpha_{\lambda 2}\})\rangle_{\mathcal{N}}, \qquad (18.4)
\end{aligned}$$

where terms such as

$$\begin{aligned}
&\langle |V(x_1, \{\alpha_{\lambda 1}\})|^2\, V^*(x_2, \{\alpha_{\lambda 1}\})\, V(x_2, \{\alpha_{\lambda 2}\})\rangle_{\mathcal{N}}, \\
&\langle V^*(x_1, \{\alpha_{\lambda 1}\})\, V(x_1, \{\alpha_{\lambda 2}\})\, V^*(x_2, \{\alpha_{\lambda 1}\})\, V(x_2, \{\alpha_{\lambda 2}\})\rangle_{\mathcal{N}},
\end{aligned}$$

etc. are equal to zero. The fifth and sixth terms in this expression generally lead to an almost periodic variation of the correlation $\langle I(x_1)\, I(x_2)\rangle_{\mathcal{N}}$ with $|\mathbf{x}_2 - \mathbf{x}_1|$ and $|t_2 - t_1|$. Assuming quasi-monochromatic beams with a spread Δk much less than the mean wave number k_0 and if $|\mathbf{x}_1 - \mathbf{x}_2| \ll 1/\Delta k$ and $|t_1 - t_2| \ll 1/c\,\Delta k$, we obtain the expression

$$\begin{aligned}
&\langle I(x_1)\, I(x_2)\rangle_{\mathcal{N}} = \langle [|V(x_1, \{\alpha_{\lambda 1}\})|^2 + |V(x_1, \{\alpha_{\lambda 2}\})|^2]^2\rangle_{\mathcal{N}} + \\
&+ 2\langle |V(x_1, \{\alpha_{\lambda 1}\})|^2\rangle_{\mathcal{N}} \langle |V(x_1, \{\alpha_{\lambda 2}\})|^2\rangle_{\mathcal{N}} \times \\
&\times \cos[(\mathbf{k}_0^{(1)} - \mathbf{k}_0^{(2)})(\mathbf{x}_2 - \mathbf{x}_1) - c(k_0^{(1)} - k_0^{(2)})(t_2 - t_1)], \qquad (18.5)
\end{aligned}$$

where $\mathbf{k}_0^{(j)}$ and $k_0^{(j)}$ refer to the jth beam ($j = 1, 2$). Thus, over a limited space-time region, the intensity correlation shows a cosinusoidal dependence on space and time, which can be interpreted both in terms of interference fringes and light beats. This has been experimentally verified by Vajnshtejn, Melechin, Mishin and Podoljak (1981). The generating function has been calculated in a closed form for this case by Klauder and Sudarshan (1968) (Sec. 10.2).

The role of interference of photons may be demonstrated as follows. Consider a two-slit experiment with the photon annihilation operator $\hat{a} = \alpha_1 \hat{a}_1 + \alpha_2 \hat{a}_2$ (α_j being propagators of single beams) on the screen of observation and let the whole field be described by the normalized state $|\rangle \equiv (|n, n-1\rangle + |n-1, n\rangle)/2^{1/2}$ (the number of photons in the beams must differ by one, otherwise the mean intensity $\langle \hat{a}^+ \hat{a}\rangle$ is zero). Then using (11.11a, b)

$$\langle |\hat{a}^+ \hat{a}| \rangle = I_1 + I_2 + (I_1 I_2)^{1/2}(\gamma + \text{c.c.}), \qquad (18.6)$$

where $I_j = |\alpha_j|^2(n - 1/2)$, $j = 1, 2$, $\gamma = n\alpha_1^* \alpha_2/(2n-1)|\alpha_1|\,|\alpha_2|$. Thus such a field in general produces interference with visibility $|\gamma| = 1/(2 - 1/n)$ and it is partially coherent; but the one-photon field ($n = 1$) provides $|\gamma| = 1$, the visibility is maximal and the field is fully coherent in the second order (only interferences of a photon with itself are possible), as with a field in the coherent state. For $n \to \infty$ the field is partially coherent with $|\gamma| = 1/2$. For further discussions we refer the reader to a paper by Walls (1977).

REVIEW OF NONLINEAR OPTICAL PHENOMENA

In this chapter we deal briefly with some fundamental optical nonlinear phenomena, such as the generation of higher harmonics, parametric generation and amplification, frequency conversion, Raman, Brillouin and hyper-Raman scattering, and multiphoton absorption and emission. A quantum statistical treatment is then given in Chapter 22 as demonstration of the application of the coherent-state methods to nonlinear optical phenomena. The treatment given here is only introductory, more detailed treatments of the traditional description of the nonlinear optical phenomena can be found in the specialized monographs and reviews by Bloembergen (1965), Baldwin (1969), Yariv (1967, 1975), Kleinman (1972), Akhmanov, Khokhlov and Sukhorukov (1972), Courtens (1972), Zernike and Midwinter (1973), Paul (1973), Svelto (1974), Allen and Eberly (1975), Shen (1976), Schubert and Wilhelmi (1978), Klyshko (1980), Kielich (1981) and Akhmanov, Dyakov and Tchirkin (1981).

19.1 General classical description

The classical phenomenological description of nonlinear optical phenomena is based on the Maxwell equations for the electric and magnetic strength vectors **E** and **H** in the nonlinear medium. By the standard method we can obtain the equivalent wave equation, for instance for **E**, in the nonlinear medium

$$\Delta \mathbf{E}(\mathbf{x}, t) - \mu_0 \varepsilon_0 \frac{\partial^2 \mathbf{E}(\mathbf{x}, t)}{\partial t^2} = \mu_0 \frac{\partial^2 \mathbf{P}(\mathbf{x}, t)}{\partial t^2} \tag{19.1a}$$

and

$$\nabla . \mathbf{E}(\mathbf{x}, t) = -\frac{1}{\varepsilon_0} \nabla . \mathbf{P}(\mathbf{x}, t), \tag{19.1b}$$

where ε_0 and μ_0 are the permittivity and permeability constants of the vacuum and it is assumed that the medium is non-magnetic and non-conducting without external charges. If the intensity of radiation is sufficiently high, the medium has a nonlinear response to the radiation and the generalized electric polarization vector **P** can be written in the frequency domain as

$$\mathbf{P}(\omega_i) = \boldsymbol{\chi}^{(1)}(\omega_i) . \mathbf{E}(\omega_i) + \sum_{j,k} \boldsymbol{\chi}^{(2)}(\omega_i = \omega_j + \omega_k) : \mathbf{E}(\omega_j)\, \mathbf{E}(\omega_k) + \\ + \sum_{j,k,l} \boldsymbol{\chi}^{(3)}(\omega_i = \omega_j + \omega_k + \omega_l) \vdots \mathbf{E}(\omega_j)\, \mathbf{E}(\omega_k)\, \mathbf{E}(\omega_l) + \ldots, \tag{19.2}$$

where $\chi^{(n)}$ is the susceptibility tensor of the $(n+1)$th order and multiple scalar products are denoted by points. Note that in the time domain (19.2) is represented by multifold convolutions. Now we can summarize the nonlinear processes of various orders classified on the basis of (19.2). Since $\mathbf{E}(t)$ and $\mathbf{P}(t)$ are real vectors, $\chi^{(n)}$ must fulfil the cross-symmetry conditions, such as $\chi^{(1)*}(\omega_j) = \chi^{(1)}(-\omega_j)$, etc.

19.2 Second-order phenomena

The second-order nonlinear optical phenomena are characterized by the second term in (19.2). Assuming the propagation to be along the z-axis and monochromatic, we obtain from (19.1a) for the parametric interaction of three waves, provided that $k\,\mathrm{d}E_i/\mathrm{d}z \gg \mathrm{d}^2E_i/\mathrm{d}z^2$ $(k = |\mathbf{k}|)$,

$$\frac{\mathrm{d}E_{1i}}{\mathrm{d}z} = -\frac{\mathrm{i}\omega_1}{2}\left(\frac{\mu_0}{\varepsilon_1}\right)^{1/2} \sum_{j,k} \chi^{(2)}_{ijk} E_{3j} E^*_{2k} \exp(\mathrm{i}\Delta kz), \tag{19.3a}$$

$$\frac{\mathrm{d}E_{2k}}{\mathrm{d}z} = -\frac{\mathrm{i}\omega_2}{2}\left(\frac{\mu_0}{\varepsilon_2}\right)^{1/2} \sum_{i,j} \chi^{(2)}_{kij} E^*_{1i} E_{3j} \exp(\mathrm{i}\Delta kz), \tag{19.3b}$$

$$\frac{\mathrm{d}E_{3j}}{\mathrm{d}z} = -\frac{\mathrm{i}\omega_3}{2}\left(\frac{\mu_0}{\varepsilon_3}\right)^{1/2} \sum_{i,k} \chi^{(2)}_{jik} E_{1i} E_{2k} \exp(-\mathrm{i}\Delta kz), \tag{19.3c}$$

where i, j, k denote the cartesian components, $\omega_j = k_j/(\mu_0\varepsilon_j)^{1/2}$ and $\Delta k = k_3^{(j)} - k_2^{(k)} - k_1^{(i)}$ represents the phase mismatch, $k_j^{(i)}$ being the ith polarization component of the wave vector of the jth wave. Further, the frequency resonance condition $\omega_3 = \omega_1 + \omega_2$ holds.

If one cannot distinguish between waves 1 and 2, this three-wave interaction reduces to the degenerate case, described by the following set of coupled equations

$$\frac{\mathrm{d}E_{1i}}{\mathrm{d}z} = -\mathrm{i}\omega_1\left(\frac{\mu_0}{\varepsilon_1}\right)^{1/2} \sum_{k,j} \chi^{(2)}_{ijk} E^*_{1k} E_{2j} \exp(\mathrm{i}\Delta kz), \tag{19.4a}$$

$$\frac{\mathrm{d}E_{2j}}{\mathrm{d}z} = -\mathrm{i}\frac{\omega_1}{2}\left(\frac{\mu_0}{\varepsilon_2}\right)^{1/2} \sum_{i,k} \chi^{(2)}_{jik} E_{1i} E_{1k} \exp(-\mathrm{i}\Delta kz), \tag{19.4b}$$

where $\Delta k = k_2^{(j)} - 2k_1^{(i)}$ and $\omega_2 = 2\omega_1$. Compared to (19.3), where the subfrequency modes are labeled as 1 and 2 and the sum-frequency mode as 3, we have denoted in the degenerate case the subfrequency mode as 1 and the sum-frequency mode as 2.

Thus equations (19.3) describe, in various channels, the process of sum-frequency generation if radiation of the frequency $\omega_3 = \omega_1 + \omega_2$ is generated from the subfrequency radiations, the splitting of the radiation of frequency ω_3 to two radiations with subfrequencies ω_1 and ω_2 (if both the modes 1 and 2 start from the vacuum fluctuations, we speak of the parametric generation process; if the signal mode 1 is amplified whereas the idler mode 2 starts from the vacuum fluctuations, we speak of the parametric amplification process), the frequency down-conversion, $\omega_2 = \omega_3 - \omega_1$, radiations of the frequencies ω_3 and ω_1 being introduced, or frequency up-conversion, $\omega_3 = \omega_1 + \omega_2$, if radiations of frequencies ω_1 and ω_2 are introduced. In a simplified description the idler mode 2 may be interpreted as a phonon mode and we can consider the Brillouin and Raman scattering. A fully quantum description of all of these processes starts with effective Hamiltonians which will be discussed in Chapter 22. Quite similarly in the degenerate case equations (19.4) describe sum-frequency generation $\omega_2 = 2\omega_1$ (pairs of red photons produce blue photons) or subharmonic generation $\omega_1 = \omega_2/2$.

One can introduce the coherence length $L_c = \pi/\Delta k$ giving the distance over which a systematic exchange of energy between the pumping radiation and the signal occurs.

From equations (19.4a, b) the following conservation law follows simply, if we suppress the polarization indices,

$$\frac{1}{g_1}\frac{\mathrm{d}|E_1|^2}{\mathrm{d}z} + \frac{4}{g_2}\frac{\mathrm{d}|E_2|^2}{\mathrm{d}z} = 0, \tag{19.5}$$

where $g_j = \omega_j \mu_0^{1/2} \chi_{ijk}^{(2)}/2\varepsilon_j^{1/2}$. The equations of motion (19.4a, b) can be solved in terms of hyperbolic or elliptic functions (Bloembergen (1965), Zernike and Midwinter (1973)).

The intensity of the second harmonics is strongly dependent on the angle of deviation from the synchronization direction determined by $\Delta \boldsymbol{k} = 0$. This synchronization condition can be fulfilled in anisotropic crystals, if the fundamental wave corresponds to the ordinary beam and the second harmonic wave to the extraordinary beam. Systematic studies of the spatial synchronization of optical beams in parametric processes have been carried out by a number of authors (e.g. Chmela (1971) and references therein) and aperture effects and effects of the intensity distribution in the pumping beams have been investigated as well (Chmela (1974)). Questions of the effectiveness of second harmonic generation in relation to the initial statistics of the pumping field have been also discussed (Crosignani, Di Porto and Solimeno (1972), Chmela (1973)).

Similar results can be derived for non-degenerate parametric processes governed by the equations of motion (19.3a – c). From them we obtain the conservation laws

$$\frac{1}{g_1}\frac{\mathrm{d}|E_1|^2}{\mathrm{d}z} = \frac{1}{g_2}\frac{\mathrm{d}|E_2|^2}{\mathrm{d}z} = -\frac{1}{g_3}\frac{\mathrm{d}|E_3|^2}{\mathrm{d}z}. \tag{19.6}$$

Also the equations of motion (19.3a – c) can be solved in terms of hyperbolic or elliptic functions (Bloembergen (1965), Zernike and Midwinter (1973)).

Consider now the case of the parametric amplifier, in the case when the pumping radiation of the frequency ω_3 is so strong that it remains practically unchanged during the nonlinear interaction. Then the solution for the subfrequency radiations has the form ($|E_3|$ is included in g, $\Delta k = 0$)

$$E_1(z) = E_1(0)\cosh(gz) - iE_2^*(0)\sinh(gz),$$
$$E_2(z) = E_2(0)\cosh(gz) - iE_1^*(0)\sinh(gz), \tag{19.7}$$

where $g_1 = g_2 = g$ is assumed to be real. If the idler mode 2 starts with zero amplitude, the signal mode 1 and the idler mode 2 are both amplified since

$$n_1(z) = |E_1(z)|^2 = |E_1(0)|^2 \cosh^2(gz) \underset{z\to\infty}{\simeq} \frac{|E_1(0)|^2}{4}\exp(2gz),$$
$$n_2(z) = |E_2(z)|^2 = |E_1(0)|^2 \sinh^2(gz) \underset{z\to\infty}{\simeq} \frac{|E_1(0)|^2}{4}\exp(2gz). \tag{19.8}$$

On the other hand, considering frequency conversion with strong pumping radiation of frequency ω_2, we obtain the periodic solution ($|E_2|$ included in g, $\Delta k = 0$)

$$E_1(z) = E_1(0)\cos(gz) - iE_3(0)\sin(gz),$$
$$E_3(z) = E_3(0)\cos(gz) - iE_1(0)\sin(gz). \tag{19.9}$$

For the up-conversion process $\omega_1 + \omega_2 \to \omega_3$ ($E_3(0) = 0$), for down-conversion $\omega_3 \to \omega_1 + \omega_2$ ($E_1(0) = 0$) and

$$n_1(z) = |E_3(0)|^2 \sin^2(gz),$$
$$n_3(z) = |E_3(0)|^2 \cos^2(gz), \tag{19.10}$$

and $n_1(z) + n_3(z) = n_3(0)$ (cf. (19.6)).

A more realistic treatment must take into account the depletion of the pumping radiation (Bloembergen (1965), Yariv (1967)). Such solutions are shown in Fig. 19.1 for the parametric amplification process and in Fig. 19.2 for frequency up- and down-conversion. Note that in the later stage of the process the inverse process starts.

The above equations also describe Brillouin or Raman scattering, in which laser radiation of the frequency ω_L is scattered by acoustical or molecular vibrations

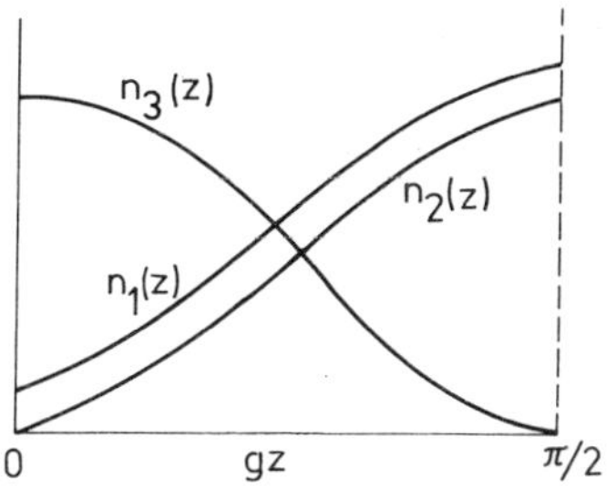

Fig. 19.1 Parametric amplification of light with frequency ω_1, ω_2 is the frequency of the idler mode and ω_3 is the frequency of pumping.

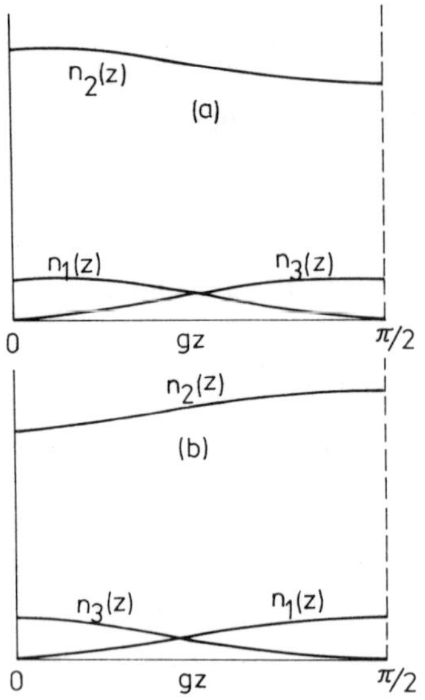

Fig. 19.2 Frequency up-conversion $\omega_1 + \omega_2 \to \omega_3$ (a) and frequency down-conversion $\omega_3 \to \omega_1 + \omega_2$ (b), ω_2 is the frequency of pumping.

(phonons) of the scattering medium possessing a typical frequency ω_V. In these processes scattered modes arise which have the Stokes and anti-Stokes frequencies

$$\omega_S = \omega_L - \omega_V, \tag{19.11a}$$

$$\omega_A = \omega_L + \omega_V, \tag{19.11b}$$

respectively. If the incident laser field is extremely strong, higher-order frequencies of the scattered modes can occur, such as $\omega_S = \omega_L - k\omega_V$, $\omega_A = \omega_L + k\omega_V$, $k \geqq 1$ being integer. For strong classical laser fields the Stokes interaction determined by (19.11a) corresponds to the parametric amplification process, whereas the anti-Stokes interaction related to (19.11b) corresponds to the frequency conversion process.

19.3 Third- and higher-order phenomena

In the same way as above, third and higher harmonic generation can be considered where $\omega_2 = k\omega_1$, $k \geq 3$ being integer; in the course of the inversion process, the kth subharmonic frequency is generated, $\omega_1 = \omega_2/k$. Also higher-order parametric processes can be considered, including the generation of sum and difference frequency fields, with more than two fields.

The important class of third-order nonlinear optical processes is represented by stimulated Raman scattering. In contrast to the optical parametric processes, where the medium is involved parametricly, stimulated Raman scattering represents a direct resonant interaction of the incident laser field with the active medium. In general, in a scattering process a photon of a frequency $\omega_1(\mathbf{k}_1)$ is absorbed, while a photon of a frequency $\omega_2(\mathbf{k}_2)$ is emitted by the medium. The radiation state changes from its initial state $|i\rangle$ to its final state $|f\rangle$, and the active scattering medium makes a transition as well. The excitation of the medium can include an entropy wave without changing the frequency of the incident light (Rayleigh scattering), pressure acoustic waves (Brillouin scattering), a phonon, magnon, plasmon or electric excitation (Raman scattering), a concentration variation (concentration scattering), etc. The transition probability is given by (16.94), which defines spontaneous scattering if $\langle \hat{n}_2 \rangle = 0$ and stimulated scattering if $\langle \hat{n}_2 \rangle \neq 0$. For stimulated scattering, the Stokes and anti-Stokes radiations are coherent and coupled, and considering an effective internal vibration frequency ω_V of atoms and molecules, the frequency conditions (19.11a, b) hold and consequently

$$2\omega_L = \omega_S + \omega_A, \tag{19.12}$$

which expresses the coupling of the laser, Stokes and anti-Stokes modes. Thus the corresponding equations of motion for the Stokes and anti-Stokes amplitudes E_S and E_A can be written in the form

$$\begin{aligned} \frac{\mathrm{d}E_S}{\mathrm{d}z} &= g_S |E_L|^2 E_S + g_{SA} E_L^2 E_A^* \exp(\mathrm{i}\Delta kz), \\ \frac{\mathrm{d}E_A}{\mathrm{d}z} &= -g_A |E_L|^2 E_A - g_{SA} E_L^2 E_S^* \exp(\mathrm{i}\Delta kz), \end{aligned} \tag{19.13}$$

where g_S, g_A and g_{SA} are the Stokes, anti-Stokes and mutual coupling constants, E_L is the complex amplitude of the laser field and $\Delta k = |2\mathbf{k}_L - \mathbf{k}_S - \mathbf{k}_A|$ characterizes the phase mismatch.

In general, in stimulated scattering the Stokes photons are amplified, whereas the anti-Stokes photons are attenuated. A more detailed discussion, based on the fully quantum theory, will be presented in Chapter 22.

With very powerful laser beams hyper-Raman scattering may be realized, which is a higher-order nonlinear process. Using two laser beams with frequencies ω_1 and ω_2, the Stokes and anti-Stokes modes have frequencies $\omega_{S,A} = \omega_1 + \omega_2 \mp \omega_V$, whereas in the degenerate case $\omega_{S,A} = 2\omega_L \mp \omega_V$.

Other interesting nonlinear optical phenomena are multiphoton absorption and emission. Consider first n-photon absorption. In the course of the n-photon absorption, the atomic system makes a transition from the ground state to an exciting state simultaneously absorbing n photons, in general of frequencies $\omega_1, \ldots, \omega_n$ (of course it may be that $\omega_1 = \ldots = \omega_n = \omega$). The transition probability is proportional, with respect to Sec. 12.1, to the normal correlation function $\Gamma_{\mathcal{N}}^{(n,n)}(x, \ldots, x) = \langle \hat{A}^{(-)n}(x) \hat{A}^{(+)n}(x) \rangle$. It equals $\langle \hat{A}^{(-)}(x) \hat{A}^{(+)}(x) \rangle^n = \langle I(x) \rangle^n$ for coherent radiation and $n! \langle I(x) \rangle^n$ for chaotic radiation; thus n-photon absorption is $n!$ times more effective for chaotic radiation than for coherent radiation, which is a consequence of the bunching effect of chaotic photons. The presence of the factor $n!$ in the transition probability of n-photon absorption for chaotic radiation, or for laser radiation with a very high number of modes, has been experimentally tested and verified by Shiga and Inamura (1967), Teich, Abrams and Gandrud (1970), Jakeman, Oliver and Pike (1968b), Kovarskii (1974), Krasiński, Chudzyński and Majewski (1974, 1976) and Glódź (1978) (cf. also Schubert and Wilhelmi (1980) for a review). It has been also experimentally shown by Le Compte, Mainfray and Manus (1974) and Le Compte, Mainfray, Manus and Sanchez (1975) that the eleventh-order ionization of Xe-atoms with laser radiation composed of 100 modes is $10^{6.9 \pm 0.3}$ times more effective than for single-mode laser radiation, which approximately corresponds to the factor 11! which is correct for chaotic radiation. In general, multiphoton absorption smoothes the fluctuations of the incident radiation. Multiphoton ionization of atoms in strong stochastic fields has been discussed by Kraynov and Todirashku (1980).

Multiphoton absorption processes are of great practical importance because they make possible the construction of optical filters which are transparent for low-power radiation and which absorb high-power radiation. Further, in combination with tunable lasers two-photon absorption is a powerful spectroscopic technique with high resolution. Multiphoton absorption is an inverse process to multiphoton emission; a transition scheme for two-photon stimulated emission is shown in Fig. 19.3.

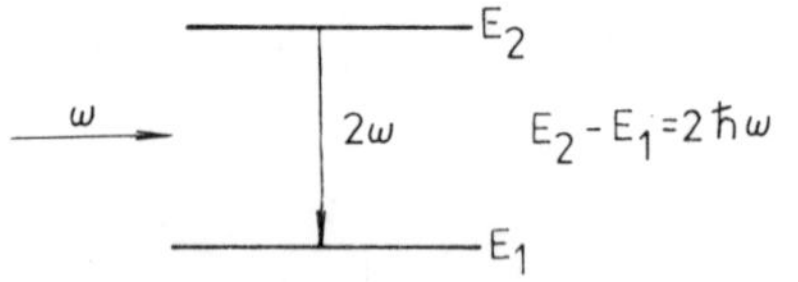

Fig. 19.3 Energy levels for two-photon stimulated emission.

Another third-order nonlinear effect is the self-focusing of laser beams. This effect is caused by the dependence of the refractive index or the permittivity on the intensity of radiation, $\varepsilon = \bar{\varepsilon} + \varepsilon_2 |\mathbf{E}|^2$, $\bar{\varepsilon}$ and $\varepsilon_2 > 0$ being constants. If a laser beam has a Gaussian intensity profile, it propagates slower in the central part than at the borders (Fig. 19.4) and therefore the beam is focused at a distance z_f. However, if the beam propagates through finite apertures, diffraction of the beam also occurs and both these effects are superimposed. If they just compensate one another, the beam propagates without change of the cross-section along great distances and it is self-trapped. Since $\Delta\varepsilon = \varepsilon - \bar{\varepsilon} = \varepsilon_2 |\mathbf{E}|^2$ has to be high enough

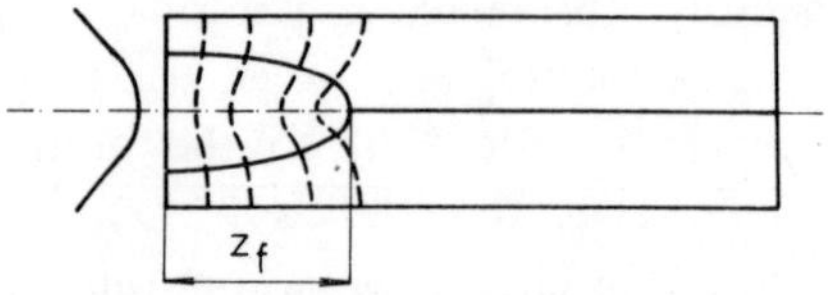

Fig. 19.4 Self-focusing of a laser beam.

in order to have an observable effect, pulsed lasers are usually used. One may distinguish two cases: i) The laser pulse is broader than a typical response time for $\Delta\varepsilon$, so that the response of $\Delta\varepsilon$ to the laser intensity variations can be considered as instantaneous and we have quasi-steady state of self-focusing; ii) If the laser pulse width is comparable with, or shorter than, the response time of $\Delta\varepsilon$, a transient self-focusing appears. In the transient case the front of the pulse remains unchanged and only the back-part of the pulse is made narrower, so that the pulse takes a horn-shaped form, propagating further without any change. The phenomenon is described by the corresponding nonlinear wave equation with $\varepsilon = \bar{\varepsilon} + \varepsilon_2 |\mathbf{E}|^2$.

The self-focusing phenomenon plays an important role in the design of high-power lasers, where optical breakdown of the active medium may be caused. On the other hand, it may be of great significance for optical communication and for the transfer of energy in narrow beams without losses.

Further details concerning self-focusing can be found in reviews by Akhmanov, Khokhlov and Sukhorukov (1972), Svelto (1974) and Shen (1976).

Recently, much effort has been devoted to the development of so-called dynamical holography or adaptive optics, which represents a combination of the holographic principle of imagery with a reference beam and nonlinear optics (Yariv (1975), Bespalov (1979), Adaptive Optics (1977) (a special issue of the Journal of Optical Society of America)).

19.4 Transient coherent optical effects

It has been recognized that there exists a close analogy between magnetic resonances and optical resonance phenomena. In general, any two-level system can be treated as a (1/2)-spin system. Magnetic transient coherent phenomena are now used routinely for relaxation studies in the radio- and micro-wave range. Similar transient resonant phenomena also exist in the optical region and have been the subject of intensive research in recent years (Courtens (1972), Nussenzveig (1973), Sargent, Scully and Lamb (1974), Slusher (1974), Allen and Eberly (1975), Butylkin, Kaplan, Khronopulo and Yakubovitch (1977), Kujawski and Eberly (1978), Schubert and Wilhelmi (1978)). The role of fluctuations in nonlinear pulse propagation has been discussed by Crosignani, Papas and Di Porto (1980). The Maxwell equations (or the equivalent wave equation) for the radiation field coupled to the Bloch equations for the two-level atomic active system, can serve as the usual basis for the description of nonlinear optical phenomena associated with the propagation of very short optical pulses through nonlinear media.

19.4.1 Self-induced transparency

It has been shown by McCall and Hahn (1967, 1969) theoretically as well as experimentally that, for sufficiently intense light pulses of a suitable shape, the

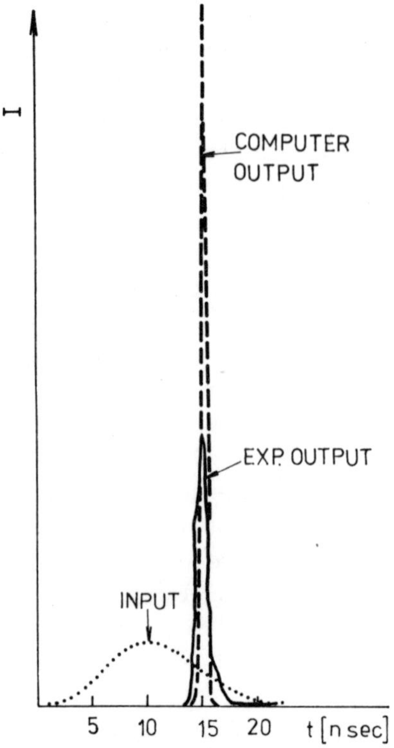

Fig. 19.5 Optical 3.5π-pulse compression by focusing (after R. E. Slusher, 1974, Progress in Optics, Vol. 12, ed. E. Wolf (North-Holland, Amsterdam), p. 53).

medium, due to nonlinear effects, can behave as completely transparent. This phenomenon is called *self-induced transparency*, and it arises when the front edge of the pulse can excite the medium and the back edge then stimulates emission. Thus the pulse propagates without any change, it is only delayed. The pulse must be short to avoid relaxation processes.

We can define the pulse area

$$\vartheta(z) = \frac{\mathscr{P}}{\hbar} \int_{-\infty}^{+\infty} \mathscr{E}(z, t)\, \mathrm{d}t, \tag{19.14}$$

where $\mathscr{E}$ represents the complex envelope function of the pulse and $\mathscr{P}$ is the matrix element of the dipole moment. Then it follows from the Maxwell and Bloch equations (McCall and Hahn (1967, 1969)) that

$$\frac{\mathrm{d}\vartheta(z)}{\mathrm{d}z} = \frac{a}{2} \sin \vartheta(z); \tag{19.15}$$

for absorbing media ($a < 0$) and weak pulses ($\vartheta \ll 1$) the well-known Lambert-Beer law is obtained,

$$[\vartheta(z)]^2 = [\vartheta(0)]^2 \exp(-|a|z). \tag{19.16}$$

The solution of (19.15) has the shape of a hyperbolic secant, which remains unchanged during the propagation through the nonlinear medium. Pulses with $\vartheta < \pi$ are attenuated to zero, whereas pulses with $2\pi > \vartheta > \pi$ are amplified to a 2π-pulse. Arbitrarily shaped pulses are split into a number of 2π-pulses, which thereafter propagate without any change. In Fig. 19.5 the compression of the 3.5π-pulse into a 2π-pulse is demonstrated (Gibbs and Slusher (1971), Slusher (1974)).

19.4.2 Photon echo

The optical photon echo is analogous to the spin echo in magnetic resonance. The principle of the phenomenon is sketched in Fig. 19.6. Application of the first resonant coherent $\pi/2$-pulse causes the rotation of the dipoles $\vec{P}$ from the direction $\vec{z}$ by 90° about $\vec{\mathscr{E}}$. The dipoles then precess around $\vec{z}$ with a frequency ω.

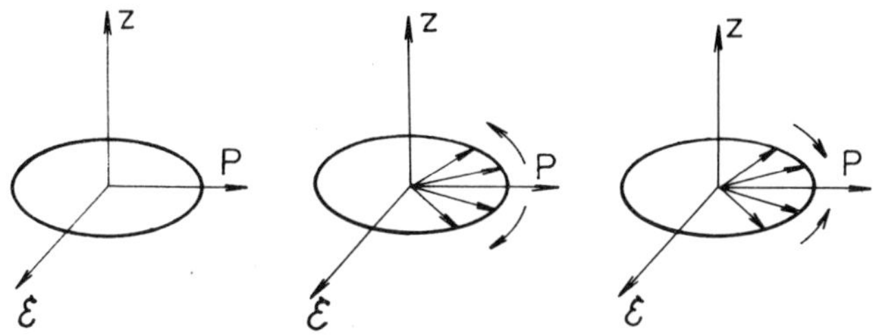

Fig. 19.6 A sketch of the photon echo phenomenon.

As a result of inhomogeneous broadening, different dipoles precess with slightly different frequencies. At a time τ after the first pulse we apply the second resonant π-pulse. Each dipole is now rotated by 180° about $\vec{\mathscr{E}}$. After the pulse is over the dipoles again precess around $\vec{z}$, but in the opposite directions. Thus at time τ after the second pulse all the dipoles create one giant dipole which radiates a giant pulse representing the photon echo.

19.4.3 Superradiance

As shown by Dicke (1954, 1964), in a system composed of N atoms, of which $N/2$ are excited, the correlation of the atoms can appear via their interaction with the electromagnetic field. The system then represents a big dipole in an atomic coherent state which spontaneously emits radiation in a cooperative way, the intensity being proportional to N^2, whereas the resultant intensity of ordinary spontaneous emission is proportional to N. The superradiant photons obey Poissonian statistics as a coherent field, with $\langle(\Delta n)^2\rangle \approx \langle n\rangle$. The statistical properties of superradiance have been discussed by Eberly and Rehler (1970) and Karczewski (1976). Subharmonic superradiance has been considered also (Sczaniecki and Buchert (1978)).

HEISENBERG-LANGEVIN AND FOKKER-PLANCK APPROACHES TO THE STATISTICAL PROPERTIES OF RADIATION INTERACTING WITH MATTER

In this chapter we introduce the Heisenberg-Langevin approach, the master equation approach, and the related generalized Fokker-Planck equation approach to the statistical properties of radiation in interaction with matter. First we consider a model of the interaction of a single-mode radiation field with an infinite reservoir boson system (the quantum theory of damping). A general formulation of the interaction of the electromagnetic field with a system of two-level atoms is also presented. The methods developed in this chapter will be applied to the propagation of radiation in random and nonlinear media in Chapters 21 and 22.

20.1 The Heisenberg-Langevin approach

Consider a one-mode radiation field of frequency ω described in the Heisenberg picture by the annihilation and creation operators $\hat{a}$ and $\hat{a}^+$ respectively, and an infinite boson system with the annihilation and creation operators $\hat{b}_j$ and $\hat{b}_j^+$ respectively and with the corresponding frequencies ψ_j. If the coupling constants of this interaction are $\varkappa_j$, then the Hamiltonian of the composite system can be written in the form

$$\hat{H} = \hat{H}_0 + \hat{H}_{\text{int}}, \tag{20.1a}$$

where

$$\hat{H}_0 = \hbar\omega(\hat{a}^+\hat{a} + \tfrac{1}{2}) + \sum_j \hbar\psi_j(\hat{b}_j^+\hat{b}_j + \tfrac{1}{2}) \tag{20.1b}$$

is the free Hamiltonian and

$$\hat{H}_{\text{int}} = \sum_j (\hbar\varkappa_j\hat{b}_j\hat{a}^+ + \hbar\varkappa_j^*\hat{b}_j^+\hat{a}) \tag{20.1c}$$

is the interaction Hamiltonian of the one-mode radiation field with the infinite reservoir system $\{\hat{b}_j\}$.

The Heisenberg equations (11.47a) for this interaction read

$$\frac{\mathrm{d}\hat{F}}{\mathrm{d}t} = \frac{1}{\mathrm{i}\hbar}[\hat{F}, \hat{H}] = -\mathrm{i}\omega\frac{\partial\hat{F}}{\partial\hat{a}}\hat{a} + \mathrm{i}\omega\hat{a}^+\frac{\partial\hat{F}}{\partial\hat{a}^+} - \\ - \mathrm{i}\frac{\partial\hat{F}}{\partial\hat{a}}\sum_j \varkappa_j\hat{b}_j + \mathrm{i}\frac{\partial\hat{F}}{\partial\hat{a}^+}\sum_j \varkappa_j^*\hat{b}_j^+, \tag{20.2a}$$

$$\frac{\mathrm{d}\hat{b}_j}{\mathrm{d}t} = -\mathrm{i}\psi_j\hat{b}_j - \mathrm{i}\varkappa_j^*\hat{a}, \tag{20.2b}$$

where $\hat{F}(\hat{a}^+, \hat{a})$ is an arbitrary operator and we have used the identities

$$[\hat{F}, \hat{a}^+\hat{a}] = [\hat{F}, \hat{a}^+]\,\hat{a} + \hat{a}^+[\hat{F}, \hat{a}] \tag{20.3}$$

and (13.31b).

If $\hat{F} = \hat{a}$, (20.2a) reduces to

$$\frac{\mathrm{d}\hat{a}}{\mathrm{d}t} = -\mathrm{i}\omega\hat{a} - \mathrm{i}\sum_j \varkappa_j \hat{b}_j. \tag{20.4}$$

Writing the solution of (20.2b) in the form

$$\hat{b}_l(t) = \hat{b}_l(0)\exp(-i\psi_l t) - \mathrm{i}\varkappa_l^* \int_0^t \hat{a}(t')\exp[\mathrm{i}\psi_l(t' - t)]\,\mathrm{d}t' \tag{20.5}$$

and substituting this into (20.4), we obtain, in the Wigner-Weisskopf approximation,

$$\frac{\mathrm{d}\hat{a}}{\mathrm{d}t} = -\mathrm{i}\left(\omega' - \mathrm{i}\frac{\gamma}{2}\right)\hat{a} + \hat{L}, \tag{20.6}$$

where we have assumed quasi-monochromatic light, $\hat{a}(t') \approx \hat{a}(t)\exp[\mathrm{i}\omega(t - t')]$, replaced the integration over the interval $(0, t)$ by the integration over $(0, \infty)$, and used the identity

$$\int_0^\infty \exp[\mathrm{i}(\omega - \psi_l)\,t]\,\mathrm{d}t = \lim_{\substack{\varepsilon > 0 \\ \varepsilon \to 0}} \frac{1}{-\mathrm{i}(\omega - \psi_l) + \varepsilon} = \frac{\mathrm{i}P}{\omega - \psi_l} + \pi\delta(\omega - \psi_l). \tag{20.7}$$

Here P denotes the Cauchy principal value of the integral, $\omega' = \omega + \Delta\omega$ and the frequency shift $\Delta\omega$ (which may usually be neglected) and the damping constant γ are determined by

$$\Delta\omega = P\int_{-\infty}^{+\infty} \frac{|\varkappa(\psi)|^2\,\varrho(\psi)}{\omega - \psi}\,\mathrm{d}\psi, \tag{20.8a}$$

$$\gamma = 2\pi\,|\varkappa(\omega)|^2\,\varrho(\omega), \tag{20.8b}$$

$\varrho(\omega)$ being the density function of the reservoir oscillators, the sum over l has been replaced by an integration over ψ (assuming that the reservoir oscillators are sufficiently dense) and the Langevin force $\hat{L}(t)$ has the form

$$\hat{L}(t) = -\mathrm{i}\sum_l \varkappa_l \hat{b}_l(0)\exp(-\mathrm{i}\psi_l t). \tag{20.9}$$

It has the following properties

$$\begin{aligned}
&\langle \hat{L}(t)\rangle = \langle \hat{L}^+(t)\rangle = 0,\\
&\langle \hat{L}^+(t)\,\hat{L}(t')\rangle = \gamma\langle n_d\rangle\,\delta(t - t'),\\
&\langle \hat{L}(t)\,\hat{L}^+(t')\rangle = \gamma(\langle n_d\rangle + 1)\,\delta(t - t'),\\
&\langle \hat{L}(t)\,\hat{L}(t')\rangle = \langle \hat{L}^+(t)\,\hat{L}^+(t')\rangle = 0,
\end{aligned} \tag{20.10}$$

provided that the reservoir spectrum is flat so that the mean number of reservoir oscillators (phonons) in the mode l is $\langle n_d \rangle = \langle \hat{b}_l^+(0)\, \hat{b}_l(0) \rangle$ independently of l and the reservoir oscillators form a chaotic system.

The Heisenberg-Langevin equation (20.6) may be solved, for instance, by the Laplace transform method, in the form

$$\hat{a}(t) = u(t)\, \hat{a}(0) + \sum_l w_l(t)\, \hat{b}_l(0), \tag{20.11}$$

where

$$u(t) = \exp(-\mathrm{i}\omega' t - \gamma t/2),$$
$$w_l(t) = \varkappa_l \exp(-\mathrm{i}\psi_l t) \frac{1 - \exp[\mathrm{i}(\psi_l - \omega')t - \gamma t/2]}{\psi_l - \omega' + \mathrm{i}\gamma/2}. \tag{20.12}$$

We can verify (again assuming a very great density of reservoir modes, and using the residue theorem) that

$$|u(t)|^2 + \sum_l |w_l(t)|^2 = 1, \tag{20.13a}$$

that is

$$\sum_l |w_l(t)|^2 = 1 - \exp(-\gamma t). \tag{20.13b}$$

Consequently the commutation rule is preserved for all times,

$$[\hat{a}(t), \hat{a}^+(t)] = \hat{1}. \tag{20.14}$$

Therefore the character of the particles is preserved, although their statistics can change with time since the radiation mode represents an open non-equilibrium system. Neglecting the Langevin force in (20.6), then $\hat{a}(t) = \hat{a}(0) \exp(-\mathrm{i}\omega' t - \gamma t/2)$ and we have the incorrect commutation rule $[\hat{a}(t), \hat{a}^+(t)] = \hat{1} \exp(-\gamma t)$, which violates the boson character of photons (it may be considered to be approximately valid if $t \ll 1/\gamma$). Since $\Delta\omega \ll \omega$ we neglect $\Delta\omega$ in the following.

From (20.11) and (20.13a, b) it follows that

$$\langle \hat{n}(t) \rangle = \langle \hat{a}^+(t)\, \hat{a}(t) \rangle = \exp(-\gamma t) \langle \hat{n}(0) \rangle + \langle n_d \rangle [1 - \exp(-\gamma t)], \tag{20.15}$$

the first term being the radiation term, the second one representing the reservoir contribution. Denoting by $\langle\ \rangle_R$ the average over the reservoir variables only, we can show that

$$\frac{\mathrm{d}\langle \hat{n}(t) \rangle_R}{\mathrm{d}t} = -\gamma \langle \hat{n}(t) \rangle_R + \gamma \langle n_d \rangle, \tag{20.16}$$

which is satisfied by (20.15), and that the following important identity holds

$$\langle \hat{L}^+(t)\, \hat{a}(t) + \hat{a}^+(t)\, \hat{L}(t) \rangle_R = \gamma \langle n_d \rangle. \tag{20.17}$$

Having found the solution (20.11), we can calculate the normal quantum characteristic function (13.72), which can be written, in view of (11.50), in the forms

$$\begin{aligned}C_{\mathcal{N}}(\beta, t) &= \mathrm{Tr}\,\{\hat{\varrho}(t)\exp(\beta\hat{a}^{+}(0))\exp(-\beta^{*}\hat{a}(0))\} = \\ &= \mathrm{Tr}\,\{\hat{\varrho}(0)\exp(\beta\hat{a}^{+}(t))\exp(-\beta^{*}\hat{a}(t))\} = \\ &= \int \Phi_{\mathcal{N}}(\xi, 0)\exp(\beta u^{*}\xi^{*} - \beta^{*}u\xi)\,\mathrm{d}^{2}\xi \times \\ &\times \prod_{\lambda}\int \exp\left(-\frac{|\eta_l|^2}{\langle n_{dl}\rangle} + \beta w_l^{*}\eta_l^{*} - \beta^{*}w_l\eta_l\right)\frac{\mathrm{d}^2\eta_l}{\pi\langle n_{dl}\rangle} = \\ &= \langle \exp[-\langle n_d\rangle(1-\exp(-\gamma t))|\beta|^2 + \beta\xi^{*}(t) - \beta^{*}\xi(t)]\rangle; \end{aligned} \tag{20.18}$$

here we have used the Glauber-Sudarshan representation (13.60) of the density matrix $\hat{\varrho}(0) = \hat{\varrho}_{\mathrm{rad}}(0)\,\hat{\varrho}_{\mathrm{reserv}}(0)$ in terms of the coherent states $|\xi\rangle$ and $|\{\eta_j\}\rangle$ as eigenstates of $\hat{a}$ and $\hat{b}_j$. Since the reservoir is a chaotic system,

$$\Phi_{\mathcal{N},\mathrm{reserv}}(\{\eta_l\}, 0) = \prod_{l}(\pi\langle n_{dl}\rangle)^{-1}\exp\left(-\frac{|\eta_l|^2}{\langle n_{dl}\rangle}\right); \tag{20.19}$$

here $\langle n_{dl}\rangle \equiv \langle n_d\rangle$ is independent of l, the angle brackets in (20.18) mean the average over the initial complex field amplitude $\xi = \xi(0)$ with probability distribution $\Phi_{\mathcal{N}}(\xi, 0)$ and

$$\xi(t) = u(t)\,\xi = \xi\exp(-\mathrm{i}\omega t - \gamma t/2). \tag{20.20}$$

The quasi-distribution $\Phi_{\mathcal{N}}(\alpha, t)$ is derived using the Fourier transformation (13.74) together with the integral (16.41),

$$\Phi_{\mathcal{N}}(\alpha, t) = [\pi\langle n_d\rangle(1-\exp(-\gamma t))]^{-1}\left\langle\exp\left[-\frac{|\alpha-\xi(t)|^2}{\langle n_d\rangle(1-\exp(-\gamma t))}\right]\right\rangle. \tag{20.21a}$$

The quasi-distribution $\Phi_{\mathcal{A}}(\alpha, t)$ related to antinormal ordering (cf. 13.68) and (13.71)) is equal to

$$\Phi_{\mathcal{A}}(\alpha, t) = \frac{1}{\pi[\langle n_d\rangle(1-\exp(-\gamma t))+1]}\left\langle\exp\left[-\frac{|\alpha-\xi(t)|^2}{\langle n_d\rangle(1-\exp(-\gamma t))+1}\right]\right\rangle. \tag{20.21b}$$

If the initial field is in the coherent state $|\xi\rangle$, the angle brackets in (20.21a, b) (also in (20.18)) can be omitted. Thus the initial coherent state $|\xi\rangle$ is attenuated in its amplitude and the quantum noise $\langle n_d\rangle(1-\exp(-\gamma t))$ increases with time to the saturation value $\langle n_d\rangle$ for $t\to\infty$, when the coherent energy of the radiation mode is transferred to the reservoir. The photocount generating function, the photocount distribution and its factorial moments are given by (17.92) with $s = 1$, (17.94) and (17.93) with $s = 1$ and with $M = 1$, $\langle n_c\rangle = |\xi(t)|^2$ and $\langle n_{ch}\rangle = \langle n_d\rangle(1-\exp(-\gamma t))$.

As another example of the application of the Heisenberg approach we consider the interaction described by the Hamiltonian (13.56). The Heisenberg equations are

$$i\hbar \frac{d\hat{a}_j(t)}{dt} = \sum_k f_{jk}(t)\, \hat{a}_k(t) + g_j(t), \tag{20.22}$$

which do not contain any creation operators. Their solution is expressed, using the perturbation technique (equation (11.55)), as

$$\hat{a}_j(t) = \sum_m S_{jm}(t)\, \hat{a}_m(0) - q_j(t), \tag{20.23}$$

where the time-dependence operator is

$$\hat{S}(t) = \hat{1} + \sum_{n=1}^{\infty} \left(-\frac{i}{\hbar}\right)^n \int_0^t dt'_1 \ldots \int_0^{t'_{n-1}} dt'_n \hat{f}(t'_1) \ldots \hat{f}(t'_n), \tag{20.24}$$

$\hat{f}$ being the matrix (f_{jk}) and

$$q_j(t) = \frac{i}{\hbar} \sum_{m,n} S_{jm}(t) \int_0^t S^+_{mn}(t')\, g_n(t')\, dt'. \tag{20.25}$$

Since the annihilation operators $\hat{a}_j(t)$ in (20.23) are dependent only on the initial annihilation operators $\hat{a}_m(0)$, the normal characteristic function is also in the normal form in the initial operators and the Fourier transform $\Phi_{\mathcal{N}}(\{\alpha_j\}, t)$ has the form of a δ-function, provided that the initial field is in a coherent state (including the vaccum state). Thus the Hamiltonian (13.56) preserves the coherent state and generates the coherent state from the vacuum state. More generally, the initial statistics of the field remain unchanged by this interaction (Webber (1968, 1969), Robl (1967, 1968)), as can be shown by substituting (20.23) into the multimode version of the normal characteristic function (13.72) and performing the Fourier transformation (13.74), giving

$$\Phi_{\mathcal{N}}(\{\alpha_j\}, t) = \Phi_{\mathcal{N}}(\{\sum_m S^+_{jm}(t)\, \alpha_m + Q_j(t)\}, 0), \tag{20.26}$$

where

$$Q_j(t) = \frac{i}{\hbar} \sum_m \int_0^t S^+_{jm}(t)\, g_m(t')\, dt'. \tag{20.27}$$

Thus, if the field is in the coherent state $|\{\beta_j\}\rangle$ for $t = 0$, $\Phi_{\mathcal{N}}(\{\alpha_j\}, 0) = \prod_j \delta(\alpha_j - \beta_j)$ and $\Phi_{\mathcal{N}}(\{\alpha_j\}, t) = \prod_j \delta(\alpha_j - \bar{\alpha}_j(t))$, where $\bar{\alpha}_j(t) = \sum_m S_{jm}(t)\, \beta_m - q_j(t)$; if $\beta_m = 0$, we see that the classical currents indeed produce a field in the coherent state. Also Gaussian distributions remain Gaussian for all times in this interaction. Making use of (14.13) we arrive at

$$\begin{aligned} P_{\mathcal{N}}(W, t) &= \frac{1}{2\pi} \int_{-\infty}^{+\infty} \exp(-isW)\, ds \int \Phi_{\mathcal{N}}(\{\alpha_j\}, t) \exp(is \sum_j |\alpha_j|^2)\, d^2\{\alpha_j\} = \\ &= \frac{1}{2\pi} \int_{-\infty}^{+\infty} \exp(-isW)\, ds \int \Phi_{\mathcal{N}}(\{\gamma_j + Q_j(t)\}, 0) \exp(is \sum_j |\gamma_j|^2)\, d^2\{\gamma_j\}, \end{aligned} \tag{20.28}$$

where we have substituted $\gamma_j = \sum_m S^+_{jm}\alpha_m$ (Det $\hat{S} = 1$, $\sum_j |\alpha_j|^2 = \sum_j |\gamma_j|^2$, $\hat{S}$ being unitary). If the currents are vanishingly small, $g_m = 0$ and $P_{\mathcal{N}}(W, t) = P_{\mathcal{N}}(W, 0)$ and consequently the photocount distribution and its factorial moments are time independent. Applying (20.28) to the distribution (17.81) which is valid for the superposition of coherent and chaotic fields, we arrive again at the superposition of the coherent field whose complex amplitude is $\beta_j - Q_j(t)$ and the chaotic field whose mean number of photons is $\langle n_{chj}\rangle$. Consequently, this interaction only changes the coherent component of the field and yields its modulation (Horák, Mišta and Peřina (1971c)).

20.2 The master equation and the generalized Fokker-Planck equation

An alternative equivalent approach developed in the framework of the Schrödinger picture can be based on the equation of motion for the density matrix,

$$i\hbar \frac{\partial \hat{\varrho}}{\partial t} = [\hat{H}_0 + \hat{H}_{\text{int}}, \hat{\varrho}]. \tag{20.29}$$

The Gaussian property of the reservoir, leading to the Markoffian $\hat{L}(t)$, as expressed in (20.10), permits us to perform two iterations when solving (20.29) for the reduced density matrix $\hat{\varrho}$ (with traced reservoir operators) and to calculate easily the involved expressions $\langle[\hat{H}_{\text{int}}, \hat{\varrho}]\rangle_R = 0\,(\langle\hat{b}_j\rangle = \langle\hat{b}_j^+\rangle = 0)$ and $\langle[\hat{H}_{\text{int}}, [\hat{H}_{\text{int}}, \hat{\varrho}]]\rangle_R$; in this way we arrive at the master equation for the reduced density matrix

$$\begin{aligned}\frac{\partial \hat{\varrho}}{\partial t} &= -i\omega([\hat{a}, \hat{\varrho}\hat{a}^+] + [\hat{a}^+, \hat{a}\hat{\varrho}]) + \frac{\gamma}{2}(\langle n_d\rangle + 1) \times \\ &\quad \times ([\hat{a}\hat{\varrho}, \hat{a}^+] + [\hat{a}, \hat{\varrho}\hat{a}^+]) + \frac{\gamma}{2}\langle n_d\rangle([\hat{a}^+\hat{\varrho}, \hat{a}] + [\hat{a}^+, \hat{\varrho}\hat{a}]) = \\ &= \left(\frac{\gamma}{2} - i\omega\right)[\hat{a}, \hat{\varrho}\hat{a}^+] - \left(\frac{\gamma}{2} + i\omega\right)[\hat{a}^+, \hat{a}\hat{\varrho}] - \\ &\quad - \frac{\gamma\langle n_d\rangle}{2}\{[\hat{a}^+, [\hat{a}, \hat{\varrho}]] + [\hat{a}, [\hat{a}^+, \hat{\varrho}]]\}.\end{aligned} \tag{20.30}$$

Performing the antinormal ordering in this equation with the help of (13.31b) and using (13.89a), we obtain the generalized Fokker-Planck equation

$$\begin{aligned}\frac{\partial \Phi_{\mathcal{N}}}{\partial t} &= -\frac{\partial}{\partial \alpha}(A_\alpha \Phi_{\mathcal{N}}) - \frac{\partial}{\partial \alpha^*}(A_{\alpha^*}\Phi_{\mathcal{N}}) + \frac{\partial^2}{\partial \alpha\, \partial \alpha^*}(2D_{\alpha^*\alpha}\Phi_{\mathcal{N}}) = \\ &= \left(\frac{\gamma}{2} + i\omega\right)\frac{\partial}{\partial \alpha}(\alpha\Phi_{\mathcal{N}}) + \left(\frac{\gamma}{2} - i\omega\right)\frac{\partial}{\partial \alpha^*}(\alpha^*\Phi_{\mathcal{N}}) + \\ &\quad + \gamma\langle n_d\rangle \frac{\partial^2 \Phi_{\mathcal{N}}}{\partial \alpha\, \partial \alpha^*}.\end{aligned} \tag{20.31a}$$

Using the Fourier transformation (13.72) we have the equation of motion for the normal characteristic function

$$\frac{\partial C_{\mathcal{N}}}{\partial t} = -\left(\frac{\gamma}{2} - \mathrm{i}\omega\right)\beta\frac{\partial C_{\mathcal{N}}}{\partial \beta} - \left(\frac{\gamma}{2} + \mathrm{i}\omega\right)\beta^*\frac{\partial C_{\mathcal{N}}}{\partial \beta^*} - \gamma\langle n_d\rangle|\beta|^2 C_{\mathcal{N}}. \tag{20.31b}$$

This equation can be solved by the standard method of characteristics in the form (20.18) and $\Phi_{\mathcal{N}}$ given by (20.21a) satisfies the Fokker-Planck equation (20.31a). Thus both these approaches are entirely equivalent.

The Fokker-Planck equation (20.31a) directly corresponds to the Heisenberg-Langevin equation (20.6), because the drift vector A_α and the diffusion constant $D_{\alpha^*\alpha}$ are determined by

$$A_\alpha = \left\langle\frac{\mathrm{d}\alpha}{\mathrm{d}t}\right\rangle_R = -\left(\frac{\gamma}{2} + \mathrm{i}\omega\right)\alpha,$$

$$2D_{\alpha^*\alpha} = \left\langle\frac{\mathrm{d}(\alpha^*\alpha)}{\mathrm{d}t}\right\rangle - \left\langle\alpha^*\frac{\mathrm{d}\alpha}{\mathrm{d}t}\right\rangle - \left\langle\frac{\mathrm{d}\alpha^*}{\mathrm{d}t}\alpha\right\rangle_R = \gamma\langle n_d\rangle, \tag{20.32}$$

whereas the diffusion constants such as $D_{\alpha\alpha}$ and $D_{\alpha^*\alpha^*}$ are zero. If the antinormal order is adopted, $2D_{\alpha\alpha^*} = \gamma(\langle n_d\rangle + 1)$.

Further details concerning the interaction of light with reservoirs can be found in papers by Senitzky (1967a, b, 1968, 1969, 1973, 1978, 1981), Mollow (1968b), Peřina, Peřinová, Mišta and Horák (1974), Agarwal (1973), Haake (1973) and Scully and Whitney (1972), among others. An alternative quantum theory of damping, involving the quadratic terms in $\hat{a}$ and $\hat{a}^+$ in the Hamiltonian, has been considered by Colegrave and Abdalla (1981). Also the quantum statistical properties of randomly modulated harmonic oscillators have been investigated (Crosignani, Di Porto and Solimeno (1969), Mollow (1970)).

20.3 The interaction of radiation with a nonlinear medium

The interaction of radiation with the atomic system of a nonlinear medium can be described by the following interaction Hamiltonian (McNeil and Walls (1974))

$$\hat{H}_{\text{int}} = \sum_j \hbar\mu^{(n)}(\mathbf{x}_j)\,\hat{c}_{2j}^+\hat{c}_{1j}^+\hat{O}^{(n)} + \text{h.c.},$$

$$\hat{O}^{(n)} = \prod_{l=1}^{m}\hat{a}_l^+\prod_{k=m+1}^{n}\hat{a}_k, \tag{20.33}$$

where h.c. means the Hermitian conjugate terms, $\hat{a}_l$ and $\hat{a}_l^+$ are again the annihilation and creation operators of a photon in the lth radiation mode, $\hat{c}_{\lambda j}$ and $\hat{c}_{\lambda j}^+$ are the annihilation and creation operators of the λth level ($\lambda = 1,2$) of the jth atom and $\mu^{(n)}(\mathbf{x}_j)$ is the coupling constant proportional to the n-photon transition matrix element. This Hamiltonian describes m emission and $n - m$ absorption events during one atomic transition. In quantum optics we are mostly interested in the properties of radiation and the atomic variables may be eliminated in the same way as the reservoir variables in the previous treatment; thus the Heisenberg-Langevin equations or the generalized Fokker-Planck equation can be derived. If virtual electronic transitions are taken into account and real transitions may be neglected (Graham (1970)), an effective Hamiltonian $\hat{H}_{\text{int,eff}}$ may be derived with the atomic variables eliminated and the nonlinear optical process is described by the Heisenberg-Langevin equations

$$\mathrm{i}\hbar \frac{\mathrm{d}\hat{a}_j}{\mathrm{d}t} = [\hat{a}_j, \hat{H}_0 + \hat{H}_{\text{int,eff}}] + \mathrm{i}\hbar \hat{L}_j, \tag{20.34}$$

where $\hat{L}_j$ are the Langevin forces arising in the elimination procedure, usually represented as Markoffian processes. If in addition real transitions are involved, the effective Hamiltonian cannot be obtained; however, one can always perform the elimination of atomic and reservoir variables in order to derive directly the Heisenberg-Langevin equations or the master equation, including only the radiation variables. Under the Markoff approximation, the master equation for the reduced density matrix $\hat{\varrho}$ can be derived in the same way as (20.30) in the form (Shen (1967), Haken (1970a, b), Agarwal (1973))

$$\begin{aligned}\frac{\partial \hat{\varrho}}{\partial t} &= K\{N_1([\hat{O}^{(n)}\hat{\varrho}, \hat{O}^{(n)+}] + [\hat{O}^{(n)}, \hat{\varrho}\hat{O}^{(n)+}]) - \\ &\quad - N_2([\hat{O}^{(n)}, \hat{O}^{(n)+}\hat{\varrho}] + [\hat{\varrho}\hat{O}^{(n)}, \hat{O}^{(n)+}])\},\end{aligned} \tag{20.35}$$

where K is a constant related to $\mu^{(n)}$ and N_1 and N_2 are the occupation numbers of the atomic levels 1 and 2 respectively, under the condition of thermal equilibrium. If $\hat{O}^{(n)} = \hat{a}(m = 0,\ n = 1)$ and $K = \gamma/2$, $N_1 = \langle n_d \rangle + 1$, $N_2 = \langle n_d \rangle$, we just arrive at the interaction part of the Fokker-Planck equation (20.30) (with the $\exp(-\mathrm{i}\omega t)$-dependence eliminated) for the damped harmonic oscillator. The same procedure as above leads to the generalized Fokker-Planck equation

$$\frac{\partial \Phi}{\partial t} = \sum_j \left[- \frac{\partial}{\partial \alpha_j}(A_j \Phi) + \sum_{i,j} \frac{\partial^2}{\partial \alpha_i \, \partial \alpha_j}(D_{ij}\Phi) \right], \tag{20.36}$$

where for instance $\alpha_1 = \alpha$, $\alpha_2 = \alpha^*$, etc., and the drift vectors $A_j(A_\alpha, A_{\alpha^*}$, etc.) and the diffusion constants $D_{ij}(D_{\alpha^*\alpha}, D_{\alpha\alpha}, D_{\alpha^*\alpha^*}$, etc.) are determined as

$$A_j = \left\langle \frac{\mathrm{d}\hat{a}_j}{\mathrm{d}t} \right\rangle_{R,\text{atoms}},$$

$$2D_{ij} = \left\langle \frac{\mathrm{d}(\hat{a}_i \hat{a}_j)}{\mathrm{d}t} \right\rangle_{R,\text{atoms}} - \left\langle \hat{a}_i \frac{\mathrm{d}\hat{a}_j}{\mathrm{d}t} \right\rangle_{R,\text{atoms}} - \left\langle \frac{\mathrm{d}\hat{a}_i}{\mathrm{d}t} \hat{a}_j \right\rangle_{R,\text{atoms}} \tag{20.37}$$

This follows from taking into account that for any operator $\hat{F}$

$$\left\langle \frac{d\hat{F}}{dt} \right\rangle \doteq \left\langle \sum_j \frac{\partial \hat{F}}{\partial \hat{a}_j} \frac{\Delta \hat{a}_j}{\Delta t} + \sum_{i>j} \frac{\partial^2 \hat{F}}{\partial \hat{a}_i \partial \hat{a}_j} \frac{\Delta \hat{a}_i \Delta \hat{a}_j}{\Delta t} + \sum_i \frac{1}{2} \frac{\partial^2 \hat{F}}{\partial \hat{a}_i^2} \frac{(\Delta \hat{a}_i)^2}{\Delta t} \right\rangle \quad (20.38)$$

and integrating by parts, we have (20.36) with

$$A_j = \left\langle \frac{\Delta \hat{a}_j}{\Delta t} \right\rangle, \qquad 2D_{ij} = \left\langle \frac{\Delta \hat{a}_i \Delta \hat{a}_j}{\Delta t} \right\rangle, \quad (20.39)$$

giving (20.37), as $\Delta(\hat{a}_i \hat{a}_j) = \hat{a}_i(t + \Delta t)\, \hat{a}_j(t + \Delta t) - \hat{a}_i(t)\, \hat{a}_j(t) = \Delta \hat{a}_i\, \Delta \hat{a}_j + (\Delta \hat{a}_i)\, \hat{a}_j + \hat{a}_i\, \Delta \hat{a}_j$. For instance, the most general single-mode generalized Fokker-Planck equation has the form

$$\frac{\partial \Phi}{\partial t} = - \frac{\partial}{\partial \alpha} (A_\alpha \Phi) - \frac{\partial}{\partial \alpha^*} (A_{\alpha^*} \Phi) + \frac{\partial^2}{\partial \alpha^2} (D_{\alpha\alpha} \Phi) + + \frac{\partial^2}{\partial \alpha^{*2}} (D_{\alpha^*\alpha^*} \Phi) + \frac{\partial^2}{\partial \alpha\, \partial \alpha^*} (2D_{\alpha^*\alpha} \Phi), \quad (20.40)$$

where $A_\alpha = \langle \Delta \hat{a}/\Delta t \rangle$, $A_{\alpha^*} = \langle \Delta \hat{a}^+/\Delta t \rangle$, $2D_{\alpha\alpha} = \langle (\Delta \hat{a})^2/\Delta t \rangle$, $2D_{\alpha^*\alpha^*} = \langle (\Delta \hat{a}^+)^2/\Delta t \rangle$, $2D_{\alpha^*\alpha} = \langle \Delta \hat{a}^+ \Delta \hat{a}/\Delta t \rangle \neq 2D_{\alpha\alpha^*} = \langle \Delta \hat{a}\, \Delta \hat{a}^+/\Delta t \rangle$.

If antinormal ordering is performed in (20.35) and if the multimode version of (13.89a), $\Phi_{\mathcal{N}}(\{\alpha_j\}) = \varrho^{(\mathcal{A})}(\{\hat{a}_j^+ \to \alpha_j^*\}, \{\hat{a}_j \to \alpha_j\})/\pi^M$, M being the number of modes, is used, the generalized Fokker-Planck equation for $\Phi_{\mathcal{N}}$ is obtained; if the normal ordering is carried out and the multimode version of (13.91), $\Phi_{\mathcal{A}}(\{\alpha_j\}) = \varrho^{(\mathcal{N})}(\{\hat{a}_j^+ \to \alpha_j^*\}, \{\hat{a}_j \to \alpha_j\})/\pi^M$ is adopted, we derive the generalized Fokker-Planck equation for $\Phi_{\mathcal{A}}$. These procedures of deriving the generalized Fokker-Planck equations will be explicitly demonstrated in the next chapters. In general we prefer to use the quasi-distribution $\Phi_{\mathcal{A}}$, which is always well behaved, whereas the Glauber-Sudarshan quasi-distribution $\Phi_{\mathcal{N}}$ does not exist usually in interaction problems. General procedures for obtaining the equations of motion for $\Phi_{\mathcal{N}}$ for general classes of Hamiltonian have been developed by Crosignani, Ganiel, Solimeno and Di Porto (1971a, b) and Graham (1973). An interesting use of the Fokker-Planck equation has been suggested by Arecchi and Politi (1980) and Arecchi, Politi and Ulivi (1982) to measure transient fluctuations, by fitting the experimental and theoretical moments.

Finally note that an alternative method for the investigation of the interaction of light and matter can be developed in terms of the Fock states. The master equation (20.35) may then be solved recursively by applying the Laplace transformation. Such a method has been developed by Scully and Lamb for the laser (see e.g. Sargent and Scully (1972)) and has been employed in two- and multi-photon absorption and Raman scattering processes (see e.g. Loudon (1973), Simaan and Loudon (1975, 1978), Simaan (1975, 1978), Gupta and Mohanty (1980), Mohanty and Gupta (1981a – c)), in the second- and third-harmonic generation (Nayak and Mohanty (1977), Nayak (1980)), and in two-photon emission (Nayak and Mohanty (1979), Gupta and Mohanty (1981)). However, the phase information about the field is lost in such formulations, unless all the off-diagonal Fock matrix elements of the density matrix are determined. Therefore we prefer to use the coherent-state technique and the generalized Fokker-Planck equation approach.

A quantum theory of propagation of the electromagnetic field, particularly of strong fields, has been developed by Gordov and Tvorogov (1978, 1980, 1984).

The present brief review of the Heisenberg-Langevin, master equation, and Fokker-Planck equation methods is sufficient for applications in this book. More extended reviews are available in the literature (Lax (1968a), Haken (1970b), Scully and Whitney (1972), Agarwal (1973), Haake (1973), Louisell (1973), Sargent, Scully and Lamb (1974), Davies (1976)).

CHAPTER 21

QUANTUM STATISTICS OF RADIATION IN RANDOM MEDIA

We can now demonstrate the general methods developed in the preceding chapters by applying them to the interaction of radiation with random media, such as the turbulent atmosphere. We present a theory treating radiation in a fully quantal way, including the self-radiation of the medium, whereas the random medium is described by matrix elements of permittivity fluctuations.

The photon statistics of radiation propagating through random media, described in the standard way using the Mandel photodetection equation (10.3a) in which the integrated intensity is randomly modulated, have been discussed and many results have been obtained for the turbulent atmosphere (Diament and Teich (1970b, 1971), Solimeno, Corti and Nicoletti (1970), Teich and Rosenberg (1971), Rosenberg and Teich (1972), Peřina and Peřinová (1972), Lachs and Laxpati (1973), Prucnal (1980)) and Gaussian media (Bertolotti, Crosignani and Di Porto (1970), Peřina and Peřinová (1972)) (for reviews see Peřina (1980, 1984)). Regarding classical treatments of the structure of turbulence, they are rather complex and many references are available (Tatarskii (1967, 1970), Beran and Ho (1969), Ho (1969), Strohbehn (1971, 1978), De Wolf (1973a–c), Furutsu and Furuhama (1973), Klyackin and Tatarskii (1973), Klyackin (1975), Rytov, Kravcov and Tatarskii (1978), Furutsu (1972, 1976), Tatarskii and Zavorotnyi (1980), Jakeman and Pusey (1980), Leader (1981), Hill and Clifford (1981), Clifford and Hill (1981), Ito and Furutsu (1982)). Particularly, the formulae for the superposition of coherent and chaotic light discussed in Sec. 17.3 have been employed to show that the increasing level of turbulence causes broadening of the photocount distribution (an increase of photon bunching) and a shift of the peak of the photocount distribution to lower n; for a saturation value of fluctuations the most probable number of photocounts is zero and the spectral information is practically lost from the photocount distribution (Diament and Teich (1971), Peřina and Peřinová (1972), Peřina, Peřinová, Teich and Diament (1973)).

21.1 The Hamiltonian for radiation in interaction with a random medium

We characterize the random medium by its fluctuating dielectric permittivity $\varepsilon(\nu, \mathbf{x}, t)$ depending on the frequency ν of radiation, on the position of a point $\mathbf{x}$ and on time t, and we assume for simplicity that the mean value of ε is $\langle\varepsilon\rangle = 1$. We set $\varepsilon = 1 + \varepsilon'$, so that ε' represents the fluctuating part of the permittivity and

$\langle \varepsilon' \rangle = 0$. We will neglect dispersion of the medium. We can then write the Hamiltonian in a volume V in the form (Louisell, Yariv and Siegman (1961), Yariv (1967), Crosignani, Di Porto and Solimeno (1971), Peřina and Peřinová (1976))

$$\hat{H}_{\text{rad}} = \frac{1}{8\pi} \int_V [\hat{\mathbf{E}}^2(\mathbf{x}, t) + \hat{\mathbf{H}}^2(\mathbf{x}, t)] \, d^3x + \frac{1}{8\pi} \int_V \varepsilon'(\mathbf{x}, t) \, \hat{\mathbf{E}}^2(\mathbf{x}, t) \, d^3x =$$

$$= \sum_l \hbar\omega_l(\hat{a}_l^+ \hat{a}_l + \tfrac{1}{2}) - \sum_{l,m} \hbar K'_{lm}(t) \, (\hat{a}_l^+ - \hat{a}_l)(\hat{a}_m^+ - \hat{a}_m), \tag{21.1}$$

where (11.1a, b) and

$$\hat{\mathbf{A}}(x) = \sum_l \left(\frac{\hbar}{2\omega_l} \right)^{1/2} \mathbf{u}_l(\mathbf{x}) \, [\hat{a}_l(t) + \hat{a}_l^+(t)] \tag{21.2}$$

have been used. The real mode functions $\mathbf{u}_l(\mathbf{x})$ satisfy the Helmholtz equation and the orthogonality condition

$$\int_V \mathbf{u}_l(\mathbf{x}) \, . \, \mathbf{u}_m(\mathbf{x}) \, d^3x = 4\pi c^2 \delta_{lm} . \tag{21.3}$$

The matrix elements of the fluctuating part of the permittivity, which are typical fluctuating quantities in the medium, are given by

$$K'_{lm}(t) = \frac{(\omega_l \omega_m)^{1/2}}{16\pi c^2} \int_V \varepsilon'(\mathbf{x}, t) \, \mathbf{u}_l(\mathbf{x}) \, . \, \mathbf{u}_m(\mathbf{x}) \, d^3x. \tag{21.4}$$

The first term in (21.1) represents the free-field radiation Hamiltonian, while the second one is the interaction Hamiltonian.

We assume for simplicity that $|\varepsilon'| \ll 1$, i.e. the fluctuations are assumed to be weak and, in addition, that the fluctuations are slow compared to optical vibrations (for a more general treatment including strong and fast fluctuations, see Peřina (1980, 1984)); then ε depends only parametrically on t through $\mathbf{x}$ (Tatarskii (1971)) and consequently the K'_{lm} are independent of t. Adiabatic behaviour of $K'_{lm}(t)$ ($|dK'_{lm}(t)/dt| \ll |K'_{lm}(t)|$) has been also considered (Peřina, Peřinová and Horák (1973a, b), Peřina, Peřinová and Mišta (1974), Peřina, Peřinová, Mišta and Horák (1974)). Further we assume weakly inhomogeneous media so that the inhomogeneities are much larger than the radiation wavelength and consequently ε depends slowly on $\mathbf{x}$, off-diagonal elements (21.4) may be neglected (i.e. fluctuations of the direction of propagation may be neglected), and the radiation modes may be considered as practically independent.

The loss mechanism may be described in the same way as was discussed in Chapter 20.

21.2 Heisenberg-Langevin equations and the generalized Fokker-Planck equation

The Heisenberg equations are

$$\mathrm{i}\frac{\mathrm{d}\hat{a}_j}{\mathrm{d}t} = (\omega + 2K'_j)\,\hat{a}_j - 2K'_j\hat{a}_j^+ + \sum_l \varkappa_{jl}\hat{b}_l^{(j)},$$

$$\mathrm{i}\frac{\mathrm{d}\hat{b}_l^{(j)}}{\mathrm{d}t} = \psi_l^{(j)}\hat{b}_l^{(j)} + \varkappa_{jl}^*\hat{a}_j, \tag{21.5}$$

$\hat{b}_l^{(j)}$ being the reservoir operators coupled to the jth radiation mode and $K'_j \equiv K'_{jj}$. As in Chapter 20 we obtain the following set of Heisenberg-Langevin equations for the radiation operators

$$\frac{\mathrm{d}\hat{a}_j}{\mathrm{d}t} = -\left(\mathrm{i}\omega_j\mathscr{K}_j^{1/2}\cosh\left(\frac{\varphi_j}{2}\right) + \frac{\gamma_j}{2}\right)\hat{a}_j + \mathrm{i}\omega_j\mathscr{K}_j^{1/2}\sinh\left(\frac{\varphi_j}{2}\right)\hat{a}_j^+ + \hat{L}_j,$$

$$\frac{\mathrm{d}\hat{a}_j^+}{\mathrm{d}t} = -\left(-\mathrm{i}\omega_j\mathscr{K}_j^{1/2}\cosh\left(\frac{\varphi_j}{2}\right) + \frac{\gamma_j}{2}\right)\hat{a}_j^+ - \mathrm{i}\omega_j\mathscr{K}_j^{1/2}\sinh\left(\frac{\varphi_j}{2}\right)\hat{a}_j + \hat{L}_j^+, \tag{21.6}$$

where we have neglected the frequency shifts $\Delta\omega_j$ and the damping constants are

$$\gamma_j = 2\pi\,|\,\varkappa_j(\omega_j\mathscr{K}_j^{1/2})\,|^2\,\varrho_j(\omega_j\mathscr{K}_j^{1/2}) \tag{21.7}$$

(see (20.8b)) and

$$\mathscr{K}_j = 1 + \frac{4K'_j}{\omega_j}; \tag{21.8}$$

we have used that $\omega_j + 2K'_j = \omega_j\mathscr{K}_j^{1/2}\cosh(\varphi_j/2)$, $2K'_j = \omega_j\mathscr{K}_j^{1/2}\sinh(\varphi_j/2)$, $\varphi_j = \ln\mathscr{K}_j$. Assuming that $\varkappa_{jl} \approx \varkappa_l$, $\mathscr{K}_j \approx \mathscr{K}$, $\omega_j \approx \omega$, $\varrho_j \approx \varrho$, all the γ_j are practically the same. The Langevin forces are expressed by (20.9).

Since the quasi-distribution $\Phi_{\mathscr{N}}(\{\alpha_j\}, t)$ (in the Schrödinger picture) appropriate to normal ordering of field operators, does not exist in general in interaction problems, at least for some time intervals, we prefer to adopt the quasi-distribution $\Phi_{\mathscr{A}}(\{\alpha_j\}, t)$, appropriate to antinormal ordering of field operators, which is always well behaved and non-negative, as discussed in Sec. 13.2. These multimode quasi-distributions are related to their characteristic functions by the multimode Fourier transformations, in analogy to (13.74) and (13.75),

$$\Phi_{\mathscr{N},\mathscr{A}}(\{\alpha_j\}, t) = \int C_{\mathscr{N},\mathscr{A}}(\{\beta_j\}, t)\prod_j \exp(\alpha_j\beta_j^* - \alpha_j^*\beta_j)\frac{\mathrm{d}^2\beta_j}{\pi^2}, \tag{21.9}$$

where the multimode normal and antinormal characteristic functions are defined, in analogy to (13.72) and (13.73), as follows

$$\begin{aligned} C_{\mathscr{N},\mathscr{A}}(\{\beta_j\}, t) &= \mathrm{Tr}\,\{\hat{\varrho}(t)\,\mathscr{N}, \mathscr{A}\prod_j \exp(\beta_j\hat{a}_j^+(0))\exp(-\beta_j^*\hat{a}_j(0))\} = \\ &= \mathrm{Tr}\,\{\hat{\varrho}(0)\,\mathscr{N}, \mathscr{A}\prod_j \exp(\beta_j\hat{a}_j^+(t))\exp(-\beta_j^*\hat{a}_j(t))\} = \\ &= \int \Phi_{\mathscr{N},\mathscr{A}}(\{\alpha_j\}, t)\prod_j \exp(-\alpha_j\beta_j^* + \alpha_j^*\beta_j)\,\mathrm{d}^2\alpha_j, \end{aligned} \tag{21.10}$$

$\mathcal{N}$ and $\mathcal{A}$ being the normal and antinormal ordering operators, $\hat{\varrho}(t)$, $\hat{a}_j(0)$ and $\hat{\varrho}(0)$, $\hat{a}_j(t)$ are the density matrices and annihilation operators in the Schrödinger and Heisenberg pictures respectively.

Making use of these tools in the same manner as in Chapter 20, we arrive at the generalized Fokker-Planck equation appropriate for propagation of radiation of arbitrary photon statistics through a random medium (Peřina and Peřinová (1975, 1976))

$$\frac{\partial \Phi_{\mathcal{A}}}{\partial t} = \left\{ \sum_j \left[\frac{\partial}{\partial \alpha_j} \left\{ \left[\frac{\gamma_j}{2} + \mathrm{i}\omega_j \mathcal{K}_j^{1/2} \cosh\left(\frac{\varphi_j}{2}\right) \right] \alpha_j - \right.\right.\right.$$
$$\left. - \mathrm{i}\omega_j \mathcal{K}_j^{1/2} \sinh\left(\frac{\varphi_j}{2}\right) \alpha_j^* \right\} + \mathrm{c.c.} + \gamma_j(\langle n_{dj}\rangle + 1) \frac{\partial^2}{\partial \alpha_j \partial \alpha_j^*} -$$
$$\left.\left. - \left(\mathrm{i}K_j' \frac{\partial^2}{\partial \alpha_j^2} + \mathrm{c.c.} \right) \right] \right\} \Phi_{\mathcal{A}}; \tag{21.11a}$$

for the corresponding antinormal characteristic function

$$\frac{\partial C_{\mathcal{A}}(\{\beta_j\}, t)}{\partial t} = \left\{ \sum_j \left[\left\{ \left[-\frac{\gamma_j}{2} + \mathrm{i}\omega_j \mathcal{K}_j^{1/2} \cosh\left(\frac{\varphi_j}{2}\right) \right] \beta_j - \right.\right.\right.$$
$$\left. - \mathrm{i}\omega_j \mathcal{K}_j^{1/2} \sinh\left(\frac{\varphi_j}{2}\right) \beta_j^* \right\} \frac{\partial}{\partial \beta_j} + \mathrm{c.c.} - \gamma_j(\langle n_{dj}\rangle + 1) |\beta_j|^2 +$$
$$\left.\left. + (\mathrm{i}K_j' \beta_j^2 + \mathrm{c.c.}) \right] \right\} C_{\mathcal{A}}(\{\beta_j\}, t). \tag{21.11b}$$

Here c.c. denotes the complex conjugate terms, α_j is the eigenvalue of $\hat{a}_j$ in the coherent state $|\{\alpha_j\}\rangle$ and $\langle n_{dj}\rangle = \langle n_d(\omega \mathcal{K}_j^{1/2})\rangle$. The drift vectors in (21.11a) are determined, as discussed in Chapter 20, by $\langle \mathrm{d}\alpha_j/\mathrm{d}t\rangle_R$, the coefficient of $\partial^2/\partial\alpha_j \partial\alpha_j^*$ is equal to $2D_j$ where the diffusion constant $D_j = \gamma_j(\langle n_{dj}\rangle + 1)/2$ from (20.10) and this term represents the reservoir contribution. The coefficient of $\partial^2/\partial\alpha_j^2$ is equal to the expression which arises from the nonlinear term $-\hbar K_j'(\hat{a}_j^2 + \hat{a}_j^{+2})$ in the Hamiltonian (21.1). It is derived from the expression $\mathrm{i}K_j'[\hat{a}_j^2 + \hat{a}_j^{+2}, \hat{\varrho}]$ in the equation of motion for the density matrix, if the normal ordering is performed by means of the identities (13.31b), $[\hat{a}_j^2, \hat{\varrho}] = \hat{a}_j[\hat{a}_j, \hat{\varrho}] + [\hat{a}_j, \hat{\varrho}]\,\hat{a}_j = \hat{a}_j\, \partial\hat{\varrho}/\partial\hat{a}_j^+ + (\partial\hat{\varrho}/\partial\hat{a}_j^+)\,\hat{a}_j$, $[\hat{a}_j^{+2}, \hat{\varrho}] = -(\partial\hat{\varrho}/\partial\hat{a}_j)\,\hat{a}_j^+ - \hat{a}_j^+\,\partial\hat{\varrho}/\partial\hat{a}_j$, c-numbers α_j and α_j^* are substituted for the operators $\hat{a}_j$ and $\hat{a}_j^+$ and the standard procedure is applied.

21.3 Solutions of the Heisenberg-Langevin equations and the generalized Fokker-Planck equation

The solution of (21.11b) can be written, adopting for instance the standard method of characteristics, in the following form (Peřina and Peřinová (1975, 1976))

$$C_{\mathcal{A}}(\{\beta_j\}, t) = \langle \prod_j \exp\left[-B_j(t)|\beta_j|^2 + C_j^*(t)\beta_j^2/2 + \right.$$
$$\left. + C_j(t)\beta_j^{*2}/2 + \xi_j^*(t)\beta_j - \xi_j(t)\beta_j^* \right]\rangle, \tag{21.12}$$

in full agreement with the Heisenberg-Langevin approach (Peřina, Peřinová and Horák (1973a, b), Peřina, Peřinová and Mišta (1974), Peřina, Peřinová, Mišta and Horák (1974)). Here $\xi_j(t)$, $B_j(t)$ and $C_j(t)$ are functions of time (the differential equations for them can be derived by substituting (21.12) into (21.11b)), which have been determined for various fluctuation regimes (Peřina and Peřinová (1975, 1976)). In general, they are rather complicated; however they have simple form in our case of slow and small fluctuations (see equation (21.21)). The Fourier transformation (21.9) provides the averaged shifted multimode Gaussian distribution

$$\Phi_{\mathscr{A}}(\{\alpha_j\}, t) = \left\langle \prod_j (\pi K_j^{1/2}(t))^{-1} \exp\left[-\frac{B_j(t)}{K_j(t)} |\alpha_j - \xi_j(t)|^2 + \right.\right. \\ \left.\left. + \frac{C_j^*(t)(\alpha_j - \xi_j(t))^2 + \text{c.c.}}{2K_j(t)} \right] \right\rangle, \tag{21.13}$$

where $K_j(t) = B_j^2(t) - |C_j(t)|^2$ and the angle brackets in (21.12) and (21.13) denote the average over the initial complex amplitudes $\xi_j(0) \equiv \xi_j$ with the initial probability distribution $\Phi_{\mathscr{N}}(\{\xi_j\}, 0)$.

We note that if the $\hat{a}_l$ for scattered radiation modes are interpreted as reservoir operators $\hat{b}_l^{(j)}$, then the K'_{jl} corresponding to these modes have the role of coupling constants $\varkappa_{jl}$ between radiation and reservoir modes, and it follows from (21.7) that $\gamma_j = 2\pi\langle K_j'^2\rangle \varrho_j = 2\pi\sigma^2\varrho_j$ ($\langle K'_{jl}\rangle = 0$ since $\langle \varepsilon' \rangle = 0$), if we assume the Gaussian probability distribution for K'_j with standard deviation σ; this means that the damping constants are proportional to the square of the standard deviation, a result obtained earlier by means of more complicated classical considerations, based on diffraction of light (Tatarskii (1967)).

Assuming that fluctuations of ε' are Gaussian, then fluctuations of ε are log-normal ($\varepsilon' \approx \ln \varepsilon$) and similarly if fluctuations of K'_{jl} are Gaussian, fluctuations of $\mathscr{K}_j$ are log-normal (for a discussion of sums of independent lognormally distributed random variables, see Barakat (1976)). Hence, the probability distribution $\bar{P}(\mathscr{K}_j)$ and the moments $\langle \mathscr{K}_j^n \rangle$ are expressed as

$$\bar{P}(\mathscr{K}_j) = \frac{1}{(2\pi)^{1/2}\sigma\mathscr{K}_j} \exp\left[-\frac{\left(\ln \mathscr{K}_j + \frac{\sigma^2}{2}\right)^2}{2\sigma^2} \right] \tag{21.14}$$

and

$$\langle \mathscr{K}_j^n \rangle = \langle \exp(n\varphi_j) \rangle = \exp[\sigma^2 n(n-1)/2], \qquad n = 0, \pm 1, \ldots, \tag{21.15}$$

where σ is the standard deviation of $\varphi = \ln \mathscr{K}$, whose saturation value in the turbulent atmosphere is about $\sigma = 3/2$ (Tatarskii (1967), Strohbehn (1971), Diament and Teich (1970b)). Obviously $\langle \mathscr{K} \rangle = \langle \exp \varphi \rangle = 1$, which expresses the energy conservation law.

Equations (21.6) can be solved in the form

$$\hat{a}_j(t) = u_j(t)\,\hat{a}_j + v_j(t)\,\hat{a}_j^+ + \sum_l (w_{jl}(t)\,\hat{b}_l^{(j)} + z_{jl}(t)\,\hat{b}_l^{(j)+}), \tag{21.16}$$

where

$$u_j(t) = \exp(-\gamma_j t/2)\left[\cos(\omega_j \mathscr{K}_j^{1/2} t) - \mathrm{i}\cosh(\varphi_j/2)\sin(\omega_j \mathscr{K}_j^{1/2} t)\right],$$
$$v_j(t) = \mathrm{i}\exp(-\gamma_j t/2)\sinh(\varphi_j/2)\sin(\omega_j \mathscr{K}_j^{1/2} t). \tag{21.17}$$

We shall not need the explicit expressions for $w_{jl}(t)$ and $z_{jl}(t)$, which satisfy the identities

$$\mathrm{i}\sum_l \left[\varkappa_{jl}^* w_{jl}\exp(\mathrm{i}\psi_l^{(j)}t) - \varkappa_{jl} w_{jl}^* \exp(-\mathrm{i}\psi_l^{(j)}t)\right] = \gamma_j, \tag{21.18a}$$

$$\sum_l \varkappa_{jl} z_{jl}\exp(-\mathrm{i}\psi_l^{(j)}t) = 0; \tag{21.18b}$$

these quantities are rather complicated (Peřina, Peřinová, Mišta and Horák (1974), Peřina and Peřinová (1976)) and we only need the expression

$$\sum_l |w_{jl}(t)|^2 \approx 1 - \exp(-\gamma_j t), \tag{21.19}$$

since $z_{jl} \approx 0$ in the optical region. Further, it follows from the commutation rule $[\hat{a}_j(t), \hat{a}_j^+(t)] = \hat{1}$ that $|u_j|^2 - |v_j|^2 + \sum_l (|w_{jl}|^2 - |z_{jl}|^2) = 1$ (and other identities follow from the other commutation rules, such as $[\hat{a}_j(t), \hat{a}_k^+(t)] = \hat{0}$ for $j \neq k$, $[\hat{a}_j(t), \hat{a}_k(t)] = \hat{0}$, etc.). Note that the second term in (21.16) is related to self-radiation of the medium, since, neglecting the reservoir terms, we have for the vacuum expectation value of the number operator $\langle 0|\hat{a}_j^+(t)\hat{a}_j(t)|0\rangle = |v_j(t)|^2$, which is the response of the medium in the absence of incident radiation. This property enables us to consider such a process as a prototype of the amplification process (Yuen (1976), Sec. 22.1). Alternatively, we may substitute the solution (21.16) into the antinormal characteristic function (21.10), perform the normal ordering in the initial operators, use the diagonal representation (13.142) of the density matrix and average over the reservoir variables with a probability distribution of the Gaussian form (20.19),

$$\Phi_{\mathscr{N}}(\{\eta_{jl}\}, 0) = \prod_{j,l} (\pi\langle n_{dj}\rangle)^{-1}\exp\left(-\frac{|\eta_{jl}|^2}{\langle n_{dj}\rangle}\right), \tag{21.20}$$

where $\langle n_{dj}\rangle = (\exp(\hbar\psi^{(j)}/KT) - 1)^{-1}$, T being the absolute temperature of the reservoir. We then arrive at the expressions (21.12) and (21.13) where

$$B_j(t) = |u_j(t)|^2 + (1 + \langle n_{dj}\rangle)(1 - \exp(-\gamma_j t)),$$
$$C_j(t) = u_j(t)\, v_j(t),$$
$$\xi_j(t) = u_j(t)\,\xi_j + v_j(t)\,\xi_j^*, \tag{21.21}$$

$\xi_j = \xi_j(0)$ being the initial amplitudes. If the incident radiation is in the coherent state $|\{\xi_j'\}\rangle$, then $\Phi_{\mathscr{N}}(\{\xi_j\}, 0) = \prod_j \delta(\xi_j - \xi_j')$ and we may omit the angle brackets in (21.12) and (21.13) substituting $\xi_j \to \xi_j'$. More generally, the form of these functions is preserved if the incident radiation can be represented by the superposition of coherent and chaotic fields (Peřina, Peřinová and Mišta (1974)). Thus in these cases the statistics of radiation propagating through the random medium are described by the multimode superposition of coherent and chaotic fields with correlated real and imaginary parts of the complex amplitudes α_j. This correlation vanishes if $C_j(t) \approx 0$ ($v_j(t) \approx 0$), so that $K_j(t) \approx B_j^2(t)$, $B_j(t) = 1 + \langle n_{dj} \rangle (1 - \exp(-\gamma_j t))$ and

$$\Phi_{\mathscr{A}}(\{\alpha_j\}, t) = \prod_j (\pi B_j(t))^{-1} \exp\left[- \frac{|\alpha_j - \xi_j(t)|^2}{B_j(t)} \right], \tag{21.22}$$

in correspondence to (20.21b). In this case $\Phi_{\mathscr{N}}(\{\alpha_j\}, t)$ exists for all times and is equal to (21.22) with $B_j(t) = \langle n_{dj} \rangle (1 - \exp(-\gamma_j t))$ and $\xi_j(t) = \xi_j \exp(-i\omega_j t - \gamma_j t/2)$, in agreement with (20.21a).

We point out that the angle brackets in (21.12) and (21.13) also mean the additional average over $\mathscr{K}_j$ or φ_j.

21.4 Photocount statistics

We define the time-dependent photon-number characteristic function $\mathrm{Tr}\,\{\hat{\varrho}(t) \exp(is\hat{n})\} = \mathrm{Tr}\,\{\hat{\varrho} \exp(is\hat{n}(t))\}$, where $\hat{n}(t) = \sum_j \hat{a}_j^+(t)\, \hat{a}_j(t)$ is the photon-number operator in the Heisenberg picture (cf. (11.24b)). Further, in correspondence to (14.13), we define the integrated intensity probability distributions

$$P_{\mathscr{N},\mathscr{A}}(W, t) = \int \Phi_{\mathscr{N},\mathscr{A}}(\{\alpha_j\}, t)\, \delta(W - \sum_j |\alpha_j|^2)\, d^2\{\alpha_j\}, \tag{21.23}$$

and in agreement with (14.12), the integrated intensity is $W = \langle \{\alpha_j\} | \hat{n} | \{\alpha_j\} \rangle = \sum_j |\alpha_j|^2$. In analogy to (14.22) we have for the normal and antinormal characteristic functions, writing the δ-function in (21.23) in the form of the Fourier integral and taking into account that $P_{\mathscr{N},\mathscr{A}}(W, t)$ and $\langle \exp(isW) \rangle_{\mathscr{N},\mathscr{A}}$ are related by the Fourier transformations (cf. (10.22) and (10.46)),

$$\langle \exp(isW) \rangle_{\mathscr{N},\mathscr{A}} = \int \Phi_{\mathscr{N},\mathscr{A}}(\{\alpha_j\}, t) \exp(is \sum_j |\alpha_j|^2)\, d^2\{\alpha_j\}. \tag{21.24}$$

If $\Phi_{\mathscr{N}}(\{\alpha_j\}, t)$ does not exist, we calculate the antinormal characteristic function $\langle \exp(isW) \rangle_{\mathscr{A}}$ with the help of $\Phi_{\mathscr{A}}(\{\alpha_j\}, t)$. However, in order to determine the photocount distribution $p(n, t)$ and its factorial moments $\langle W^k \rangle_{\mathscr{N}}$ with the help of equations (10.12) and (10.13) we need to have the normal characteristic generating function $\langle \exp(isW) \rangle_{\mathscr{N}}$, which is determined by the simple substitution

$$\langle \exp(isW) \rangle_{\mathscr{N}} = (1 + is)^{-M} \left\langle \exp\left(\frac{isW}{1 + is} \right) \right\rangle_{\mathscr{A}} \tag{21.25}$$

following from (16.64) for $s_1 = -1$ and $s_2 = 1$; we remember that M denotes the number of degrees of freedom.

Sometimes we are not interested in the explicit forms of the quasi-distributions, but only in the photocount statistics. Then the non-existence of $\Phi_{\mathcal{N}}$ may be avoided by a direct determination of the generating function $\langle \exp(-\lambda W)\rangle_{\mathcal{N}}$ ($\lambda = -\mathrm{i}s$) from the characteristic function $C_{\mathcal{N}}(\{\beta_j\}, t)$ (Rockower and Abraham (1978), Peřina (1979)). Substituting $\Phi_{\mathcal{N}}(\{\alpha_j\}, t)$ from (21.9) into $\langle \exp(-\lambda W)\rangle_{\mathcal{N}}$ given in (21.24), changing the order of the integrations and performing the integration over $\{\alpha_j\}$, we arrive at the relation

$$\langle \exp(-\lambda W)\rangle_{\mathcal{N}} = (\pi\lambda)^{-M} \int \exp\left(-\frac{1}{\lambda}\sum_j |\beta_j|^2\right) C_{\mathcal{N}}(\{\beta_j\}, t)\, \mathrm{d}^2\{\beta_j\}. \quad (21.26)$$

This relation holds regardless of whether $\Phi_{\mathcal{N}}$ exists or not; this can be proved (Peřina (1979)) using the same relation between $\langle \exp(-\lambda W)\rangle_{\mathcal{A}}$ and $C_{\mathcal{A}}(\{\beta_j\}, t)$, following in the same way as (21.26), and making use of (21.25). Further, note that there are always such small $\lambda(-\mathrm{i}s)$ that (21.24) and (21.26) exist (still at $\lambda = 1$, $\langle \exp(-W)\rangle_{\mathcal{N}} = \pi^M \Phi_{\mathcal{A}}(\{0\}, t)$, cf. (13.71) and (13.75)); then for all values of λ these generating functions are defined by analytic continuation, as discussed in Sec. 10.2 (see also Peřina (1979)).

Making use of the quasi-distribution (21.13) in (21.24) we obtain

$$\langle \exp(\mathrm{i}sW)\rangle_{\mathcal{A}} = \Big\langle \prod_j (1 - \mathrm{i}sE_j)^{-1/2} (1 - \mathrm{i}sF_j)^{-1/2} \times \times \exp\left[\frac{\mathrm{i}sA_{1j}}{1 - \mathrm{i}sE_j} + \frac{\mathrm{i}sA_{2j}}{1 - \mathrm{i}sF_j}\right]\Big\rangle \quad (21.27a)$$

and, substituting (21.27a) into (21.25), we finally have

$$\langle \exp(\mathrm{i}sW)\rangle_{\mathcal{N}} = \Big\langle \prod_j [1 - \mathrm{i}s(E_j - 1)]^{-1/2} [1 - \mathrm{i}s(F_j - 1)]^{-1/2} \times \times \exp\left[\frac{\mathrm{i}sA_{1j}}{1 - \mathrm{i}s(E_j - 1)} + \frac{\mathrm{i}sA_{2j}}{1 - \mathrm{i}s(F_j - 1)}\right]\Big\rangle, \quad (21.27b)$$

where

$$E_j = B_j - |C_j|, \qquad F_j = B_j + |C_j|,$$
$$A_{1,2j} = \frac{1}{2}\left[|\xi_j(t)|^2 \mp \frac{1}{2|C_j|}(\xi_j^2(t)\, C_j^* + \text{c.c.})\right]. \quad (21.28)$$

Of course, the same expressions are obtained from (21.26), where $C_{\mathcal{N}}(\{\beta_j\}, t)$ is given by (21.12) with $B_j(t) \to B_j(t) - 1$ (cf. (13.71)). The angle brackets in (21.27a, b) denote the average over ξ_j and $\mathcal{K}_j$ with the probability distributions $\Phi_{\mathcal{N}}(\{\xi_j\}, 0)$ and $\bar{P}(\mathcal{K}_j)$. The expressions in the angle brackets of (21.27a, b) are the generating functions for the Laguerre polynomials, as seen from (17.63a, b). The quantities $E_j - 1$ and $F_j - 1$ play the role of the "mean numbers of chaotic photons" (the subtraction of unity from E_j and F_j in the normal generating function eliminates the contribution of the physical vacuum), while the quantities $A_{1,2j}$, related to the incident field, are playing the role of the "mean numbers of coherent photons".

The deviation of the expression in the angle brackets of (21.27b) from the Poissonian generating function reflects the change of the photon statistics caused by the dynamics of the process, whereas the average over the initial complex amplitudes, represented by the angle brackets, leads to an additional change of the photon statistics with respect to the initial state of the field (Srinivas (1978)). The character of the dynamical change of the photon statistics depends on the sign of the "mean number of chaotic photons" $E_j - 1 = B_j - |C_j| - 1$ ($F_j - 1 = B_j + |C_j| - 1$ is always non-negative). If it is positive, as for instance in the case of propagation of radiation through random media, the uncertainty (bunching effect) of photons is higher than that for the Poissonian statistics, corresponding to the coherent state. If it is negative, the uncertainty is reduced below that for the coherent state and antibunching of photons and sub-Poisson statistics may occur, as is the case in nonlinear optical processes (Chapter 22).

The photocount distribution $p(n, t)$ and its factorial moments $\langle W^k \rangle_{\mathcal{N}}$ can be obtained as in Sec. 17.3 (Peřina, Peřinová and Horák (1973b)). However, in the following we consider a simplified case in the spirit of the approximate formulae discussed in Sec. 17.3, where all the E_j and F_j are the same, i.e. we assume that all modes are equally damped, only the mean frequency is considered and also all $\mathcal{K}_j$ are assumed to fluctuate uniformly ($\mathcal{K}_j \approx \varepsilon(t)$ in weakly inhomogeneous media). This provides the simplified generating function

$$\langle \exp(\mathrm{i}sW) \rangle_{\mathcal{N}} = \Big\langle [1 - \mathrm{i}s(E - 1)]^{-M/2} [1 - \mathrm{i}s(F - 1)]^{-M/2} \times \times \exp\left[\frac{\mathrm{i}sA_1}{1 - \mathrm{i}s(E - 1)} + \frac{\mathrm{i}sA_2}{1 - \mathrm{i}s(F - 1)}\right]\Big\rangle, \tag{21.29}$$

where

$$E = B - |C|, \qquad F = B + |C|,$$
$$A_{1,2} = \frac{1}{2}\left[\sum_j^M |\xi_j(t)|^2 \mp \frac{1}{2|C|}\Big(C^* \sum_j^M \xi_j^2(t) + \text{c.c.}\Big)\right]; \tag{21.30a}$$

this generating function may be further approximated when the average over the initial random phases is performed, giving approximately

$$A_{1,2} = \tfrac{1}{2}(|u| \mp |v|)^2 W_0, \tag{21.30b}$$

where $W_0 = \Sigma_j^M |\xi_j|^2$ is the integrated intensity of the incident radiation.

From (21.29) we obtain for the mean number of photons

$$\langle n(t) \rangle = \langle W(t) \rangle_{\mathcal{N}} = \frac{\mathrm{d}}{\mathrm{d}(\mathrm{i}s)} \langle \exp(\mathrm{i}sW) \rangle_{\mathcal{N}} \Big|_{\mathrm{i}s = 0}$$
$$= \langle |u|^2 + |v|^2 \rangle \langle W_0 \rangle + M\langle |v|^2 \rangle + M\langle n_d \rangle (1 - \exp(-\gamma t)); \tag{21.31}$$

the angle brackets mean now the average over W_0 with the probability distribution $P_{\mathcal{N}}(W_0)$, and over $\mathcal{K}$ with the probability distribution $\bar{P}(\mathcal{K})$.

Applying the identities (17.63a, b) to (21.29), we arrive at the following photocount distribution and its factorial moments

$$p(n,t) = \left\langle (EF)^{-M/2}(1 - F^{-1})^n \exp\left(-\frac{A_1}{E} - \frac{A_2}{F}\right) \times \right.$$
$$\times \sum_{k=0}^{n} \frac{1}{\Gamma(k + M/2)\,\Gamma(n - k + M/2)} \left(\frac{1 - E^{-1}}{1 - F^{-1}}\right)^k \times$$
$$\left. \times L_k^{M/2-1}\left(-\frac{A_1}{E(E-1)}\right) L_{n-k}^{M/2-1}\left(-\frac{A_2}{F(F-1)}\right)\right\rangle, \tag{21.32a}$$

$$\langle W^k \rangle_{\mathcal{N}} = k! \left\langle (F-1)^k \sum_{l=0}^{k} \frac{1}{\Gamma(l + M/2)\,\Gamma(k - l + M/2)} \left(\frac{E-1}{F-1}\right)^l \times \right.$$
$$\left. \times L_l^{M/2-1}\left(-\frac{A_1}{E-1}\right) L_{k-l}^{M/2-1}\left(-\frac{A_2}{F-1}\right)\right\rangle; \tag{21.32b}$$

here

$$E, F - 1 = |v|(|v| \mp |u|) + \langle n_d \rangle (1 - \exp(-\gamma t)). \tag{21.33}$$

The corresponding integrated intensity probability distributions $P_{\mathcal{N},\mathcal{A}}(W, t)$ can be calculated from $\langle \exp(isW) \rangle_{\mathcal{N},\mathcal{A}}$ by means of the Fourier transformation (Peřina, Peřinová and Horák (1973b)).

The first term in (21.31) represents the response to the incident radiation and it is zero if the incident field is zero, $\langle W_0 \rangle = 0$. The second term represents the self-radiation of the medium and it is non-zero even when $\langle W_0 \rangle = 0$. The third term is the reservoir contribution. If the incident radiation is sufficiently strong, quantum effects may be neglected; then $E - 1, F - 1 \approx \langle n_d \rangle (1 - \exp(-\gamma t))$ and we obtain the photocount distribution and its factorial moments as given in (17.94) and (17.93) with $s = 1$ averaged over $\mathcal{K}$, corresponding to the photocount generating function in the form (17.92) with $s = 1$; here $\langle n_c \rangle = (|u|^2 + |v|^2) W_0$ and $\langle n_{ch} \rangle = M \langle n_d \rangle (1 - \exp(-\gamma t))$. If $\gamma = \langle n_d \rangle = 0$, i.e. if the damping mechanism is neglected, we obtain the standard photodetection equation (10.3a) with $W \to W_0 \mathcal{L}$, where $\mathcal{L} = |u|^2 + |v|^2$, additionally averaged over $\mathcal{K}$.

The photocount distribution $p(n, z/c)$ (z being the distance travelled in the random medium) for radiation propagating through the turbulent atmosphere is shown in Fig. 21.1 for (a) initially chaotic light ($\langle n_c \rangle = 0$, $\langle n_{ch} \rangle = 20$, $\gamma = \Gamma T = 0$), (b) initially coherent light ($\langle n_c \rangle = 20$, $\langle n_{ch} \rangle = 0$ independently of γ). Curves *a* are for $\sigma = 0$, curves *b* are for an intermediate level of fluctuations with $\sigma = 1/2$ and curves *c* for a saturated level of fluctuations with $\sigma = 3/2$. The dotted curves represent the values obtained from the above quantum dynamical description of propagation of radiation through the turbulent atmosphere, the solid and dashed curves correspond to the phenomenological descriptions using the modulation of the photon number or of the intensity suggested by Tatarskii (1971) and Diament and Teich (1970b). More details can be found in original papers (Peřina, Peřinová, Teich and Diament (1973), Peřina, Peřinová, Diament and Teich (1975)) and

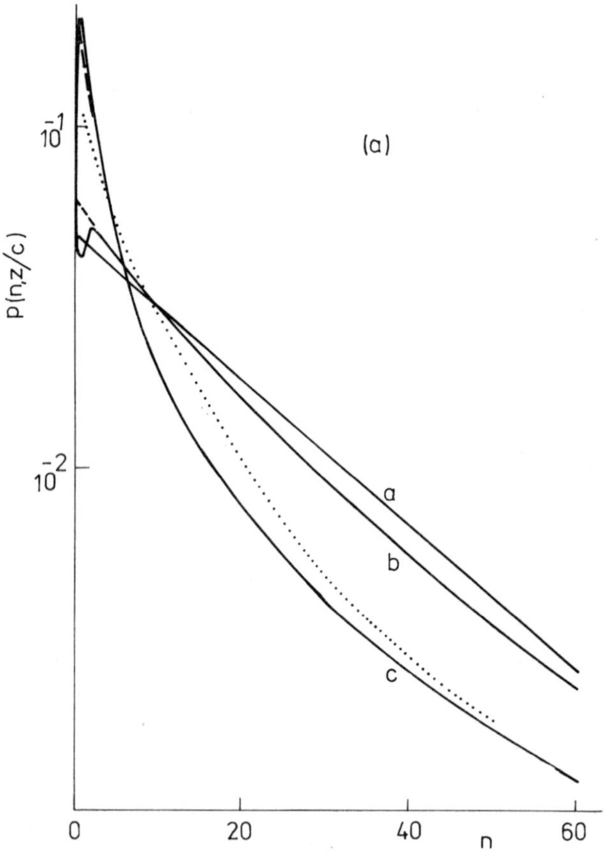

in reviews (Peřina (1980, 1984)). Thus the phenomenological descriptions may be considered as effective descriptions of radiation in lossless random media. As we have noted the photocount distribution is broadened and shifted to lower *n* with increasing level of fluctuations.

Such a behaviour of the photocount distributions of radiation in random media, as shown in Fig. 21.1, has been also obtained by Estes, Kuppenheimer and Narducci (1970) in a quantum statistical analysis of randomly modulated laser beams, by Bufton, Iyer and Taylor (1977) in connection with the scintillation statistics caused by atmospheric turbulence and speckle in satellite laser ranging and by Churnside and McIntyre (1978) in a study of the signal current probability distribution for optical heterodyne receivers in the turbulent atmosphere. This behaviour has been experimentally observed by Bluemel, Narducci and Tuft (1972) using the log-normal distribution for intensity fluctuations of scattered light by a rotating ground glass, by Bertolotti (1974) using nematic liquid crystals with the voltage varied to randomize the light, by Churnside and McIntyre (1978) who used an optical heterodyne receiver operating in the presence of clear air turbulence in a 1.6 km propagation path through the open atmosphere, and

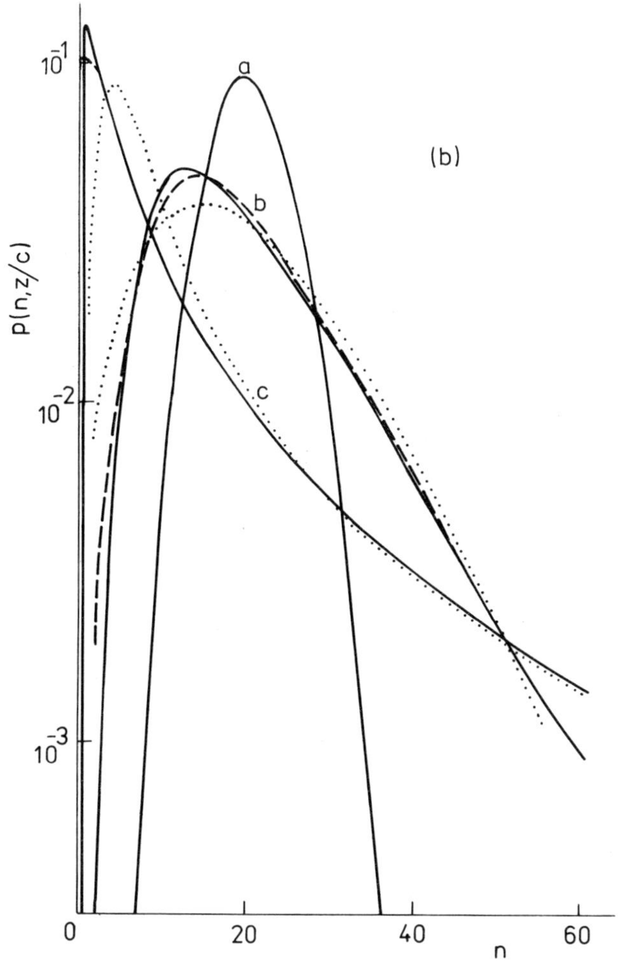

Fig. 21.1 A comparison of the photocount distributions based on the Tatarskii (solid curves) and the Diament-Teich (dashed curves) descriptions, for $\sigma = 0$ (curves *a*), $\sigma = 1/2$ (curves *b*) and $\sigma = 3/2$ (curves *c*) and (a) $\langle n_c \rangle = 0$, $\langle n_{ch} \rangle = 20$, $\gamma = 0$, (b) $\langle n_c \rangle = 20$, $\langle n_{ch} \rangle = 0$, independently of M and γ. The dotted curves represent values based on the quantum dynamical description. When only a solid curve is shown, the solid and dashed curves coincide (after J. Peřina, V. Peřinová, M. C. Teich and P. Diament, 1973, *Phys. Rev.* **A 7**, 1732, and J. Peřina, V. Peřinová, P. Diament and M. C. Teich, 1975, *Czech. J. Phys.* **B 25**, 483).

by Parry (1981) who reported measurements of atmospheric turbulence that induced intensity fluctuations in a laser beam propagating through a 1.125 km path in the atmosphere. In the last measurements the so-called *K*-distribution was used (Parry, Pusey, Jakeman and McWhirter (1978)). Similar results were obtained by Phillips and Andrews (1981).

Finally we note that if we neglect losses ($\gamma = \langle n_d \rangle = 0$), $|u|^2 + |v|^2 = 1 + 2|v|^2$ ($|u|^2 - |v|^2 = 1$ from the commutation rule) and hence the enhancement factor is generally connected to the uncertainty (the quantum noise component) in the photocount distribution, determined by $M(E - 1 + F - 1)/2 = M|v|^2$, i.e. the quantum noise is generally increasing together with an increase in the level of amplification (Mollow and Glauber (1967a, b) and Chapter 22).

21.5 Speckle phenomenon

When coherent laser light is reflected from a rough surface or when it propagates through a medium with random refractive index fluctuations, we can observe a typical granular pattern called the *speckle phenomenon.* This phenomenon depends on the coherence of the incident light and the structure of the rough surface or of the random medium.

The statistical properties of the speckle patterns may be in general very complicated. However, their description is quite analogous to the description of the photocount statistics of radiation and therefore most results from this field can be directly transferred to investigations of the statistical properties of the speckle. If the number of scatterers is high, so that the central limit theorem of probability theory can be applied, the negative exponential intensity distribution is appropriate to describe the statistics of the speckle. For smaller numbers of scatterers deviations from Gaussian statistics may occur (Crosignani, Di Porto and Bertolotti (1975)). For the sum of M speckles the probability distribution (10.46) is established. The intensity probability distributions for superposed multifold speckle patterns, in general partially polarized, with a coherent background, can also be obtained (Peřina (1977), cf. Sec. 17.3). These formulae are discussed in greater detail by Peřina (1977), Saleh and Irshid (1979) and Peřina and Horák (1981).

If we define the contrast C of the speckle pattern as $C = \langle(\Delta I)^2\rangle_{\mathcal{N}}^{1/2}/\langle I\rangle_{\mathcal{N}}$, then from (10.48)

$$C = M^{-1/2}, \tag{21.34}$$

which means that the contrast of the speckle pattern decreases with increase of M. Such a dependence of the intensity probability distribution on M and C represented by (21.34) has been verified experimentally by Ohtsubo and Asakura (1977). A connection of these quantities to the structure of the rough surface may also be established (e.g. Welford (1977)). For further studies we can recommend reviews of the statistical properties of the speckle by Goodman (1975), Dainty (1975, 1976) and Zardecki (1978), where further references may be found. Finally note that the statistics of the speckle fields in terms of the Stokes parameters have been considered by Fercher and Steeger (1981) and Steeger and Fercher (1982), and the role of speckle in optical fibres has been discussed by Daino, Marchis and Piazzolla (1980).

QUANTUM STATISTICS OF RADIATION IN NONLINEAR MEDIA

We continue our demonstration of the general methods of investigation of the statistical properties of radiation by applying them to nonlinear optical processes, as discussed in Chapter 19. We particularly pay attention to optical parametric processes and Raman scattering. A more detailed treatment of this subject can be found in the monograph by this author (Peřina (1984), Chapter 9).

22.1 Optical parametric processes with classical pumping

The simplest non-trivial nonlinear case is the degenerate parametric process with classical pumping, which is mathematically fully analogous to the single-mode quantum model for propagation of radiation through random media, discussed in Chapter 21. Assuming the pumping radiation to be so strong that it may be considered as classical radiation, the renormalized Hamiltonian for second subharmonic generation (degenerate parametric amplification) reads

$$\hat{H}_{\text{rad}} = \hbar\omega\hat{a}^+\hat{a} - \tfrac{1}{2}\hbar g(\hat{a}^2 \exp(\mathrm{i}\,2\omega t - \mathrm{i}\varphi) + \text{h.c.}); \tag{22.1}$$

here, for simplicity, phase matching is assumed so that the coupling constant $g > 0$ is real, the frequency of pumping is 2ω and its phase is φ (here as well as in (22.6) the pumping amplitude is included in the coupling constant g). Losses may be included in the same manner as in Chapters 20 and 21. The corresponding Heisenberg-Langevin equation is

$$\frac{\mathrm{d}\hat{a}}{\mathrm{d}t} = -(\mathrm{i}\omega + \gamma/2)\,\hat{a} + \mathrm{i}g\hat{a}^+ \exp(-\mathrm{i}\,2\omega t + \mathrm{i}\varphi) + \hat{L}, \tag{22.2}$$

where the Langevin force $\hat{L}$ is given by (20.9) and the generalized Fokker-Planck equation can be written in the form

$$\frac{\partial \Phi_{\mathscr{A}}}{\partial t} = \left\{(\mathrm{i}\omega + \gamma/2)\frac{\partial}{\partial\alpha}\alpha + \text{c.c.} + \gamma(\langle n_d\rangle + 1)\frac{\partial^2}{\partial\alpha\,\partial\alpha^*} + \right.$$
$$\left. + \left[\mathrm{i}g \exp(\mathrm{i}\,2\omega t - \mathrm{i}\varphi)\left(\alpha\frac{\partial}{\partial\alpha^*} + \frac{1}{2}\frac{\partial^2}{\partial\alpha^{*2}}\right) + \text{c.c.}\right]\right\}\Phi_{\mathscr{A}}, \tag{22.3}$$

where α is an eigenvalue of $\hat{a}$ for the coherent state $|\alpha\rangle$. The solution of the

Heisenberg-Langevin equation (22.2) together with its Hermitian conjugate equation has the form (21.16), and if losses are neglected

$$u(t) = \cosh(gt)\exp(-\mathrm{i}\omega t),$$
$$v(t) = \mathrm{i}\sinh(gt)\exp(-\mathrm{i}\omega t + \mathrm{i}\varphi). \quad (22.4)$$

The characteristic function $C_{\mathscr{A}}$ and the quasi-distribution $\Phi_{\mathscr{A}}$ are determined by equations (21.12) and (21.13), with the appropriate generating function (21.29), the photocount distribution (21.32a), and its factorial moments (21.32b) with $M = 1$, where the functions $B(t)$ and $C(t)$ can be obtained in closed forms (Mišta, Peřinová, Peřina and Braunerová (1977)).

Neglecting losses and making the correspondence to equation (13.149a) for the two-photon coherent states, we see that $u = \mu^*$ and $v = -\nu$, and both the processes are described in the same way. This makes it possible to show the full mathematical equivalence of the two-photon coherent state technique with the results obtained for the Hamiltonian (22.1) (Yuen (1976), Mišta, Peřinová, Peřina and Braunerová (1977)). This shows that the two-photon coherent states are generated rather by the degenerate parametric amplification process with classical pumping, than by two-photon stimulated emission, unless the atomic variables are described classically.

Since $(B-1)^2 - |C|^2 = (\cosh^2(gt) - 1)^2 - \cosh^2(gt)\sinh^2(gt) = -\sinh^2(gt) < 0$, which is the expression involved in the Fourier transformation to obtain $\Phi_{\mathscr{N}}(\alpha, t)$, this quasi-distribution does not exist at all for $t > 0$ (on the other hand, $\Phi_{\mathscr{A}}(\alpha, t)$ always exists, since $B^2 - |C|^2 = \cosh^2(gt) > 0$).

For the mean number of photons and for the variance of the integrated intensity we obtain (Stoler (1974), Mišta and Peřina (1977b, c))

$$\langle \hat{n}(t)\rangle = \langle W(t)\rangle_{\mathscr{N}} = \langle \hat{a}^+(t)\,\hat{a}(t)\rangle = \frac{\mathrm{d}}{\mathrm{d(is)}}\langle \exp(\mathrm{is}W)\rangle_{\mathscr{N}}\bigg|_{\mathrm{is}=0} =$$
$$= [\cosh(2gt) + \sinh(2gt)\sin(2\vartheta - \varphi)]\,|\xi|^2 + \sinh^2(gt), \quad (22.5a)$$

$$\langle(\Delta W(t))^2\rangle_{\mathscr{N}} = \langle \hat{a}^{+2}(t)\,\hat{a}^2(t)\rangle - \langle \hat{a}^+(t)\,\hat{a}(t)\rangle^2 =$$
$$= \frac{\mathrm{d}^2}{\mathrm{d(is)}^2}\langle \exp(\mathrm{is}W)\rangle_{\mathscr{N}}\bigg|_{\mathrm{is}=0} - \left[\frac{\mathrm{d}}{\mathrm{d(is)}}\langle \exp(\mathrm{is}W)\rangle_{\mathscr{N}}\bigg|_{\mathrm{is}=0}\right]^2 =$$
$$= \sinh(gt)\left[\frac{\sinh(3gt) - \sinh(gt)}{2} + 2|\xi|^2\sinh(3gt) + \right.$$
$$\left. + 2|\xi|^2\cosh(3gt)\sin(2\vartheta - \varphi)\right], \quad (22.5b)$$

where, for simplicity, we do not consider the reservoir and ϑ is the phase of the initial complex amplitude ξ of the incident coherent signal field. The last term in (22.5a) and the first term in (22.5b) represent the quantum noise contributions (arising from the physical vacuum fluctuations) since they are independent of the incident field. It has been shown by Stoler (1974) that if the phases are so related that $2\vartheta - \varphi = -\pi/2$, maximal antibunching of photons occurs, i.e. $\langle(\Delta W(t))^2\rangle_{\mathcal{N}} < 0$. In this case the photocount distribution $p(n, t)$ is narrower than the Poisson distribution corresponding to the coherent state. This is true for a certain time interval after the switching on of the interaction. In Fig. 22.1 we can see the time behaviour of the mean intensity (the mean number of photons) in the antibunching and sub-Poisson regime. The attenuation of the incident field initially in the coherent state stops at $t \approx 6 \times 10^{-5}$ s, in approximate agreement with the region of antibunching seen in Fig. 22.2, where $\langle W^2\rangle_{\mathcal{N}}/\langle W\rangle_{\mathcal{N}}^2 - 1 < 0$. In this figure the reduced factorial

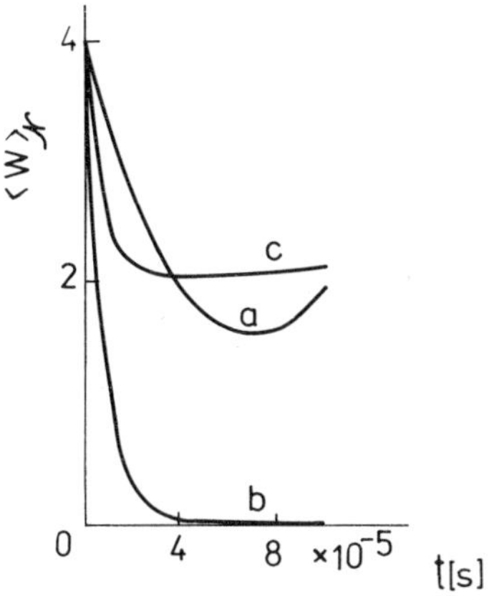

Fig. 22.1 The time behaviour of the mean integrated intensity for $|\xi| = 2, 2\vartheta - \varphi = -\pi/2, g = 10^4\ \mathrm{s}^{-1}$; curves $a-c$ are for $\gamma = \langle n_d\rangle = 0$; $\gamma = 10^5\ \mathrm{s}^{-1}$, $\langle n_d\rangle = 0$ and $\gamma = 10^5\ \mathrm{s}^{-1}$, $\langle n_d\rangle = 2$ respectively (after L. Mišta, V. Peřinová, J. Peřina and Z. Braunerová, 1977, *Acta Phys. Pol.* **A 51**, 739).

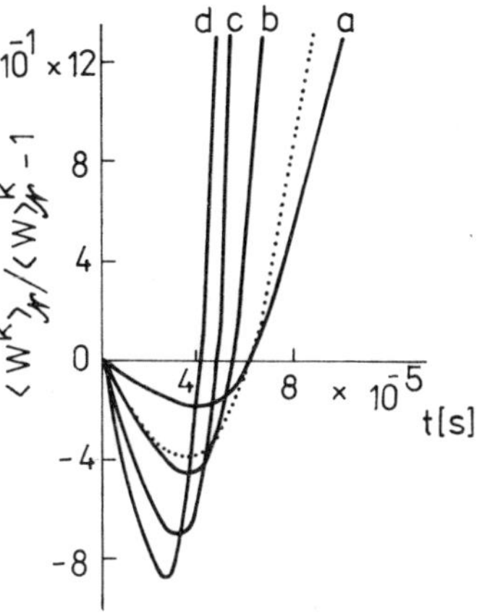

Fig. 22.2 The reduced factorial moments $\langle W^k\rangle_{\mathcal{N}}/\langle W\rangle_{\mathcal{N}}^k - 1, k = 2 - 5$ (solid curves $a-d$ respectively) for $|\xi| = 2$, $2\vartheta - \varphi = -\pi/2$, $g = 10^4\ \mathrm{s}^{-1}$ and $\gamma = \langle n_d\rangle = 0$. The dotted curve represents the quantity $\langle(\Delta W)^2\rangle_{\mathcal{N}}/\langle W\rangle_{\mathcal{N}}$ (after L. Mišta, V. Peřinová, J. Peřina and Z. Brauncrová, 1977, *Acta Phys. Pol.* **A 51**, 739).

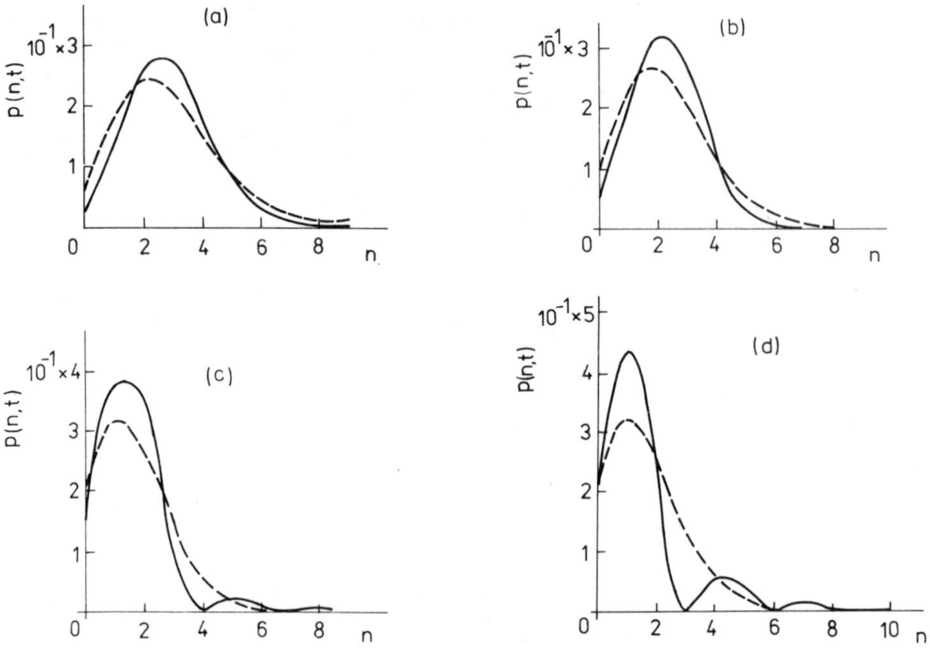

Fig. 22.3 The photocount distribution for the degenerate parametric amplifier, $|\xi| = 2$, $2\vartheta - \varphi = -\pi/2$, $g = 10^4\ s^{-1}$, $\gamma = \langle n_d \rangle = 0$ and (a) $t = 2 \times 10^{-5}$ s, (b) $t = 3 \times 10^{-5}$ s, (c) $t = 6 \times 10^{-5}$ s and (d) $t = 8 \times 10^{-5}$ s. The full curves represent the actual distributions, the dashed curves are the corresponding Poisson distributions (after L. Mišta, V. Peřinová, J. Peřina and Z. Braunerová, 1977, *Acta Phys. Pol.* **A 51**, 739).

moments $\langle W^k \rangle_{\mathcal{N}} / \langle W \rangle_{\mathcal{N}}^k - 1$ are shown for $k = 2 - 5$ (solid curves); the dotted curve represents the quantity $\langle (\Delta W)^2 \rangle_{\mathcal{N}} / \langle W \rangle_{\mathcal{N}}$, with its minimum about -0.4 (it is equal to -1 for the Fock state). Antibunching reflects itself also in the higher-order moments. However, the curves corresponding to various values of k have not a common point of intersection with the t-axis and thus the initially coherent field cannot be coherent again at any later time. (In Brillouin scattering it may happen that the field returns to the coherent state again during the interaction (Pieczonková (1982a, b)).) However, at a certain instant the field may be coherent to fourth order. This is a typical property of fields having no classical analogues (for classical fields coherence to all orders follows from second- and fourth-order coherence (Titulaer and Glauber (1965), Sec. 12.2)). Further we observe that the amplification is related to an increase in quantum noise, and the decrease of fluctuations in n below the level appropriate to the coherent state is usually related to the attenuation of the field. The asymptotic values of the curves shown are $\langle W^k \rangle_{\mathcal{N}} / \langle W \rangle_{\mathcal{N}}^k - 1 = (2k-1)!! - 1$ $(= 2$ for $k = 2)$, i.e. chaotic fluctuations are enhanced for long-time intervals (cf. (10.37)). Note that the entropy properties of "superchaotic" fields have been investigated by Sotskii and Glazatchev (1981).

In Fig. 22.3 we see the time development of the photocount distribution including the antibunching regime ($2\vartheta - \varphi = -\pi/2$) for $g = 10^4\ \mathrm{s}^{-1}$, neglecting losses ($\gamma = \langle n_d \rangle = 0$) and with $|\xi|^2 = 4$ for the initial intensity: $t = 2 \times 10^{-5}$ s (a), 3×10^{-5} s (b), 6×10^{-5} s (c) and 8×10^{-5} s (d). The full curves show the actual distribution, the dashed curves represent the Poisson distribution with the same $\langle W(t) \rangle_{\mathcal{N}}$ appropriate to the coherent state ($p(n, t)$ are always shown as continuous curves for simplicity although they are defined for n integer only). The figures (a) and (b) clearly demonstrate sub-Poisson statistics of photons when the actual distributions are narrower than the corresponding Poisson distributions with the same $\langle W(t) \rangle_{\mathcal{N}}$. For later times a typical oscillating behaviour is reached, reflecting the appearance of competing states. Antibunching and sub-Poisson statistics of photons can also occur if losses are included, but with increasing $\langle n_d \rangle$ they are rapidly smoothed out (Mišta, Peřinová, Peřina and Braunerová (1977)). In the generating function (21.27b) sub-Poisson statistics of photons are shown by a negative "mean number of chaotic photons" $E - 1 = B - |C| - 1$.

Some efforts have been made to enhance photon antibunching using interference effects, by Bandilla and Ritze (1979).

A proposal for an experiment to demonstrate photon antibunching, based on degenerate parametric amplification with classical pumping, has been given by Stoler (1974). In the first nonlinear crystal radiation of frequency 2ω is generated and it is used as the pumping radiation in the second crystal, where the fundamental wave of frequency ω is introduced as the initial signal. If $2\vartheta - \varphi = -\pi/2$, the initial signal is attenuated at the beginning of the interaction and antibunching occurs in the signal mode.

The non-degenerate version of the parametric generation ($\xi_1 = \xi_2 = 0$) or amplification ($\xi_1 \neq 0$, $\xi_2 = 0$) process with classical pumping is described by the renormalized Hamiltonian

$$\hat{H}_{\mathrm{rad}} = \hbar\omega_1 \hat{a}_1^+ \hat{a}_1 + \hbar\omega_2 \hat{a}_2^+ \hat{a}_2 - \hbar g(\hat{a}_1 \hat{a}_2 \exp(\mathrm{i}\omega t - \mathrm{i}\varphi) + \mathrm{h.c.}), \tag{22.6}$$

assuming phase matching again (the coupling constant g is real) and $\omega = \omega_1 + \omega_2$. The corresponding Heisenberg-Langevin equations are

$$\frac{\mathrm{d}\hat{a}_1}{\mathrm{d}t} = -\mathrm{i}\omega_1 \hat{a}_1 + \mathrm{i}g\hat{a}_2^+ \exp(-\mathrm{i}\omega t + \mathrm{i}\varphi) + \hat{L}_1,$$

$$\frac{\mathrm{d}\hat{a}_2}{\mathrm{d}t} = -\mathrm{i}\omega_2 \hat{a}_2 + \mathrm{i}g\hat{a}_1^+ \exp(-\mathrm{i}\omega t + \mathrm{i}\varphi) + \hat{L}_2, \tag{22.7}$$

together with the Hermitian conjugate equations, where the Langevin forces are given by (20.9) for every mode. If one neglects losses, the solution of (22.7) has the form

$$\hat{a}_1(t) = \exp(-\mathrm{i}\omega_1 t)\,[\hat{a}_1 \cosh(gt) + \mathrm{i}\hat{a}_2^+ \sinh(gt) \exp(\mathrm{i}\varphi)],$$

$$\hat{a}_2(t) = \exp(-\mathrm{i}\omega_2 t)\,[\hat{a}_2 \cosh(gt) + \mathrm{i}\hat{a}_1^+ \sinh(gt) \exp(\mathrm{i}\varphi)]. \tag{22.8}$$

The generalized Fokker-Planck equation for this process can be obtained as previously (see Graham (1968a, b), Mišta and Peřina (1978), Peřinová (1981), Peřina (1984)).

Calculating $C_{\mathcal{N}}(\beta_1, 0, t)$ and $\Phi_{\mathcal{N}}(\alpha_1, t)$ by the use of (22.8), we find that the initial field $|\xi_1, 0\rangle$ is transformed by this process to the superposition of signal $\xi_1(t) = \xi_1 \cosh(gt)\exp(-i\omega_1 t)$ and noise $B_1(t) - 1 = \sinh^2(gt)$, with $\Phi_{\mathcal{A}}$ given by (21.22), that is

$$\Phi_{\mathcal{N}}(\alpha_1, t) = [\pi(B_1(t) - 1)]^{-1} \exp\left[-\frac{|\alpha_1 - \xi_1(t)|^2}{B_1(t) - 1}\right], \tag{22.9}$$

and the peak of this Gaussian distribution, whose standard deviation $\sinh^2(gt)$ increases with time, moves along the spiral $\xi_1(t) = \xi_1 \cosh(gt)\exp(-i\omega_1 t)$ (Mollow and Glauber (1967a, b), Mišta (1969), Mišta and Peřina (1978)). For the mean number of photons in the first signal mode we obtain the expression

$$\begin{aligned}\langle \hat{a}_1^+(t)\,\hat{a}_1(t)\rangle &= \langle \hat{a}_1^+ \hat{a}_1\rangle \cosh^2(gt) + \langle \hat{a}_2 \hat{a}_2^+\rangle \sinh^2(gt) + \\ &+ \frac{i}{2}\sinh(2gt)(\hat{a}_1^+ \hat{a}_2^+ \exp(i\varphi) - \text{c.c.}) \underset{t\to\infty}{\simeq} \\ &\underset{t\to\infty}{\simeq} \frac{1}{4}(\langle n_1\rangle + \langle n_2\rangle + 1)\exp(2gt),\end{aligned} \tag{22.10}$$

provided that the initial phases are uncertain; this is in agreement with (19.8) ($\langle n_2\rangle = 0$, $\langle n_1\rangle \gg 1$, so the quantum noise $\exp(2gt)/4$ can be neglected). We observe again that the amplification of the field is related to an increase of uncertainty.

The quasi-distribution $\Phi_{\mathcal{A}}$ has again the form of a multidimensional superposition of coherent and chaotic radiation, averaged over the initial amplitudes, and the normal generating function, the photocount distribution, and its factorial moments, can be expressed by equations (21.29) and (21.32a, b) with $M = 2$. The non-existence of the Glauber-Sudarshan quasi-distribution $\Phi_{\mathcal{N}}(\alpha_1, \alpha_2, t)$ in this process (Mollow and Glauber (1967b)) is related to possible anticorrelation between the signal and idler modes 1 and 2, $\langle \Delta W_1 \Delta W_2\rangle_{\mathcal{N}} < 0$, as follows from the following expression (Mišta and Peřina (1977b, c))

$$\begin{aligned}\langle(\Delta W)^2\rangle_{\mathcal{N}} &= \langle(\Delta W_1)^2\rangle_{\mathcal{N}} + \langle(\Delta W_2)^2\rangle_{\mathcal{N}} + 2\langle \Delta W_1 \Delta W_2\rangle_{\mathcal{N}} = \\ &= 2\sinh(gt)\left[\frac{\sinh(3gt) - \sinh(gt)}{2} + \sinh(3gt)(|\xi_1|^2 + |\xi_2|^2) + \right.\\ &\left. + 2\cosh(3gt)|\xi_1||\xi_2|\sin(\varphi_1 + \varphi_2 - \varphi)\right];\end{aligned} \tag{22.11}$$

here φ_j are the phases of the initial complex amplitudes ξ_j of the coherent field, $\langle(\Delta W)^2\rangle_{\mathcal{N}} = \langle \mathcal{N}\hat{n}^2(t)\rangle - \langle \hat{n}(t)\rangle^2$, $\hat{n}(t) = \hat{a}_1^+(t)\,\hat{a}_1(t) + \hat{a}_2^+(t)\,\hat{a}_2(t)$, $\mathcal{N}$ being the normal ordering operator in $\hat{a}_j(t)$ and $\hat{a}_j^+(t)$. The expression (22.11) generalizes the Stoler expression (22.5b) for second subharmonic generation, which is obtained by putting $\xi_1 = \xi_2 = \xi$ and dividing this expression by 2; here the simultaneous detection of both the modes is assumed. In analogy to the degenerate case, antibunching is maximal when $\varphi_1 + \varphi_2 - \varphi = -\pi/2$ and is due to the coupling of modes. It is a purely quantum phenomenon, as discussed by Paul and Brunner (1981). The time behaviour of the reduced factorial moments and of the photocount distributions has been found to be quite similar to that in the degenerate process (Mišta and Peřina (1978), for reviews see Peřina (1980, 1984)). The effect of pumping fluctuations has been investigated by Peřinová (1981) and Peřinová and Peřina (1981). It has been found that antibunching and sub-Poisson effect are successively missing with increasing level of fluctuations of the pumping radiation.

The renormalized Hamiltonian for frequency up-conversion ($\xi_1 = 0$) and down-conversion ($\xi_2 = 0$) can be written as

$$\begin{aligned} \hat{H}_{\text{rad}} &= \hbar\omega_1 \hat{a}_1^+ \hat{a}_1 + \hbar\omega_2 \hat{a}_2^+ \hat{a}_2 + \hbar(g\hat{a}_1\hat{a}_2^+ \exp(\mathrm{i}\omega t - \mathrm{i}\varphi) + \text{h.c.}), \\ \omega_1 &= \omega_2 + \omega, \end{aligned} \tag{22.12}$$

leading to the Heisenberg equations

$$\begin{aligned} \frac{\mathrm{d}\hat{a}_1}{\mathrm{d}t} &= -\mathrm{i}\omega_1\hat{a}_1 - \mathrm{i}g^*\hat{a}_2 \exp(-\mathrm{i}\omega t + \mathrm{i}\varphi), \\ \frac{\mathrm{d}\hat{a}_2}{\mathrm{d}t} &= -\mathrm{i}\omega_2\hat{a}_2 - \mathrm{i}g\hat{a}_1 \exp(\mathrm{i}\omega t - \mathrm{i}\varphi), \end{aligned} \tag{22.13}$$

whose solution is

$$\begin{aligned} \hat{a}_1(t) &= \exp(-\mathrm{i}\omega_1 t)\left[\hat{a}_1 \cos(|g|t) - \mathrm{i}\frac{g^*}{|g|}\exp(\mathrm{i}\varphi)\,\hat{a}_2 \sin(|g|t)\right], \\ \hat{a}_2(t) &= \exp(-\mathrm{i}\omega_2 t)\left[\hat{a}_2 \cos(|g|t) - \mathrm{i}\frac{g}{|g|}\exp(-\mathrm{i}\varphi)\,\hat{a}_1 \sin(|g|t)\right]. \end{aligned} \tag{22.14}$$

Losses can be taken into account in the same way as before. Thus, compared to the parametric generation and amplification processes, the solution for the frequency converter is periodic. If, for instance, the second mode starts from the vacuum state, then

$$\begin{aligned} \langle \hat{a}_1^+(t)\,\hat{a}_1(t)\rangle &= \langle n_1\rangle \cos^2(|g|t), \\ \langle \hat{a}_2^+(t)\,\hat{a}_2(t)\rangle &= \langle n_1\rangle \sin^2(|g|t), \end{aligned} \tag{22.15}$$

and $\langle \hat{a}_1^+(t)\,\hat{a}_1(t)\rangle + \langle \hat{a}_2^+(t)\,\hat{a}_2(t)\rangle = \langle n_1\rangle$, $\langle n_1\rangle$ being the initial mean number of photons in mode 1; this is in agreement with (19.10). Since in (22.14) there are the initial annihilation operators only, the normal characteristic function $C_{\mathcal{N}}(\beta_1, \beta_2, t)$ is also normally ordered in the initial operators and therefore $\Phi_{\mathcal{N}}(\alpha_1, \alpha_2, t)$ has the form of the δ-function and the initial statistics are conserved by this process (Mišta (1969)). Assuming g real (the phase-matching condition is fulfilled), we can obtain anticorrelation of modes 1 and 2 from (22.14), $\langle \Delta W_1 \Delta W_2 \rangle_{\mathcal{N}} = -|\xi_1|^2 |\xi_2|^2 \sin^2(2gt)/2$ regardless of the intensity level of the field, provided that the field is initially in the coherent state $|\xi_1, \xi_2\rangle$ and the pumping phase is uniformly distributed in the interval $(0, 2\pi)$. However, $\langle(\Delta W)^2\rangle_{\mathcal{N}} = \langle(\Delta W_1)^2\rangle_{\mathcal{N}} + \langle(\Delta W_2)^2\rangle_{\mathcal{N}} + 2\langle \Delta W_1 \Delta W_2\rangle_{\mathcal{N}} = 0$, no bunching or antibunching (which is purely quantum effect) can occur. However, if $g(t)$ is the coupling Markoffian stochastic function, antibunching can occur (Kryszewski and Chrostowski (1977), Mielniczuk (1979), Srinivasan and Udayabaskaran (1979), Mielniczuk and Chrostowski (1981)).

22.2 Interaction of three single-mode boson quantum fields

If we consider the third pumping mode in the Hamiltonian (22.6) also to be a quantum mode and if we remove the assumption of phase matching, we may write the Hamiltonian for the three-mode interaction in the general form (Graham (1968a, b, 1970))

$$\hat{H}_{\text{rad}} = \sum_{j=1}^{3} \hbar\omega_j \hat{a}_j^+ \hat{a}_j - \hbar(g\hat{a}_1\hat{a}_2\hat{a}_3^+ + \text{h.c.}), \tag{22.16}$$

where $\omega_3 = \omega_1 + \omega_2$. This Hamiltonian describes, in fully quantum manner, sum and difference frequency generation, and frequency conversion, as well as superradiative emission (Walls and Barakat (1970), Nussenzveig (1973)). The Heisenberg-Langevin equations are

$$\frac{d\hat{a}_1}{dt} = -(i\omega_1 + \gamma_1/2)\,\hat{a}_1 + ig^*\hat{a}_2^+\hat{a}_3 + \hat{L}_1,$$

$$\frac{d\hat{a}_2}{dt} = -(i\omega_2 + \gamma_2/2)\,\hat{a}_2 + ig^*\hat{a}_1^+\hat{a}_3 + \hat{L}_2,$$

$$\frac{d\hat{a}_3}{dt} = -(i\omega_3 + \gamma_3/2)\,\hat{a}_3 + ig\hat{a}_1\hat{a}_2 + \hat{L}_3, \tag{22.17}$$

with the corresponding Fokker-Planck equation

$$\frac{\partial \Phi_{\mathcal{A}}}{\partial t} = \left\{ \sum_{j=1}^{3} \left[(i\omega_j + \gamma_j/2)\frac{\partial}{\partial \alpha_j}\alpha_j + \text{c.c.} + \gamma_j(\langle n_{dj}\rangle + 1) \times \right.\right.$$
$$\left. \times \frac{\partial^2}{\partial \alpha_j \partial \alpha_j^*} \right] + \left[ig^* \left(\alpha_1^*\alpha_2^* \frac{\partial}{\partial \alpha_3^*} - \alpha_1^*\alpha_3 \frac{\partial}{\partial \alpha_2} - \right.\right.$$
$$\left.\left.\left. - \alpha_2^*\alpha_3 \frac{\partial}{\partial \alpha_1} - \alpha_3 \frac{\partial^2}{\partial \alpha_1 \partial \alpha_2} \right) + \text{c.c.} \right] \right\} \Phi_{\mathcal{A}}. \tag{22.18}$$

From (22.17), neglecting losses, the following conservation laws are obtained

$$\frac{d}{dt}(\langle \hat{a}_1^+\hat{a}_1\rangle + \langle \hat{a}_3^+\hat{a}_3\rangle) = 0,$$

$$\frac{\mathrm{d}}{\mathrm{d}t}(\langle \hat{a}_2^+ \hat{a}_2 \rangle + \langle \hat{a}_3^+ \hat{a}_3 \rangle) = 0,$$

$$\frac{\mathrm{d}}{\mathrm{d}t}(\langle \hat{a}_1^+ \hat{a}_1 \rangle - \langle \hat{a}_2^+ \hat{a}_2 \rangle) = 0; \tag{22.19}$$

they correspond to the conservation laws (19.6).

The short-time solutions (up to $|gt|^2$) are easily obtained for the set of equations (22.17), taking into account that $\hat{a}_j(t) \approx \hat{a}_j(0) + \hat{a}'_j(0)\,t + \hat{a}''_j(0)\,t^2/2$, giving, if the losses are neglected (Agrawal and Mehta (1974), Peřina (1976)),

$$\hat{a}_1(t) = \exp(-\mathrm{i}\omega_1 t)\left[\hat{a}_1 + \mathrm{i}g^* t \hat{a}_2^+ \hat{a}_3 - \frac{|g|^2 t^2}{2} \hat{a}_1 (\hat{a}_2^+ \hat{a}_2 - \hat{a}_3^+ \hat{a}_3)\right],$$

$$\hat{a}_2(t) = \exp(-\mathrm{i}\omega_2 t)\left[\hat{a}_2 + \mathrm{i}g^* t \hat{a}_1^+ \hat{a}_3 - \frac{|g|^2 t^2}{2} \hat{a}_2 (\hat{a}_1^+ \hat{a}_1 - \hat{a}_3^+ \hat{a}_3)\right],$$

$$\hat{a}_3(t) = \exp(-\mathrm{i}\omega_3 t)\left[\hat{a}_3 + \mathrm{i}gt \hat{a}_1 \hat{a}_2 - \frac{|g|^2 t^2}{2} \hat{a}_3 (\hat{a}_1 \hat{a}_1^+ + \hat{a}_2^+ \hat{a}_2)\right], \tag{22.20}$$

where $\hat{a}_j \equiv \hat{a}_j(0)$. One can verify that $[\hat{a}_j(t), \hat{a}_k^+(t)] \approx \hat{1}\delta_{jk}$ up to $(|g|\,t)^2$.

We can now apply the general methods of determining the quantum statistical properties, as explained in Chapter 20. Either we can substitute the solutions (22.20) in the characteristic functions $C_{\mathscr{N}}$ or $C_{\mathscr{A}}$ and use the Baker-Hausdorff identity (the assumptions for its use are fulfilled up to $(|g|\,t)^2$) to put them into the normal form with respect to the initial operators, or we can directly solve the Fokker-Planck equation (22.18) or the corresponding equation for the characteristic function, for which a special iterative procedure has been developed (Peřinová and Peřina (1978b)). In these ways, neglecting losses for simplicity, we arrive at

$$\begin{aligned} C_{\mathscr{N}}(\{\beta_j\}, t) = \langle \exp\{ &- \sum_{j=1}^{3} B_j(t)\,|\beta_j|^2 + \sum_{j=1}^{3} [\tfrac{1}{2}C_j^*(t)\,\beta_j^2 + \text{c.c.}] + \\ &+ [D_{12}(t)\,\beta_1^*\beta_2^* + D_{13}(t)\,\beta_1^*\beta_3^* + D_{23}(t)\,\beta_2^*\beta_3^* + \bar{D}_{13}(t)\,\beta_1\beta_3^* + \\ &+ \bar{D}_{23}(t)\,\beta_2\beta_3^* + \text{c.c.}] + \sum_{j=1}^{3} [\beta_j \xi_j^*(t) - \beta_j^* \xi_j(t)]\}\rangle, \end{aligned} \tag{22.21}$$

where

$$\begin{aligned} &B_j(t) = (1 - \delta_{j3})\,|g|^2 t^2 |\xi_3|^2, \qquad \mathrm{j} = 1, 2, 3, \\ &C_j(t) = 0, \qquad j = 1, 2, 3, \\ &D_{12}(t) = [\mathrm{i}g^* t \xi_3 - \tfrac{1}{2}|g|^2 t^2 \xi_1 \xi_2] \exp(-\mathrm{i}\omega_3 t), \\ &D_{j3}(t) = -\tfrac{1}{2}|g|^2 t^2 \xi_j \xi_3 \exp[-\mathrm{i}(\omega_j + \omega_3)t], \qquad j = 1, 2, \\ &\bar{D}_{13}(t) = \bar{D}_{23}(t) = 0, \end{aligned} \tag{22.22a}$$

$$\begin{aligned} &\xi_1(t) = \exp(-\mathrm{i}\omega_1 t)\,[\xi_1 + \mathrm{i}g^* t \xi_2^* \xi_3 + \tfrac{1}{2}|g|^2 t^2 \xi_1 (|\xi_3|^2 - |\xi_2|^2)], \\ &\xi_2(t) = \exp(-\mathrm{i}\omega_2 t)\,[\xi_2 + \mathrm{i}g^* t \xi_1^* \xi_3 + \tfrac{1}{2}|g|^2 t^2 \xi_2 (|\xi_3|^2 - |\xi_1|^2)], \\ &\xi_3(t) = \exp(-\mathrm{i}\omega_3 t)\,[\xi_3 + \mathrm{i}gt \xi_1 \xi_2 - \tfrac{1}{2}|g|^2 t^2 \xi_3 (1 + |\xi_1|^2 + |\xi_2|^2)]. \end{aligned} \tag{22.22b}$$

The quasi-distribution $\Phi_{\mathscr{A}}$ has again the form of the averaged, shifted, multidimensional Gaussian distribution (Peřinová and Peřina (1978c)).

Alternative methods for circumventing the impossibility of solving the Heisenberg-Langevin equations (22.17) and the generalized Fokker-Planck equation (22.18) in closed form have been based on numerical calculations (Walls and Barakat (1970), Walls and Tindle (1972), Mostowski and Rzazewski (1978)), making use of the commutativity of the free and the interaction Hamiltonians. Some operator linearization methods have been suggested by Katriel and Hummer (1981), making it possible to obtain linearized closed form solutions for a class of nonlinear optical processes, including the optical parametric processes and four-wave mixing. Nevertheless, the above short-time approximation approach is simple and enables us to derive explicitly short-time quantum statistics within the interaction time $t = z/c$ in a systematic way (z being the distance travelled by the beam in the active crystal (Shen (1967)), which is indeed very small in experimental arrangements with a propagating wave. Such solutions provide information for more detailed computations and make it possible to determine, in a straightforward way, all the possible states of the field exhibiting anti-correlation, antibunching, correlation, bunching and full coherence.

Fluctuations in single modes and correlations of fluctuations among them are easily obtained from the normal characteristic function (see (22.21))

$$\begin{aligned}\langle(\Delta W_j)^2\rangle_{\mathscr{N}} &= \langle \hat{a}_j^{+2}(t)\,\hat{a}_j^2(t)\rangle - \langle \hat{a}_j^+(t)\,\hat{a}_j(t)\rangle^2 = \\ &= \left.\frac{\partial^4 C_{\mathscr{N}}}{\partial\beta_j^2\,\partial(-\beta_j^*)^2}\right|_{\{\beta_j\}=0} - \left[\left.\frac{\partial^2 C_{\mathscr{N}}}{\partial\beta_j\,\partial(-\beta_j^*)}\right|_{\{\beta_j\}=0}\right]^2 = \\ &= \langle B_j^2 + |C_j|^2 + 2B_j|\xi_j(t)|^2 + [C_j\xi_j^{*2}(t) + \text{c.c.}]\rangle + \\ &\quad + \langle(B_j + |\xi_j(t)|^2)^2\rangle - \langle B_j + |\xi_j(t)|^2\rangle^2,\end{aligned} \tag{22.23a}$$

$$\begin{aligned}\langle\Delta W_j\,\Delta W_k\rangle_{\mathscr{N}} &= \langle \hat{a}_j^+(t)\,\hat{a}_k^+(t)\,\hat{a}_j(t)\,\hat{a}_k(t)\rangle - \langle \hat{a}_j^+(t)\,\hat{a}_j(t)\rangle\langle \hat{a}_k^+(t)\,\hat{a}_k(t)\rangle = \\ &= \left.\frac{\partial^4 C_{\mathscr{N}}}{\partial\beta_j\,\partial(-\beta_j^*)\,\partial\beta_k\,\partial(-\beta_k^*)}\right|_{\{\beta_j\}=0} - \left.\frac{\partial^2 C_{\mathscr{N}}}{\partial\beta_j\,\partial(-\beta_j^*)}\frac{\partial^2 C_{\mathscr{N}}}{\partial\beta_k\,\partial(-\beta_k^*)}\right|_{\{\beta_j\}=0} = \\ &= \langle|D_{jk}|^2 + |\bar{D}_{jk}|^2 + [D_{jk}\xi_j^*(t)\,\xi_k^*(t) - \bar{D}_{jk}\xi_j(t)\,\xi_k^*(t) + \text{c.c.}]\rangle + \\ &\quad + \langle(B_j + |\xi_j(t)|^2)(B_k + |\xi_k(t)|^2)\rangle - \\ &\quad - \langle B_j + |\xi_j(t)|^2\rangle\langle B_k + |\xi_k(t)|^2\rangle, \qquad j \neq k.\end{aligned} \tag{22.23b}$$

If the field is initially coherent, the last two terms in (22.23a, b) are cancelled.

Applying this to (22.21) and (22.22a, b) we arrive at

$$\begin{aligned}&\langle(\Delta W_j)^2\rangle_{\mathscr{N}} = 2|g|^2 t^2 |\xi_j|^2 |\xi_3|^2, \qquad j = 1, 2, \\ &\langle(\Delta W_3)^2\rangle_{\mathscr{N}} = 0, \\ &\langle\Delta W_1\,\Delta W_2\rangle_{\mathscr{N}} = \mathrm{i}g^* t\xi_1^*\xi_2^*\xi_3 + \text{c.c.} + \\ &+ |g|^2 t^2[(1 + 2|\xi_1|^2 + 2|\xi_2|^2)|\xi_3|^2 - |\xi_1|^2|\xi_2|^2], \\ &\langle\Delta W_j\,\Delta W_3\rangle_{\mathscr{N}} = -|g|^2 t^2 |\xi_j|^2 |\xi_3|^2, \qquad j = 1, 2\end{aligned} \tag{22.24a}$$

and

$$\begin{aligned}\langle(\Delta W)^2\rangle_{\mathcal{N}} &= \langle(\Delta W_1)^2\rangle_{\mathcal{N}} + \langle(\Delta W_2)^2\rangle_{\mathcal{N}} + 2\langle\Delta W_1\,\Delta W_2\rangle_{\mathcal{N}} = \\ &= 4\,|g|\,|t|\,|\xi_1|\,|\xi_2|\,|\xi_3|\sin(\varphi_1 + \varphi_2 + \psi - \varphi_3) + \\ &+ 2\,|g|^2 t^2\,[(3(|\xi_1|^2 + |\xi_2|^2) + 1)\,|\xi_3|^2 - |\xi_1|^2\,|\xi_2|^2],\end{aligned} \tag{22.24b}$$

provided that the field is initially in the coherent state $|\xi_1, \xi_2, \xi_3\rangle$; also in (22.24b) losses are neglected. In parametric amplification $\xi_2 = 0$ and $\langle(\Delta W)^2\rangle_{\mathcal{N}} = 6\,|g|^2 \times t^2(|\xi_1|^2 + 1/3)\,|\xi_3|^2 > 0$ and in parametric generation $\xi_1 = \xi_2 = 0$ and $\langle(\Delta W)^2\rangle_{\mathcal{N}} = 2\,|g|^2 t^2\,|\xi_3|^2 > 0$. From (22.24a) we see that the subfrequency modes are subject to quantum noise (bunching of photons increases) in the interaction, while the sum-frequency mode is coherent in this approximation. Further, there is phase-dependent anticorrelation between the subfrequency signal and idler modes 1 and 2, which is maximal if $\varphi_1 + \varphi_2 + \psi - \varphi_3 = -\pi/2$, where φ_j are again the phases of ξ_j and ψ is the phase of g. This phase condition is in agreement with that of Sec. 22.1 for classical pumping. There always exists anticorrelation in the course of the sum-frequency generation with $\xi_3 = 0$, given by $\langle\Delta W_1\,\Delta W_2\rangle_{\mathcal{N}} = -\,|g|^2 t^2\,|\xi_1|^2\,|\xi_2|^2$, independently of the initial phases. This is a typical higher-order quantum term in the intensities, which cannot be obtained if the pumping mode is treated as classical. We also see from (22.24a) that anticorrelation always appears between the subfrequency and sum-frequency modes at the beginning of the interaction, in consequence of the coupling of modes. Proceeding to higher powers of t, one can show that the sum-frequency mode 3 exhibits antibunching

$$\langle(\Delta W_3)^2\rangle_{\mathcal{N}} = -\tfrac{2}{3}\,|g|^6 t^6\,|\xi_1|^4\,|\xi_2|^4, \tag{22.25}$$

provided that $\xi_3 = 0$, in close analogy to the similar result of Kozierowski and Tanaś (1977) and Kielich, Kozierowski and Tanaś (1978) for the second harmonic generation (see equation (22.30d)). In the course of parametric amplification ($\xi_2 = 0$), there is anticorrelation $\langle\Delta W_1\,\Delta W_3\rangle_{\mathcal{N}} = -\,|g|^2 t^2\,|\xi_1|^2\,|\xi_3|^2$ and $\langle\Delta W_2\,\Delta W_3\rangle_{\mathcal{N}} = 0$. Anticorrelation may also occur if some modes are initially chaotic. For example, if the idler mode 2 is initially chaotic (it may represent an effective mode of molecular and atomic vibrations in Raman scattering) with the mean number of photons or phonons $\langle n_{ch2}\rangle$, we obtain, in the case of sum-frequency generation ($\xi_3 = 0$), the enhanced anticorrelation (Trung and Schütte (1978), Peřinová and Peřina (1978b))

$$\langle\Delta W_1\,\Delta W_2\rangle_{\mathcal{N}} = -\,|g|^2 t^2\,|\xi_1|^2\,\langle n_{ch2}\rangle\,(1 + \langle n_{ch2}\rangle). \tag{22.26}$$

It should be noted that a random mode may create an enhancement of anticorrelation between different modes because fluctuations in them may be opposite as a consequence of randomness; however $\langle(\Delta W_j)^2\rangle_{\mathcal{N}} < 0$ may only occur as resulting from quantum noise terms arising from the commutators (Paul and Brunner (1981)).

Note that questions of saturation, lost in the short-time approximation, have been discussed by Graham (1968a, b, 1970, 1973), Graham and Haken (1968), Oliver and Bendjaballah (1980) and Abraham (1980). Methods for possible enhancement of antibunching in the non-degenerate parametric process (with classical pumping), using interference, have been suggested by Bandilla and Ritze (1980a). Also higher-order parametric processes have been treated (Peřinová, Peřina and Knesel (1977), Graham (1973)), but the results obtained are more complicated.

Proposals to generate quantum fields exhibiting photon antibunching and sub-Poisson statistics have been given by Stoler (1974), based on the degenerate parametric amplifier with classical pumping, and by Paul and Brunner (1980) and Chmela, Horák and Peřina (1981), based on the non-degenerate parametric amplifier with classical pumping. Paul and Brunner (1980) also investigated memory effects caused by dispersion in the nonlinear medium. However, they neglected the initial correlations of radiation modes in the second stage of the device, which were included by Chmela et al. (1981). Paul and Brunner (1980) proposed to use a c.w. picosecond laser emitting light of frequency ω_3, which enters a nonlinear crystal in which both signal and idler waves are generated, both with low efficiency. Subsequently the latter waves are attenuated and the phase difference is changed by π. Then in the second crystal the sum-frequency radiation is generated (the incident signal and idler waves are attenuated) and photon antibunching or sub-Poisson statistics may be registered either by means of the Hanbury Brown-Twiss coincidence technique, antibunching being demonstrated by an excess of delayed coincidences, or by photocount measurements ($p(n)$ being narrower than the Poisson distribution). Chmela et al. (1981) suggested using two powerful lasers as sources of intense coherent subfrequency radiations at ω_1 and ω_2, respectively. Each of these subfrequency light beams is optically divided into two beams, one of these being strong and the other weak. In the first stage the intense sum-frequency wave at ω_3 is generated by the two subfrequency waves at ω_1 and ω_2 ($\omega_1 + \omega_2 = \omega_3$); in the second stage the nonlinear interaction between the strong sum-frequency wave at ω_3 (pumping wave) and the two weak subfrequency waves at ω_1 and ω_3, which have not been employed in any nonlinear process in the first stage, takes place. The optical paths of both the strong and weak waves at ω_j ($j = 1, 2$), interacting in the first and the second stages respectively, must be equal to ensure the initial condition for sum-frequency generation and for antibunching in the subfrequency modes ($\varphi_1 + \varphi_2 - \varphi = -\pi/2$) in the second stage of the experiment. In this way the initial uncorrelation of the beams at the second stage is achieved.

The effect of pump coherence on frequency conversion and parametric amplification has been further investigated by Crosignani, Di Porto, Ganiel, Solimeno and Yariv (1972). A quantum theory of light propagation in amplifying media has been considered by Foerster and Glauber (1971).

22.3 Second and higher harmonic and subharmonic generation

If we cannot distinguish between the signal and idler modes, the degenerate version of the Hamiltonian (22.16) is appropriate,

$$\hat{H}_{\text{rad}} = \hbar\omega_1 \hat{a}_1^+ \hat{a}_1 + \hbar\omega_2 \hat{a}_2^+ \hat{a}_2 - \hbar(g\hat{a}_1^2 \hat{a}_2^+ + \text{h.c.}), \tag{22.27}$$

where $\omega_2 = 2\omega_1$. Just as in Sec. 22.2 the Heisenberg-Langevin and Fokker-Planck equations may be obtained. The Heisenberg-Langevin equations are, in analogy to (22.17),

$$\begin{aligned} \frac{\mathrm{d}\hat{a}_1}{\mathrm{d}t} &= -(\mathrm{i}\omega_1 + \gamma_1/2)\,\hat{a}_1 + \mathrm{i}2g^*\hat{a}_1^+\hat{a}_2 + \hat{L}_1, \\ \frac{\mathrm{d}\hat{a}_2}{\mathrm{d}t} &= -(\mathrm{i}\omega_2 + \gamma_2/2)\,\hat{a}_2 + \mathrm{i}g\hat{a}_1^2 + \hat{L}_2, \end{aligned} \tag{22.28}$$

from which the following conservation law follows if losses are neglected again (see (19.5)),

$$\frac{\mathrm{d}}{\mathrm{d}t}(\langle \hat{a}_1^+ \hat{a}_1 \rangle + 2\langle \hat{a}_2^+ \hat{a}_2 \rangle) = 0. \tag{22.29}$$

As in Sec. 22.2 we can obtain the short-time solutions of (22.28) and of the corresponding Fokker-Planck equation (Peřinová and Peřina (1978a), see also Peřina (1984)). For fluctuations in the subharmonic and the second harmonic modes and for the correlation of fluctuations between them we obtain in the same way as before for the initially coherent field (again neglecting losses, for simplicity)

$$\begin{aligned} \langle (\Delta W_1)^2 \rangle_{\mathcal{N}} &= \mathrm{i}\,2g^* t \xi_1^{*2} \xi_2 + \text{c.c.} + 2|g|^2 t^2 (2|\xi_2|^2 + 12|\xi_1|^2|\xi_2|^2 - |\xi_1|^4), \\ \langle (\Delta W_2)^2 \rangle_{\mathcal{N}} &= 0, \\ \langle \Delta W_1\, \Delta W_2 \rangle_{\mathcal{N}} &= -4|g|^2 t^2 |\xi_1|^2 |\xi_2|^2. \end{aligned} \tag{22.30a}$$

Thus the subharmonic mode 1 exhibits antibunching if $2\varphi_1 - \varphi_2 + \psi = -\pi/2$, in agreement with the results of Stoler (1974) and Sec. 22.1. For second harmonic generation ($\xi_2 = 0$) we have antibunching independent of the initial phases, $\langle (\Delta W_1)^2 \rangle_{\mathcal{N}} = -2|g|^2 t^2 |\xi_1|^4$, which needs the quantum description of both the modes, as being the higher-order effect in the intensity. This has been used by Chmela et al. (1981) as the basis of a proposed cascade experiment for observing antibunching in the fundamental mode with the successive filtering of the second-harmonic mode; for this, the use of optical fibres may be suitable. More generally, for the kth harmonic we have

$$\langle (\Delta W_1)^2 \rangle_{\mathcal{N}} = -k(k-1)|g|^2 t^2 |\xi_1|^{2k}. \tag{22.30b}$$

In the course of the second subharmonic generation ($\xi_1 = 0$) we have photon bunching in the fundamental mode, $\langle (\Delta W_1)^2 \rangle_{\mathcal{N}} = \langle (\Delta W)^2 \rangle_{\mathcal{N}} = 4|g|^2 t^2 |\xi_2|^2 > 0$. Between modes 1 and 2 there is in general anticorrelation at the beginning of the interaction, whereas in the course of the second harmonic or subharmonic generation these modes are uncorrelated in this approximation. More generally for the kth-order process (Hofman (1980))

$$\langle \Delta W_1 \, \Delta W_2 \rangle_{\mathcal{N}} = -(k!)^2 |g|^2 t^2 |\xi_1|^2 |\xi_2|^2 \sum_{s=0}^{k-2} \frac{|\xi_1|^{2s}}{s!\,(k-s-1)!\,(s+1)!}. \tag{22.30c}$$

The second or the kth harmonic mode is coherent in this approximation. This has been verified experimentally by Clark, Estes and Narducci (1970), Akhmanov, Tchirkin and Tunkin (1970) and Akhmanov and Tchirkin (1971). However, in higher powers of t antibunching occurs in the course of second harmonic generation ($\xi_2 = 0$) (Kozierowski and Tanaś (1977), Kielich, Kozierowski and Tanaś (1978)),

$$\langle (\Delta W_2)^2 \rangle_{\mathcal{N}} = -\tfrac{8}{3} |g|^6 t^6 |\xi_1|^8. \tag{22.30d}$$

This result is in agreement with a numerical solution by Walls and Tindle (1972). If the pump exceeds some critical value, self-pulsing behaviour can occur (McNeil, Drummond and Walls (1978)).

Further investigation of the interaction between the fundamental and the second harmonic waves inside a Fabry-Perot cavity with external coherent driving fields was carried out by Drummond, McNeil and Walls (1979, 1980, 1981), who found also that antibunching of the fundamental wave in the steady state may appear and some bistable operation is possible. A review has been given by Walls, Drummond and McNeil (1981).

An experiment based on nonlinear optical phenomena is to be performed to detect photon antibunching and sub-Poisson behaviour as basic properties of quantum fields having no classical analogues. However, Wagner, Kurowski and Martienssen (1979) performed a simulation experiment using an analogue device and a computer, employing an electro-optical filter. Fig. 22.4 shows the logarithmic probability $P_{\mathcal{N}}(W)$, which is equal to $p(n)$ for strong fields, (a) in the fundamental component and (b) in the second harmonic wave. Their measurement provides an analogue demonstration of antibunching in second harmonic generation.

Higher harmonics and subharmonic generation has also been discussed using the Heisenberg-Langevin and the Fokker-Planck approaches (Peřinová, Peřina and Knesel (1977), Peřina, Peřinová and Knesel (1977), Kielich, Kozierowski and Tanaś (1978), Kielich (1981)).

A correction to antibunching in second subharmonic generation with classical pumping, derived from a perturbative quantum treatment of the pumping field, has been obtained by Neumann and Haug (1979) and Milburn and Walls (1981). Photon statistics of the second – and third – harmonic generation in gaseous systems, adopting the Fock-state technique in the density matrix treatment and taking into account atomic motion, have been discussed by Nayak and Mohanty (1977) and Nayak (1980). Reviews of results concerning the antibunching phenomenon have been published recently by Loudon (1980), Kozierowski (1981),

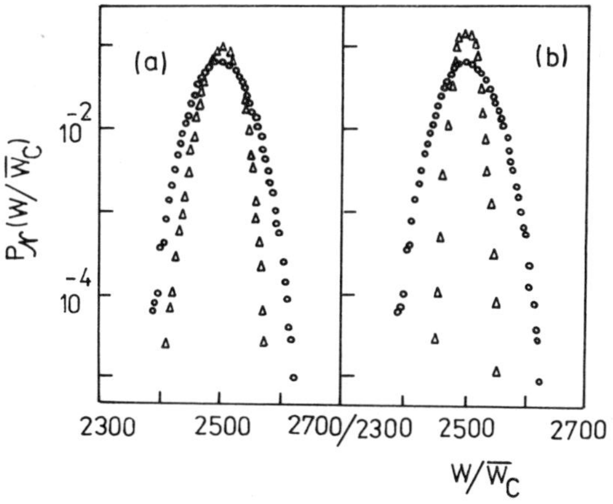

Fig. 22.4 Measured intensity probability distributions in a simulation experiment giving evidence for antibunching (a) in the fundamental wave, (b) in the second harmonic wave; the initial distributions are represented by circles and the final distributions by triangles; $\overline{W}_C$ is the mean intensity within one coherence area (after J. Wagner, P. Kurowski and W. Martienssen, 1979, *Z. Phys.* **B 33**, 391).

Kielich (1981) and Paul (1982). Methods used for the description of the interaction of N two-level atoms with the electromagnetic field have been exploited by Orszag (1979) to find also solutions for a parametric amplifier, a frequency converter, Brillouin and Raman scattering. Semiclassical phenomenological methods of describing radiation statistics and coherence properties in nonlinear optical processes have been developed and applied by Tunkin and Tchirkin (1970), Akhmanov and Tchirkin (1971) and Akhmanov, Dyakov and Tchirkin (1981).

Further results concerning the quantum statistical and coherence properties of radiation in optical parametric processes are available in the literature (Mollow and Glauber (1967a, b), Mišta (1969), Tucker and Walls (1969), Mishkin and Walls (1969), Smithers and Lu (1974), Graham (1968a, b), Graham and Haken (1968), Walls and Barakat (1970), Echtermeyer (1971), Dewael (1975), Trung and Schütte (1977, 1978)) and they have been reviewed in greater detail (Peřina (1984)).

22.4 Raman scattering

Consider Raman scattering of intense classical laser light with complex amplitude e_L by an infinite Markoffian system of phonons, as expressed by the renormalized Hamiltonian (Walls (1973), Peřina (1981a, b))

$$\hat{H} = \hbar\omega_L \hat{a}_L^+ \hat{a}_L + \hbar\omega_S \hat{a}_S^+ \hat{a}_S + \hbar\omega_A \hat{a}_A^+ \hat{a}_A + \sum_l \hbar\omega_{Vl} \hat{a}_{Vl}^+ \hat{a}_{Vl} - \\ - \sum_l (\hbar g_l e_L \hat{a}_S^+ \hat{a}_{Vl}^+ + \hbar\varkappa_l^* e_L \hat{a}_{Vl} \hat{a}_A^+ + \text{h.c.}), \tag{22.31}$$

where the first four terms represent the free Hamiltonians of the laser (L), Stokes (S), anti-Stokes (A) and vibration phonon (V) modes, with the corresponding annihilation operators $\hat{a}_L$, $\hat{a}_S$, $\hat{a}_A$, $\hat{a}_{Vl}$ and with frequencies ω_L, ω_S, ω_A, ω_{Vl} respectively; the remaining terms are the interaction terms between laser photons, scattered Stokes and anti-Stokes photons and phonons, g_l and $\varkappa_l$ being the Stokes and anti-Stokes coupling constants. The Heisenberg equations read

$$\frac{d\hat{a}_S}{dt} = -i\omega_S\hat{a}_S + i\sum_l g_l e_L \hat{a}^+_{Vl},$$

$$\frac{d\hat{a}_A}{dt} = -i\omega_A\hat{a}_A + i\sum_l \varkappa^*_l e_L \hat{a}_{Vl},$$

$$\frac{d\hat{a}_{Vl}}{dt} = -i\omega_{Vl}\hat{a}_{Vl} + ig_l e_L \hat{a}^+_S + i\varkappa_l e^*_L \hat{a}_A. \tag{22.32}$$

Introducing the frequency mismatch difference $\Delta = \omega_L - (\omega_S + \omega_A)/2$ and the new operators from the relation

$$\hat{a}_j(t) = \hat{A}_j(t)\exp(-i\omega_j t - i\Delta t), \qquad j = S, A, \tag{22.33}$$

after the elimination of chaotic phonon operators $\hat{a}_{Vl}$ we obtain in the Wigner-Weisskopf approximation in the same manner as in Chapter 20, the following set of equations of motion

$$\frac{d}{dt}\begin{pmatrix}\hat{A}_S\\ \hat{A}^+_A\end{pmatrix} = \begin{pmatrix}\frac{1}{2}\gamma_S|E_L|^2 + i\Delta & \frac{1}{2}\gamma_{SA}E^2_L\\ -\frac{1}{2}\gamma^*_{SA}E^{*2}_L & -\frac{1}{2}\gamma_A|E_L|^2 - i\Delta\end{pmatrix}\begin{pmatrix}\hat{A}_S\\ \hat{A}^+_A\end{pmatrix} + \begin{pmatrix}E_L\hat{L}_S\\ E^*_L\hat{L}^+_A\end{pmatrix}, \tag{22.34}$$

where $\gamma_S = 2\pi|g|^2\varrho$, $\gamma_A = 2\pi|\varkappa|^2\varrho$, $\gamma_{SA} = 2\pi g\varkappa^*\varrho$ are the gain, damping and the mutual damping constants of modes S and A ($|\gamma_{SA}|^2 = \gamma_S\gamma_A$), ϱ being the phonon density function. At frequency resonance $\Delta = 0$ and only such ω_{Vl} are strongly coupled to modes S and A that $\mp\omega_{Vl} \approx \omega_{S,A} - \omega_L$, that is

$$2\omega_L \approx \omega_S + \omega_A, \tag{22.35}$$

in agreement with (19.12); further we have introduced $E_L = e_L\exp(i\omega_L t)$ and the Langevin forces

$$\hat{L}_S(t) = i\sum_l g_l\hat{a}^+_{Vl}(0)\exp\left[i\left(\frac{\omega_S - \omega_A}{2} + \omega_{Vl}\right)t\right],$$

$$\hat{L}_A(t) = i\sum_l \varkappa^*_l\hat{a}_{Vl}(0)\exp\left[-i\left(\frac{\omega_S - \omega_A}{2} + \omega_{Vl}\right)t\right] \tag{22.36}$$

satisfy the Markoffian conditions

$$\langle\hat{L}_S(t)\,\hat{L}^+_S(t')\rangle_R = \gamma_S\langle n_V\rangle\,\delta(t - t'),$$

$$\langle\hat{L}_A(t)\,\hat{L}^+_A(t')\rangle_R = \gamma_A(\langle n_V\rangle + 1)\,\delta(t - t'),$$

$$\tfrac{1}{2}\langle\hat{L}_S(t)\,\hat{L}_A(t') + \hat{L}_A(t')\,\hat{L}_S(t)\rangle_R = -\gamma_{SA}(\langle n_V\rangle + \tfrac{1}{2})\,\delta(t - t'),$$

$$\langle\hat{L}_S(t)\,\hat{L}^+_A(t')\rangle_R = \langle\hat{L}^+_S(t)\,\hat{L}_A(t')\rangle_R = 0, \tag{22.37}$$

where $\langle\ \rangle_R$ denotes the average over the phonon reservoir system and $\langle n_V \rangle$ is the mean number of phonons. The corresponding generalized Fokker-Planck equation is derived directly from the Heisenberg-Langevin equations (22.34) and from the properties of the Langevin forces expressed in (22.37) in the standard way (Chapter 20), or using the master equation (20.35) with $N_1 = \langle n_V \rangle + 1$, $N_2 = \langle n_V \rangle$, $\hat{O}^{(n)} = (2\pi\varrho)^{1/2}(gE_L\hat{A}_S^+ + \varkappa E_L^*\hat{A}_A)$ (Peřina (1981a)),

$$\begin{aligned}\frac{\partial \Phi_{\mathscr{A}}}{\partial t} = &-\left[\left(\frac{\gamma_S}{2}|E_L|^2 + \mathrm{i}\Delta\right)\frac{\partial}{\partial \alpha_S}(\alpha_S \Phi_{\mathscr{A}}) + \text{c.c.}\right] + \\ &+\left[\left(\frac{\gamma_A}{2}|E_L|^2 - \mathrm{i}\Delta\right)\frac{\partial}{\partial \alpha_A}(\alpha_A \Phi_{\mathscr{A}}) + \text{c.c.}\right] - \\ &-\left[\frac{\gamma_{SA}}{2}E_L^2\left(\alpha_A^*\frac{\partial \Phi_{\mathscr{A}}}{\partial \alpha_S} - \alpha_S^*\frac{\partial \Phi_{\mathscr{A}}}{\partial \alpha_A}\right) + \text{c.c.}\right] + \\ &+\gamma_S|E_L|^2\langle n_V\rangle\frac{\partial^2 \Phi_{\mathscr{A}}}{\partial \alpha_S\,\partial \alpha_S^*} + \gamma_A|E_L|^2(\langle n_V\rangle + 1)\frac{\partial^2 \Phi_{\mathscr{A}}}{\partial \alpha_A\,\partial \alpha_A^*} - \\ &-\left[\gamma_{SA}E_L^2(\langle n_V\rangle + \tfrac{1}{2})\frac{\partial^2 \Phi_{\mathscr{A}}}{\partial \alpha_S\,\partial \alpha_A} + \text{c.c.}\right],\end{aligned} \tag{22.38}$$

where α_j are eigenvalues of $\hat{A}_j$ ($j = S, A$) in the coherent state $|\alpha_S, \alpha_A\rangle$.

For the normal characteristic function we obtain

$$\begin{aligned}C_{\mathscr{N}}(\beta_S, \beta_A, t) = \langle \exp\{&-B_S(t)|\beta_S|^2 - B_A(t)|\beta_A|^2 + \\ &+[D_{SA}(t)\beta_S^*\beta_A^* + \text{c.c.}] + [\beta_S\xi_S^*(t) + \beta_A\xi_A^*(t) - \text{c.c.}]\}\rangle,\end{aligned} \tag{22.39a}$$

where, assuming for simplicity $\Delta = 0$ ($2\omega_L = \omega_S + \omega_A$),

$$\xi_S(t) = U_S(t)\,\xi_S + V_S(t)\,\xi_A^*,$$

$$\xi_A(t) = U_A(t)\,\xi_A + V_A(t)\,\xi_S^*,$$

$$U_S(t) = \frac{1}{\gamma_S - \gamma_A}(\gamma_S \exp(xt) - \gamma_A) \geqq 0,$$

$$U_A(t) = \frac{1}{\gamma_S - \gamma_A}(\gamma_S - \gamma_A \exp(xt)),$$

$$V_S(t) = -V_A(t) = \frac{(\gamma_S\gamma_A)^{1/2}}{\gamma_S - \gamma_A}(\exp(xt) - 1)\exp(\mathrm{i}\,2\varphi_L + \mathrm{i}\psi_S - \mathrm{i}\psi_A),$$

$$x = \tfrac{1}{2}(\gamma_S - \gamma_A)|E_L|^2,$$

$$\begin{aligned}B_S(t) &= |V_S(t)|^2 + (\langle n_V\rangle + 1)\sum_l |W_{Sl}(t)|^2 = \\ &= \frac{1}{(\gamma_S - \gamma_A)^2}\{\gamma_S^2(\exp(2xt) - 1) + 2\gamma_S\gamma_A(1 - \exp(xt))\} + \\ &+ \frac{\gamma_S\langle n_V\rangle}{\gamma_S - \gamma_A}(\exp(2xt) - 1) \geqq 0,\end{aligned}$$

$$B_A(t) = | V_A(t) |^2 + \langle n_V \rangle \sum_l | W_{Al}(t) |^2 =$$

$$= \frac{\gamma_S \gamma_A}{(\gamma_S - \gamma_A)^2} (\exp(xt) - 1)^2 + \frac{\gamma_A \langle n_V \rangle}{\gamma_A - \gamma_S} (1 - \exp(2xt)) \geqq 0,$$

$$D_{SA}(t) = V_S(t) U_A(t) + (\langle n_V \rangle + 1) \sum_l W_{Sl}(t) W_{Al}(t) = \frac{(\gamma_S \gamma_A)^{1/2}}{\gamma_A - \gamma_S} \times$$

$$\times \left\{ \frac{(\exp(xt) - 1)(\gamma_A - \gamma_S \exp(xt))}{\gamma_A - \gamma_S} + \langle n_V \rangle (\exp(2xt) - 1) \right\} \times$$

$$\times \exp(\mathrm{i}\, 2\varphi_L + \mathrm{i}\psi_S - \mathrm{i}\psi_A), \tag{22.39b}$$

φ_L, ψ_S and ψ_A being the phases of E_L, g and $\varkappa$ respectively; and the solution of the Heisenberg-Langevin equations has the form

$$\hat{A}_S(t) = U_S(t)\, \hat{a}_S + V_S(t)\, \hat{a}_A^+ + \sum_l W_{Sl}(t)\, \hat{a}_{Vl}^+(0),$$

$$\hat{A}_A(t) = U_A(t)\, \hat{a}_A + V_A(t)\, \hat{a}_S^+ + \sum_l W_{Al}(t)\, \hat{a}_{Vl}(0), \tag{22.39c}$$

where $\hat{a}_j = \hat{a}_j(0) = \hat{A}_j(0)$ and U_S, U_A, V_S, V_A, W_{Sl} and W_{Al} are time-dependent functions (Peřina (1981a, b)). The angle brackets in (22.39a) denote the average over the initial complex amplitudes ξ_S and ξ_A, with the probability function $\Phi_{\mathcal{N}}(\xi_S, \xi_A, 0)$.

One can show that

$$\mathscr{D} = B_S B_A - | D_{SA} |^2 = - \frac{\gamma_S \gamma_A}{(\gamma_S - \gamma_A)^2} (\exp(xt) - 1)^2 < 0, \tag{22.40}$$

that is the quasi-distribution $\Phi_{\mathcal{N}}(\alpha_S, \alpha_A, t)$ never exists for $t > 0$, if the fields are in the coherent state at $t = 0$. If $\Delta \neq 0$, the analytic expression for $\mathscr{D}$ is more complicated; however for small Δ the determinant $\mathscr{D}$ is practically independent of $\langle n_V \rangle$ and it is negative; for large values of Δ and small $\langle n_V \rangle$, $\mathscr{D} < 0$; in these cases $\Phi_{\mathcal{N}}$ does not exist. If Δ and $\langle n_V \rangle \gg 1$, then $\mathscr{D} > 0$ and $\Phi_{\mathcal{N}}$ does exist. If the field is initially coherent, $\Phi_{\mathscr{A}}$ has the form of the shifted multidimensional Gaussian distribution involving correlation between modes and correlation between the real and imaginary parts of the complex amplitudes α_S and α_A. Therefore the statistics of the superposition of signal and noise in the generalized quantum form are appropriate for the statistical description of the scattering process under discussion.

The normal generating function can be obtained, either using (21.25) having determined the antinormal generating function $\langle \exp(\mathrm{i}sW) \rangle_{\mathscr{A}}$, or using (21.26) directly, leading to

$$\langle \exp(-\lambda W) \rangle_{\mathcal{N}} = \left(1 - \frac{\lambda}{\lambda_1}\right)^{-1} \left(1 - \frac{\lambda}{\lambda_2}\right)^{-1} \times$$

$$\times \exp\left(\frac{\lambda \mathscr{A}}{1 - \dfrac{\lambda}{\lambda_1}} + \frac{\lambda \mathscr{B}}{1 - \dfrac{\lambda}{\lambda_2}} \right), \tag{22.41a}$$

where

$$\lambda_{1,2} = \frac{-(B_S + B_A) \pm [(B_S + B_A)^2 - 4\mathscr{D}]^{1/2}}{2\mathscr{D}},$$

$$\mathscr{A} = \frac{1}{\mathscr{D}(\lambda_1 - \lambda_2)}\left[|\xi_S(t)|^2\left(B_A + \frac{1}{\lambda_1}\right) + |\xi_A(t)|^2\left(B_S + \frac{1}{\lambda_1}\right) - \right.$$
$$\left. - (D_{SA}\xi_S^*(t)\,\xi_A^*(t) + \text{c.c.})\right],$$

$$\mathscr{B} = \frac{1}{\mathscr{D}(\lambda_2 - \lambda_1)}\left[|\xi_S(t)|^2\left(B_A + \frac{1}{\lambda_2}\right) + |\xi_A(t)|^2\left(B_S + \frac{1}{\lambda_2}\right) - \right.$$
$$\left. - (D_{SA}\xi_S^*(t)\,\xi_A^*(t) + \text{c.c.})\right], \tag{22.41b}$$

provided that the Stokes and anti-Stokes radiation modes are initially coherent. Further, it holds that $\lambda_1 < 0$, $\lambda_2 > 0$ provided that $\mathscr{D} < 0$; if $\mathscr{D} > 0$, then λ_1, $\lambda_2 < 0$. Since $(-\lambda_1^{-1})$ and $(-\lambda_2^{-2})$ are playing the role of the "mean numbers of chaotic photons", this provides an insight into the occurrence of photon antibunching and sub-Poisson statistics in Raman scattering $(-\lambda_2 < 0)$ and the non-existence of the quasi-distribution $\Phi_{\mathscr{N}}(\mathscr{D} < 0)$.

Applying (17.63a, b) to (22.41a), we arrive at the photocount distribution and its factorial moments

$$p(n, t) = \frac{p(0, t)}{(1 - \lambda_1)^n} \sum_{m=0}^{n} \frac{1}{m!(n-m)!}\left(\frac{1-\lambda_1}{1-\lambda_2}\right)^m L_m^0\left(-\frac{\mathscr{B}\lambda_2^2}{\lambda_2 - 1}\right) \times$$
$$\times L_{n-m}^0\left(-\frac{\mathscr{A}\lambda_1^2}{\lambda_1 - 1}\right), \tag{22.42a}$$

$$\langle W^k \rangle_{\mathscr{N}} = \frac{(-1)^k k!}{\lambda_2^k} \sum_{l=0}^{k} \frac{(\lambda_2/\lambda_1)^l}{l!(k-l)!} L_l^0(-\mathscr{A}\lambda_1)\, L_{k-l}^0(-\mathscr{B}\lambda_2), \tag{22.42b}$$

where

$$p(0, t) = \langle \exp(-W) \rangle_{\mathscr{N}} = \left(1 - \frac{1}{\lambda_1}\right)^{-1}\left(1 - \frac{1}{\lambda_2}\right)^{-1} \times$$
$$\times \exp\left(\frac{\mathscr{A}}{1 - \dfrac{1}{\lambda_1}} + \frac{\mathscr{B}}{1 - \dfrac{1}{\lambda_2}}\right). \tag{22.42c}$$

For the mean and variance of the integrated intensity we obtain

$$\langle W \rangle_{\mathscr{N}} = \langle W_S \rangle_{\mathscr{N}} + \langle W_A \rangle_{\mathscr{N}} = -(\mathscr{A} + \mathscr{B}) - \left(\frac{1}{\lambda_1} + \frac{1}{\lambda_2}\right) =$$
$$= |\xi_S(t)|^2 + |\xi_A(t)|^2 + B_S + B_A, \tag{22.43a}$$

$$\langle (\Delta W)^2 \rangle_{\mathscr{N}} = \langle (\Delta W_S)^2 \rangle_{\mathscr{N}} + \langle (\Delta W_A)^2 \rangle_{\mathscr{N}} + 2\langle \Delta W_S\, \Delta W_A \rangle_{\mathscr{N}} =$$
$$= \frac{2\mathscr{A}}{\lambda_1} + \frac{2\mathscr{B}}{\lambda_2} + \frac{1}{\lambda_1^2} + \frac{1}{\lambda_2^2}, \tag{22.43b}$$

where

$$\langle(\Delta W_S)^2\rangle_{\mathcal{N}} = B_S^2 + 2B_S\,|\,\xi_S(t)\,|^2,$$
$$\langle(\Delta W_A)^2\rangle_{\mathcal{N}} = B_A^2 + 2B_A\,|\,\xi_A(t)\,|^2,$$
$$\langle\Delta W_S\,\Delta W_A\rangle_{\mathcal{N}} = |\,D_{SA}\,|^2 + (D_{SA}\xi_S^*(t)\,\xi_A^*(t) + \text{c.c.}) \qquad (22.43c)$$

are fluctuations in the Stokes and anti-Stokes modes and the correlation of fluctuations in these modes. The terms B_S^2, B_A^2 and $|\,D_{SA}\,|^2$ in (22.43c) are independent of the initial field and represent quantum noise related to vacuum fluctuations, arising from the commutators.

It is evident from (22.39a) that separately both the Stokes and anti-Stokes modes have the behaviour characteristic of the superposition of the signal $\xi_j(t)$ and noise $B_j(t)$ $(j = S, A)$ provided that these modes are initially coherent. If moreover $\varkappa = 0$, i.e. $\gamma_A = 0$ (only the Stokes interaction is switched on) or $g = 0$, i.e. $\gamma_S = 0$ (only the anti-Stokes interaction is switched on), then $D_{SA} = 0$ and from (22.39b)

$$B_S(t) = [\exp(\gamma_S\,|\,E_L\,|^2\,t) - 1](\langle n_V\rangle + 1),$$
$$\xi_S(t) = \xi_S \exp\left(\frac{\gamma_S}{2}\,|\,E_L\,|^2\,t\right), \qquad (22.44a)$$

or

$$B_A(t) = [1 - \exp(-\gamma_A\,|\,E_L\,|^2\,t)]\,\langle n_V\rangle,$$
$$\xi_A(t) = \xi_A \exp\left(-\frac{\gamma_A}{2}\,|\,E_L\,|^2\,t\right), \qquad (22.44b)$$

that is the Stokes mode is amplified and the quantum noise in it increases exponentially for later times, whereas the anti-Stokes mode is attenuated and its quantum noise is saturated by the value $\langle n_V\rangle$ in the course of stimulated scattering.

In Fig. 22.5 we see the time development of the fluctuations $\langle(\Delta W_{S,A})^2\rangle_{\mathcal{N}}$, $\langle\Delta W_S\,\Delta W_A\rangle_{\mathcal{N}}$ and $\langle(\Delta W)^2\rangle_{\mathcal{N}}$ subject to the initial phase condition $2\varphi_L - \varphi_S - \varphi_A + \psi_S - \psi_A = n2\pi$ $(n = 0, 1, \ldots)$, where φ_L, φ_S, φ_A, ψ_S and ψ_A are the phases of E_L, ξ_S, ξ_A, g and $\varkappa$ respectively. The anticorrelation of the modes S and A reflects itself in the competition of these modes, as demonstrated by the time-dependence of the integrated intensities $\langle W_S\rangle_{\mathcal{N}}$ and $\langle W_A\rangle_{\mathcal{N}}$ in Fig. 22.6. The reduced factorial moments and the photocount distribution giving evidence for antibunching can be obtained in the same way as in parametric processes (Peřina (1981a, b)).

An analogous quantum description of Raman scattering with damping has been proposed by Germey, Schütte and Tiebel (1981). A more detailed discussion, including the dynamics of phonons in Raman and Brillouin scattering (Pieczonková and Peřina (1981), Pieczonková (1982a, b)), a fully quantum treatment (Walls (1970), Simaan (1975), Szlachetka, Kielich, Peřina and Peřinová (1979, 1980a, b), Gupta

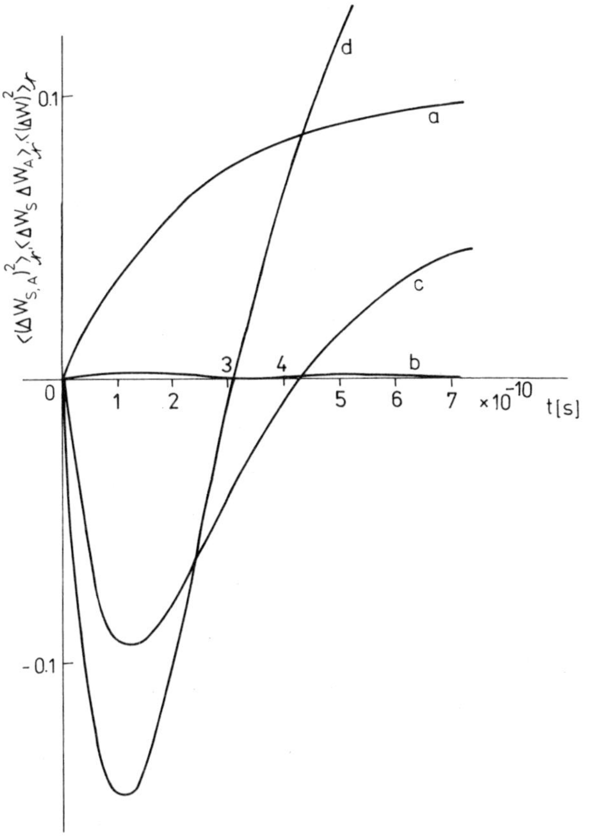

Fig. 22.5 The time development of mode fluctuations $\langle(\Delta W_S)^2\rangle_{\mathcal{N}}$ (curve *a*) and $\langle(\Delta W_A)^2\rangle_{\mathcal{N}}$ (curve *b*) and their correlation $\langle\Delta W_S \Delta W_A\rangle_{\mathcal{N}}$ (curve *c*) and of the field fluctuations $\langle(\Delta W)^2\rangle_{\mathcal{N}} = \langle(\Delta W_S)^2\rangle_{\mathcal{N}} + \langle(\Delta W_A)^2\rangle_{\mathcal{N}} + 2\langle\Delta W_S \Delta W_A\rangle_{\mathcal{N}}$ (curve *d*) for $\gamma_S = 100\ \mathrm{s}^{-1}$, $\gamma_A = 10^4\ \mathrm{s}^{-1}$, $|E_L|^2 = 10^6$, $|\xi_S| = |\xi_A| = 2^{1/2}$, $\Delta = 1 - 10^6$ (all the curves coincide), $\langle n_V\rangle = 0$, under the initial phase condition $2\varphi_L - \varphi_S - \varphi_A + \psi_S - \psi_A = n\,2\pi$ (after J. Peřina, 1981, *Opt. Acta* **28**, 1529).

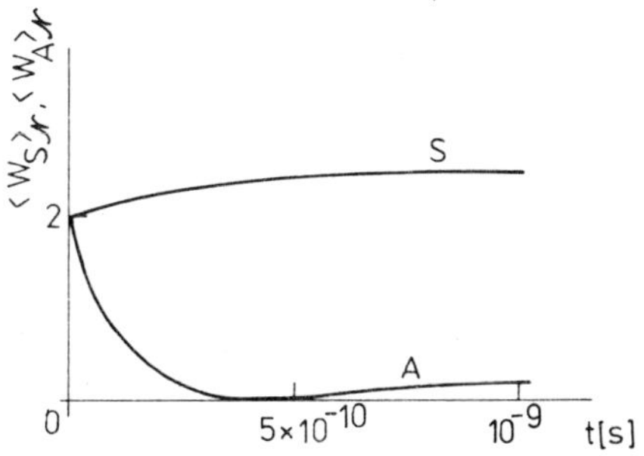

Fig. 22.6 The time development of the mean integrated intensities $\langle W_S\rangle_{\mathcal{N}}$ and $\langle W_A\rangle_{\mathcal{N}}$ in the Stokes and anti-Stokes modes for the same values of parameters as in Fig. 22.5 (after J. Peřina, 1981, *Opt. Acta* **28**, 325).

and Mohanty (1980), Mohanty and Gupta (1981a, b)) and hyper-Raman scattering (Peřinová, Peřina, Szlachetka and Kielich (1979a, b), Szlachetka, Kielich, Peřinová and Peřina (1980), Simaan (1978), Tänzler and Schütte (1981a, b), Mohanty and Gupta (1981c)), has been given by this author in a special monograph (Peřina (1984)).

22.5 Multiphoton absorption and emission

Multiphoton absorption is described by the master equation (20.35) if $N_2 = 0$, $m = 0$ and $n = k$. Thus for the k-photon absorption of single-mode radiation we have the master equation for the reduced density matrix in the form

$$\frac{\partial \hat{\varrho}}{\partial t} = -\frac{\mu}{2}\left[\hat{a}^{+k}\hat{a}^{k} - 2\hat{a}^{k}\hat{\varrho}\hat{a}^{+k} + \hat{\varrho}\hat{a}^{+k}\hat{a}^{k}\right], \tag{22.45}$$

where $\mu = 2KN_1 > 0$. If we multiply this equation by the number operator $\hat{a}^{+}\hat{a}$ and take the trace, we obtain

$$\frac{\mathrm{d}}{\mathrm{d}t}\langle \hat{a}^{+}\hat{a}\rangle = -k\mu\langle \hat{a}^{+k}\hat{a}^{k}\rangle, \tag{22.46}$$

which points out the strong dependence of the multiphoton absorption on the statistical properties of the light. If $k = 1$, then $\langle \hat{a}^{+}(t)\,\hat{a}(t)\rangle = \exp(-\mu t)\,\langle \hat{a}^{+}(0)\,\hat{a}(0)\rangle$, which is the standard result. For radiation in a coherent state

$$\frac{\mathrm{d}}{\mathrm{d}t}|\alpha|^2 = -k\mu(|\alpha|^2)^k \tag{22.47a}$$

and for chaotic radiation

$$\frac{\mathrm{d}}{\mathrm{d}t}\langle \hat{a}^{+}\hat{a}\rangle = -kk!\mu\langle \hat{a}^{+}\hat{a}\rangle^k, \tag{22.47b}$$

reflecting the fact that this process is $k!$-times more effective for chaotic light than for coherent light. Similar conclusions are true also for multiphoton emission, Raman scattering (Simaan (1975)), second-harmonic generation (Crosignani, Di Porto and Solimeno (1972), Chmela (1973)) and other nonlinear optical processes.

A more general connection can be found (Chmela (1979a – e, 1981)) between the efficiency of nonlinear optical processes and intermodal correlations. For instance, the correlation between subfrequency modes leads to sum-frequency generation; but it acts against subfrequency generation, while anticorrelation between them acts in the opposite way. The correlation between the subfrequency and sum-frequency modes supports difference-frequency generation, while the anticorrelation between them supports sum-frequency generation.

For the solution of (22.45) the generating function technique has been used (Tornau and Bach (1974), Bandilla and Ritze (1976a)). The quantity $\langle(\Delta n)^2\rangle/\langle n\rangle$, which is time-dependent, is always less than unity for two- and higher-photon absorption of coherent light, which shows that antibunching and sub-Poisson effect occur in this process. This phenomenon is qualitatively explained by taking into account that pairs of photons are more probably absorbed, which produces "holes" in the light beam and this leads to antibunching of the photons.

The photon statistics of higher-order absorption processes have been further discussed by Paul, Mohr and Brunner (1976), Mohr and Paul (1978), Voigt, Bandilla and Ritze (1980), Zubairy and Yeh (1980), Voigt and Bandilla (1981) and Mohr (1981). The quantity $\langle(\Delta n(t))^2\rangle/\langle n(t)\rangle$ as a function of $\langle n(t)\rangle/\langle n(0)\rangle$ for $k = 1, 2, 3, 5$ and 10 is shown in Fig. 22.7, demonstrating photon antibunching also for higher-order absorption; we see that for one-photon process there is indeed no change of photon statistics (Mohr and Paul (1978)). The sub-Poissonian distributions for two- and three-photon absorbers have been demonstrated by Voigt, Bandilla and Ritze (1980).

General solutions of the master equation for multiphoton absorption have been found by Zubairy and Yeh (1980). In Figs. 22.8a, b we see the transient behaviour of the photon-number distribution $p(n, t)$ for $\mu t = 0, 0.1, 0.2$ and ∞ for two-photon absorption and initially coherent light (Fig. 22.8a) and chaotic light (Fig. 22.8b)

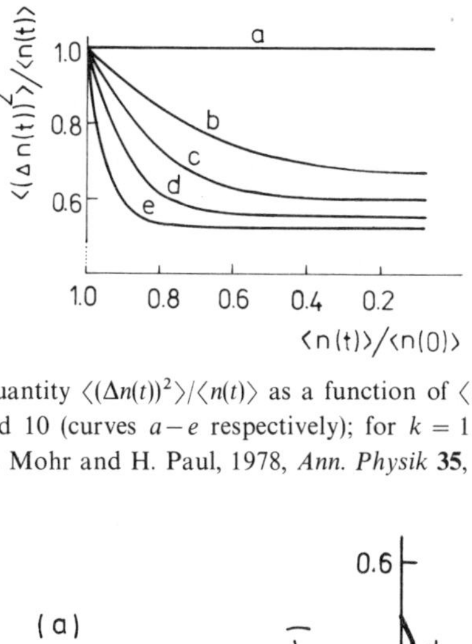

Fig. 22.7 The change of the quantity $\langle(\Delta n(t))^2\rangle/\langle n(t)\rangle$ as a function of $\langle n(t)\rangle/\langle n(0)\rangle$ due to k-photon absorption for $k = 1, 2, 3, 5$ and 10 (curves $a-e$ respectively); for $k = 1$ the statistics are conserved (after U. Mohr and H. Paul, 1978, *Ann. Physik* **35**, 461).

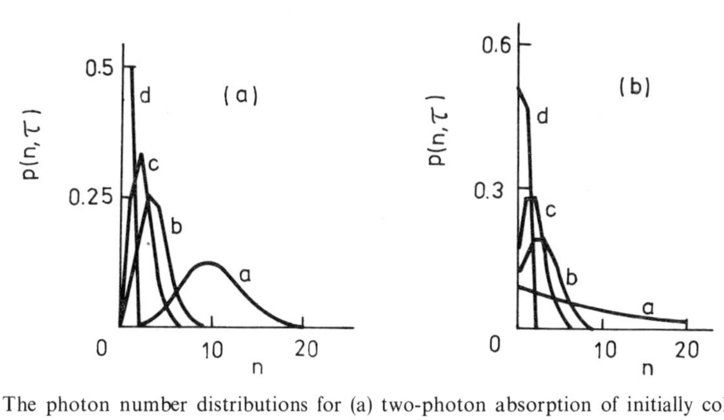

Fig. 22.8 The photon number distributions for (a) two-photon absorption of initially coherent light, (b) two-photon absorption of initially chaotic light; $\langle n(0)\rangle = 10$, curves $a-d$ are for $\tau = \mu t = 0, 0.1, 0.2$ and ∞ respectively (after M. S. Zubairy and J. J. Yeh, 1980, *Phys. Rev.* **A 21**, 1624).

if $\langle n(0)\rangle = 10$ (Zubairy and Yeh (1980)). Very similar results have been obtained by these authors for three- and four-photon absorption. The transient photon-number distributions for two-photon absorption have been discussed also by Simaan and Loudon (1978).

Using the master equation approach single- as well as multimode multiphoton absorption have been investigated in a number of papers (McNeil and Walls (1974), Simaan and Loudon (1975, 1978), Simaan (1979)). The effect of partial coherence of the driving fields in two-photon absorbers in the resonator has been considered by Chaturvedi, Drummond and Walls (1977). The use of interference to enhance photon antibunching in two-photon absorbed beams has been suggested by Bandilla and Ritze (1980b, 1981). Photon statistics for a laser with an intercavity two-photon absorber have been reported by Bandilla and Ritze (1976b), Bandilla (1978) and Hildred (1980).

The process of multiphoton emission is described in a similar way by the master equation (20.35), where $N_1 = 0$ and $n = m = k$, i.e. for the k-photon emission we have

$$\frac{\partial \hat{\varrho}}{\partial t} = -\frac{\mu}{2}\left[\hat{a}^k\hat{a}^{+k}\hat{\varrho} - 2\hat{a}^{+k}\hat{\varrho}\hat{a}^k + \hat{\varrho}\hat{a}^k\hat{a}^{+k}\right], \tag{22.48}$$

where $\mu = 2KN_2 > 0$. In analogy to (22.46) we obtain for the k-photon emission

$$\frac{\mathrm{d}}{\mathrm{d}t}\langle \hat{a}\hat{a}^+\rangle = k\mu\langle \hat{a}^k\hat{a}^{+k}\rangle, \tag{22.49}$$

showing that antinormal ordering of the field operators is appropriate to describe emission (see equations (11.8a, b)); it is also evident that multiphoton emission is strongly dependent on the statistical properties of light and $\langle \hat{a}(t)\,\hat{a}^+(t)\rangle = \exp(\mu t)\,\langle \hat{a}(0)\,\hat{a}^+(0)\rangle$ for $k = 1$.

Further discussions of two- and multi-photon emission and of two-photon lasers have been provided by McNeil and Walls (1975a–c), Yuen (1976), Helstrom (1979a), Nayak and Mohanty (1979), Gupta and Mohanty (1981), Golubev (1979), Schubert and Vogel (1981), Carusotto (1980), Zubairy and Yeh (1980), Sharma and Brescausin (1981), Reid, McNeil and Walls (1981), Sczaniecki (1980, 1982), Bandilla and Voigt (1982) and Zubairy (1982). In particular McNeil and Walls (1975a, b), Nayak and Mohanty (1979) and Gupta and Mohanty (1981) demonstrated that the two- and three-photon lasers may provide sharper photon-number distributions than one-photon lasers.

22.6 Resonance fluorescence

Although resonance fluorescence is a single-photon phenomenon, it is of increasing theoretical and experimental importance, since it was used by Kimble, Dagenais and Mandel (1977) to exhibit photon antibunching for the first time. Later more precise experiments have been performed by Kimble, Dagenais and Mandel (1978), Dagenais and Mandel (1978), Leuchs, Rateike and Walther (1979) (see Walls (1979)) and Cresser, Häger, Leuchs, Rateike and Walther (1982). Sub-Poisson light was observed by Short and Mandel (1983) and in Franck-Hertz-Teich-Saleh experiment (Teich and Saleh (1985)).

Carmichael and Walls (1976), Kimble and Mandel (1976) and Cohen-Tannoudji (1977) predicted that the fourth-order correlation function of light emitted by a single atom undergoing resonance fluorescence exhibits photon antibunching. The normal and time ordered intensity correlation function can be written in the factorized form

$$\langle \mathscr{T} \mathscr{N} \hat{I}(t) \hat{I}(t + \tau) \rangle = \langle \hat{I} \rangle \langle \hat{I}(\tau) \rangle_G, \tag{22.50}$$

where $\mathscr{T}$ and $\mathscr{N}$ are the operators of time and normal ordering respectively, $\hat{I} = \hat{A}^{(-)} \hat{A}^{(+)}$ is the intensity operator, $\langle \hat{I} \rangle$ is the steady-state mean intensity and $\langle \hat{I}(\tau) \rangle_G$ is the mean intensity of light that is radiated by an atom driven by an external field at time τ if it starts in the ground state at $t = 0$. As an atom cannot radiate in its ground state, it follows that $\langle \hat{I}(\tau) \rangle_G$ always starts from zero at $\tau = 0$ and then grows with τ and reaches its steady state value $\langle \hat{I} \rangle$ after a time long compared with the natural life-time. Denoting the Rabi frequency as $\Omega = 2\mathscr{P}\mathscr{E}/h$ ($\mathscr{P}$ being the atomic dipole matrix element and $\mathscr{E}$ the driving field amplitude), then (Carmichael and Walls (1976), Kimble and Mandel (1976, 1977), Cohen-Tannoudji (1977))

$$\langle \hat{I}(\tau) \rangle_G = \langle \hat{I} \rangle [1 + \lambda(\tau)], \tag{22.51}$$

where $\lambda(\tau)$ represents the normalized correlation function of the fluctuations from both the detectors of the Hanbury Brown-Twiss correlation arrangement,

$$\lambda(\tau) = \frac{\langle \mathscr{T} \mathscr{N} \hat{I}_1(t) \hat{I}_2(t + \tau) \rangle}{\langle \hat{I}_1(t) \rangle \langle \hat{I}_2(t + \tau) \rangle} - 1. \tag{22.52}$$

Kimble, Dagenais and Mandel (1977) and Dagenais and Mandel (1978) used an atomic beam of sodium atoms optically pumped in order to prepare a pure two-level system. The atomic beam was irradiated at right angles by a highly stabilized dye laser tuned on resonance with the $3^2S_{1/2}$, $F = 2$, $m_F = 2$ to $3^2P_{3/2}$, $F = 3$, $m_F = 3$ transition in sodium. The intensity of the atomic beam was reduced so that on average no more than one atom is present in the observation region at a time. The fluorescent light from a small observation volume is obtained in a direction orthogonal to both the atomic and laser beams. Further the usual Hanbury Brown-Twiss coincidence arrangement was used, i.e. the fluorescent light was divided by a splitter and the arrival of photons in each beam was detected by two photomultipliers. The pulses from the two detectors were fed to the start and stop inputs of a time-to-digital converter where the time intervals between start and stop pulses were digitized in units of 0.5 ns and stored. The number of events $n(\tau)$ stored at address τ was a measure of the joint probability density $\eta^2\gamma_{\mathcal{N}}^{(2,2)}(\tau)$ of separation of photons by τ seconds (η being the photodetector efficiency).

The results of the measurements of Dagenais and Mandel (1978) are shown in Figs. 22.9a, b for $\Omega/\beta = 3.3$ and 1.4, where β is half of the Einstein A-coefficient for the transition. Fig. 22.9a represents the unnormalized data; the growing part of the curve violates the classical inequality $\gamma^{(2,2)}(\tau) \leqq \gamma^{(2,2)}(0)$ and demonstrates the antibunching phenomenon in resonance fluorescence radiation. If the data are normalized (and also scattered light is eliminated and transit time effects and atom number fluctuations are taken into account), we obtain Fig. 22.9b giving clear evidence for photon antibunching in resonance fluorescence ($\lambda(\tau)$ in (22.52) equals -1 for $\tau = 0$, so that $1 + \lambda(0)$ is zero, whereas for classical systems it is maximal). Also the measurements by Cresser, Häger, Leuchs, Rateike and Walther (1982) provide clear evidence for photon antibunching in fluorescent light and demonstrate the effect of the smoothing of this phenomenon with increasing atomic fluctuations.

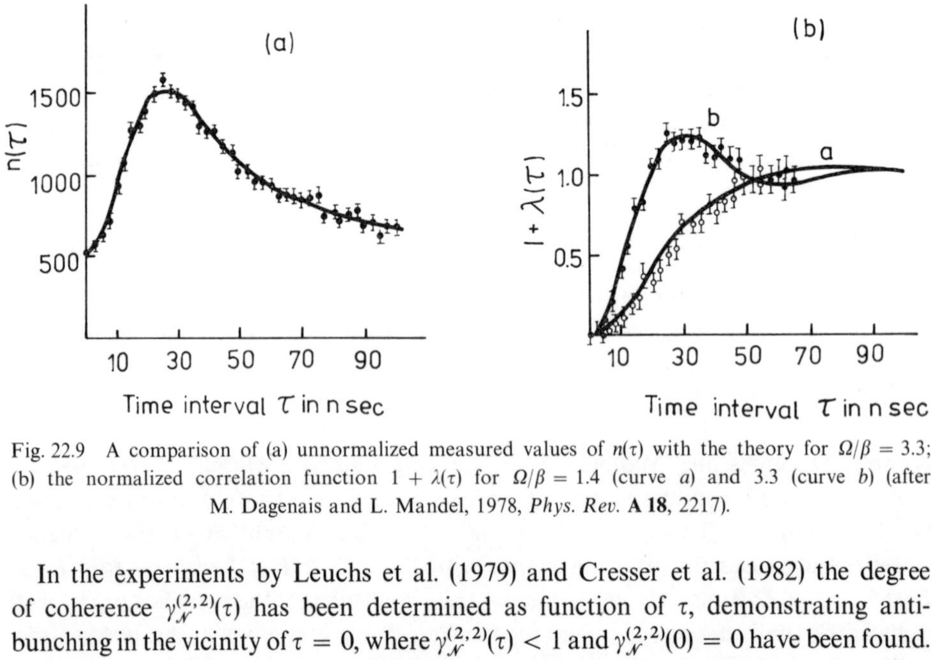

Fig. 22.9 A comparison of (a) unnormalized measured values of $n(\tau)$ with the theory for $\Omega/\beta = 3.3$; (b) the normalized correlation function $1 + \lambda(\tau)$ for $\Omega/\beta = 1.4$ (curve a) and 3.3 (curve b) (after M. Dagenais and L. Mandel, 1978, *Phys. Rev.* **A 18**, 2217).

In the experiments by Leuchs et al. (1979) and Cresser et al. (1982) the degree of coherence $\gamma_{\mathcal{N}}^{(2,2)}(\tau)$ has been determined as function of τ, demonstrating antibunching in the vicinity of $\tau = 0$, where $\gamma_{\mathcal{N}}^{(2,2)}(\tau) < 1$ and $\gamma_{\mathcal{N}}^{(2,2)}(0) = 0$ have been found.

Thus it has been experimentally demonstrated that fluorescent photons from one atom exhibit antibunching in time, which may be regarded as a reflection of the fact that the atom makes a quantum jump to the ground state in the process of emission of a photon and is unable to radiate again immediately afterwards. No classical system can exhibit such a behaviour. Thus the photon coincidence technique makes it possible to test the behaviour of the atom and to observe the antibunching effect which is a direct manifestation of the quantum nature of light, in agreement with predictions of quantum electrodynamics.

The photon-number distribution of resonance fluorescence radiation has been discussed by Mandel (1979b), Cook (1980, 1981) and Smirnov and Trochin (1981). Mandel (1979b) and Smirnov and Trochin (1981) clearly demonstrated that the photon-number distribution for fluorescent photons from one atom is indeed narrower than the corresponding Poisson distribution, as in optical parametric processes, Raman and Brillouin scattering and multiphoton absorption. This was observed by Short and Mandel (1983).

With respect to the basic importance of these experiments an extensive discussion appeared in the literature concerning the effect of atom number fluctuations, which smooth out the antibunching phenomenon (Jakeman, Pike, Pusey and Vaugham (1977), Carmichael, Drummond, Meystre and Walls (1978), Carmichael, Drummond, Walls and Meystre (1980), Schubert, Süsse, Vogel, Welsch and Wilhelmi (1982)).

The use of reduced quantum fluctuations in resonance fluorescence to produce squeezed states (Loudon (1984)) for the detection of gravitational waves has been suggested by Walls and Zoller (1981). A review of the theory of intensity dependent resonance light scattering and resonance fluorescence has been published by Mollow (1981).

CHAPTER 23

CONCLUSIONS

In this book we have studied the coherence properties of optical fields by classical and quantum methods and the relation between the classical and quantum methods of describing coherence has been investigated by using the formalism of the coherent states and the Glauber-Sudarshan representation of the density matrix. Particular attention has been paid to the detection of optical fields and many theoretical results have been shown to be in very good agreement with experimental results obtained by a number of authors. The topics of the first edition of this book are supplemented by the discussion of radiometry with light fields of arbitrary states of coherence, and the general coherent-state methods are demonstrated, using the Heisenberg-Langevin and Fokker-Planck approaches to the interaction of radiation with matter, by the propagation of radiation through random and nonlinear media.

In conclusion it is worthwhile to point out a few other topics which are undergoing development at the present time and some possible applications of the methods of the theory of coherence in other branches of physics. As mentioned in the Introduction, very interesting applications of photocount statistics have been provided in biophysics and psychophysics by Teich and McGill (1976), Teich et al. (1978, 1982a, b) and Prucnal and Teich (1982). Further the coherent-state technique has been applied to the solid state, superfluidity, superconductivity, ferromagnetism, etc. by Carruthers and Dy (1966), Carruthers and Nieto (1968), López (1967), Langer (1968, 1969), Rezende and Zagury (1969) (see also Man'ko (1972)) and Malkin and Man'ko (1979), among others. The process of γ-emission by stimulated annihilation of electron–positron pairs has been discussed by Bertolotti and Sibilia (1979), Sibilia and Bertolotti (1981), Peřinová, Peřina, Bertolotti and Sibilia (1982), who discovered anticorrelation and antibunching of the γ-photons. The possibility of taking into account the structure of sources of light has been discussed by Zardecki (1969b), in continuation to papers by Halpern and Doermann (1937) and Doermann and Halpern (1939). There is great significance in the experimental field of generation of radiation with antibunching of photons, useful for optical communication (Yuen and Shapiro (1978, 1979)) and in the detection of gravitational waves using optical interferometers (Caves (1981), Walls and Zoller (1981), Loudon (1981), Walls and Milburn (1981)). The photon statistics of radiation in free electron lasers have been considered by Becker et al. (1982), Peřina, Peřinová, Křepelka, Lukš, Sibilia and Bertolotti (1983), Sibilia, Bertolotti, Peřinová, Peřina and Lukš (1983).

The methods and results of this book demonstrate that the theory of coherence of the electromagnetic field is also fruitful in other branches of physics, including high-energy physics (Biyajima (1984), Blažek (1984)). This is particularly true for the coherent-state formalism which permits a formulation of quantum electrodynamics in a form applicable to systems with a large number of particles in cooperative states, since such a formulation has a classical limit and retains phase information of the field; and all the quantum properties of the systems are conserved in such formulations.

REFERENCES

Ablekov, V. K., V. S. Avdyrevskii, Yu. N. Babaev, S. A. Koljadin, A. V. Frolov and V. A. Fulov, 1980, *Proc. Acad. Sci. USSR* **251**, 1098.

Ablekov, V. K., Yu. N. Babaev, S. A. Koljadin and A. V. Frolov, 1980, *Proc. Acad. Sci. USSR* **250**, 90.

Ablekov, V. K., Yu. N. Babaev, S. A. Koljadin, Yu. P. Syrich and A. V. Frolov, 1981, *Proc. Acad. Sci. USSR* **261**, 609.

Ablekov, V. K., P. I. Zubkov and A. V. Frolov, 1976, Optical and Opto-Electronical Data Processing (Mashinostroyeniye, Moscow), in Russian.

Abraham, N. B., 1980, *Phys. Rev.* **A 21**, 1595.

Adaptive Optics, 1977, *J. Opt. Soc. Am.* **67**, No 3.

Agarwal, G. S., 1969, *Phys. Rev.* **178**, 2025.

Agarwal, G. S., 1970, *Phys. Rev.* **A 1**, 1445.

Agarwal, G. S., 1973, Progress in Optics, Vol. 11, ed. E. Wolf (North-Holland, Amsterdam), p. 1.

Agarwal, G. S. and E. Wolf, 1968a, *Phys. Lett.* **26 A**, 485.

Agarwal, G. S. and E. Wolf, 1968b, *Phys. Rev. Lett.* **21**, 180.

Agarwal, G. S. and E. Wolf, 1968c, *Phys. Rev. Lett.* **21**, 656 (E).

Agarwal, G. S. and E. Wolf, 1970, *Phys. Rev.* **D 2**, 2161, 2187 and 2206.

Agrawal, G. P. and C. L. Mehta, 1974, *J. Phys.* **A 7**, 607.

Akhiezer, A. I. and V. B. Berestetsky, 1965, Quantum Electrodynamics (Interscience, New York).

Akhmanov, S. A., Yu. E. Dyakov and A. S. Tchirkin, 1981, Introduction to Statistical Radiophysics and Optics (Nauka, Moscow), in Russian.

Akhmanov, S. A., R. V. Khokhlov and A. P. Sukhorukov, 1972, Laser Handbook, Vol. 2, eds. F. T. Arecchi and E. O. Schulz-Dubois (North-Holland, Amsterdam), p. 1151.

Akhmanov, S. A. and A. S. Tchirkin, 1971, Statistical Phenomena in Nonlinear Optics (Lomonosov Univ., Moscow), in Russian.

Akhmanov, S. A., A. S. Tchirkin and V. G. Tunkin, 1970, *Opto-Electronics* **2**, No 2.

Aldridge, M. D., 1969, *J. Appl. Phys.* **40**, 1720.

Allen, L. and J. H. Eberly, 1975, Optical Resonance and Two-Level Atoms (J. Wiley, New York).

Aoki, T., 1977, *Phys. Rev.* **A 16**, 2432.

Arecchi, F. T., 1965, *Phys. Rev. Lett.* **15**, 912.

Arecchi, F. T., 1969, Quantum Optics, ed. R. J. Glauber (Acad. Press, New York).

Arecchi, F. T., M. Asdente and A. M. Ricca, 1976, *Phys. Rev.* **14**, 383.

Arecchi, F. T., A. Berné and P. Burlamacchi, 1966, *Phys. Rev. Lett.* **16**, 32.

Arecchi, F. T., A. Berné and A. Sona, 1966, *Phys. Rev. Lett.* **17**, 260.

Arecchi, F. T., A. Berné, A. Sona and P. Burlamacchi, 1966, *IEEE J. Quant. Electr.* **QE-2**, 341.

Arecchi, F. T., E. Courtens, R. Gilmore and H. Thomas, 1972, *Phys. Rev.* **A 6**, 2211.

Arecchi, F. T., E. Courtens, R. Gilmore and H. Thomas, 1973, Coherence and Quantum Optics, eds. L. Mandel and E. Wolf (Plenum Press, New York), p. 191.

Arecchi, F. T. and V. Degiorgio, 1971, *Phys. Rev.* **A 3**, 1108.

Arecchi, F. T. and V. Degiorgio, 1972, Laser Handbook, Vol. 1, eds. F. T. Arecchi and E. O. Schulz-Dubois (North-Holland, Amsterdam), p. 191.

Arecchi, F. T., V. Degiorgio and B. Querzola, 1967, *Phys. Rev. Lett.* **19**, 1168.

Arecchi, F. T., E. Gatti and A. Sona, 1966, *Phys. Lett.* **20**, 27.
Arecchi, F. T., M. Giglio and A. Sona, 1967, *Phys. Lett.* **25 A**, 341.
Arecchi, F. T. and A. Politi, 1980, *Phys. Rev. Lett.* **45**, 1219.
Arecchi, F. T., A. Politi and L. Ulivi, 1982, *Phys. Lett.* **87 A**, 333.
Arecchi, F. T., G. S. Rodari and A. Sona, 1967, *Phys. Lett.* **25 A**, 59.
Armstrong, J. A. and A. W. Smith, 1967, Progress in Optics, Vol. 6, ed. E. Wolf (North-Holland, Amsterdam), p. 211.
Asakura, T., 1974a, *Phys. Lett.* **47 A**, 101.
Asakura, T., 1974b, *Nouv. Rev. Opt.* **5**, 169.
Asakura, T. and H. Mishina, 1974, *Phys. Lett.* **47 A**, 443.
Bakut, P. A., K. N. Sviridov and N. D. Ustinov, 1981, *Opt. Spectr.* (USSR) **51**, 1056.
Baldwin, G. C., 1969, An Introduction to Nonlinear Optics (Plenum Press, New York).
Baltes, H. P., 1977, *Appl. Phys.* **12**, 221.
Baltes, H. P., J. Geist and A. Walther, 1978, Inverse Source Problems, ed. H. P. Baltes (Springer, Berlin), p. 119.
Baltes, H. P., B. Steinle and G. Antes, 1976, *Opt. Comm.* **18**, 242.
Bandilla, A., 1978, Quantum Optics, eds. J. Heldt and J. Czub (Univ. Toruń and Gdańsk), p. 20.
Bandilla, A. and H. H. Ritze, 1976a, *Ann. Physik* **33**, 207.
Bandilla, A. and H. H. Ritze, 1976b, *Opt. Comm.* **19**, 169.
Bandilla, A. and H. H. Ritze, 1979, *Opt. Comm.* **28**, 126.
Bandilla, A. and H. H. Ritze, 1980a, *Opt. Comm.* **34**, 190.
Bandilla, A. and H. H. Ritze, 1980b, *Opt. Comm.* **32**, 195.
Bandilla, A. and H. H. Ritze, 1981, *Ann. Physik* **38**, 123.
Bandilla, A. and H. Voigt, 1982, Quantum statistics of light after saturated two-photon emission process and the photon statistics of a two-photon laser, *Opt. Comm.*, preprint, ZOS Berlin.
Barakat, R., 1963, *J. Opt. Soc. Am.* **53**, 317.
Barakat, R., 1966, *J. Opt. Soc. Am.* **56**, 739.
Barakat, R., 1976, *J. Opt. Soc. Am.* **66**, 211.
Barakat, R., 1977, *Opt. Comm.* **23**, 147.
Barakat, R., 1980, *J. Opt. Soc. Am.* **70**, 688.
Barakat, R., 1981, *Opt. Comm.* **38**, 159.
Barakat, R. and J. Blake, 1980, *Phys. Rep.* **60**, 225.
Barashev, P. P., 1970a, *J. Exp. Theor. Phys.* (USSR) **59**, 1318.
Barashev, P. P., 1970b, *Phys. Lett.* **32 A**, 291.
Barashev, P. P., 1976, *Opt. Spectr.* (USSR) **40**, 349.
Bastiaans, M. J., 1977, *Opt. Acta* **24**, 261.
Bastiaans, M. J., 1981, *Opt. Acta* **28**, 1215.
Becker, W., M. O. Scully and M. S. Zubairy, 1982, *Phys. Rev. Lett.* **48**, 475.
Bédard, G., 1966a, *Phys. Rev.* **151**, 1038.
Bédard, G., 1966b, *Phys. Lett.* **21**, 32.
Bédard, G., 1967a, *J. Opt. Soc. Am.* **57**, 1201.
Bédard, G., 1967b, *Phys. Lett.* **24 A**, 613.
Bédard, G., 1967c, *Proc. Phys. Soc.* **90**, 131.
Bédard, G., 1967d, *Phys. Rev.* **161**, 1304.
Bédard, G., J. C. Chang and L. Mandel, 1967, *Phys. Rev.* **160**, 1496.
Bénard, C. M., 1969, *C. R. Acad. Sci. Paris* **268**, 1504.
Bénard, C. M., 1970a, Quantum Optics, eds. S. M. Kay and A. Maitland (Acad. Press, London), p. 535.
Bénard, C., 1970b, *Phys. Rev.* **A 2**, 2140.
Bénard, C., 1975, *J. Math. Phys.* **16**, 710.
Bendjaballah, C., 1969, *C. R. Acad. Sci. Paris* **268**, 1719.
Bendjaballah, C., 1971, *C. R. Acad. Sci. Paris* **272**, 1244.
Bendjaballah, C., 1979, *J. Appl. Phys.* **50**, 62.

Bendjaballah, M. C. and F. Perrot, 1970, *C. R. Acad. Sci. Paris* **271**, 1085.
Bendjaballah, C. and F. Perrot, 1971, *Opt. Comm.* **3**, 21.
Bendjaballah, C. and F. Perrot, 1973, *J. Appl. Phys.* **44**, 5130.
Beran, M. and P. Corson, 1965, *J. Math. Phys.* **6**, 271.
Beran, M., J. DeVelis and G. Parrent, 1967, *Phys. Rev.* **154**, 1224.
Beran, M. J. and T. L. Ho, 1969, *J. Opt. Soc. Am.* **59**, 1134.
Beran, M. and G. B. Parrent, 1962, *J. Opt. Soc. Am.* **52**, 98.
Beran, M. and G. B. Parrent, 1963, *Nuovo Cim.* **27**, 1049.
Beran, M. and G. B. Parrent, 1964, Theory of Partial Coherence (Prentice-Hall, Englewood Cliffs, New Jersey).
Berek, M., 1926a, *Z. Phys.* **36**, 675 and 824.
Berek, M., 1926b, *Z. Phys.* **37**, 287.
Berek, M., 1926c, *Z. Phys.* **40**, 420.
Bertero, M., P. Boccacci and E. R. Pike, 1982, *Opt. Acta* **29**, 1599.
Bertero, M. and E. R. Pike, 1982, *Opt. Acta* **29**, 727.
Bertolotti, M., 1974, Photon Correlation and Light Beating Spectroscopy, eds. H. Z. Cummins and E. R. Pike (Plenum Press, New York).
Bertolotti, M., 1983, Masers and Lasers (A. Hilger, Bristol).
Bertolotti, M., B. Crosignani and P. Di Porto, 1970, *J. Phys.* **A 3**, L 37.
Bertolotti, M., B. Crosignani, P. Di Porto and D. Sette, 1966, *Phys. Rev.* **150**, 1054.
Bertolotti, M., B. Crosignani, P. Di Porto and D. Sette, 1967a, *Phys. Rev.* **157**, 146.
Bertolotti, M., B. Crosignani, P. Di Porto and D. Sette, 1967b, *Z. Phys.* **205**, 129.
Bertolotti, M., B. Crosignani, P. Di Porto and D. Sette, 1969, *J. Phys.* **A 2**, 126 and 473.
Bertolotti, M. and C. Sibilia, 1979, *Appl. Phys.* **19**, 127.
Bertolotti, M. and C. Sibilia, 1980, *Phys. Rev.* **A 21**, 234.
Bertrand, P. P. and E. A. Mishkin, 1967, *Phys. Lett.* **25 A**, 204.
Bespalov, V. I., 1979, Inversion of Wave Front of Optical Radiation in Nonlinear Media (Acad. Sci. USSR, Gorkii), in Russian.
Bhatnagar, G. S., R. S. Sirohi and S. K. Sharma, 1971, *Opt. Comm.* **3**, 269.
Bialynicka-Birula, Z., 1968, *Phys. Rev.* **173**, 1207.
Biyajima, M., 1984, *Phys. Lett.* **B 137**, 225; **B 139**, 93; **B 143**, 463.
Blake, J. and R. Barakat, 1973, *J. Phys.* **A 6**, 1196.
Blake, J. and R. Barakat, 1976, *Opt. Comm.* **16**, 303.
Blanc-Lapierre, A. and P. Dumontet, 1955, *Rev. Opt.* **34**, 1.
Blažek, M., 1979, *Acta Phys. Slov.* **29**, 3.
Blažek, M., 1984, *Czech. J. Phys.* **B 34**, 838.
Bloembergen, N., 1965, Nonlinear Optics (W. A. Benjamin, New York).
Bluemel, V., L. M. Narducci and R. A. Tuft, 1972, *J. Opt. Soc. Am.* **62**, 1309.
Bogolyubov, N. N., B. V. Medvedev and M. K. Polivanov, 1958, Questions of the Theory of Dispersion Relations (Gos. Izd. Fiz. Mat. Lit., Moscow), in Russian.
Bogolyubov, N. N. and D. V. Shirkov, 1959, Introduction to the Theory of Quantized Fields (Interscience, New York).
Boileau, E. and B. Picinbono, 1968, *J. Opt. Soc. Am.* **58**, 1238.
Bonifacio, R., L. M. Narducci and E. Montaldi, 1966, *Phys. Rev. Lett.* **16**, 1125.
Born, M. and E. Wolf, 1965, Principles of Optics (Pergamon, Oxford).
Bothe, W., 1927, *Z. Phys.* **41**, 345.
Bourret, R. C., 1960, *Nuovo Cim.* **18**, 347.
Brannen, E., H. I. S. Ferguson and W. Wehlau, 1958, *Can. J. Phys.* **36**, 871.
Brevik, I. and E. Suhonen, 1968, *Phys. Norveg.* **3**, 135.
Brevik, I. and E. Suhonen, 1970, *Nuovo Cim.* **65 B**, 187.
Brown, R. Hanbury, 1964, *Sky and Telesc.* **28**, 64.
Brown, R. Hanbury and R. Q. Twiss, 1956a, *Nature* (Lond.) **177**, 27.

Brown, R. Hanbury and R. Q. Twiss, 1956b, *Nature* (Lond.) **178**, 1046.
Brown, R. Hanbury and R. Q. Twiss, 1956c, *Nature* (Lond.) **178**, 1447.
Brown, R. Hanbury and R. Q. Twiss, 1957a, *Proc. Roy. Soc.* (A) **242**, 300.
Brown, R. Hanbury and R. Q. Twiss, 1957b, *Proc. Roy. Soc.* (A) **243**, 291.
Brown, R. Hanbury and R. Q. Twiss, 1958, *Proc. Roy. Soc.* (A) **248**, 199 and 222.
Brunner, W., H. Paul and G. Richter, 1965, *Ann. Physik* **16**, 343.
Bufton, J. L., R. S. Iyer and L. S. Taylor, 1977, *Appl. Opt.* **16**, 2408.
Bures, J., C. Delisle and A. Zardecki, 1971, *Can. J. Phys.* **49**, 3064.
Bures, J., C. Delisle and A. Zardecki, 1972a, *Can. J. Phys.* **50**, 760.
Bures, J., C. Delisle and A. Zardecki, 1972b, *Can. J. Phys.* **50**, 1307.
Bures, J., C. Delisle and A. Zardecki, 1972c, *Phys. Rev.* **A 6**, 2237.
Burge, R. E., M. A. Fiddy, A. H. Greenaway and G. Ross, 1974, *J. Phys.* **D 7**, L 65.
Burge, R. E., M. A. Fiddy, A. H. Greenaway and G. Ross, 1976, *Proc. Roy. Soc. Lond.* **A 350**, 191.
Butylkin, V. S., A. E. Kaplan, Yu. G. Khronopulo and E. I. Yakubovitch, 1977, Resonant Interactions of Light with Matter (Nauka, Moscow), in Russian.
Cahill, K. E., 1965, *Phys. Rev.* **138**, B 1566.
Cahill, K. E., 1969, *Phys. Rev.* **180**, 1244.
Cahill, K. E. and R. J. Glauber, 1969, *Phys. Rev.* **177**, 1857 and 1882.
Campagnoli, G. and G. Zambotti, 1968, *Nuovo Cim.* **57 A**, 468.
Cantor, B. I. and M. C. Teich, 1975, *J. Opt. Soc. Am.* **65**, 786.
Cantrell, C. D., 1970, *Phys. Rev.* **A 1**, 672.
Cantrell, C. D., 1971, *Phys. Rev.* **A 3**, 728.
Cantrell, C. D. and J. R. Fields, 1973, *Phys. Rev.* **A 7**, 2063.
Cantrell, C. D. and W. A. Smith, 1971, *Phys. Lett.* **37 A**, 167.
Carmichael, H. J., P. Drummond, P. Meystre and D. F. Walls, 1978, *J. Phys.* **A 11**, L 121.
Carmichael, H. J., P. D. Drummond, D. F. Walls and P. Meystre, 1980, *Opt. Acta* **27**, 581.
Carmichael, H. J. and D. F. Walls, 1974, *Phys. Rev.* **A 8**, 2686.
Carmichael, H. J. and D. F. Walls, 1976, *J. Phys.* **B 9**, L 43 and 1199.
Carruthers, P. and K. S. Dy, 1966, *Phys. Rev.* **147**, 214.
Carruthers, P. and M. M. Nieto, 1968, *Rev. Mod. Phys.* **40**, 411.
Carter, W. H., 1977, *Appl. Opt.* **16**, 558.
Carter, W. H., 1980, *J. Opt. Soc. Am.* **70**, 1067.
Carter, W. H. and M. Bertolotti, 1978, *J. Opt. Soc. Am.* **68**, 329.
Carter, W. H. and E. Wolf, 1973, *J. Opt. Soc. Am.* **63**, 1619.
Carter, W. H. and E. Wolf, 1975, *J. Opt. Soc. Am.* **65**, 1067.
Carter, W. H. and E. Wolf, 1977, *J. Opt. Soc. Am.* **67**, 785.
Carter, W. H. and E. Wolf, 1981a, *Opt. Acta* **28**, 227.
Carter, W. H. and E. Wolf, 1981b, *Opt. Acta* **28**, 245.
Carusotto, S., 1970, *Nuovo Cim.* **70 B**, 73.
Carusotto, S., 1980, *Opt. Acta* **27**, 1567.
Carusotto, S., G. Fornaca and E. Polacco, 1967, *Phys. Rev.* **157**, 1207.
Carusotto, S., G. Fornaca and E. Polacco, 1968, *Phys. Rev.* **165**, 1391.
Caves, C. M., 1981, *Phys. Rev.* **D 23**, 1693.
Chand, P., 1979, *Nuovo Cim.* **50 B**, 17.
Chang, R. F., R. W. Detenbeck, V. Korenman, C. O. Alley and U. Hochuli, 1967, *Phys. Lett.* **25 A**, 272.
Chang, R. F., V. Korenman, C. O. Alley and R. W. Detenbeck, 1969, *Phys. Rev.* **178**, 612.
Chang, R. F., V. Korenman and R. W. Detenbeck, 1968, *Phys. Lett.* **26 A**, 417.
Chaturvedi, S., P. Drummond and D. F. Walls, 1977, *J. Phys.* **A 10**, L 187.
Chmela, P., 1971, *Acta Univ. Palack. Olom.* **33**, 253.
Chmela, P., 1973, *Czech. J. Phys.* **B 23**, 884.
Chmela, P., 1974, *Czech. J. Phys.* **B 24**, 1 and 506; *Acta Univ. Palack. Olom.* **45**, 5.
Chmela, P., 1979a, *Opt. Quant. Electr.* **11**, 103.

Chmela, P., 1979b, *Czech. J. Phys.* **B 29**, 129.
Chmela, P., 1979c, *Opt. Quant. Electr.* **11**, 287.
Chmela, P., 1979d, *Opt. Appl.* **9**, 223.
Chmela, P., 1979e, *Acta Phys. Pol.* **A 55**, 945.
Chmela, P., 1981, *Czech. J. Phys.* **B 31**, 977 and 999.
Chmela, P., R. Horák and J. Peřina, 1981, *Opt. Acta* **28**, 1209.
Chu, B., 1974, Laser Light Scattering (Acad. Press, New York).
Chung, J. C., J. C. Huang and N. B. Abraham, 1980, *Phys. Rev.* **A 22**, 1018.
Churnside, J. H. and C. M. McIntyre, 1978, *Appl. Opt.* **17**, 2141 and 2148.
Clark, W. G., L. E. Estes and L. M. Narducci, 1970, *Phys. Lett.* **33 A**, 517.
Clifford, S. F. and R. J. Hill, 1981, *J. Opt. Soc. Am.* **71**, 112.
Cohen-Tannoudji, C., 1977, Frontiers in Laser Spectroscopy, eds. R. Balian, S. Haroche and S. Liberman (North-Holland, Amsterdam).
Cohen-Tannoudji, C. and A. Kastler, 1966, Progress in Optics, Vol. 5, ed. E. Wolf (North-Holland, Amsterdam), p. 1.
Colegrave, R. K. and M. S. Abdalla, 1981, *Opt. Acta* **28**, 495.
Collet, E. and E. Wolf, 1979, *J. Opt. Soc. Am.* **69**, 942.
Collet, E. and E. Wolf, 1980, *Opt. Comm.* **32**, 27.
Cook, R. J., 1980, *Opt. Comm.* **35**, 347.
Cook, R. J., 1981, *Phys. Rev.* **A 23**, 1243.
Cook, R. J., 1982, *Phys. Rev.* **A 26**, 2754.
Cornacchio, J. V. and K. A. Farnham, 1966, *Nuovo Cim.* **42 B**, 108.
Cornacchio, J. V. and R. P. Soni, 1965, *Nuovo Cim.* **38**, 1169.
Corti, M. and V. Degiorgio, 1974, *Opt. Comm.* **11**, 1.
Corti, M. and V. Degiorgio, 1976a, *Phys. Rev.* **A 14**, 1475.
Corti, M. and V. Degiorgio, 1976b, Recent Advances in Optical Physics, eds. B. Havelka and J. Blabla (Soc. Czech. Math. Phys., Prague), p. 59.
Corti, M., V. Degiorgio and F. T. Arecchi, 1973, *Opt. Comm.* **8**, 329.
Courtens, E., 1972, Laser Handbook, Vol. 2, eds. F. T. Arecchi and E. O. Schulz-Dubois (North-Holland, Amsterdam), p. 1259.
Cresser, J. D., J. Häger, G. Leuchs, M. Rateike and H. Walther, 1982, Dissipative Systems in Quantum Optics, Topics in Current Physics, Vol. 27, ed. R. Bonifacio (Springer, Berlin), p. 21.
Crosignani, B., P. Di Porto and M. Bertolotti, 1975, Statistical Properties of Scattered Light (Acad. Press, New York).
Crosignani, B., P. Di Porto, U. Ganiel, S. Solimeno and A. Yariv, 1972, *IEEE J. Quant. Electr.* **QE-8**, 731.
Crosignani, B., P. Di Porto and S. Solimeno, 1968a, *Phys. Lett.* **27 A**, 568.
Crosignani, B., P. Di Porto and S. Solimeno, 1968b, *Phys. Lett.* **28 A**, 271.
Crosignani, B., P. Di Porto and S. Solimeno, 1969, *Phys. Rev.* **186**, 1342.
Crosignani, B., P. Di Porto and S. Solimeno, 1971, *Phys. Rev.* **D 3**, 1729.
Crosignani, B., P. Di Porto and S. Solimeno, 1972, *J. Phys.* **A 5**, L 119.
Crosignani, B., U. Ganiel, S. Solimeno and P. Di Porto, 1971a, *Phys. Rev. Lett.* **26**, 1130.
Crosignani, B., U. Ganiel, S. Solimeno and P. Di Porto, 1971b, *Phys. Rev.* **A 4**, 1570.
Crosignani, B., C. H. Papas and P. Di Porto, 1980, *Opt. Lett.* **5**, 467.
Dagenais, M. and L. Mandel, 1978, *Phys. Rev.* **A 18**, 2217.
Daino, B., G. De Marchis and S. Piazzolla, 1980, *Opt. Acta* **27**, 1151.
Dainty, J. C., 1975, Laser Speckle and Related Phenomena (Springer, Berlin).
Dainty, J. C., 1976, Progress in Optics, Vol. 14, ed. E. Wolf (North-Holland, Amsterdam), p. 1.
Davidson, F., 1969, *Phys. Rev.* **185**, 446.
Davidson, F. and L. Mandel, 1967, *Phys. Lett.* **25 A**, 700.
Davidson, F. and L. Mandel, 1968, *Phys. Lett.* **27 A**, 579.
Davies, E. B., 1976, Quantum Theory of Open Systems (Acad. Press, New York).
Delone, N. B. and A. V. Masalov, 1980, *Opt. Quant. Electr.* **12**, 291.

Demtchuk, M. I. and M. A. Ivanov, 1981, Statistical One-Quantum Method in Optico-Physical Experiment (Izd. BGU, Minsk), in Russian.
De Santis, P., F. Gori, G. Guattari and C. Palma, 1979, *Opt. Comm.* **29**, 256.
Dewael, P., 1975, *J. Phys.* **A 8**, 1614.
De Wolf, D. A., 1973a, *J. Opt. Soc. Am.* **63**, 171.
De Wolf, D. A., 1973b, *J. Opt. Soc. Am.* **63**, 657.
De Wolf, D. A., 1973c, *J. Opt. Soc. Am.* **63**, 1249.
Dialetis, D., 1967, *J. Math. Phys.* **8**, 1641.
Dialetis, D., 1969a, *J. Phys.* **A 2**, 229.
Dialetis, D., 1969b, *J. Opt. Soc. Am.* **59**, 74.
Dialetis, D. and E. Wolf, 1967, *Nuovo Cim.* **47**, 113.
Diament, P. and M. C. Teich, 1969, *J. Opt. Soc. Am.* **59**, 661.
Diament, P. and M. C. Teich, 1970a, *J. Opt. Soc. Am.* **60**, 682.
Diament, P. and M. C. Teich, 1970b, *J. Opt. Soc. Am.* **60**, 1489.
Diament, P. and M. C. Teich, 1971, *Appl. Opt.* **10**, 1664.
Dicke, R. H., 1954, *Phys. Rev.* **93**, 99.
Dicke, R. H., 1964, Quantum Electronics, eds. N. Bloembergen and P. Grivet (Dunod et Cie., Paris), p. 35.
Dirac, P. A. M., 1958, Principles of Quantum Mechanics, 4th Edition (Clarendon, Oxford).
Doermann, F. W. and O. Halpern, 1939, *Phys. Rev.* **55**, 486.
Drummond, P. D. and C. W. Gardiner, 1980, *J. Phys.* **A 13**, 2353.
Drummond, P. D., K. J. McNeil and D. F. Walls, 1979, *Opt. Comm.* **28**, 255.
Drummond, P. D., K. J. McNeil and D. F. Walls, 1980, *Opt. Acta* **27**, 321.
Drummond, P. D., K. J. McNeil and D. F. Walls, 1981, *Opt. Acta* **28**, 211.
Eberly, J. H. and A. Kujawski, 1967a, *Phys. Lett.* **24 A**, 426.
Eberly, J. H. and A. Kujawski, 1967b, *Phys. Rev.* **155**, 10.
Eberly, J. H. and A. Kujawski, 1972, *Acta Phys. Pol.* **A 41**, 259.
Eberly, J. H. and N. E. Rehler, 1970, *Phys. Rev.* **A 2**, 1607.
Eberly, J. H. and K. Wódkiewicz, 1977, *J. Opt. Soc. Am.* **67**, 1252.
Echtermeyer, B., 1971, *Z. Phys.* **246**, 225.
Edwards, S. F. and G. B. Parrent, 1959, *Opt. Acta* **6**, 367.
Einstein, A., 1909, *Phys. Z.* **10**, 185 and 817.
Einstein, A., 1912, La theorie du rayonnment et les quanta, Instituts Solvay, Brussels, Conseil de Physique 1er 1911, eds. P. Langevin and L. de Broglie (Gauthier-Villars, Paris), p. 407.
Estes, L. E., J. D. Kuppenheimer and L. M. Narducci, 1970, *Phys. Rev.* **A 1**, 710.
Fano, U., 1954, *Phys. Rev.* **93**, 121.
Fano, U., 1957, *Rev. Mod. Phys.* **29**, 74.
Farina, J. D., L. M. Narducci and E. Collet, 1980, *Opt. Comm.* **32**, 203.
Fercher, A. F. and P. F. Steeger, 1981, *Opt. Acta* **28**, 443.
Ferwerda, H. A., 1978, Inverse Source Problems, ed. H. P. Baltes (Springer, Berlin), p. 13.
Ferwerda, H. A. and B. J. Hoenders, 1974, *Optik* **40**, 14.
Ferwerda, H. A. and B. J. Hoenders, 1975, *Opt. Acta* **22**, 25 and 35.
Feynman, R. P., 1948, *Rev. Mod. Phys.* **20**, 367.
Feynman, R. P. and A. R. Hibbs, 1965, Quantum Mechanics and Path Integrals (McGraw-Hill, New York).
Fiddy, M. A. and A. H. Greenaway, 1979, *Opt. Comm.* **29**, 270.
Fillmore, G. L., 1969, *Phys. Rev.* **182**, 1384.
Fillmore, G. L. and G. Lachs, 1969, *IEEE Trans. Inform. Theory* **IT-15**, 465.
Fleck, J. A., 1966a, *Phys. Rev.* **149**, 309 and 322.
Fleck, J. A., 1966b, *Phys. Rev.* **152**, 278.
Foerster, T. von and R. J. Glauber, 1971, *Phys. Rev.* **A 3**, 1484.
Forrester, A. T., R. A. Gudmundsen and P. O. Johnson, 1955, *Phys. Rev.* **99**, 1691.
Francon, M., 1966, Diffraction. Coherence in Optics (Pergamon, Oxford).

Francon, M. and S. Mallick, 1967, Progress in Optics, Vol. 6, ed. E. Wolf (North-Holland, Amsterdam), p. 71.
Francon, M. and S. Slansky, 1965, Cohérence en optique (Centre Nat. Rech. Sci., Paris).
Fray, S., F. A. Johnson, R. Jones, T. P. McLean and E. R. Pike, 1967, *Phys. Rev.* **153**, 357.
Freed, C. and H. A. Haus, 1965, *Phys. Rev. Lett.* **15**, 943.
Freed, C. and H. A. Haus, 1966, *IEEE J. Quant. Electr.* **QE-2**, 190.
Friberg, A. T., 1978a, Coherence and Quantum Optics IV, eds. L. Mandel and E. Wolf (Plenum Press, New York), p. 449.
Friberg, A. T., 1978b, *J. Opt. Soc. Am.* **68**, 1281.
Friberg, A. T., 1979a, *J. Opt. Soc. Am.* **69**, 192.
Friberg, A. T., 1979b, Appl. Opt. Coherence, Proc. SPIE, Vol. 194, ed. W. H. Carter, p. 55 and 71.
Friberg, A. T., 1981a, Optics in Four Dimensions – 1980, eds. M. A. Machado and L. M. Narducci (Am. Inst. Phys., New York), p. 313.
Friberg, A. T., 1981b, *Opt. Acta* **28**, 261.
Friberg, A. T. and E. Wolf, 1983, *Opt. Acta* **30**, 1417.
Frieden, B. R., 1971, Progress in Optics, Vol. 9, ed. E. Wolf (North-Holland, Amsterdam), p. 311.
Fürth, R., 1928a, *Z. Phys.* **48**, 323.
Fürth, R., 1928b, *Z. Phys.* **50**, 310.
Furutsu, K., 1972, *J. Opt. Soc. Am.* **62**, 240.
Furutsu, K., 1976, *J. Math. Phys.* **17**, 1252.
Furutsu, K. and Y. Furuhama, 1973, *Opt. Acta* **20**, 707.
Gabor, D., 1946, *J. Inst. Electr. Engrs.* **93**, 429.
Gabor, D., 1956, Proc. Symp. Astr. Optics and Rel. Subjects, ed. Z. Kopal (North-Holland, Amsterdam), p. 17.
Gabor, D., 1961, Progress in Optics, Vol. 1, ed. E. Wolf (North-Holland), p. 109.
Gagliardi, R. M. and S. Karp, 1976. Optical Communications (J. Wiley, New York).
Gamo, H., 1963a, Electromagnetic Theory and Antennas, Part 2, ed. E. C. Jordan (Macmillan, New York), p. 801.
Gamo, H., 1963b, *J. Appl. Phys.* **34**, 875.
Gamo, H., 1964, Progress in Optics, Vol. 3, ed. E. Wolf (North-Holland), p. 187.
Gase, R. and M. Schubert, 1982, *Opt. Acta* **29**, 1331.
Gelfand, J. M. and P. A. Minlos, 1954, *Proc. Acad. Sci. USSR* **97**, 209.
Gelfand, I. M. and G. E. Shilov, 1964, Generalized Functions, Vol. I (Acad. Press, New York).
Gelfand, I. M. and N. Ya. Vilenkin, 1964, Generalized Functions, Vol. 4 (Acad. Press, New York).
Gelfand, I. M. and A. M. Yaglom, 1956, *Usp. Math. Nauk* **11**, 77.
Germey, K., 1963, *Ann. Physik* **10**, 141.
Germey, K., F. J. Schütte and R. Tiebel, 1981, *Ann. Physik* **38**, 80.
Ghielmetti, F., 1964, *Phys. Lett.* **12**, 210.
Ghielmetti, F., 1976, *Nuovo Cim.* **35 B**, 243.
Gibbs, H. M. and R. E. Slusher, 1971, *Appl. Phys. Lett.* **18**, 505.
Glauber, R. J., 1963a, *Phys. Rev.* **130**, 2529.
Glauber, R. J., 1963b, *Phys. Rev.* **131**, 2766.
Glauber, R. J., 1963c, *Phys. Rev. Lett.* **10**, 84.
Glauber, R. J., 1964, Quantum Electronics, eds. N. Bloembergen and P. Grivet (Dunod et Cie., Paris, Columbia Univ., New York), p. 111.
Glauber, R. J., 1965, Quantum Optics and Electronics, eds. C. DeWitt, A. Blandin and C. Cohen-Tannoudji (Gordon and Breach, New York), p. 144.
Glauber, R. J., 1966a, Physics of Quantum Electronics, eds. P. L. Kelley, B. Lax and P. E. Tannenwald (McGraw-Hill, New York), p. 788.
Glauber, R. J., 1966b, *Phys. Lett.* **21**, 650.
Glauber, R. J., 1967, Proc. Symp. Modern Optics (Polytech. Press, New York), p. 1.
Glauber, R. J., 1969, Quantum Optics, ed. R. J. Glauber (Acad. Press, New York).

Glauber, R. J., 1970, Quantum Optics, eds. S. M. Kay and A. Maitland (Acad. Press, London), p. 53.
Glauber, R. J., 1972, Laser Handbook, Vol. 1, eds. F. T. Arecchi and E. O. Schulz-Dubois (North-Holland, Amsterdam), p. 1.
Glódź, M., 1978, *Acta Phys. Pol.* **A 54**, 213.
Gnutzmann, U., 1969, *Z. Phys.* **222**, 283.
Gnutzmann, U., 1970, *Z. Phys.* **233**, 380.
Golay, M. J. E., 1961, *Proc. IRE* **49**, 958.
Goldberger, M. L., H. W. Lewis and K. M. Watson, 1963, *Phys. Rev.* **132**, 2764.
Goldberger, M. L., H. W. Lewis and K. M. Watson, 1966, *Phys. Rev.* **142**, 25.
Goldberger, M. L. and K. M. Watson, 1964, *Phys. Rev.* **134**, B 919.
Goldberger, M. L. and K. M. Watson, 1965, *Phys. Rev.* **137**, B 1396.
Golubev, Yu. M., 1979, *Opt. Spectr.* (USSR) **46**, 3 and 398.
Goodman, J. W., 1975, Laser Speckle and Related Phenomena, ed. J. C. Dainty (Springer, Berlin), p. 9.
Gordov, E. P. and S. D. Tvorogov, 1978, The Quantum Theory of Propagation of the Electromagnetic Field (Nauka, Novosibirsk), in Russian.
Gordov, E. P. and S. D. Tvorogov, 1980, *Phys. Rev.* **D 22**, 908.
Gordov, E. P. and S. D. Tvorogov, 1984, Method of Semiclassical Representation of Quantum Theory (Nauka, Novosibirsk), in Russian.
Gradshteyn, I. S. and I. M. Ryzhik, 1965, Table of Integrals, Series and Products (Acad. Press, New York).
Graham, R., 1968a, *Z. Phys.* **210**, 319.
Graham, R., 1968b, *Z. Phys.* **211**, 469.
Graham, R., 1970, Quantum Optics, eds. S. M. Kay and A. Maitland (Acad. Press, London), p. 489.
Graham, R., 1973, Springer Tracts in Modern Optics, Vol. 66, ed. G. Höhler (Springer, Berlin), p. 1.
Graham, R., 1974, Progress in Optics, Vol. 12, ed. E. Wolf (North-Holland, Amsterdam), p. 233.
Graham, R. and H. Haken, 1968, *Z. Phys.* **210**, 276.
Greenaway, A. H., 1977, *Opt. Lett.* **1**, 10.
Gupta, P. S. and B. K. Mohanty, 1980, *Czech. J. Phys.* **B 30**, 1127.
Gupta, P. S. and B. K. Mohanty, 1981, *Opt. Acta* **28**, 521.
Gusev, V. G. and B. I. Pojzner, 1982, Quantum Radiophysics (Laboratory excercises) (Izd. Tomsk Univ., Tomsk), in Russian.
Güttinger, W., 1966, *Fortschr. Phys.* **14**, 483.
Haake, F., 1973, Springer Tracts in Modern Physics, Vol. 66, ed. G. Höhler (Springer, Berlin), p. 98.
Haig, N. D. and R. M. Sillitto, 1968, *Phys. Lett.* **28 A**, 463.
Haken, H., 1967, Dynamical Processes in Solid State Optics, Part I, eds. R. Kubo and H. Kamimura (Syokabo, Tokyo, W. A. Benjamin, New York), p. 168.
Haken, H., 1970a, Quantum Optics, eds. S. M. Kay and A. Maitland (Acad. Press, London), p. 201.
Haken, H., 1970b, Handbuch der Physik, Vol. XXV/2c (Springer, Berlin).
Haken, H., 1972, Laser Handbook, Vol. 1, eds. F. T. Arecchi and E. O. Schulz-Dubois (North-Holland, Amsterdam), p. 115.
Haken, H., 1978, Synergetics. An Introduction. Nonequilibrium Phase Transtitions and Selforganization in Physics, Chemistry and Biology, 2nd Edition (Springer, Berlin).
Haken, H., H. Risken and W. Weidlich, 1967, *Z. Phys.* **206**, 355.
Halpern, O. and G. W. Doermann, 1937, *Phys. Rev.* **52**, 937.
Harwit, M., 1960, *Phys. Rev.* **120**, 1551.
Helstrom, C. W., 1972, Progress in Optics, Vol. 10, ed. E. Wolf (North-Holland, Amsterdam), p. 284.
Helstrom, C. W., 1976, Quantum Detection and Estimation Theory (Acad. Press, New York).
Helstrom, C. W., 1979a, *IEEE Trans. Inform. Theory* **IT-25**, 69.
Helstrom, C. W., 1979b, *Opt. Comm.* **28**, 363.
Hempstead, R. D. and M. Lax, 1967, *Phys. Rev.* **161**, 350.
Hildred, G. P., 1980, *Opt. Acta* **27**, 1621.
Hill, R. J. and S. F. Clifford, 1981, *J. Opt. Soc. Am.* **71**, 675.
Ho, T. L., 1969, *J. Opt. Soc. Am.* **59**, 385.

Hoenders, B. J., 1978, Inverse Source Problems, ed. H. P. Baltes (Springer, Berlin), p. 41.
Hofman, M., 1980, *Acta Univ. Palack. Olom.* **65**, 35.
Holliday, D., 1964, *Phys. Lett.* **8**, 250.
Holliday, D. and M. L. Sage, 1964, *Ann. Phys.* **29**, 125.
Holliday, D. and M. L. Sage, 1965, *Phys. Rev.* **138**, B 485.
Holý, V., 1980, *Phys. Stat. Sol.* (b) **101**, 575.
Holý, V., 1982, *Phys. Stat. Sol.* (b) **111**, 341; **112**, 161.
Hopkins, H. H., 1951, *Proc. Roy. Soc.* (A) **208**, 263.
Hopkins, H. H., 1953, *Proc. Roy. Soc.* (A) **217**, 408.
Hopkins, H. H., 1957a, *J. Opt. Soc. Am.* **47**, 508.
Hopkins, H. H., 1957b, *Proc. Roy. Soc.* **B 70**, 1002.
Horák, R., 1969a, *Opt. Acta* **16**, 111.
Horák, R., 1969b, *Czech. J. Phys.* **B 19**, 827.
Horák, R., 1971, *Czech. J. Phys.* **B 21**, 7.
Horák, R., L. Mišta and J. Peřina, 1971a, *J. Phys.* **A 4**, 231.
Horák, R., L. Mišta and J. Peřina, 1971b, *Czech. J. Phys.* **B 21**, 614.
Horák, R., L. Mišta and J. Peřina, 1971c, *Phys. Lett.* **35 A**, 400.
Ingarden, R. S., 1965, *Fortschr. Phys.* **13**, 755.
Ito, S. and K. Furutsu, 1982, *J. Opt. Soc. Am.* **72**, 760.
Jacquinot, P., 1960, *Rept. Progr. Phys.* **23**, 267.
Jaiswal, A. K. and G. S. Agarwal, 1969, *J. Opt. Soc. Am.* **59**, 1446.
Jaiswal, A. K. and C. L. Mehta, 1969, *Phys. Rev.* **186**, 1355.
Jaiswal, A. K. and C. L. Mehta, 1970, *Phys. Rev.* **A 2**, 168.
Jakeman, E., 1981, *Opt. Acta* **28**, 435.
Jakeman, E., C. J. Oliver and E. R. Pike, 1968a, *J. Phys.* **A 1**, 406.
Jakeman, E., C. J. Oliver and E. R. Pike, 1968b, *J. Phys.* **A 1**, 497.
Jakeman, E., C. J. Oliver and E. R. Pike, 1971, *Phys. Lett.* **34 A**, 101.
Jakeman, E., C. J. Oliver, E. R. Pike, M. Lax and M. Zwanziger, 1970, *J. Phys.* **A 3**, L 52.
Jakeman, E. and E. R. Pike, 1968, *J. Phys.* **A 1**, 128.
Jakeman, E. and E. R. Pike, 1969, *J. Phys.* **A 2**, 115.
Jakeman, E., E. R. Pike, P. N. Pusey and J. M. Vaugham, 1977, *J. Phys.* **A 10**, L 257.
Jakeman, E. and P. N. Pusey, 1980, Inverse Scattering Problems in Optics, ed. H. P. Baltes (Springer, Berlin), p. 73.
Jannson, T., 1980, *J. Opt. Soc. Am.* **70**, 1544.
Janossy, L., 1957, *Nuovo Cim.* **6**, 125.
Janossy, L., 1959, *Nuovo Cim.* **12**, 369.
Javan, A., E. A. Ballik and W. L. Bond, 1962, *J. Opt. Soc. Am.* **52**, 96.
Johnson, F. A., R. Jones, T. P. McLean and E. R. Pike, 1966, *Phys. Rev. Lett.* **16**, 589.
Johnson, F. A., T. P. McLean and E. R. Pike, 1966, Physics of Quantum Electronics, eds. P. L. Kelley, B. Lax and P. E. Tannenwald (McGraw-Hill, New York).
Jordan, F. T. and F. Ghielmetti, 1964, *Phys. Rev. Lett.* **12**, 607.
Kahn, F. D., 1958, *Opt. Acta* **5**, 93.
Kano, Y., 1962, *Nuovo Cim.* **23**, 328.
Kano, Y., 1964a, *J. Phys. Soc. Jap.* **19**, 1555.
Kano, Y., 1964b, *Ann. Phys.* **30**, 127.
Kano, Y., 1965, *J. Math. Phys.* **6**, 1913.
Kano, Y., 1966, *Nuovo Cim.* **43**, 1.
Kano, Y., 1976, *Phys. Lett*, **56 A**, 7.
Kano, Y. and E. Wolf, 1962, *Proc. Phys. Soc.* **80**, 1273.
Karczewski, B., 1963a, *Phys. Lett.* **5**, 191.
Karczewski, B., 1963b, *Nuovo Cim.* **30**, 906.

Karczewski, B., 1976, Recent Advances in Optical Physics, eds. B. Havelka and J. Blabla (Soc. Czech. Math. Phys., Prague), p. 53.

Kastler, A., 1964, Quantum Electronics, eds. N. Bloembergen and P. Grivet (Dunod and Cie., Paris), p. 3.

Katriel, J. and D. G. Hummer, 1981, *J. Phys.* **A 14**, 1211.

Keller, E. F., 1965, *Phys. Rev.* **139**, B 202.

Kelley, P. L. and W. H. Kleiner, 1964, *Phys. Rev.* **136**, A 316.

Khalfin, L. A., 1960, *Proc. Acad. Sci. USSR* **132**, 1051.

Khurgyn, J. L. and V. P. Yakovlev, 1971, Finite Functions in Physics and Technique (Nauka, Moscow), in Russian.

Kiedroń, P., 1980, *Opt. Appl.* **10**, 253.

Kiedroń, P., 1981, *Optik* **59**, 303.

Kielich, S., 1981, Molecular Nonlinear Optics (Nauka, Moscow), in Russian.

Kielich, S., M. Kozierowski and R. Tanaś, 1978, Coherence and Quantum Optics IV, eds. L. Mandel and E. Wolf (Plenum Press, New York), p. 511.

Kimble, H. J., M. Dagenais and L. Mandel, 1977, *Phys. Rev. Lett.* **39**, 691.

Kimble, H. J., M. Dagenais and L. Mandel, 1978, *Phys. Rev.* **A 18**, 201.

Kimble, H. J. and L. Mandel, 1976, *Phys. Rev.* **A 13**, 2123.

Kimble, H. J. and L. Mandel, 1977, *Phys. Rev.* **A 15**, 689.

Kintner, E. C. and R. M. Sillitto, 1973, *Opt. Acta* **20**, 721.

Kitazima, I., 1974, *Opt. Comm.* **10**, 137.

Klauder, J. R., 1960, *Ann. Phys.* **11**, 123.

Klauder, J. R., 1966, *Phys. Rev. Lett.* **16**, 534.

Klauder, J. R., J. McKenna and D. G. Currie, 1965, *J. Math. Phys.* **6**, 734.

Klauder, J. R. and E. C. G. Sudarshan, 1968, Fundamentals of Quantum Optics (W. A. Benjamin, New York).

Kleinman, D. A., 1972, Laser Handbook, Vol. 2, eds. F. T. Arecchi and E. O. Schulz-Dubois (North-Holland, Amsterdam), p. 1229.

Klyackin, V. I., 1975, Statistical Description of Dynamical Systems with Fluctuating Parameters (Nauka, Moscow), in Russian.

Klyackin, V. I. and V. I. Tatarskii, 1973, *Usp. Phys. Nauk* **110**, 499.

Klyshko, A. N., 1980, Photons and Nonlinear Optics (Nauka, Moscow), in Russian.

Kohler, D. and L. Mandel, 1970, *J. Opt. Soc. Am.* **60**, 280.

Kohler, D. and L. Mandel, 1973, *J. Opt. Soc. Am.* **63**, 126.

Koňák, Č., P. Štěpánek, L. Dvořák, Z. Kupka, J. Křepelka and J. Peřina, 1982, *Opt. Acta* **29**, 1105.

Korenman, V., 1965, *Phys. Rev. Lett.* **14**, 293.

Korenman, V., 1966, *Ann. Phys.* **39**, 72.

Korenman, V., 1967, *Phys. Rev.* **154**, 1233.

Kovarskii, V. A., 1974, Multiphoton Transitions (Shtiinca, Kishinev), in Russian.

Kozierowski, M., 1981, *Kvant. Electr.* (USSR) **8**, 1157.

Kozierowski, M. and R. Tanaś, 1977, *Opt. Comm.* **21**, 229.

Krasiński, J., S. Chudzyński and W. Majewski, 1974, *Opt. Comm.* **12**, 304.

Krasiński, J., S. Chudzyński and W. Majewski, 1976, Recent Advances in Optical Physics, eds. B. Havelka and J. Blabla (Soc. Czech. Math. Phys., Prague), p. 323.

Kraynov, V. P. and S. S. Todirashku, 1980, *J. Exp. Theor. Phys.* (USSR) **79**, 69.

Krivoshlykov, S. G. and I. N. Sissakian, 1979, *Opt. Quant. Electr.* **11**, 393.

Krivoshlykov, S. G. and I. N. Sissakian, 1980a, *Opt. Quant. Electr.* **12**, 463.

Krivoshlykov, S. G. and I. N. Sissakian, 1980b, *Kvant. Electr.* (USSR) **7**, 553.

Krivoshlykov, S. G. and I. N. Sissakian, 1983, *Kvant. Electr.* (USSR) **10**, 735.

Kryszewski, S. and J. Chrostowski, 1977, *J. Phys.* **A 10**, L 261.

Kujawski, A., 1966, *Nuovo Cim.* **44**, 326.

Kujawski, A., 1968, *Acta Phys. Pol.* **34**, 957.

Kujawski, A., 1969, *Bull. Acad. Pol. Sci.* **17**, 467 and 839.

Kujawski, A. and J. H. Eberly, 1978, Coherence and Quantum Optics IV, eds. L. Mandel and E. Wolf (Plenum Press, New York), p. 989.
Kuriksha, A. K., 1973, Quantum Optics and Optical Location (Sov. Radio, Moscow), in Russian.
Lachs, G., 1965, *Phys. Rev.* **138**, B 1012.
Lachs, G., 1967, *J. Appl. Phys.* **38**, 3439.
Lachs, G., 1971, *J. Appl. Phys.* **42**, 602.
Lachs, G. and S. R. Laxpati, 1973, *J. Appl. Phys.* **44**, 3332.
Lambropoulos, P., 1968, *Phys. Rev.* **168**, 1418.
Lambropoulos, P., C. Kikuchi and R. K. Osborn, 1966, *Phys. Rev.* **144**, 1081.
Landau, L. D. and E. M. Lifshitz, 1964, Statistical Physics, 2nd Ed. (Nauka, Moscow), in Russian.
Langer, J. S., 1968, *Phys. Rev.* **167**, 183.
Langer, J. S., 1969, *Phys. Rev.* **184**, 219.
Laue, M., 1907, *Ann. Physik* (Leipzig) **23**, 1.
Lawton, W., 1981, *J. Opt. Soc. Am.* **71**, 1519.
Lax, M., 1967a, Dynamical Processes in Solid State Optics, Part I, eds. R. Kubo and H. Kamimura (W. A. Benjamin, New York), p. 195.
Lax, M., 1967b, *Phys. Rev.* **157**, 213.
Lax, M., 1968a, Fluctuation and Coherence Phenomena in Classical and Quantum Physics. Statistical Physics, Phase Transitions and Superconductivity, eds. M. Chrétien, E. P. Gross and S. Deser (Gordon and Breach, New York).
Lax, M., 1968b, *Phys. Rev.* **172**, 350.
Lax, M. and W. H. Louisell, 1967, *IEEE J. Quant. Electr.* **QE-3**, 47.
Lax, M. and Yuen, H., 1968, *Phys. Rev.* **172**, 362.
Lax, M. and M. Zwanziger, 1970, *Phys. Rev. Lett*, **24**, 937.
Lax, M. and M. Zwanziger, 1973, *Phys. Rev.* **A 7**, 750.
Laxpati, S. R. and G. Lachs, 1972, *J. Appl. Phys.* **43**, 4773.
Leader, J. C., 1981, *J. Opt. Soc. Am.* **71**, 542.
Le Compte, C., G. Mainfray and C. Manus, 1974, *Phys. Rev. Lett.* **32**, 265.
Le Compte, C., G. Mainfray, C. Manus and F. Sanchez, 1975, *Phys. Rev.* **A 11**, 1009.
Ledinegg, E., 1967, *Z. Phys.* **205**, 25.
Ledinegg, E. and E. Schachinger, 1983, *Phys. Rev.* **A 27**, 2555.
Lehmberg, R. H., 1968, *Phys. Rev.* **167**, 1152.
Leuchs, G., M. Rateike and H. Walther, 1979, see Walls, D. F., 1979, *Nature* **280**, 451.
Linfoot, E. H., 1956, Proc. Symp. Astr. Optics and Rel. Subj., ed. Z. Kopal (North-Holland, Amsterdam), p. 38.
Lipsett, M. S. and L. Mandel, 1963, *Nature* **199**, 553.
Lipsett, M. S. and L. Mandel, 1964, Quantum Electronics, eds. N. Bloembergen and P. Grivet (Dunod et Cie, Paris, Columbia Univ., New York), p. 1271.
López, A., 1967, *Phys. Lett.* **25 A**, 83.
Loudon, R., 1973, The Quantum Theory of Light (Clarendon, Oxford); second edition 1981.
Loudon, R., 1980, *Rep. Progr. Phys.* **43**, 913.
Loudon, R., 1981, *Phys. Rev. Lett.* **47**, 815.
Loudon, R., 1984, *Opt. Comm.* **49**, 24.
Louisell, W. H., 1964, Radiation and Noise in Quantum Electronics (McGraw-Hill, New York).
Louisell, W. H., 1970, Quantum Optics, eds. S. M. Kay and A. Maitland (Acad. Press, London), p. 177.
Louisell, W. H., 1973, Quantum Statistical Properties of Radiation (J. Wiley, New York).
Louisell, W. H., A. Yariv and A. E. Siegman, 1961, *Phys. Rev.* **124**, 1646.
Lukš, A., 1976, *Czech. J. Phys.* **B 26**, 1095.
Lyons, J. and G. J. Troup, 1970, *Phys. Lett.* **32 A**, 352.
Magill, P. J. and R. P. Soni, 1966, *Phys. Rev. Lett.* **16**, 911.
Magyar, G. and L. Mandel, 1963, *Nature* **198**, 255.

Magyar, G. and L. Mandel, 1964, Quantum Electronics, eds. N. Bloembergen and P. Grivet (Dunod et Cie., Paris, Columbia Univ., New York), p. 1247.
Malkin, I. A. and V. I. Man'ko, 1979, Dynamical Symmetries and Coherent States of Quantum Systems (Nauka, Moscow), in Russian.
Mandel, L., 1958, *Proc. Phys. Soc.* (Lond.) **72**, 1037.
Mandel, L., 1959, *Proc. Phys. Soc.* (Lond.) **74**, 233.
Mandel, L., 1961a, *J. Opt. Soc. Am.* **51**, 1342.
Mandel, L., 1961b, *J. Opt. Soc. Am.* **51**, 797.
Mandel, L., 1962, *J. Opt. Soc. Am.* **52**, 1335.
Mandel, L., 1963a, Progress in Optics, Vol. 2, ed. E. Wolf (North-Holland, Amsterdam), p. 181.
Mandel, L., 1963b, *Proc. Phys. Soc.* **81**, 1104.
Mandel, L., 1963c, *Phys. Lett.* **7**, 117.
Mandel, L., 1964a, *Phys. Lett.* **10**, 166.
Mandel, L., 1964b, *Phys. Rev.* **134**, A 10.
Mandel, L., 1964c, *Phys. Rev.* **136**, B 1221.
Mandel, L., 1964d, Quantum Electronics, eds. N. Bloembergen and P. Grivet (Dunod et Cie., Paris, Columbia Univ., New York), p. 101.
Mandel, L., 1965, *Phys. Rev.* **138**, B 753.
Mandel, L., 1966a, *Phys. Rev.* **144**, 1071.
Mandel, L., 1966b, *Phys. Rev.* **152**, 438.
Mandel, L., 1967, Proc. Symp. Modern Optics (Polytech. Press, New York), p. 143.
Mandel, L., 1979a, *Phys. Rev.* **A 20**, 1590.
Mandel, L., 1979b, *Opt. Lett.* **4**, 205.
Mandel, L., 1981a, *Opt. Comm.* **36**, 87.
Mandel, L., 1981b, *Opt. Acta* **28**, 1447.
Mandel, L., 1983, *Phys. Rev.* **A 28**, 929.
Mandel, L. and D. Meltzer, 1969, *Phys. Rev.* **188**, 198.
Mandel, L., E. C. G. Sudarshan and E. Wolf, 1964, *Proc. Phys. Soc.* **84**, 435.
Mandel, L. and E. Wolf, 1961a, *J. Opt. Soc. Am.* **51**, 815.
Mandel, L. and E. Wolf, 1961b, *Phys. Rev.* **124**, 1696.
Mandel, L. and E. Wolf, 1962, *Proc. Phys. Soc.* (Lond.) **80**, 894.
Mandel, L. and E. Wolf, 1965, *Rev. Mod. Phys.* **37**, 231.
Mandel, L. and E. Wolf, 1966, *Phys. Rev.* **149**, 1033.
Mandel, L. and E. Wolf, 1976, *J. Opt. Soc. Am.* **66**, 529.
Mandel, L. and E. Wolf, 1981, *Opt. Comm.* **36**, 247.
Mandel, L. and E. Wolf, eds., 1984, Coherence and Quantum Optics V (Plenum, New York).
Mandelstam, L. I., 1947, Collected papers, Vol. 2, p. 388.
Mandelstam, L. I., 1948, Collected papers, Vol. 1, p. 229.
Man'ko, V. I. (ed.), 1972, Coherent States in Quantum Theory (Mir, Moscow), in Russian.
Marathay, A. S., 1966, *J. Opt. Soc. Am.* **56**, 619.
Marathay, A. S., 1982, Elements of Optical Coherence Theory (J. Wiley, New York).
Marchand, E. W. and E. Wolf, 1972, *Opt. Comm.* **6**, 305.
Marchand, E. W. and E. Wolf, 1974a, *J. Opt. Soc. Am.* **64**, 1219.
Marchand, E. W. and E. Wolf, 1974b, *J. Opt. Soc. Am.* **64**, 1273.
Maréchal, A. and M. Francon, 1960, Diffraction. Structures des images (Ed. Rev. Opt. Théor. Inst., Paris).
Martienssen, W. and E. Spiller, 1964, *Am. J. Phys.* **32**, 919.
Martienssen, W. and E. Spiller, 1966a, *Phys. Rev. Lett.* **16**, 531.
Martienssen, W. and E. Spiller, 1966b, *Phys. Rev.* **145**, 285.
Martínez - Herrero, R., 1979, *Nuovo Cim.* **54 B**, 205.
Martínez - Herrero, R., 1980, *Opt. Lett.* **5**, 502.
Martínez - Herrero, R., 1981, *Opt. Acta* **28**, 1151.

Martínez - Herrero, R., 1982, *Opt. Acta* **29**, 1255.
Martínez - Herrero, R. and A. Durán, 1981, *Opt. Acta* **28**, 65.
Martínez - Herrero, R. and P. M. Mejías, 1981, *Opt. Comm.* **37**, 234.
Martínez - Herrero, R. and P. M. Mejías, 1982a, *Opt. Acta* **29**, 187.
Martínez - Herrero, R. and P. Mejías, 1982b, *J. Opt. Soc. Am.* **72**, 131 and 765.
McCall, S. L. and E. L. Hahn, 1967, *Phys. Rev. Lett.* **18**, 908.
McCall, S. L. and E. L. Hahn, 1969, *Phys. Rev.* **183**, 457.
McGill, W. J., 1967, *J. Math. Psych.* **4**, 351.
McLean, T. P. and E. R. Pike, 1965, *Phys. Lett.* **15**, 318.
McMillan, J. L., R. M. Sillitto and W. Sillitto, 1979, *Opt. Acta* **26**, 1125.
McNeil, K. J., P. D. Drummond and D. F. Walls, 1978, *Opt. Comm.* **27**, 292.
McNeil, K. J. and D. F. Walls, 1974, *J. Phys.* **A 7**, 617.
McNeil, K. J. and D. F. Walls, 1975a, *J. Phys.* **A 8**, 104.
McNeil, K. J. and D. F. Walls, 1975b, *J. Phys.* **A 8**, 111.
McNeil, K. J. and D. F. Walls, 1975c, *Phys. Lett.* **51 A**, 233.
Mehta, B. L., 1974, *Nouv. Rev. Opt.* **5**, 95.
Mehta, C. L., 1963, *Nuovo Cim.* **28**, 401.
Mehta, C. L., 1964, *J. Math. Phys.* **5**, 677.
Mehta, C. L., 1965, *Nuovo Cim.* **36**, 202.
Mehta, C. L., 1966, *Nuovo Cim.* **45**, 280.
Mehta, C. L., 1967, *Phys. Rev. Lett.* **18**, 752.
Mehta, C. L., 1968, *J. Opt. Soc. Am.* **58**, 1233.
Mehta, C. L., 1970, Progress in Optics, Vol. 8., ed. E. Wolf (North-Holland, Amsterdam), p. 373.
Mehta, C. L., P. Chand, E. C. G. Sudarshan and R. Vedam, 1967, *Phys. Rev.* **157**, 1198.
Mehta, C. L. and S. Gupta, 1975, *Phys. Rev.* **A 11**, 1634.
Mehta, C. L. and A. K. Jaiswal, 1970, *Phys. Rev.* **A** 2, 2570.
Mehta, C. L. and L. Mandel, 1967, Electromagnetic Wave Theory, Part 2 (Pergamon, Oxford), p. 1069.
Mehta, C. L. and E. C. G. Sudarshan, 1965, *Phys. Rev.* **138**, B 274.
Mehta, C. L. and E. C. G. Sudarshan, 1966, *Phys. Lett.* **22**, 574.
Mehta, C. L. and E. Wolf, 1964, *Phys. Rev.* **134**, A 1143 and A 1149.
Mehta, C. L. and E. Wolf, 1967a, *Phys. Rev.* **157**, 1183 and 1188.
Mehta, C. L. and E. Wolf, 1967b, *Phys. Rev.* **161**, 1328.
Mehta, C. L., E. Wolf and A. P. Balachandran, 1966, *J. Math. Phys.* **7**, 133.
Meltzer, D., W. Davis and L. Mandel, 1970, *Appl. Phys. Lett.* **17**, 242.
Meltzer, D. and L. Mandel, 1970, *Phys. Rev. Lett.* **25**, 1151.
Meltzer, D. and L. Mandel, 1971, *Phys. Rev.* **A 3**, 1763.
Messiah, A., 1961, Quantum Mechanics, Vol. I (J. Wiley, New York).
Messiah, A., 1962, Quantum Mechanics, Vol. II (J. Wiley, New York).
Michelson, A. A., 1890, *Phil. Mag.* **30**, 1.
Mielniczuk, W. J., 1979, *Opt. Acta* **26**, 1115.
Mielniczuk, W. J. and J. Chrostowski, 1981, *Phys. Rev.* **A 23**, 1382.
Milburn, G. and D. F. Walls, 1981, *Opt. Comm.* **39**, 401.
Miller, M. M. and E. A. Mishkin, 1966, *Phys. Rev.* **152**, 1110.
Miller, M. M. and E. A. Mishkin, 1967a, *Phys. Lett.* **24 A**, 188.
Miller, M. M. and E. A. Mishkin, 1967b, *Phys. Rev.* **164**, 1610.
Millet, J. and W. Usselio-La-Verna, 1970, *Opt. Comm.* **2**, 12.
Millet, J. and B. Varnier, 1969, *Opt. Comm.* **1**, 211.
Misell, D. L., 1973, *J. Phys.* **D 6**, 2200 and 2217.
Misell, D. L., R. E. Burge and A. H. Greenaway, 1974, *Nature* **247**, 401.
Misell, D. L. and A. H. Greenaway, 1974, *J. Phys.* **D 7**, 832 and 1660.
Mishkin, E. A. and D. F. Walls, 1969, *Phys. Rev.* **185**, 1618.

Mišta, L., 1967, *Phys. Lett.* **25 A**, 646.
Mišta, L., 1969, *Czech. J. Phys.* **B 19**, 443.
Mišta, L., 1971, *J. Phys.* **A 4**, L 73.
Mišta, L., 1973, *Czech. J. Phys.* **B 23**, 715.
Mišta, L., 1981, *Acta Univ. Palack. Olom.* **69**, 47.
Mišta, L. and J. Peřina, 1971, *Opt. Comm.* **2**, 441.
Mišta, L. and J. Peřina, 1977a, *Czech. J. Phys.* **B 27**, 373.
Mišta, L. and J. Peřina, 1977b, *Acta Phys. Pol.* **A 52**, 425.
Mišta, L. and J. Peřina, 1977c, *Czech. J. Phys.* **B 27**, 831.
Mišta, L. and J. Peřina, 1978, *Czech. J. Phys.* **B 28**, 392.
Mišta, L., J. Peřina and Z. Braunerová, 1973, *Opt. Comm.* **9**, 113.
Mišta, L., J. Peřina and V. Peřinová, 1971, *Phys. Lett.* **35 A**, 197.
Mišta, L., V. Peřinová, J. Peřina and Z. Braunerová, 1977, *Acta Phys. Pol.* **A 51**, 739.
Mohanty, B. K. and P. S. Gupta, 1981a, *Czech. J. Phys.* **B 31**, 275.
Mohanty, B. K. and P. S. Gupta, 1981b, *Czech. J. Phys.* **B 31**, 1083.
Mohanty, B. K. and P. S. Gupta, 1981c, *Czech. J. Phys.* **B 31**, 857.
Mohr, U., 1981, *Ann. Physik* **38**, 143.
Mohr, U. and H. Paul, 1978, *Ann. Physik* **35**, 461.
Mollow, B. R., 1968a, *Phys. Rev.* **168**, 1896.
Mollow, B. R., 1968b, *Phys. Rev.* **175**, 1555.
Mollow, B. R., 1970, *Phys. Rev.* **A 2**, 1477.
Mollow, B. R., 1981, Progress in Optics, Vol. 19, ed. E. Wolf (North-Holland, Amsterdam), p. 1.
Mollow, B. R. and R. J. Glauber, 1967a, *Phys. Rev.* **160**, 1076.
Mollow, B. R. and R. J. Glauber, 1967b, *Phys. Rev.* **160**, 1097.
Möller, B., 1968, *Opt. Acta* **15**, 223.
Morawitz, H., 1965, *Phys. Rev.* **139**, A 1072.
Morawitz, H., 1966, *Z. Phys.* **195**, 20.
Morgan, B. L. and L. Mandel, 1966, *Phys. Rev. Lett.* **16**, 1012.
Morse, P. M. and H. Feshbach, 1953, Methods of Theoretical Physics, Vol. I (McGraw-Hill, New York).
Mostowski, J. and K. Rzażewski, 1978, *Phys. Lett* **66 A**, 275.
Moyal, J. E., 1949, *Proc. Cambr. Phil. Soc.* **45**, 99.
Muskhelishvili, N. I., 1953, Singular Integral Equations (Noordhoff, Groningen).
Nakajima, N. and T. Asakura, 1982, *Optik* **60**, 289.
Nayak, N., 1980, *IEEE J. Quant. Electr.* **QE-16**, 843.
Nayak, N. and B. K. Mohanty, 1977, *Phys. Rev.* **A 15**, 1173.
Nayak, N. and B. K. Mohanty, 1979, *Phys. Rev.* **A 19**, 1204.
Nayyar, V. P. and N. K. Verma, 1978, *Appl. Opt.* **17**, 2176.
Neumann, R. and H. Haug, 1979, *Opt.* Comm. **31**, 267.
Nieto, M. M. and L. M. Simmons, 1978, *Phys. Rev. Lett.* **41**, 207.
Nieto, M. M. and L. M. Simmons, 1979, *Phys. Rev.* **D 20**, 1332 and 1342.
Nieto, M. M., L. M. Simmons and V. P. Gutschick, 1981, *Phys. Rev.* **D 23**, 927.
Nieto-Vesperinas, M., 1980, *Optik* **56**, 377.
Nieto-Vesperinas, M., 1982, *Optik* **62**, 87.
Nieto-Vesperinas, M. and O. Hignette, 1979, *Opt. Pura y Apl.* **12**, 175.
Nussenzveig, H. M., 1967, *J. Math. Phys.* **8**, 561.
Nussenzveig, H. M., 1972, Causality and Dispersion Relations (Acad. Press, London).
Nussenzveig, H. M., 1973, Introduction to Quantum Optics (Gordon and Breach, London).
Ohtsubo, J. and T. Asakura, 1977, *Appl. Opt.* **16**, 1742.
Oliver, G. and C. Bendjaballah, 1980, *Phys. Rev.* **A 22**, 630.
O'Neill, E. L., 1963, Introduction to Statistical Optics (Addison-Wesley, Reading).
O'Neill, E. L. and A. Walther, 1963, *Opt. Acta* **10**, 33.
Orszag, M., 1979, *J. Phys.* **A 12**, 2205, 2225 and 2233.

Paley, R. E. and N. Wiener, 1934, Fourier Transforms in Complex Domain (Amer. Math. Soc., New York).
Pancharatnam, S., 1963, *Proc. Ind. Acad. Sci.* **57**, 218 and 231.
Parrent, G. B., 1959a, *J. Opt. Soc. Am.* **49**, 787.
Parrent, G. B., 1959b, *Opt. Acta* **6**, 285.
Parrent, G. B., 1961, *J, Opt. Soc. Am.* **51**, 143.
Parrent, G. B. and P. Roman, 1960, *Nuovo Cim.* **15**, 370.
Parry, G., 1981, *Opt. Acta* **28**, 715.
Parry, G., P. N. Pusey, E. Jakeman and J. G. McWhirter, 1978, Coherence and Quantum Optics IV, eds. L. Mandel and E. Wolf (Planum Press, New York), p. 351.
Paul, H., 1964, *Ann. Physik* **14**, 147.
Paul, H., 1966, *Fortschr. Phys.* **14**, 141.
Paul, H., 1967, *Ann. Physik* **19**, 210.
Paul, H., 1969, Lasertheorie I, II (Akademie-Verlag, Berlin).
Paul, H., 1973, Nichtlineare Optik I, II (Akademie-Verlag, Berlin).
Paul, H., 1974, *Fortschr. Phys.* **22**, 657.
Paul, H., 1976, Recent Advances in Optical Physics, eds. B. Havelka and J. Blabla (Soc. Czech. Math. Phys., Prague), p. 67.
Paul, H., 1982, *Rev. Mod. Phys.* **54**, 1061.
Paul, H. and W. Brunner, 1980, *Opt. Acta* **27**, 263.
Paul, H. and W. Brunner, 1981, *Ann. Physik* **38**, 89.
Paul, H., W. Brunner and G. Richter, 1963, *Ann. Physik* **12**, 325.
Paul, H., U. Mohr and W. Brunner, 1976, *Opt. Comm.* **17**, 145.
Pearl, P. and G. J. Troup, 1968, *Phys. Lett.* **27 A**, 560.
Perelomov, A. M., 1977, *Usp. Phys. Nauk* **123**, 23.
Peřina, J., 1963a, *Opt. Acta* **10**, 333.
Peřina, J., 1963b, *Opt. Acta* **10**, 337.
Peřina, J., 1963c, Proc. Symp. Interkamera, ed. J. Morávek (SNTL, Prague), p. 139.
Peřina, J., 1965a, *Phys. Lett.* **14**, 34.
Peřina, J., 1965b, *Phys. Lett.* **15**, 129.
Peřina, J., 1965c, *Acta Univ. Palack. Olom.* **18**, 49.
Peřina, J., 1965d, *Phys. Lett.* **19**, 195.
Peřina, J., 1966, *Czech. J. Phys.* **B 16**, 907.
Peřina, J., 1967a, *Czech. J. Phys.* **B 17**, 1086.
Peřina, J., 1967b, *Phys. Lett.* **24 A**, 333.
Peřina, J., 1968a, *Acta Univ. Palack. Olom.* **27**, 227.
Peřina, J., 1968b, *Czech. J. Phys.* **B 18**, 197.
Peřina, J., 1969, *Opt. Acta* **16**, 289.
Peřina, J., 1970, Quantum Optics, eds. S. M. Kay and A. Maitland (Acad. Press, London), p. 513.
Peřina, J., 1971, *Czech. J. Phys.* **B 21**, 731.
Peřina, J., 1972, Coherence of Light (Van Nostrand, London).
Peřina, J., 1974, Coherence of Light (Mir, Moscow), in Russian.
Peřina, J., 1975, Theory of Coherence (SNTL, Prague), in Czech.
Peřina, J., 1976, *Czech. J. Phys.* **B 26**, 140.
Peřina, J., 1977, *Acta Phys. Pol.* **A 52**, 559.
Peřina, J., 1979, *Opt. Acta* **26**, 821.
Peřina, J., 1980, Progress in Optics, Vol. 18, ed. E. Wolf (North-Holland, Amsterdam), p. 127.
Peřina, J., 1981a, *Opt. Acta* **28**, 325.
Peřina, J., 1981b, *Opt. Acta* **28**, 1529.
Peřina, J., 1984, Quantum Statistics of Linear and Nonlinear Optical Phenomena (D. Reidel, Dordrecht in coedition with SNTL, Prague).
Peřina, J. and R. Horák, 1969a, *J. Phys.* **A 2**, 702.
Peřina, J. and R. Horák, 1969b, *Opt. Comm.* **1**, 91.

Peřina, J. and R. Horák, 1970, *Czech. J. Phys.* **B 20**, 149.
Peřina, J. and R. Horák, 1981, *Opt. Quant. Electr.* **13**, 345.
Peřina, J. and L. Mišta, 1968a, *Phys. Lett.* **27 A**, 217.
Peřina, J. and L. Mišta, 1968b, *Czech. J. Phys.* **B 18**, 697.
Peřina, J. and L. Mišta, 1969, *Ann. Physik* **22**, 372.
Peřina, J. and L. Mišta, 1974, *Opt. Acta* **21**, 329.
Peřina, J. and V. Peřinová, 1965, *Opt. Acta* **12**, 333.
Peřina, J. and V. Peřinová, 1969, *Opt. Acta* **16**, 309.
Peřina, J. and V. Peřinová, 1971, *Phys. Lett.* **35 A**, 283.
Peřina, J. and V. Peřinová, 1972, *Czech. J. Phys.* **B 22**, 1085.
Peřina, J. and V. Peřinová, 1975, *Czech. J. Phys.* **B 25**, 605.
Peřina, J. and V. Peřinová, 1976, *Czech. J. Phys.* **B 26**, 489.
Peřina, J., V. Peřinová and Z. Braunerová, 1977, *Opt. Appl.* **VII**/3, 79.
Peřina, J., V. Peřinová, P. Diament and M. C. Teich, 1975, *Czech. J. Phys.* **B 25**, 483.
Peřina, J., V. Peřinová and R. Horák, 1973a, *Czech. J. Phys.* **B 23**, 975.
Peřina, J., V. Peřinová and R. Horák, 1973b, *Czech. J. Phys.* **B 23**, 993.
Peřina, J., V. Peřinová and L. Knesel, 1977, *Acta Phys. Pol.* **A 51**, 725.
Peřina, J., V. Peřinová and J. Koďousek, 1984, *Opt. Comm.* **49**, 210.
Peřina, J., V. Peřinová, J. Křepelka, A. Lukš, C. Sibilia and M. Bertolotti, 1983, *Opt. Acta* **30**, 959.
Peřina, J., V. Peřinová, G. Lachs and Z. Braunerová, 1973, *Czech. J. Phys.* **B 23**, 1008.
Peřina, J., V. Peřinová and L. Mišta, 1971, *Opt. Comm.* **3**, 89.
Peřina, J., V. Peřinová and L. Mišta, 1972, *Opt. Acta* **19**, 579.
Peřina, J., V. Peřinová and L. Mišta, 1974, *Czech. J. Phys.* **B 24**, 482.
Peřina, J., V. Peřinová, L. Mišta and R. Horák, 1974, *Czech. J. Phys.* **B 24**, 374.
Peřina, J., V. Peřinová, C. Sibilia and M. Bertolotti, 1984, *Opt. Comm.* **49**, 285.
Peřina, J., V. Peřinová, M. C. Teich and P. Diament, 1973, *Phys. Rev.* **A 7**, 1732.
Peřina, J. and J. Tillich, 1966, *Acta Univ. Palack. Olom.* **21**, 153.
Peřinová, V., 1969, *Čas. pěst. mat.* **94**, 253 and 297.
Peřinová, V., 1981, *Opt. Acta* **28**, 747.
Peřinová, V. and J. Peřina, 1978a, *Czech. J. Phys.* **B 28**, 306.
Peřinová, V. and J. Peřina, 1978b, *Czech. J. Phys.* **B 28**, 1183.
Peřinová, V. and J. Peřina, 1978c, *Czech. J. Phys.* **B 28**, 1196.
Peřinová, V. and J. Peřina, 1981, *Opt. Acta* **28**, 769.
Peřinová, V., J. Peřina, M. Bertolotti and C. Sibilia, 1982, *Opt. Acta* **29**, 131.
Peřinová, V., J. Peřina and L. Knesel, 1977, *Czech. J. Phys.* **B 27**, 487.
Peřinová, V., J. Peřina, P. Szlachetka and S. Kielich, 1979a, *Acta Phys. Pol.* **A 56**, 267.
Peřinová, V., J. Peřina, P. Szlachetka and S. Kielich, 1979b, *Acta Phys. Pol.* **A 56**, 275.
Petrov, A. Z., 1961, Einstein Spaces (GIFML, Moscow), in Russian.
Pfleegor, R. L. and L. Mandel, 1967a, *Phys. Lett.* **24 A**, 766.
Pfleegor, R. L. and L. Mandel, 1967b, *Phys. Rev.* **159**, 1084.
Pfleegor, R. L. and L. Mandel, 1968, *J. Opt. Soc. Am.* **58**, 946.
Phillips, D. T., H. Kleiman and S. P. Davis, 1967, *Phys. Rev.* **153**, 113.
Phillips, R. L. and L. C. Andrews, 1981, *J. Opt. Soc. Am.* **71**, 1440.
Picard, R. H. and C. R. Willis, 1965, *Phys. Rev.* **139**, A 10.
Picinbono, B., 1967, Proc. Symp. Modern Optics (Polytech. Press, New York), p. 167.
Picinbono, B., 1969, *Phys. Lett.* **29 A**, 614.
Picinbono, B., 1971, *Phys. Rev.* **A 4**, 2398.
Picinbono, B. and E. Boileau, 1968, *J. Opt. Soc. Am.* **58**, 784.
Picinbono, B. and M. Rousseau, 1970, *Phys. Rev.* **A 1**, 635.
Pieczonková, A., 1982a, *Opt. Acta* **29**, 1509.
Pieczonková, A., 1982b, *Czech. J. Phys.* **B 32**, 831.
Pieczonková, A. and J. Peřina, 1981, *Czech. J. Phys.* **B 31**, 837.

Pike, E. R., 1969, *Riv. Nuovo Cim.* (spec. issue) **1**, 277.
Pike, E. R., 1970, Quantum Optics, eds. S. M. Kay and A. Maitland (Acad. Press, London), p. 127.
Pike, E. R. and E. Jakeman, 1974, Advances in Quantum Electronics, Vol. 2, ed. D. W. Goodwin (Acad. Press, London), p. 1.
Piovoso, M. J. and L. P. Bolgiano, 1967, *Proc. IEEE* **55**, 1519.
Potechin, V. A. and V. N. Tatarinov, 1978, Theory of Coherence of Electromagnetic Field (Svyaz, Moscow), in Russian.
Prucnal, P. R., 1980, *Appl. Opt.* **19**, 3611.
Prucnal, P. R. and M. C. Teich, 1979, *J. Opt. Soc. Am.* **69**, 539.
Prucnal, P. R. and M. C. Teich, 1982, *Biol. Cybern.* **43**, 87.
Přikryl, I. and C. M. Vest, 1983, *Appl. Opt.* **22**, 2844.
Purcell, E. M., 1956, *Nature* **178**, 1449.
Radcliffe, J. M., 1971, *J. Phys.* **A 4**, 313.
Radloff, W., 1968, *Phys. Lett.* **27 A**, 366.
Radloff, W., 1971, *Ann. Physik* **26**, 178.
Rebka, G. A. and R. V. Pound, 1957, *Nature* **180**, 1035.
Reed, I. S., 1962, *IRE Trans. Inform. Theory* **IT-8**, 194.
Reid, M., K. J. McNeil and D. F. Walls, 1981, *Phys. Rev.* **A 24**, 2029.
Reynolds, G. T., K. Spartalian and D. B. Scarl, 1969, *Nuovo Cim.* **61**, 355.
Rezende, S. M. and N. Zagury, 1969, *Phys. Lett.* **29 A**, 47 and 616.
Richter, G., W. Brunner and H. Paul, 1964, *Ann. Physik* **14**, 239.
Risken, H., 1965, *Z. Phys.* **186**, 85.
Risken, H., 1966, *Z. Phys.* **191**, 302.
Risken, H., 1968, *Fortschr. Phys.* **16**, 261.
Risken, H., 1970, Progress in Optics, Vol. 8, ed. E. Wolf (North-Holland, Amsterdam), p. 239.
Risken, H. and H. D. Vollmer, 1967, *Z. Phys.* **204**, 240.
Robl, H. R., 1967, *Phys. Lett.* **24 A**, 288.
Robl, H. R., 1968, *Phys. Rev.* **165**, 1426.
Rocca, F., 1967, *J. Physique* **28**, 113.
Rockower, E. B. and N. B. Abraham, 1978, *J. Phys.* **A 11**, 1879.
Rockower, E. B., N. B. Abraham and S. R. Smith, 1978, *Phys. Rev.* **A 17**, 1100.
Roman, P., 1959, *Nuovo Cim.* **13**, 974.
Roman, P., 1961a, *Nuovo Cim.* **20**, 759.
Roman, P., 1961b, *Nuovo Cim.* **22**, 1005.
Roman, P. and A. S. Marathay, 1963, *Nuovo Cim.* **30**, 1452.
Roman, P. and E. Wolf, 1960, *Nuovo Cim.* **17**, 462 and 477.
Ronchi, L., 1972, Laser Handbook, Vol. 1, eds. F. T. Arecchi and E. O. Schulz-Dubois (North-Holland, Amsterdam), p. 151.
Rosenberg, S. and M. C. Teich, 1972, *J. Appl. Phys.* **43**, 1256.
Rosenfeld, L., 1958, Niels Bohr and Development of Physics, ed. W. Pauli.
Ross, G., M. A. Fiddy, M. Nieto-Vesperinas and M. W. L. Wheeler, 1977, *Optik* **49**, 71.
Ross, G. and M. Nieto-Vesperinas, 1981, *Opt. Acta* **28**, 77.
Rousseau, M., 1969a, *J. Phys.* **30**, 675.
Rousseau, M., 1969b, *C. R. Acad. Sci. Paris* **268**, 1477.
Ruggieri, N. F., D. O. Cummings and G. Lachs, 1972, *J. Appl. Phys.* **43**, 1118.
Rytov, S. M., Yu. A. Kravcov and V. I. Tatarskii, 1978, Introduction to Statistical Radiophysics, Vol. II (Nauka, Moscow), in Russian.
Saleh, B. E. A., 1975a, *J. Appl. Phys.* **46**, 943.
Saleh, B. E. A., 1975b, *Appl. Phys.* **8**, 269.
Saleh, B. E. A., 1978, Photoelectron Statistics (Springer, Berlin).
Saleh, B. E. A. and M. Irshid, 1979, *Opt. Quant. Electr.* **11**, 479.

Saleh, B. E. A., D. Stoler and M. C. Teich, 1982, Coherence and photon statistics of optical fields generated by Poisson random emissions, preprint; *Phys. Rev.* **A 27** (1983) 360.
Saleh, B. E. A., J. T. Tavolacci and M. C. Teich, 1981, *IEEE J. Quant. Electr.* **QE-17**, 2341.
Saleh, B. E. A. and M. C. Teich, 1982, *Proc. IEEE* **70**, 229.
Sarfatt, J., 1963, *Nuovo Cim.* **27**, 1119.
Sargent, M., III and M. O. Scully, 1972, Laser Handbook, Vol. 1, eds. F. T. Arecchi and E. O. Schulz-Dubois (North-Holland, Amsterdam), p. 45.
Sargent, M., M. O. Scully and W. E. Lamb, Jr., 1974, Laser Physics (Addison-Wesley, Reading).
Saxton, W. O., 1974, *J. Phys.* **D 7**, L 63.
Schmeidler, W., 1956, Über symmetrische algebraische Integralgleichungen (Helsinki, Ann. Finn. Akad. Wissen, *Math. Phys.* **I**, 220).
Schmidt-Weinmar, N. G., 1978, Inverse Source Problems, ed. H. Baltes (Springer, Berlin), p. 83.
Schmidt-Weinmar, N. G., B. Steinle and H. P. Baltes, 1978/79, *Optik* **52**, 205.
Schrödinger, E., 1927, *Naturwissensch.* **14**, 644.
Schubert, M., K. E. Süsse, W. Vogel, D. G. Welsch and B. Wilhelmi, 1982, *Kvant. Elektr.* (USSR) **9**, 495.
Schubert, M. and W. Vogel, 1981, *Opt. Comm.* **36**, 164.
Schubert, M. and B. Wilhelmi, 1976, Recent Advances in Optical Physics, eds. B. Havelka and J. Blabla (Soc. Czech. Math. Phys., Prague), p. 225.
Schubert, M. and B. Wilhelmi, 1978, Einführung in die nichtlineare Optik, Teil II (Teubner, Leipzig).
Schubert, M. and B. Wilhelmi, 1980, Progress in Optics, Vol. 17, ed. E. Wolf (North-Holland, Amsterdam), p. 163.
Schweber, S. S., 1961, An Introduction to Relativistic Quantum Field Theory (Row, Peterson and Co. Evanston, Ill., Elmsford, New York).
Scully, M. O. and W. E. Lamb, 1966, *Phys. Rev. Lett.* **16**, 853.
Scully, M. O. and W. E. Lamb, 1967, *Phys. Rev.* **159**, 208.
Scully, M. O. and W. E. Lamb, 1968, *Phys. Rev.* **166**, 246.
Scully, M. O. and W. E. Lamb, 1969, *Phys. Rev.* **179**, 368.
Scully, M. O. and K. G. Whitney, 1972, Progress in Optics, Vol. 10, ed. E. Wolf (North-Holland, Amsterdam), p. 89.
Sczaniecki, L., 1980, Bistability of multi-photon lasers, Proc. Inter. Conf. Lasers (New Orleans).
Sczaniecki, L., 1982, *Opt. Acta* **29**, 69.
Sczaniecki, L. and J. Buchert, 1978, *Opt. Comm.* **27**, 463.
Selloni, A., 1980, Inverse Scattering Problems in Optics, ed. H. P. Baltes (Springer, Berlin), p. 117.
Selloni, A., P. Schwendimann, A. Quattropani and H. P. Baltes, 1978, *J. Phys.* **A 11**, 1427.
Senitzky, I. R., 1958, *Phys. Rev.* **111**, 3.
Senitzky, I. R., 1962, *Phys. Rev.* **127**, 1638.
Senitzky, I. R., 1967a, *Phys. Rev.* **161**, 165.
Senitzky, I. R., 1967b, *Phys. Rev.* **155**, 1387.
Senitzky, I. R., 1968, *Phys. Rev.* **174**, 1588.
Senitzky, I. R., 1969, *Phys. Rev.* **183**, 1069.
Senitzky, I. R., 1973, *Phys. Rev. Lett.* **31**, 955.
Senitzky, I. R., 1978, Progress in Optics, Vol. 16, ed. E. Wolf (North-Holland, Amsterdam), p. 413.
Senitzky, I. R., 1981, *Phys. Rev. Lett.* **47**, 1503.
Series, G. W., 1970, Quantum Optics, eds. S. M. Kay and A. Maitland (Acad. Press, London), p. 395.
Shannon, C. E., 1949, Mathematical Theory of Communications (Univ. Illinois).
Shapiro, J. H., 1980, *Opt. Lett.* **5**, 351.
Shapiro, J. H., H. P. Yuen and J. A. Machado Mata, 1979, *IEEE Trans. Infor. Theory* **IT-25**, 179.
Sharma, M. P. and L. M. Brescausin, 1981, *Phys. Rev.* **A 23**, 1893.
Shen, Y. R., 1967, *Phys. Rev.* **155**, 921.
Shen, Y. R., 1976, *Rev. Mod. Phys.* **48**, 1.
Shepherd, T. J., 1981, *Opt. Acta* **28**, 567.
Sheremetyev, A. G., 1971, Statistical Theory of Laser Communication (Svyaz, Moscow), in Russian.

Shiga, F. and S. Inamura, 1967, *Phys. Lett.* **25 A**, 706.
Short, R. and L. Mandel, 1983, *Phys. Rev. Lett.* **51**, 384.
Sibilia, C. and M. Bertolotti, 1981, *Opt. Acta* **28**, 503.
Sibilia, C., M. Bertolotti, V. Peřinová, J. Peřina and A. Lukš, 1983, *Phys. Rev.* **A 28**, 328.
Sillitto, R. M., 1963, *Proc. Roy. Soc. Edinburgh* **A 66**, 93.
Sillitto, R. M., 1968, *Phys. Lett.* **27 A**, 624.
Simaan, H. D., 1975, *J. Phys.* **A 8**, 1620.
Simaan, H. D., 1978, *J. Phys.* **A 11**, 1799.
Simaan, H. D., 1979, *Opt. Comm.* **31**, 21.
Simaan, H. D. and R. Loudon, 1975, *J. Phys.* **A 8**, 539 and 1140.
Simaan, H. D. and R. Loudon, 1978, *J. Phys.* **A 11**, 435.
Slusher, R. E., 1974, Progress in Optics, Vol. 12, ed. E. Wolf (North-Holland, Amsterdam), p. 53.
Smirnov, D. F. and A. S. Trochin, 1981, *J. Exp. Theor. Phys.* (USSR) **81**, 1597.
Smith, A. W. and J. A. Armstrong, 1966a, *Phys. Rev. Lett.* **16**, 1169.
Smith, A. W. and J. A. Armstrong, 1966b, *Phys. Lett.* **19**, 650.
Smithers, M. E. and E. Y. C. Lu, 1974, *Phys. Rev.* **A 10**, 1874.
Solimeno, S., E. Corti and B. Nicoletti, 1970, *J. Opt. Soc. Am.* **60**, 1245.
Solimeno, S., P. Di Porto and B. Crosignani, 1969, *J. Math. Phys.* **10**, 1922.
Soroko, L. M., 1971, Fundamentals of Holography and Coherent Optics (Nauka, Moscow), in Russian.
Sotskii, B. A. and B. I. Glazatchev, 1981, *Opt. Spectr.* (USSR) **50**, 1057.
Spence, J. C. H., 1974, *Opt. Acta* **21**, 835.
Srinivas, M. D., 1978, Coherence and Quantum Optics IV, eds. L. Mandel and E. Wolf (Plenum Press, New York), p. 885.
Srinivas, M. D. and E. B. Davies, 1981, *Opt. Acta* **28**, 981.
Srinivasan, S. K., 1974, *Phys. Lett.* **50 A**, 277.
Srinivasan, S. K., 1978, *J. Phys.* **A 11**, 2333.
Srinivasan, S. K. and M. Gururajan, 1981, *J. Math. Phys. Sci.* **15**, 297.
Srinivasan, S. K. and S. Udayabaskaran, 1979, *Opt. Acta* **26**, 1535.
Starikov, A., 1982, *J. Opt. Soc. Am.* **72**, 1538.
Staseľko, D. I., V. B. Voronin and A. G. Smirnov, 1973, *Opt. Spectr.* (USSR) **34**, 561.
Steeger, P. T. and A. E. Fercher, 1982, *Opt. Acta* **29**, 1395.
Steel, W. H., 1957, *J. Opt. Soc. Am.* **47**, 405.
Steinle, B. and H. P. Baltes, 1977, *J. Opt. Soc. Am.* **67**, 241.
Stoler, D., 1970, *Phys. Rev.* **D 1**, 3217.
Stoler, D., 1971, *Phys. Rev.* **D 4**, 1925 and 2309.
Stoler, D., 1972, *Phys. Lett.* **38 A**, 433.
Stoler, D., 1974, *Phys. Rev. Lett.* **33**, 1397.
Stoler, D., 1975, *Phys. Rev.* **D 11**, 1975, 3033.
Streifer, W., 1966, *J. Opt. Soc. Am.* **56**, 1481.
Strohbehn, J. W., 1971, Progress in Optics, Vol. 9, ed. E. Wolf (North-Holland, Amsterdam), p. 73.
Strohbehn, J. W., ed., 1978, Laser Beam Propagation in the Atmosphere (Springer, Berlin).
Strong, J. and G. A. Vanasse, 1959, *J. Opt. Soc. Am.* **49**, 844.
Sudarshan, E. C. G., 1963a, *Phys. Rev. Lett.* **10**, 277.
Sudarshan, E. C. G., 1963b, Proc. Symp. Opt. Masers (J. Wiley, New York), p. 45.
Svelto, O., 1974, Progress in Optics, Vol. 12, ed. E. Wolf (North-Holland, Amsterdam), p. 1.
Szlachetka, P., S. Kielich, J. Peřina and V. Peřinová, 1979, *J. Phys.* **A 12**, 1921.
Szlachetka, P., S. Kielich, J. Peřina and V. Peřinová, 1980a, *J. Molec. Spectr.* **61**, 281.
Szlachetka, P., S. Kielich, J. Peřina and V. Peřinová, 1980b, *Opt. Acta* **27**, 1609.
Szlachetka, P., S. Kielich, V. Peřinová and J. Peřina, 1980, Proc. EKON-78, eds. S. Kielich, F. Kaczmarek and T. Bancewicz (Univ. Mickiewicz, Poznań), p. 281.
Tänzler, W. and F. J. Schütte, 1981a, *Ann. Physik* **38**, 73.
Tänzler, W. and F. J. Schütte, 1981b, *Opt. Comm.* **37**, 447.

Tatarskii, V. I., 1967, Propagation of Waves in a Turbulent Atmosphere (Nauka, Moscow), in Russian.

Tatarskii, V. I., 1970, Propagation of Short Waves in a Medium with Random Inhomogeneities in Approximation of Markoff Random Process (Inst. Phys. Atmosph., Moscow), in Russian.

Tatarskii, V. I., 1971, *J. Exp. Theor. Phys.* **61**, 1822.

Tatarskii, V. I. and V. U. Zavorotnyi, 1980, Progress in Optics, Vol. 18, ed. E. Wolf (North-Holland, Amsterdam), p. 204.

Teich, M. C., 1969, *Appl. Phys. Lett.* **14**, 201.

Teich, M. C., 1977, Nonlinear Heterodyne Detection, Topics in Applied Physics, Vol. 19, ed. R. J. Keyes (Springer, Berlin), p. 229.

Teich, M. C., 1981, *Appl. Opt.* **20**, 2457.

Teich, M. C., R. L. Abrams and W. B. Gandrud, 1970, *Opt. Comm.* **2**, 206.

Teich, M. C. and P. Diament, 1969, *J. Appl. Phys.* **40**, 625.

Teich, M. C., L. Matin and B. I. Cantor, 1978, *J. Opt. Soc. Am.* **68**, 386.

Teich, M. C. and W. J. McGill, 1976, *Phys. Rev. Lett.* **36**, 754.

Teich, M. C., P. R. Prucnal, G. Vannucci, M. E. Breton and W. J. McGill, 1982a, *J. Opt. Soc. Am.* **72**, 419.

Teich, M. C., P. R. Prucnal, G. Vannucci, M. E. Breton and W. J. McGill, 1982b, *Biol. Cybern.* **44**, 157.

Teich, M. C. and S. Rosenberg, 1971, *Opto-electronics* **3**, 63.

Teich, M. C. and B. E. A. Saleh, 1981, *Phys. Rev.* **A 24**, 1651.

Teich, M. C. and B. E. A. Saleh, 1982, *Opt. Lett.* **7**, 365.

Teich, M. C. and B. E. A. Saleh, 1985, Observation of sub-Poisson Franck-Hertz light at 253.7 nm, *J. Opt. Soc. Am.* **B 2**, 275.

Teich, M. C., B. E. A. Saleh and J. Peřina, 1984, *J. Opt. Soc. Am.* **B 1**, 366.

Teich, M. C. and G. Vannucci, 1978, *J. Opt. Soc. Am.* **68**, 1338.

Teich, M. C. and G. J. Wolga, 1966, *Phys. Rev. Lett.* **16**, 625.

Ter Haar, D., 1961, *Rep. Prog. Phys.* **24**, 304.

Thomas, G. M., 1971/72, *Proc. Roy. Soc. Edinburgh* **(A) 70**, 27.

Thompson, B. J., 1969, Progress in Optics, Vol. 7, ed. E. Wolf (North-Holland, Amsterdam), p. 171.

Titchmarsh, E. C., 1948, Introduction to the Theory of Fourier Integrals, 2nd ed. (Oxford).

Titulaer, U. M. and R. J. Glauber, 1965, *Phys. Rev.* **140**, B 676.

Titulaer, U. M. and R. J. Glauber, 1966, *Phys. Rev.* **145**, 1041.

Toll, J., 1956, *Phys. Rev.* **104**, 1760.

Toraldo di Francia, G., 1955, *J. Opt. Soc. Am.* **45**, 497.

Toraldo di Francia, G., 1966, *Opt. Acta* **13**, 323.

Toraldo di Francia, G., 1969, *J. Opt. Soc. Am.* **59**, 799.

Toraldo di Francia, G., 1970, Quantum Optics, eds. S. M. Kay and A. Maitland (Acad. Press, London), p. 323.

Tornau, N. and A. Bach, 1974, *Opt. Comm.* **11**, 46.

Tornau, N. and B. Echtermeyer, 1973, *Ann. Physik* **29**, 289.

Trias, A., 1977, *Phys. Lett.* **61 A**, 149.

Trifonov, D. A. and V. N. Ivanov, 1977, *Phys. Lett.* **64 A**, 269.

Troup, G. J., 1965, *Proc. Phys. Soc.* **86**, 39.

Troup, G. J., 1966, *Nuovo Cim.* **42**, 79.

Troup, G. J., 1967, Optical Coherence Theory (Methuen, London).

Troup, G. J., 1968, *Phys. Lett.* **28 A**, 251.

Trung, T. V. and F. J. Schütte, 1977, *Ann. Physik* **34**, 262.

Trung, T. V. and F. J. Schütte, 1978, *Ann. Physik* **35**, 216.

Tucker, J. and D. F. Walls, 1969, *Phys. Rev.* **178**, 2036.

Tunkin, V. G. and A. S. Tchirkin, 1970, *J. Exp. Theor. Phys.* **58**, 191.

Twiss, R. Q., 1969, *Opt. Acta* **16**, 423.

Twiss, R. Q. and A. G. Little, 1959, *Austr. J. Phys.* **12**, 77.

Twiss, R. Q., A. G. Little and R. Hanbury Brown, 1957, *Nature* **180**, 324.

Vajnshtejn, L. A., V. N. Melechin, S. A. Mishin and E. R. Podoljak, 1981, *J. Exp. Theor. Phys.* (USSR) **81**, 2000.

Vanasse, G. A. and H. Sakai, 1967, Progress in Optics, Vol. 6, ed. E. Wolf (North-Holland, Amsterdam), p. 259.

Van Cittert, P. H., 1934, *Physica* **1**, 201.

Van Cittert, P. H., 1939, *Physica* **6**, 1129.

Van Kampen, N. G., 1953, *Phys. Rev.* **89**, 1072.

Vannucci, G. and M. C. Teich, 1981, *J. Opt. Soc. Am.* **71**, 164.

Verdet, E., 1869, Leçon d'Optique Physique I (Imprim. Impér., Paris), p. 106.

Vinson, J. F., 1971, Optische Kohärenz (Akad., Berlin).

Voigt, H. and A. Bandilla, 1981, *Ann. Physik* **38**, 137.

Voigt, H., A. Bandilla and H. H. Ritze, 1980, *Z. Phys.* **B 36**, 295.

Volterra, V., 1959, Theory of Functionals and of Integral and Integro-Differential Equations (Dover, New York).

Wagner, J., P. Kurowski and W. Martienssen, 1979, *Z. Phys.* **B 33**, 391.

Walker, J. G., 1981, *Opt. Acta* **28**, 735.

Walls, D. F., 1970, *Z. Phys.* **237**, 224.

Walls, D. F., 1973, *J. Phys.* **A 6**, 496.

Walls, D. F., 1977, *Am. J. Phys.* **45**, 952.

Walls, D. F., 1979, *Nature* **280**, 451.

Walls, D. F., 1983, *Nature* **306**, 141.

Walls, D. F. and R. Barakat, 1970, *Phys. Rev.* **A 1**, 446.

Walls, D. F., P. D. Drummond and K. J. McNeil, 1981, Optical Bistability, eds. C. M. Bowden, M. Ciftan and H. R. Robl (Plenum Press, New York), p. 51.

Walls, D. F. and G. J. Milburn, 1981, Lectures presented at Summer School on Quantum Optics and Experimental General Relativity (Bad Windsheim).

Walls, D. F. and C. T. Tindle, 1972, *J. Phys.* **A 5**, 534.

Walls, D. F. and P. Zoller, 1981, *Phys. Rev. Lett.* **47**, 709.

Walther, A., 1963, *Opt. Acta* **10**, 41.

Walther, A., 1968, *J. Opt. Soc. Am.* **58**, 1256.

Walther, A., 1973, *J. Opt. Soc. Am.* **63**, 1622.

Wang, M. C. and G. E. Uhlenbeck, 1945, *Rev. Mod. Phys.* **17**, 323.

Webber, J. C., 1968, *Phys. Lett.* **27 A**, 5.

Webber, J. C., 1969, *Can. J. Phys.* **47**, 363.

Weidlich, W., H. Risken and H. Haken, 1967, *Z. Phys.* **201**, 396.

Welford, W. T., 1977, *Opt. Quant. Electr.* **9**, 269.

Whittaker, E. and G. N. Watson, 1940, A Course of Modern Analysis, 4th ed. (Cambridge Univ. Press).

Wiener, N., 1928, *J. Math. and Phys.* **7**, 109.

Wiener, N., 1929, *J. Franklin Inst.* **207**, 525.

Wiener, N., 1930, *Acta Math.* **55**, 117.

Wigner, E., 1932, *Phys. Rev.* **40**, 749.

Willis, C. R., 1966, *Phys. Rev.* **147**, 406.

Wolf, E., 1954a, *Proc. Roy. Soc.* **A 225**, 96.

Wolf, E., 1954b, *Nuovo Cim.* **12**, 884.

Wolf, E., 1955, *Proc. Roy. Soc.* **A 230**, 246.

Wolf, E., 1956, Proc. Symp. Astr. Optics and Rel. Subj., ed. Z. Kopal (North-Holland, Amsterdam), p. 177.

Wolf, E., 1957, *Phil. Mag.* (8) **2**, 351.

Wolf, E., 1958, *Proc. Phys. Soc.* **71**, 257.

Wolf, E., 1959, *Nuovo Cim.* **13**, 1165.

Wolf, E., 1960, *Proc. Phys. Soc.* **76**, 424.

Wolf, E., 1962, *Proc. Phys. Soc.* **80**, 1269.

Wolf, E., 1963a, Proc. Symp. Opt. Masers (J. Wiley, New York), p. 29.

Wolf, E., 1963b, *Phys. Lett.* **3**, 166.
Wolf, E., 1964, Quantum Electronics, eds. N. Bloembergen and P. Grivet (Dunod et Cie., Paris), p. 13.
Wolf, E., 1965, *Jap. J. Appl. Phys.* **4**, Suppl. I, p. 1.
Wolf, E., 1966, *Opt. Acta* **13**, 281.
Wolf, E., 1978, *J. Opt. Soc. Am.* **68**, 1597.
Wolf, E., 1981, Optics in Four Dimensions – 1980, eds. M. A. Machado and L. M. Narducci (Am. Inst. Phys., New York), p. 42.
Wolf, E., 1982, *J. Opt. Soc. Am.* **72**, 343.
Wolf, E., 1983, *Opt. Lett.* **8**, 250.
Wolf, E. and G. S. Agarwal, 1969, Polarization, Matière et Rayonnement (Soc. Franc. Phys., Press. Univ. France), p. 541.
Wolf, E. and G. S. Agarwal, 1984, *J. Opt. Soc. Am.* **A 1**, 541.
Wolf, E. and W. H. Carter, 1975, *Opt. Comm.* **13**, 205.
Wolf, E. and W. H. Carter, 1976, *Opt. Comm.* **16**, 297.
Wolf, E. and E. Collet, 1978, *Opt. Comm.* **25**, 293.
Wolf, E. and A. J. Devaney, 1981, *Opt. Lett.* **6**, 168.
Wolf, E., A. J. Devaney and J. T. Foley, 1981, Optics in Four Dimensions – 1980, eds. M. A. Machado and L. M. Narducci (Am. Inst. Phys., New York), p. 123.
Wolf, E. and C. L. Mehta, 1964, *Phys. Rev. Lett.* **13**, 705.
Yariv, A., 1967, Quantum Electronics and Nonlinear Optics (J. Wiley, New York).
Yariv, A., 1975, Quantum Electronics (J. Wiley, New York).
Yuen, H. P., 1975, *Phys. Lett.* **51 A**, 1.
Yuen, H. P., 1976, *Phys. Rev.* **A 13**, 2226.
Yuen, H. P. and J. H. Shapiro, 1978, *IEEE Trans. Infor. Theor.* **IT-24**, 657.
Yuen, H. P. and J. H. Shapiro, 1979, *Opt. Lett.* **4**, 334.
Zardecki, A., 1969a, *J. Math. Phys.* **11**, 224.
Zardecki, A., 1969b, *Acta Phys. Pol.* **35**, 271.
Zardecki, A., 1971, *Can. J. Phys.* **49**, 1724.
Zardecki, A., 1974, *J. Phys.* **A 7**, 2198.
Zardecki, A., 1978, Inverse Source Problems, ed. H. P. Baltes (Springer, Berlin), p. 155.
Zardecki, A., J. Bures and C. Delisle, 1972, *Phys. Rev.* **A 6**, 1209.
Zardecki, A., C. Delisle and J. Bures, 1972, *Opt. Comm.* **5**, 298.
Zardecki, A., C. Delisle and J. Bures, 1973, Coherence and Quantum Optics, eds. L. Mandel and E. Wolf (Plenum Press, New York), p. 259.
Zernike, F., 1938, *Physica* **5**, 785.
Zernike, F. and J. E. Midwinter, 1973, Applied Nonlinear Optics (J. Wiley, New York).
Zubairy, M. S., 1982, *Phys. Lett.* **87 A**, 162.
Zubairy, M. S. and J. J. Yeh, 1980, *Phys. Rev.* **A 21**, 1624.

INDEX